MICRO LIFE

DK SMITHSONIAN

MICRO LIFE

MIRACLES OF THE MINIATURE WORLD REVEALED

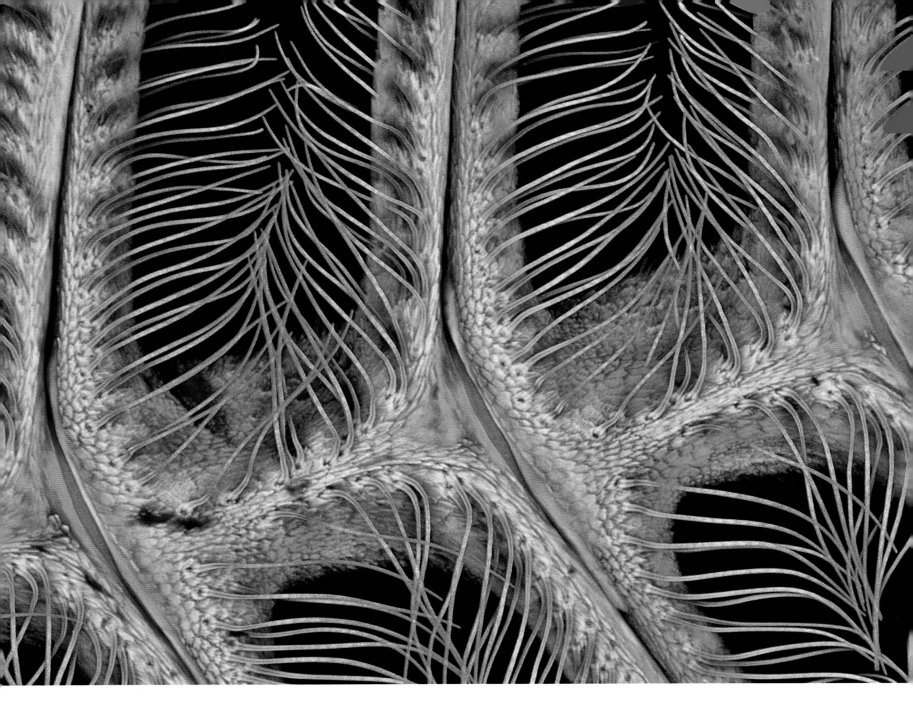

 Penguin Random House

DK LONDON

Senior Editor Dr. Rob Houston
US Editor Karyn Gerhard
Editors Jemima Dunne, Tim Harris, Annie Moss, Steve Setford, Hannah Westlake
Production Editors Andy Hilliard, Gillian Reid
Senior Production Controller Meskerem Berhane
Managing Editor Angeles Gavira Guerrero
Associate Publishing Director Liz Wheeler
Publishing Director Jonathan Metcalf

Senior Art Editor Ina Stradins
Designers Simon Murrell, Francis Wong
Illustrator Phil Gamble
Senior Jacket Designer Akiko Kato
Jacket Design Development Manager Sophia MTT
Managing Art Editor Michael Duffy
Art Director Karen Self
Design Director Phil Ormerod

DK DELHI

Senior Editor Dharini Ganesh
Editor Ishita Jha
Senior DTP Designer Jagtar Singh
Project Picture Researcher Aditya Katyal
Senior Managing Editor Rohan Sinha
Pre-production Manager Balwant Singh
Editorial Head Glenda R. Fernandes

Project Art Editor Anjali Sachar
Senior Jacket Designer Suhita Dharamjit
DTP Designers Jaypal Singh Chauhan, Rakesh Kumar
Picture Research Manager Taiyaba Khatoon
Managing Art Editor Sudakshina Basu
Production Manager Pankaj Sharma
Design Head Malavika Talukder

First American Edition, 2021
Published in the United States by DK Publishing
1450 Broadway, Suite 801, New York, NY 10018

Copyright © 2021 Dorling Kindersley Limited DK,
a Division of Penguin Random House LLC
21 22 23 24 25 10 9 8 7 6 5 4 3 2 1
001–316673–Oct/2021

All rights reserved.
Without limiting the rights under the copyright reserved above, no part of this publication may be reproduced, stored in or introduced into a retrieval system, or transmitted, in any form, or by any means (electronic, mechanical, photocopying, recording, or otherwise), without the prior written permission of the copyright owner.
Published in Great Britain by Dorling Kindersley Limited

A catalog record for this book is available from the Library of Congress.
ISBN 978-0-7440-3956-6

DK books are available at special discounts when purchased in bulk for sales promotions, premiums, fund-raising, or educational use. For details, contact: DK Publishing Special Markets, 1450 Broadway, New York, NY 10018 or SpecialSales@dk.com

Printed in China

For the curious
www.dk.com

This book was made with Forest Stewardship Council ™ certified paper—one small step in DK's commitment to a sustainable future. For more information go to www.dk.com/our-green-pledge

Contributors

Derek Harvey (lead author) is a naturalist with a particular interest in evolutionary biology, who studied Zoology at the University of Liverpool. He has taught a generation of biologists, and has led student expeditions to Costa Rica, Madagascar, and Australasia.

Dr. Elizabeth Wood is a marine biological consultant, and Coral Reef Conservation Specialist at the Marine Conservation Society. She is a diver and underwater photographer with particular research interests in reef ecology, fish behavior, and the biology of corals.

Michael Scott is a natural history writer, conservationist, and former broadcaster. He has contributed to the DK publications *Earth Matters*, *Oceans*, and *The Natural History Book*.

Tom Jackson is a science writer who has written more than 100 books and contributed to many more over the past 20 years. Tom studied zoology at Bristol University, UK, and has worked as a zookeeper and a conservationist.

Dr. Bea Perks studied zoology then gained a PhD in clinical pharmacology. She has been a science writer for 20 years, contributing articles to *New Scientist* and *Nature* magazines.

Half-title page Mineralized cell wall of a diatom (*Amphora* sp.), scanning electron micrograph
Title page Predatory mite, scanning electron micrograph
Above Moth antenna, confocal laser-scanning light micrograph
Contents page Rotifers and a desmid (single-celled alga), confocal laser-scanning light micrograph

Consultants

Mark Viney is Professor of Zoology at the University of Liverpool. He studies the biology of parasitic nematode worms and the immunology of wild mammals. He was a student at Imperial College, London and the Liverpool School of Tropical Medicine, and was previously based at the Universities of Edinburgh and Bristol.

Dr. Richard Kirby is an independent marine scientist. A former Royal Society university research fellow, his interests are the plankton and the food web it supports. In order to engage the public with plankton science he founded the global citizen science Secchi Disk study.

Dr. Kim Dennis-Bryan is a zoologist. She began her career studying fossil fish at London's Natural History Museum, before becoming an Open University lecturer specializing in natural sciences. She has written for and consulted on many science books, including DK's *Animal*, *Ocean*, and *Prehistoric Life*.

 S M I T H S O N I A N

Established in 1846, the Smithsonian Institution—the world's largest museum and research complex—includes 19 museums and galleries and the National Zoological Park. The total number of artifacts, works of art, and specimens in the Smithsonian's collections is estimated at 156 million, the bulk of which is contained in the National Museum of Natural History, which holds more than 126 million specimens and objects. The Smithsonian is a renowned research center, dedicated to public education, national service, and scholarship in the arts, sciences, and history.

contents

introduction

- **10** scale in the micro world
- **12** types of living organisms
- **14** macro life up close

getting nourishment

- **18** solar-powered microbes
- **20** absorbing light
- **22** root hairs
- **24** topping up nitrogen
- **26** fixing nitrogen
- **28** bacteria
- **30** *Escherichia coli*
- **32** damaging a host
- **34** absorbing food
- **36** *Penicillium*
- **38** engulfing prey
- **40** microbial predator
- **42** stinging cells
- **44** feeding on particles
- **46** connecting mouthparts
- **48** insect mouthparts
- **50** blood sucker
- **52** drinking sap
- **54** rasping food
- **56** venomous pincers
- **58** predigesting prey
- **60** food into the bloodstream
- **62** living in the gut
- **64** spinning webs
- **66** neither animal nor plant
- **68** eating wood

powering the body

- **72** energy release
- **74** using oxygen
- **76** mineral energy
- **78** fermentation energy
- **80** poisoned by oxygen
- **82** gas exchange
- **84** breathing tubes
- **86** the gills of insects
- **88** mammal lungs
- **90** carrying oxygen
- **92** circulatory system
- **94** leaf pores
- **96** keeping warm

sensing and responding

- **100** sensing the environment
- **102** antennae
- **104** sensing taste
- **106** hearing sound
- **108** sensory cells in the ear
- **110** judging distance
- **112** compound eyes
- **114** producing color
- **116** iridescence
- **118** producing light
- **120** color changers
- **122** nerve cells
- **124** coordinating behavior

moving

- **128** beating hairs
- **130** *Paramecium*
- **132** crawling cells
- **134** swimming with hairs
- **136** combs of cilia
- **138** simple muscles
- **140** contracting muscles
- **142** overcoming friction
- **144** smooth rowing
- **146** controlling buoyancy
- **148** tube feet
- **150** living at the surface
- **152** backswimmers
- **154** clinging feet
- **156** telescopic legs
- **158** jointed legs
- **160** catapulting
- **162** jumping with legs
- **164** insect wings
- **166** stabilizing flight
- **168** tiniest fliers
- **170** flight feather
- **172** hitchhiking mites

supporting and protecting

- **176** a cell's internal skeleton
- **178** diatoms
- **180** microscopic shells
- **182** silica skeletons
- **184** cellulose armor
- **186** sponge spicules
- **188** shedding water
- **190** chemical defenses
- **192** skeleton on the outside
- **194** staying hidden
- **196** echinoderm skeleton
- **198** shark skin
- **200** vertebrate skeletons
- **202** mammal hair
- **204** cell walls
- **206** supporting stems
- **208** leaf surfaces
- **210** insect stingers
- **212** irritating hairs
- **214** stinging hairs
- **216** internal defense

reproducing

- **220** sabotaging cells
- **222** coronavirus
- **224** swarming bacteria
- **226** asexual reproduction
- **228** fertilizing an egg
- **230** fungus reproduction
- **232** alternating generations
- **234** sex in flowering plants
- **236** pollen grains
- **238** bees
- **240** hidden pollination
- **242** boom and bust
- **244** escaping starvation
- **246** competing for mates
- **248** parental care

growing and changing

- **252** colonies of cells
- **254** cell division
- **256** developing embryo
- **258** insect eggs
- **260** how ferns grow
- **262** seeds
- **264** seed germination
- **266** simple plants
- **268** growing up as plankton
- **270** growing in steps
- **272** long and short lifespans

habitats and lifestyles

- **276** ubiquitous bacteria
- **278** surviving extremes
- **280** tardigrades
- **282** surviving cold
- **284** hanging on
- **286** marine plankton
- **288** copepods
- **290** pond microorganisms
- **292** freshwater communities
- **294** nematodes
- **296** recycling matter
- **298** between sand grains
- **300** forming galls
- **302** mosquitoes
- **304** living on skin
- **306** living in hair
- **308** gut communities
- **310** infecting blood cells
- **312** brain parasites
- **314** plant-fungus partnership
- **316** part fungus, part alga
- **318** photosynthetic helpers

- **320** classification
- **400** glossary
- **406** index
- **414** acknowledgments

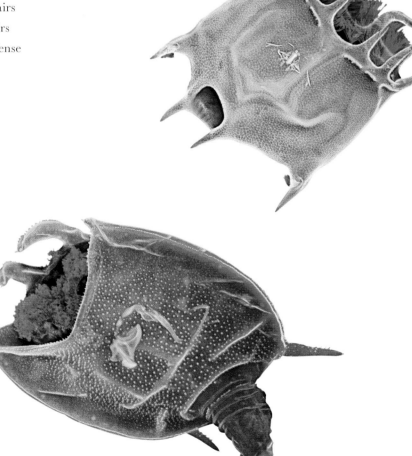

foreword

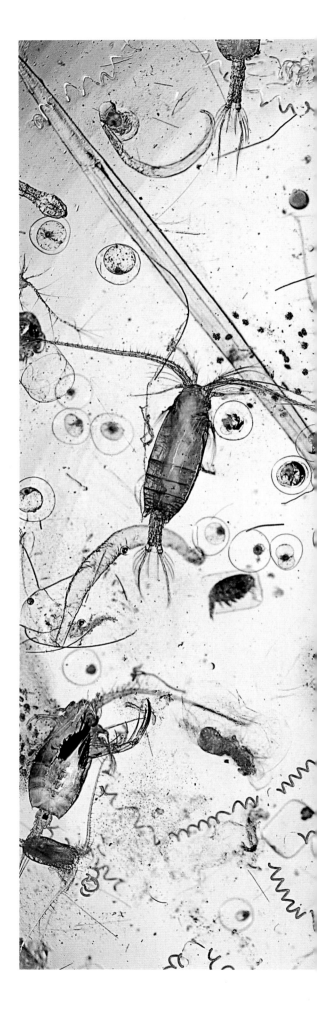

Elephants are big, dinosaurs were bigger, blue whales are the biggest ever... but is size really important, do big things make the world go round, and is being big really so impressive? And do you have to be big to be beautiful?

Put simply, no. As we take a fantastic voyage into the micro-life of our planet, the profound and unimaginable beauty previously hidden from our "big" eyes is revealed—the rainbow wefts of wafting cilia, the exquisite symmetry of diatoms' skeletons, the "op art" of a fly's compound eyes.

But beauty is more than skin deep in the microscopic world. Here lie the origins of life itself, of our life, and all its complex history and ongoing interrelationships still form the essential fabric of our whole planet's ecology. Here, the multitudes of the minute are pictured using remarkable microscopic technologies and each of their little stories told. And it's both surprising and amazing that we live our lives among this invisible horde of littler things and never appreciate or understand their fundamental importance. So shrink your world and dive into another, denied to us by scale alone. Meet your neighbors—they are all around you, on you, and inside you—viruses, bacteria, other microbes—some bad, many good. See inside your organs, peer inside cells, discover the methods of insect senses, and how pollen, seeds, and spores work. Come face to face with a tardigrade, a mite, and the curious, beady eyes of a springtail.

This spectacular book is a portal into a world within our world—one we are too big to know but clever enough to have unveiled.

CHRIS PACKHAM
NATURALIST, BROADCASTER, WRITER,
PHOTOGRAPHER, AND CONSERVATIONIST

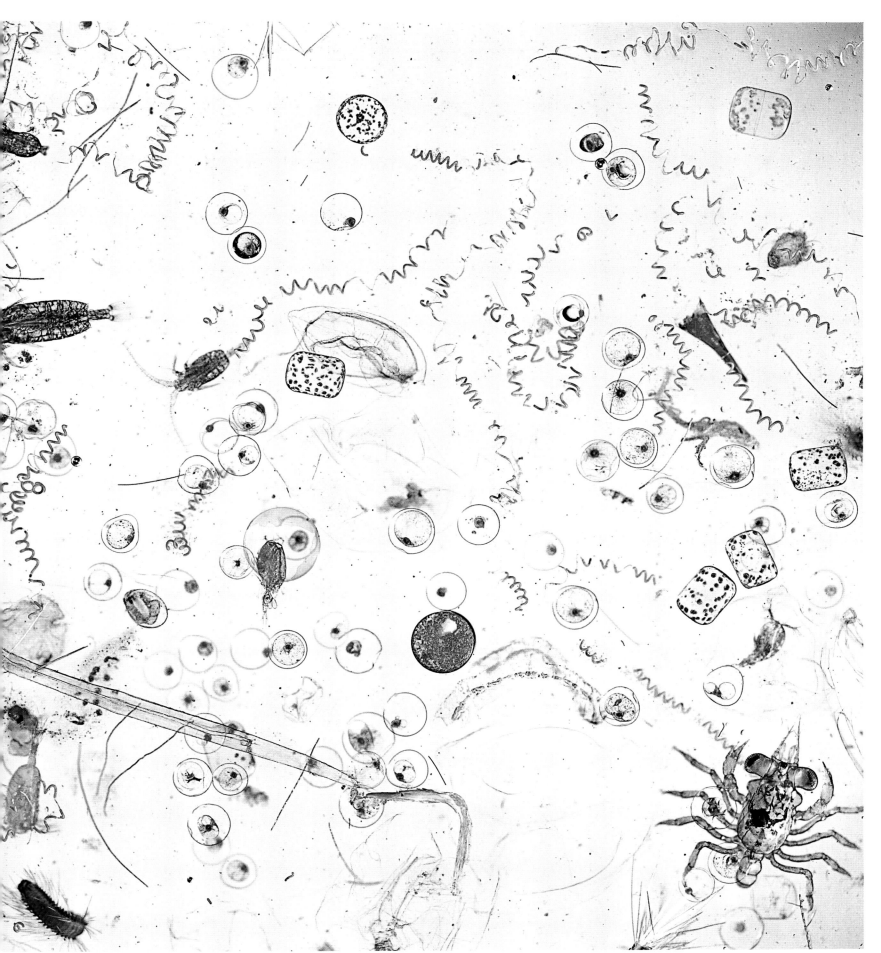

CONTENTS OF ONE DIP OF A HAND NET IN THE OCEAN, INCLUDING COPEPOD CRUSTACEANS, A CRAB LARVA, AND SINGLE-CELLED ALGAE; MAGNIFICATION × 20

Fully grown caterpillar is just 13 mm (1/2 in) long

TYPES OF MICROSCOPY
Light microscopes, which use glass lenses to focus light, can magnify to a certain limit. The wavelength of light limits resolution at high magnification, so they cannot resolve anything smaller than 2 μm, the size of a small bacterium. Electron microscopes use electron beams with much smaller wavelengths, focused with electromagnetic "lenses," to magnify further. Unlike light microscopes, electron microscopes always image dead specimens, partly because electrons transmit best in a vacuum.

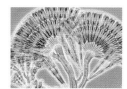

LM
Light micrographs (LM) are enhanced with many techniques. This one uses "phase contrast" to highlight the specimen.

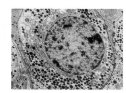

TEM
Transmission electron micrographs (TEMs) are made by transmitting electrons through thin sections of a specimen.

SEM
Scanning electron micrographs (SEMs) scan electron beams over solid samples to give a 3-D effect.

Invisible detail
This micrograph of the caterpillar of a common blue butterfly (*Polyommatus icarus*), which is no bigger than a child's fingernail, was made with a scanning electron microscope. To capture an image like this, a dead specimen is dehydrated and coated with metal to scatter electrons onto a photographic plate. As in all electron micrographs in this book, the fixed wavelength of the electron beam produced a monochrome image, and false colors were added.

scale in the micro world

Most life on Earth is too small for us to see with the naked eye—the minimum focal length of our eyes is too long and their resolution too poor. However, the size range of micro-life is vast: a dust mite may be just visible to humans, but it is 5,000 times bigger than a virus. Microscopes are needed to magnify this unseen world. Those that focus light using glass lenses can enlarge objects by up to 1,000 times, making the dust mite as big as this page. Electron microscopes go even further—up to a million times—allowing us to explore inside single cells.

House dust mite (*Dermatophagoides pteronyssinus*) viewed by scanning electron microscope

Alga (*Lepocinclis acus*) viewed by light microscope

Bacterium viewed by transmission electron microscope

Geometric shapes of virus seen with computer generated imagery (CGI)

Small, smaller, and smallest
The tiniest animals, such as mites, are smaller than 0.04 in (1 mm). Most complex single-celled organisms, such as algae and amoebas, are tenths of a millimeter across, but bacteria are typically 10–100 times smaller, at 1–2 μm (micrometers, or millionths of a meter) long. Viruses are 10 times smaller still.

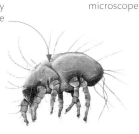
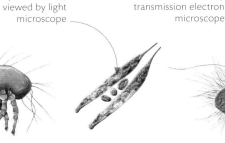
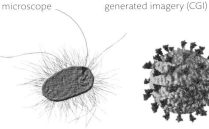

ANIMAL	SINGLE-CELLED ORGANISM	BACTERIUM	VIRUS
Dust mite	Euglenoid alga	*Escherichia coli*	SARS-CoV-2
0.3 mm long	0.17 mm long	1.4 μm long	0.1 μm across
x 80	x 232	x 12,880	x 250,000

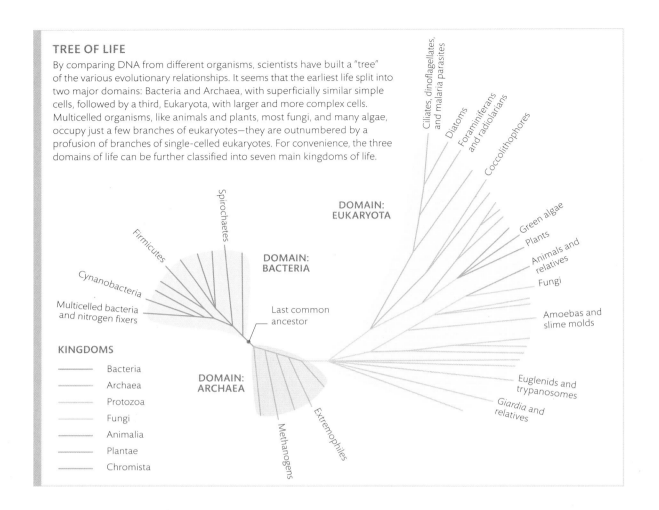

TREE OF LIFE
By comparing DNA from different organisms, scientists have built a "tree" of the various evolutionary relationships. It seems that the earliest life split into two major domains: Bacteria and Archaea, with superficially similar simple cells, followed by a third, Eukaryota, with larger and more complex cells. Multicelled organisms, like animals and plants, most fungi, and many algae, occupy just a few branches of eukaryotes—they are outnumbered by a profusion of branches of single-celled eukaryotes. For convenience, the three domains of life can be further classified into seven main kingdoms of life.

KINGDOMS
- Bacteria
- Archaea
- Protozoa
- Fungi
- Animalia
- Plantae
- Chromista

types of living organisms

All living things on Earth are products of millions of years of evolution and linked by a common ancestry. Many of those belonging to the most ancient and deepest divisions can only be seen through a microscope. The simplest single-celled microbes, such as bacteria, look deceptively uniform, but their genetic and chemical makeup is so diverse that, in evolutionary terms, they can be as different as plants and animals. Many of the more complex microorganisms are related in ways scientists are only beginning to understand. There are more than 1.2 million known species of life on Earth, but most of the millions of species awaiting discovery are part of the micro-world.

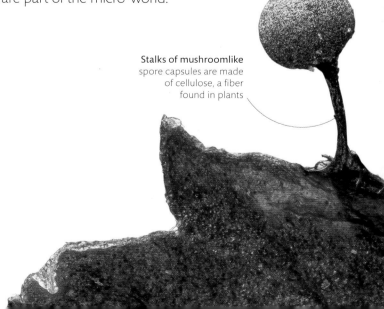

Stalks of mushroomlike spore capsules are made of cellulose, a fiber found in plants

Oddball mechanism
Plasmodial slime mold, *Lamproderma arcyrioides*, is a microorganism that defies easy classification. Slime mold DNA indicates that it is neither fungus, plant, nor animal, but occupies one of many other branches on the evolutionary tree. Like a fungus, it grows mushroomlike spore capsules, but its spores grow into single-celled amoebas that move using filaments akin to those of animal muscle.

CELLULAR COMPLEXITY

Bacteria and Archaea are described as prokaryotes: they have the smallest, simplest cells and do not develop into complex multicelled bodies. Their DNA is present as bundled rings within the cell. All other living things, including those of other microorganisms, plants, and animals, are eukaryotes—they have larger, more complex cells. Their linear molecules of DNA are packaged into a compartment, or nucleus, and thicken into threads called chromosomes when the cells divide. They have many other cellular compartments that perform different functions—such as mitochondria that respire and, in some, chloroplasts that photosynthesize.

BACTERIAL OR ARCHAEAN CELL
- Ribosome
- Genetic material, or DNA
- Small hairs, called pili
- Beating hair, or flagellum

CELL OF A EUKARYOTE
- Endoplasmic reticulum, covered in ribosomes
- Vacuole
- Lysosome
- Ribosome
- Golgi complex
- Mitochondrion
- Nucleus

Slimy filaments release enzymes that digest prey—just like a fungus

Split capsule releases spores that form amoebas, which—like many animals—prey on other small organisms

Animal connective tissue

The bulk of an animal's body is made up of connective tissues, typically composed of specialized cells embedded in a background substance, or matrix, that packs spaces between other tissues. Dense fibrous tissue contains a supporting framework of the protein collagen, while adipose tissue, with its fat deposits, serves as an energy store and helps insulate. The bone of a supporting skeleton is reinforced with hard mineral, and blood transports materials around the body.

Collagen fibers are produced by tissue's fibroblast cells

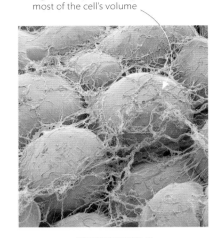

DENSE FIBROUS TISSUE
Human skin dermis
SEM, x1,000

Large fat droplets take up most of the cell's volume

ADIPOSE TISSUE
Human subcutaneous fat
SEM, x290

Supporting struts are made of the mineral calcium phosphate

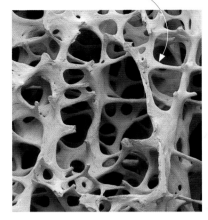

SPONGY BONE
Human long bone
SEM, x11

Other animal tissue

Epithelial tissue grows into thin layers—either around the body, as part of skin, or lining the cavities of hollow internal organs. It has specialized features in different locations in the body: ciliated epithelium in the airways, with beating cilia or hairs, wafts particles away from the lungs, while glands in the gut epithelium secrete digestive juices. Muscles and nerves are made up of tissues that carry electrical charges, which trigger contractions or send impulses.

Mucus-secreting goblet cells (brown) occur between ciliated cells (blue)

CILIATED EPITHELIUM
Lining of human trachea (windpipe)
SEM, x800

Gastric pits are lined with secretory cells

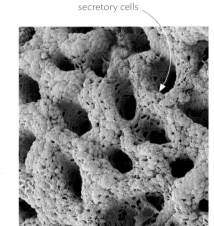

GLANDULAR EPITHELIUM
Lining of human stomach
SEM, x35

Bundles of muscle cells are packed with protein filaments that help cause contraction

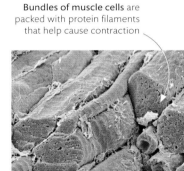

SKELETAL MUSCLE
Human limb muscle
SEM, x270

Plant tissue

Three main tissue types develop in plants: the epidermis, which forms the surface lining, or "skin;" vascular tissue made up of bundles of transport pipes (xylem and phloem); and ground tissue in between them. Ground tissue becomes specialized in different parts of the plant. In leaves, as palisade mesophyll, it is packed with chloroplasts that carry out photosynthesis, while in roots or underground tubers, it may be filled with stored starch grains.

Slitlike stoma, or "breathing" pore, is bordered by two guard cells

EPIDERMIS WITH STOMATA (PORES)
Lily leaf
LM, x200

Xylem pipes carry water and minerals

Phloem pipes carry soluble food, such as sugar

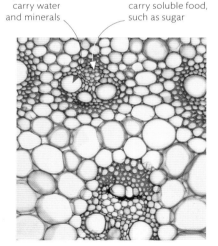

VASCULAR TISSUE
Corn stem
LM, x100

Palisade cells contain light-absorbing chloroplasts (green)

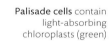
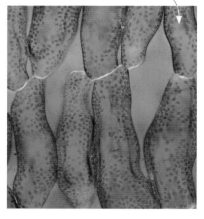

PALISADE MESOPHYLL
Tulip leaf
LM, x200

macro life up close

Micro life is often single-celled, but organisms, right up to the largest "macro life," are made of microscopic cells: scientists estimate there are about 30 trillion in an adult human body, and many times more in a whale or giant tree. Each animal or plant has kinds of cells adapted to perform particular tasks that keep the body alive. Cells that work together form aggregates, known as tissues—such as muscle in an animal, or the light-absorbing layers in a plant's leaf—that look distinctive when viewed under a microscope. The "macro life" pages in this book show these microscopic details of larger organisms.

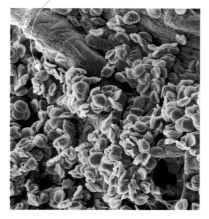

Red blood cells contain oxygen-carrying hemoglobin

BLOOD
Human peripheral (circulating) blood
SEM, x800

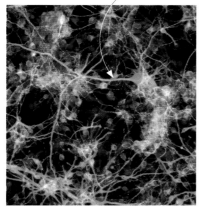

Long nerve fibers transmit electrical nerve impulses

NEURAL STEM CELLS
Cultured from rat stem cells
LM, x100

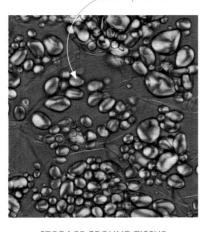

Starch grains are plant's main carbohydrate store

STORAGE GROUND TISSUE
Potato tuber
LM, x125

Leaf blade of *Lepanthes forceps* is composed of photosynthetic tissues sandwiched between upper and lower surface epidermis

Organ level of development

As multicellular bodies grow and develop, the tissues unite to form complex structures called organs. Some *Lepanthes* orchids of the Colombian cloud forest have some very tiny ones—their flowers are no more than 1/10 in (3 mm) long, and sprout beneath coin-sized leaves. Both leaves and flowers are fully formed organs, each made up of at least a dozen different kinds of tissues.

Tiny flower of *Lepanthes cercion* contains reproductive tissue that produces eggs or pollen

getting nourishment

Some living organisms—plants, algae, some bacteria—make food from simple chemicals, using energy from sunlight or from within the chemicals themselves. The organic matter they generate is food for the fungi, animals, and many other microbes that form the rest of the global food web.

solar-powered microbes

Sunlight is the ultimate source of energy for all photosynthetic life on Earth and the creatures that depend on them, and it directly powers organisms like plants and algae that make carbohydrate food from water and carbon dioxide. Called photosynthesis, this process first evolved in aquatic microbes called cyanobacteria more than 3 billion years ago, and similar organisms are still found today in bodies of water, ranging from mighty oceans to tiny puddles. In shallow bays that are too salty for more complex, faster-growing competitors, they form extraordinary "living rocks" called stromatolites.

Ancient life
Up to 3 ft (1 m) across, stromatolites in Shark Bay, Australia result from thousands of years of rock production by billions of microbes.

Stromatolites grow in shallow waters in the Bahamas and Australia where there is plenty of sunlight, even at high tide

Diatoms are larger, more complex cells than cyanobacteria

Threads of slimy mucilage secreted by cyanobacteria cause calcium carbonate to form from sea water, which provides structural support for the colony

Light-powered community
Stromatolites are built primarily by cyanobacteria, a type of bacterium. But since stromatolites first evolved, the thin, living "biofilms" that cover their sunlit surfaces have come to include more complex photosynthetic microbes, including diatoms. The cyanobacteria continue to produce most of the threads of sticky slime, which causes calcium carbonate from sea water to harden into limestone. SEM, x2,750.

OXYGEN FROM PHOTOSYNTHESIS

During photosynthesis, oxygen is released as a byproduct. When stromatolites dominated the primordial oceans billions of years ago, photosynthesis by cyanobacteria oxygenated the atmosphere. This oxygenation produced most of the oxygen in the atmosphere today, and is thought to have triggered the Cambrian Explosion—the rapid evolution of complex life.

Diatom cell walls are hardened by secreted silica, and many are strengthened by ridges

Filamentous cyanobacteria have a simple internal structure

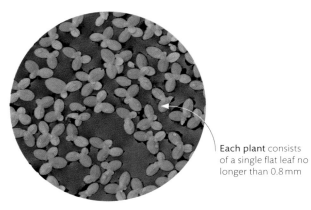

Each plant consists of a single flat leaf no longer than 0.8mm

Rapid reproducer
The world's smallest seed plant is watermeal duckweed (*Wolffia* sp.), consisting of a single leaf. It can also reproduce rapidly by budding, generating a million new plants a month.

Harnessing light energy
Unlike simple photosynthesizers like cyanobacteria (see pp.18–19), plants and algae photosynthesize with chloroplasts. The chloroplasts (shown as green) in a root of a greater duckweed (*Spirodella* sp.) have stacks of darker membranes, called thylakoids, which carry the chlorophyll pigment for absorbing light energy. Each cell also has a nucleus (blue), and mitochondria (yellow). TEM, x10,000.

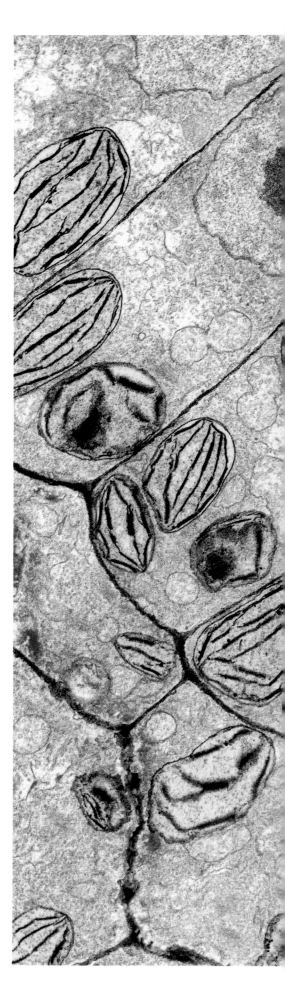

absorbing light

To live and grow, a plant must trap enough light to make sufficient food by photosynthesis. It does this using microscopic solar panels in structures called chloroplasts. Most plants carry their solar panels in many broad leaves. Floating duckweeds, however, grow just one or two tiny leaves each. But by reproducing fast, duckweeds can create a vast colonial "super-leaf." In the billions of chloroplasts, a green chemical called chlorophyll absorbs light energy, changes it into chemical energy, and converts it into food.

FLOATING SOLAR PANELS

Spirodella duckweed has leaves with air spaces that aid buoyancy. Its dangling roots absorb not only minerals but also light filtering down through the water. *Wolffia*'s tiny leaves lack air spaces and roots, representing the most extreme reduction in the body of any seed-bearing plant.

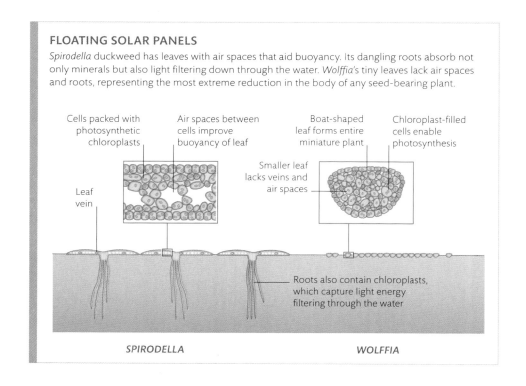

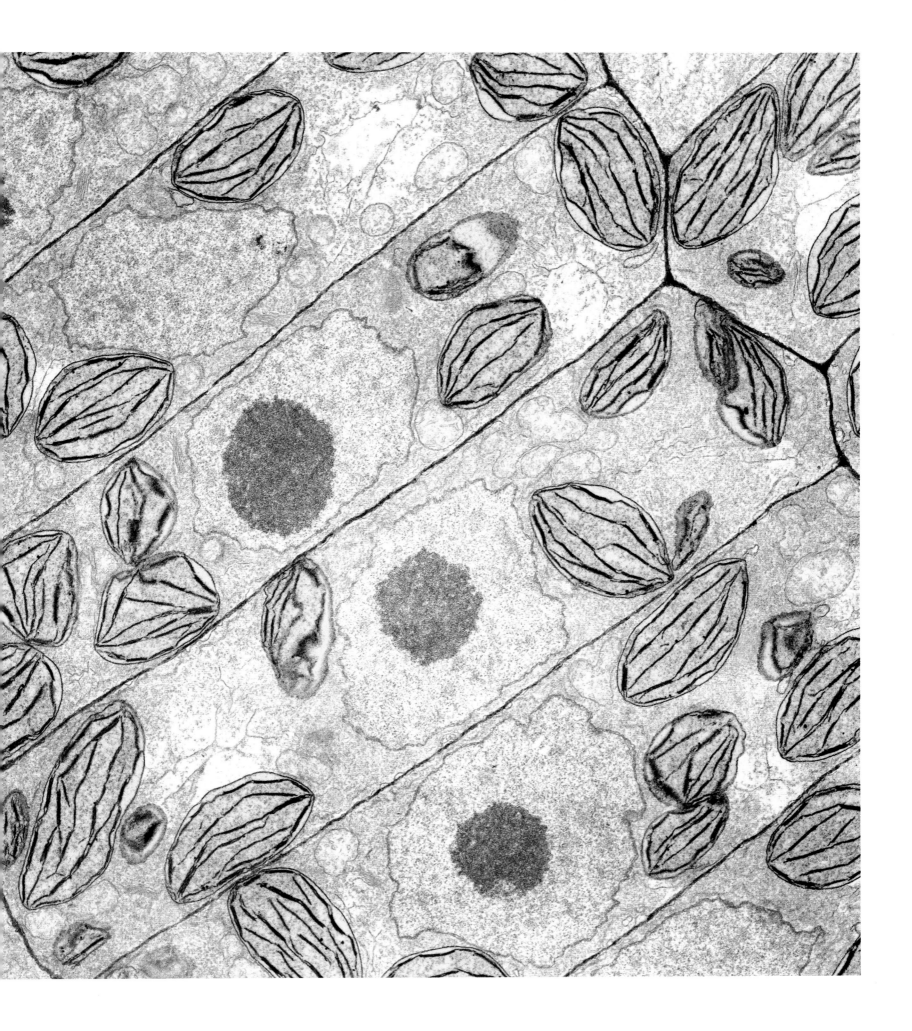

MACROLIFE

root hairs

Root hairs form a band of hairlike growths on the surface of a plant root, behind the root tip. These root hairs are extensions of the epidermis cells that form the root's surface layer. They greatly increase the surface area of the root, boosting its ability to gather water and minerals from the soil. All vascular plants—including ferns, conifers, and flowering plants of all kinds—use root hairs in this way. Nonvascular plants, such as mosses, do not grow roots but nevertheless collect minerals using hairlike rhizoids, which are developmentally similar to the root hairs of larger plants. Each root hair is a single cell in which the cell's contents are elongated to create a very long, narrow, tubelike structure. As a result, root hairs are extremely fragile and easily damaged by any kind of movement.

Large main root of mature plant functions as food storage

RADISH

Harvesting nutrients
To secure the plant's future, the dense matting of hairs on the root of a young radish (*Raphanus sativus*) must not only collect water but also take in minerals. The growing root tip lacks root hairs because root hairs are too delicate to force their way through soil.

Root cap protects growing tip as it pushes through soil

HOW ROOT HAIRS ABSORB WATER
A root hair takes in water through the process of osmosis, which forces water to move across the cell's outer membrane and into the cell. The osmotic force arises because there is a greater concentration of dissolved substances in the cell's watery cytoplasm than in the soil moisture outside the cell. This difference in concentration draws water into the cell. The water molecules are small enough to move freely across the cell membrane. Ions of dissolved minerals are too large to cross the cell membrane. Instead, they must be "pumped" into the cell by a process called active transport, using energy released by respiration inside the cell.

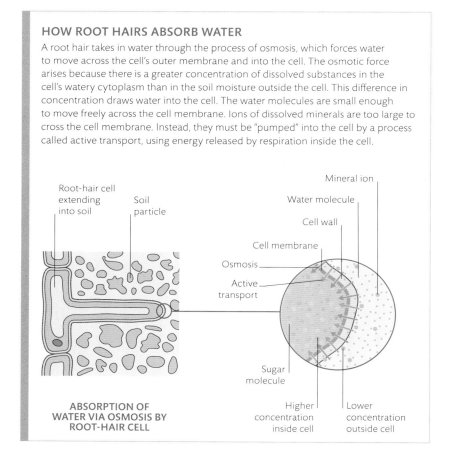

ABSORPTION OF WATER VIA OSMOSIS BY ROOT-HAIR CELL

Growing part of root absorbs little water or nutrients

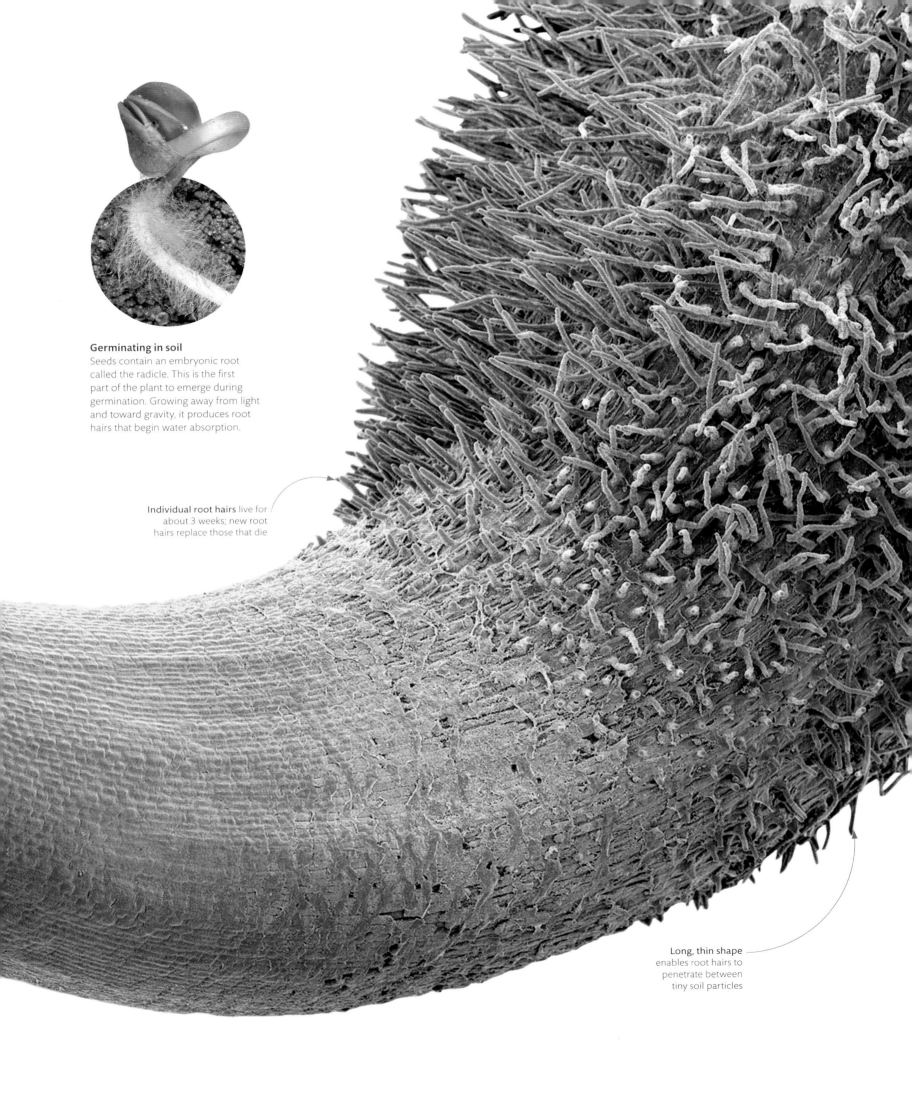

Germinating in soil
Seeds contain an embryonic root called the radicle. This is the first part of the plant to emerge during germination. Growing away from light and toward gravity, it produces root hairs that begin water absorption.

Individual root hairs live for about 3 weeks; new root hairs replace those that die

Long, thin shape enables root hairs to penetrate between tiny soil particles

PREPARING TO DIGEST

The hairlike tentacles on a sundew plant respond to the touch and movement of a struggling insect by curling inward, and eventually smother the victim with droplets of slime, called mucilage. The insect usually dies by suffocation as the slime blocks its spiracles (breathing holes), or it may die through exhaustion from struggling to break free.

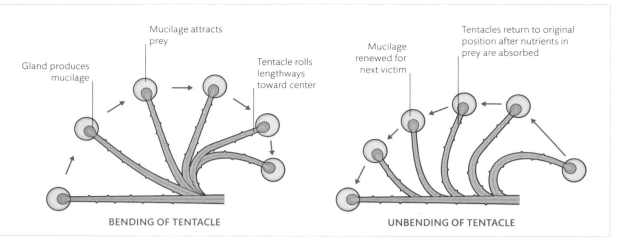

BENDING OF TENTACLE

UNBENDING OF TENTACLE

Glued to its fate

A Cape sundew (*Drosera capensis*) traps a blowfly with leaves that work like flypaper: tentacles tipped with sugary secretions lure the victim, which gets caught and dies in the sticky droplets. The leaf and its tentacles coil around the insect as digestive juices break it down.

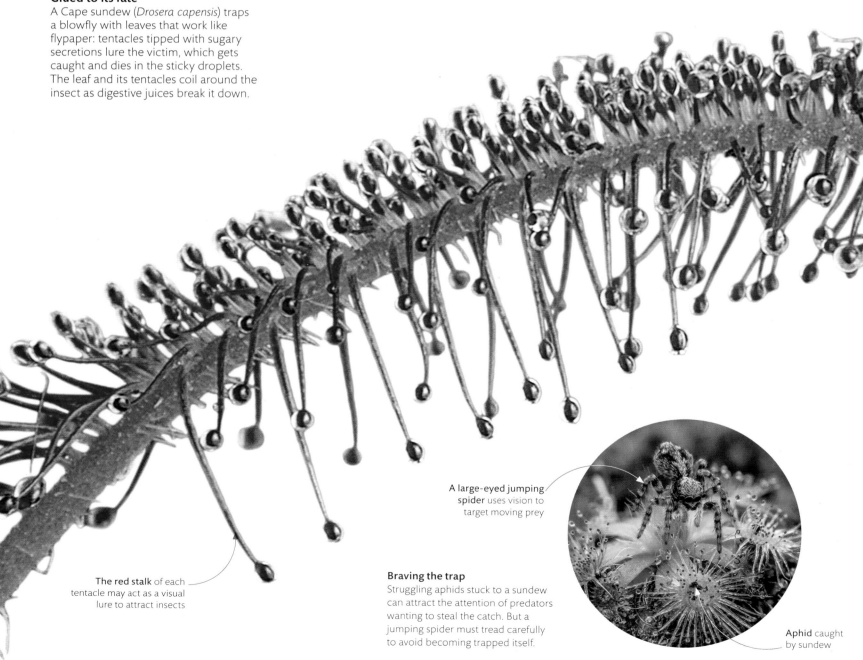

The red stalk of each tentacle may act as a visual lure to attract insects

A large-eyed jumping spider uses vision to target moving prey

Braving the trap

Struggling aphids stuck to a sundew can attract the attention of predators wanting to steal the catch. But a jumping spider must tread carefully to avoid becoming trapped itself.

Aphid caught by sundew

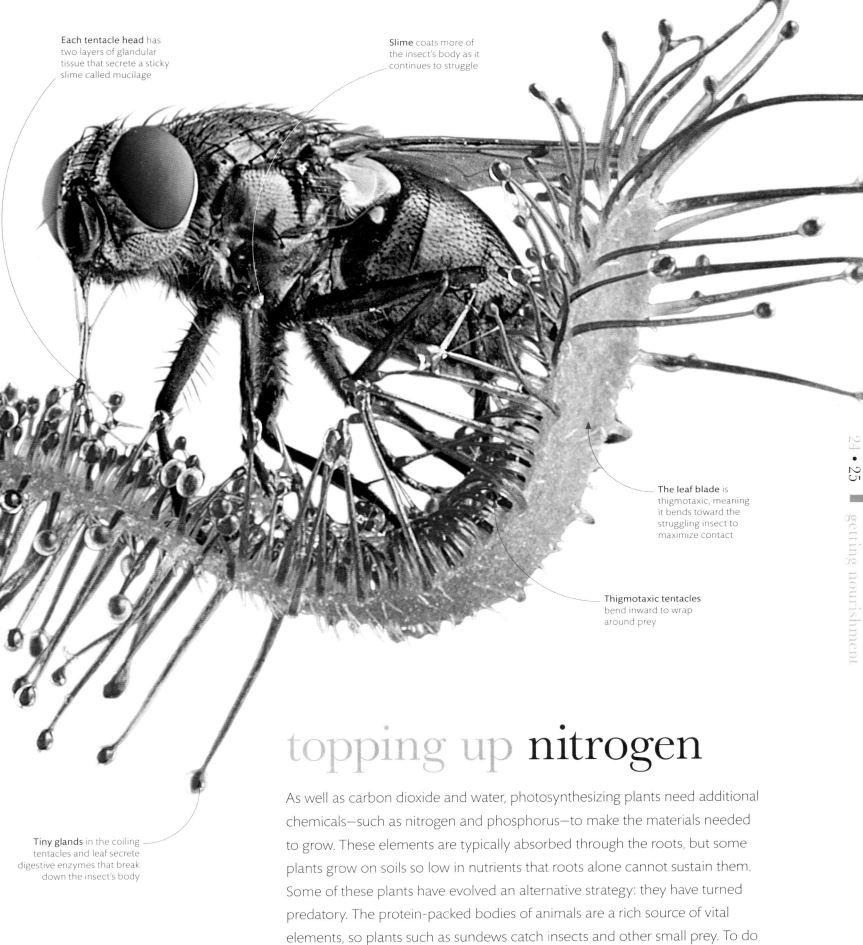

Each tentacle head has two layers of glandular tissue that secrete a sticky slime called mucilage

Slime coats more of the insect's body as it continues to struggle

The leaf blade is thigmotaxic, meaning it bends toward the struggling insect to maximize contact

Thigmotaxic tentacles bend inward to wrap around prey

Tiny glands in the coiling tentacles and leaf secrete digestive enzymes that break down the insect's body

topping up nitrogen

As well as carbon dioxide and water, photosynthesizing plants need additional chemicals—such as nitrogen and phosphorus—to make the materials needed to grow. These elements are typically absorbed through the roots, but some plants grow on soils so low in nutrients that roots alone cannot sustain them. Some of these plants have evolved an alternative strategy: they have turned predatory. The protein-packed bodies of animals are a rich source of vital elements, so plants such as sundews catch insects and other small prey. To do this, they use a trap to catch the moving target and then digest the solid body.

fixing nitrogen

All living things need nitrogen to make protein and other vital nitrogen-containing materials. More than 75 percent of air is pure gaseous nitrogen, yet few organisms are able to take nitrogen straight from the atmosphere in this simple, unreactive state. Animals and decomposers obtain nitrogen from their food, while plants absorb it as the mineral nitrate from the soil. Some plants have evolved a special relationship with certain microbes that can convert atmospheric nitrogen into protein in a process called "fixing." Legumes—such as clovers, peas, and beans—boost their nitrogen supply by nurturing nitrogen-fixing soil bacteria within their roots.

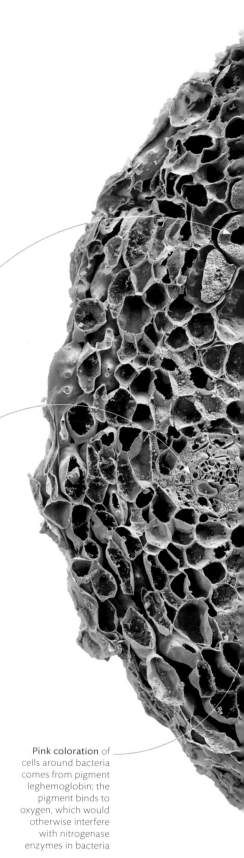

Pockets of nitrogen-fixing *Rhizobium* bacteria (blue) are packed with nitrogenase enzyme that converts atmospheric nitrogen into ammonium compounds, from which they—and the plant—make amino acids and protein

Vascular bundles (white) contain the plant's transport vessels; these transfer nitrogen compounds to the plant and sugars to the nodule

Pink coloration of cells around bacteria comes from pigment leghemoglobin; the pigment binds to oxygen, which would otherwise interfere with nitrogenase enzymes in bacteria

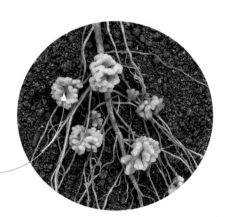

Nodule formation
Root nodules of a pea (*Pisum sativum*) grow when nitrogen-fixing bacteria invade the root hairs, multiply, and swell to form nodules that tap into the roots' transport vessels.

Nodules—here grouped as clusters—enclose the bacteria within a wall produced by the plant

NITROGEN CYCLE

The way nitrogen passes through an ecosystem depends on a wide variety of bacteria that are capable of using nitrogen in all its different forms. While nitrogen-fixers take nitrogen directly from the air, other bacteria produce the nitrate mineral in the soil or help recycle nitrogen back into the atmosphere.

KEY
- Complex organic nitrogen in organisms
- Simple inorganic nitrogen in air
- Simple inorganic nitrogen in soil

Plants use simple inorganic nitrogen (nitrate) from soil to make complex organic nitrogen, such as protein

Animals eat food containing protein and store complex organic nitrogen in their bodies

Soil bacteria break down the complex organic nitrogen of dead organisms into nitrate. This is called nitrification

Leguminous plants cultivate nitrogen-fixing bacteria in root nodules; the bacteria supply them with extra nitrogen

Nitrogen-fixing bacteria use atmospheric nitrogen to make organic nitrogen, such as protein

Bacteria in poorly aerated soil use nitrate instead of oxygen to release energy; nitrogen is released into the air as a byproduct. This is called denitrification

SIMPLIFIED NITROGEN CYCLE

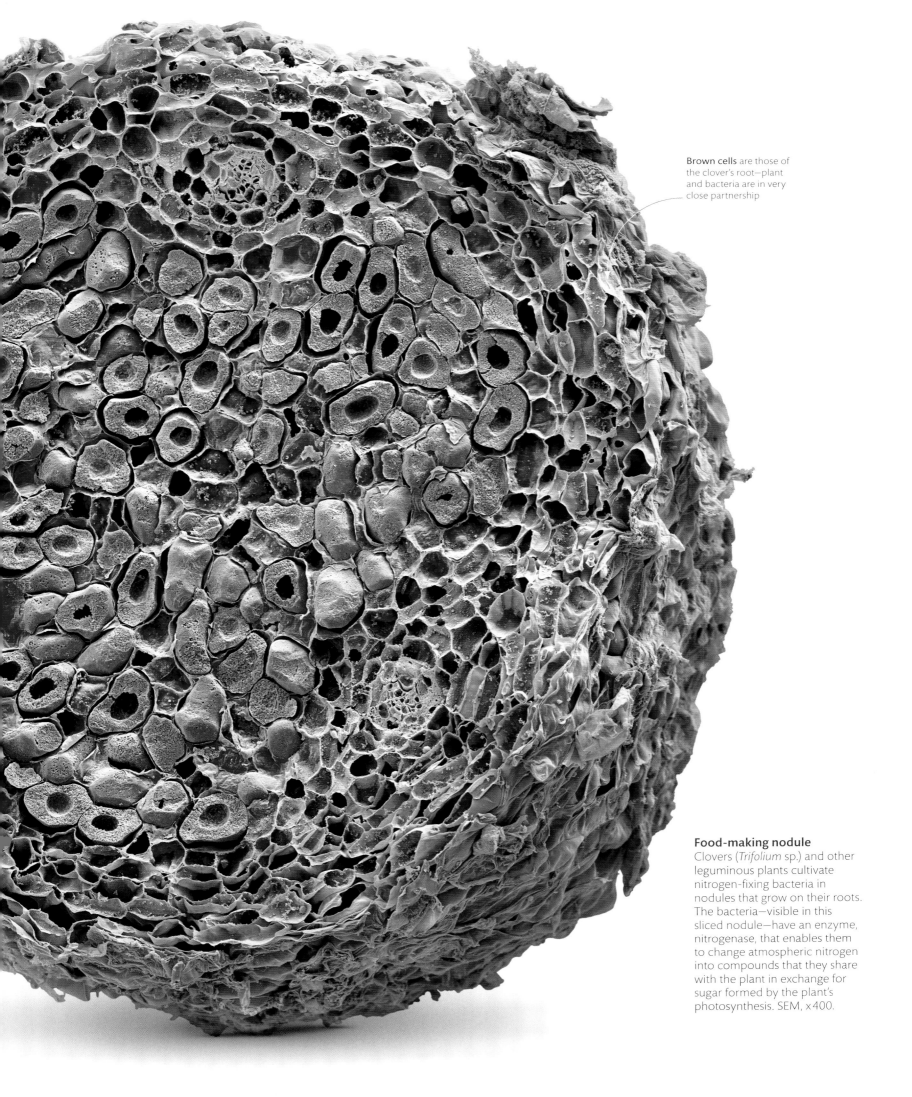

Brown cells are those of the clover's root—plant and bacteria are in very close partnership

Food-making nodule
Clovers (*Trifolium* sp.) and other leguminous plants cultivate nitrogen-fixing bacteria in nodules that grow on their roots. The bacteria—visible in this sliced nodule—have an enzyme, nitrogenase, that enables them to change atmospheric nitrogen into compounds that they share with the plant in exchange for sugar formed by the plant's photosynthesis. SEM, x400.

Spheres

A spherical bacterium is described as a coccus. Cocci can be found living singularly or in pairs called diplococci. Groupings of bacteria form when cells divide but stick together. Spherical bacteria that form chains of cells are called streptococci, while those bacteria that make more amorphous clusters, called staphylococci, form because the cells divide in multiple axes at once. One such bacterium is *Staphylococcus aureus*, which is commonly found on human skin.

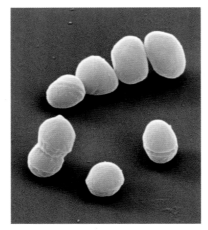

COCCUS
Enterococcus faecalis

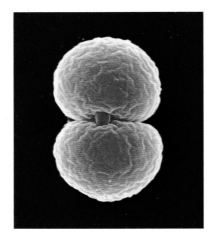

DIPLOCOCCUS
Neisseria gonorrhoeae

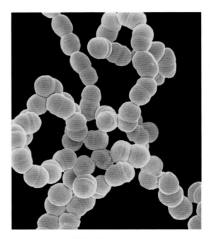

STREPTOCOCCUS
Streptococcus pyogenes

Rods

A bacterium that has a rod or cylindrical shape is called a bacillus. Bacilliform bacteria include *Escherichia coli* (see pp.30–31), a gut bacterium that can cause food poisoning. As with cocci, streptobacillus bacteria exist in chains, however some streptobacilli, known as palisades, form clusters side by side. Staphylobacilli are found in jumbled clusters similar to a bunch of grapes. Coccobacillus bacteria have very short rods, which can easily be mistaken for cocci.

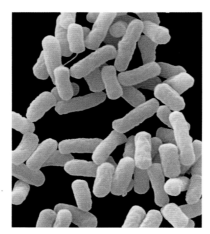

BACILLUS
Escherichia coli

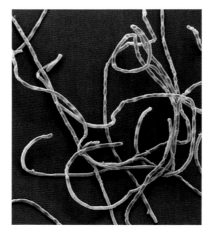

STREPTOBACILLUS
Bacillus anthracis

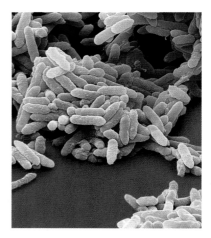

STREPTOBACILLUS PALISADES
Aquaspirillum

Complex shapes

Bacteria can adopt a wider range of shapes beyond rods and spheres. These include filaments, where a rod-shaped cell elongates without dividing, and spirals, which have more complex cell structures (see pp.32–33). There are two types of spiral-shaped bacteria—spirillum bacteria, which have rigid cells, and spirochetes, which are thinner and more flexible. Some bacteria have a curved bean shape, known as a vibrio.

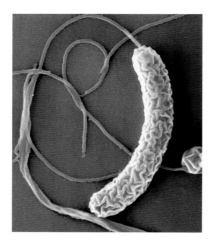

VIBRIO
Vibrio cholerae

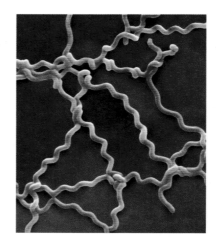

SPIROCHETE
Leptospira interrogans

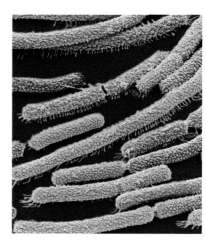

FILAMENT
Bacillus megaterium

bacteria

Bacteria are one of the simplest forms of life. They have lived for at least 4 billion years, were the first organisms to consume food, and—along with Archaea—were the only form of life for the first 1.5–2 billion years. Bacteria have diversified a great deal over billions of years. Each single-celled bacterium is only a few micrometers long and can live anywhere that it is able to feed and respire. Some make their own food using energy in light or minerals, while others absorb ready-made organic food. The wide variety of bacteria differ in their shape, and the way they form chains and clusters.

STAPHYLOCOCCUS
Staphylococcus aureus

COCCOBACILLUS
Brucella abortus

SPIRILLUM
Helicobacter pylori

Cluster of cells
Bifidobacterium spp., one of the most important types of bacterium in the human gut, is acquired by infants through breast feeding. It forms clusters as the cells divide and multiply along different axes in a characteristic "Y" shape. The cells therefore extend forward and back in disordered chains.

Cell branches apart into irregular "Y" shape as it divides

Bacterial cell in process of dividing in two

Bifidobacterium lacks flagella, so is unable to propel itself through its environment

E. coli, or *Escherichia coli* to give it its full name, is a bacterium commonly found in the gut that has a mixed reputation. On one hand, some virulent strains are occasionally implicated in serious—and potentially fatal—food poisoning cases. On the other, *E. coli* is an important model organism, a status it shares with baker's yeast, the lab mouse, and the *Drosophila* fruit fly. A model organism is a species that is easy to maintain and study

spotlight *Escherichia coli*

in a laboratory and is used in scientific investigations to inform our understanding in areas such as developmental biology, genetics, infection, and disease.

Most people have a more intimate—and beneficial relationship with *E. coli*. The bacterium lives in the lower intestine of all warm-blooded animals, including humans. It makes up about 0.1 percent of the human microbiota, which translates into about 100 billion *E. coli* per person. *E. coli* enters the gut by fecal–oral transmission, but once present it generally does more good than harm by colonizing the intestine and helping to prevent infection by pathogenic, or disease-causing, bacteria. There is little free oxygen in the gut, and *E. coli* is able to survive very well in the anaerobic conditions there. It consumes a wide range of chemicals, releasing energy from them by fermentation. One of the byproducts of *E. coli* metabolism is menaquinone, better known as vitamin K2, which is an essential component of both the blood-clotting system and healthy bones (and which can also be obtained from foods, in common with other vitamins).

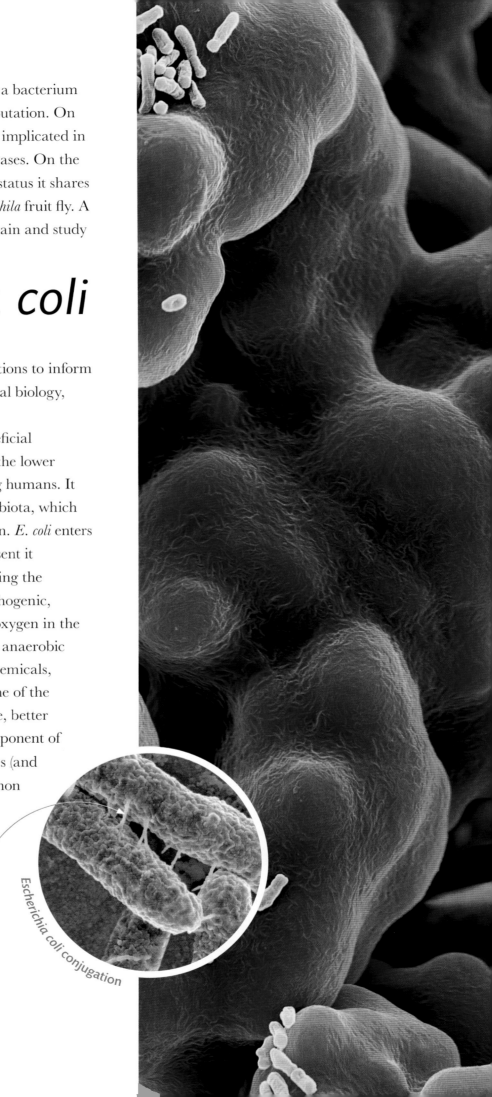

Threadlike pili connect *E. coli* cells so they can transfer DNA

Escherichia coli conjugation

Intestinal bacteria
E. coli can be seen as yellow "rods" in this *in vitro* sample. Inside the body, the 2-micrometer-long cells double their numbers by asexual cell division roughly every 90 minutes. To boost genetic diversity, *E. coli* uses a process called conjugation, in which rings of DNA called plasmids pass between cells. SEM, x7,500.

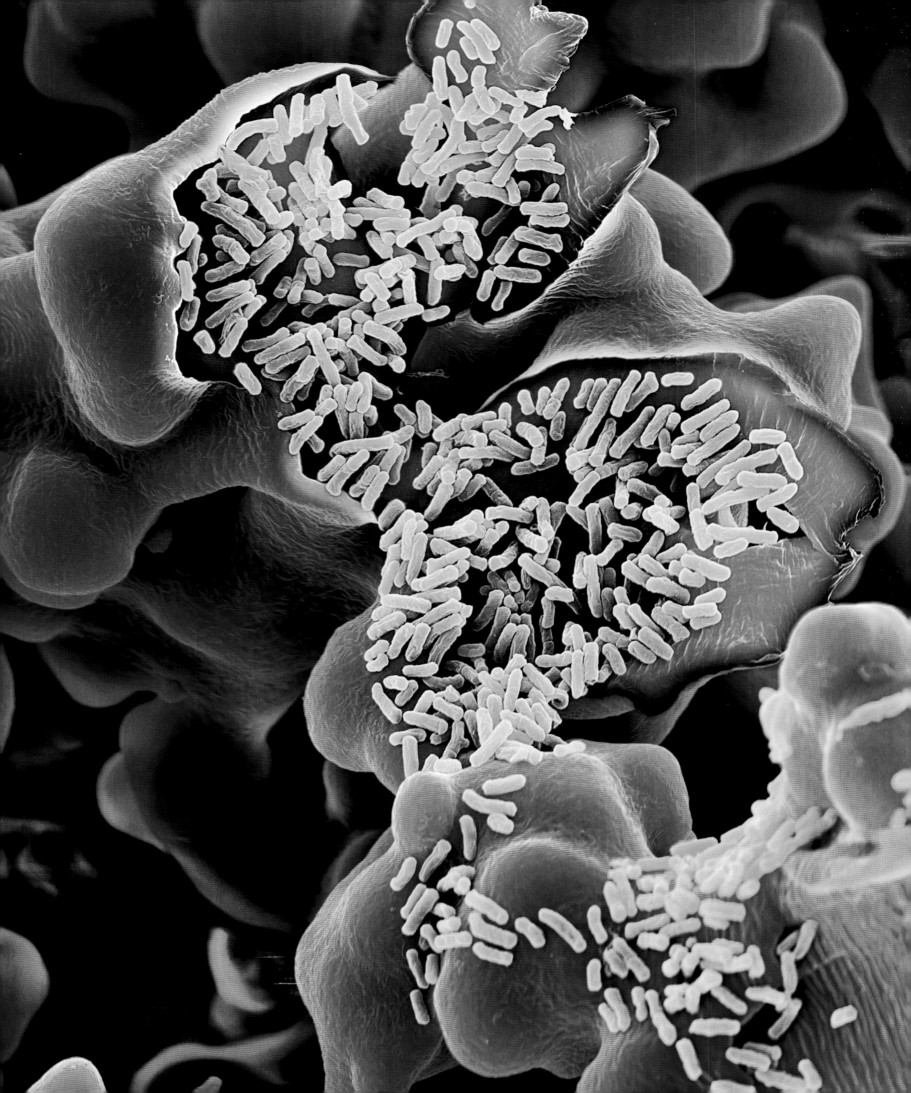

damaging a host

Bacteria can get nourishment in one of two ways. Some bacteria, called autotrophs, produce their own food like plants and algae do (see pp.20–21), while others—known as heterotrophs—gain nutrients by consuming other organisms, or their products. Heterotrophic bacteria secrete enzymes to break down their food, but in some harmful bacteria these enzymes can be used to gain entry into a host or evade the immune system. These destructive digestive enzymes can damage the cells of the host, causing disease. Some of the most dangerous bacterial diseases, such as anthrax, exert their effect in this way, while others may even be implicated in causing cancer.

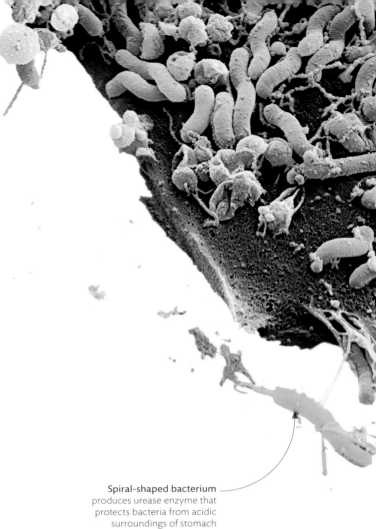

Spiral-shaped bacterium produces urease enzyme that protects bacteria from acidic surroundings of stomach

GRAM-NEGATIVE AND GRAM-POSITIVE BACTERIA
Gram staining, a method that uses a purple dye to stain cells, helps to categorize bacteria. Gram-positive bacteria have cell walls that absorb the dye, while Gram-negative bacteria have an extra membrane that limits absorbency. This also makes Gram-negative bacteria more resistant to antibiotics.

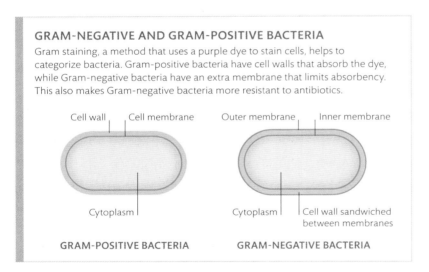

SPIRAL-SHAPED BACTERIA
Some bacteria that populate mammal stomachs have one of two types of spiral structure. Spirilliums, like *Helicobacter pylori*, are rigid with external flagella, while spirochetes are flexible spirals with internal flagella. These spiral arrangements give the bacteria a twisting motion as the cells move.

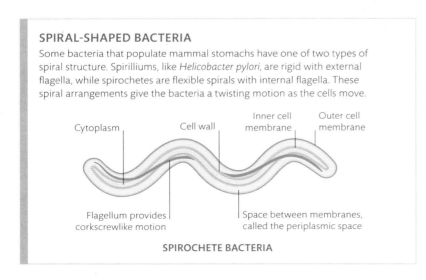

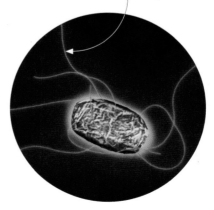

Rotating flagella help *Salmonella* swim through environment

Salmonella
The bacterium *Salmonella* sp. infects its host's intestines, and causes food poisoning. It is a "Gram-negative" bacterium (left) with a double membrane that makes it resistant to treatment.

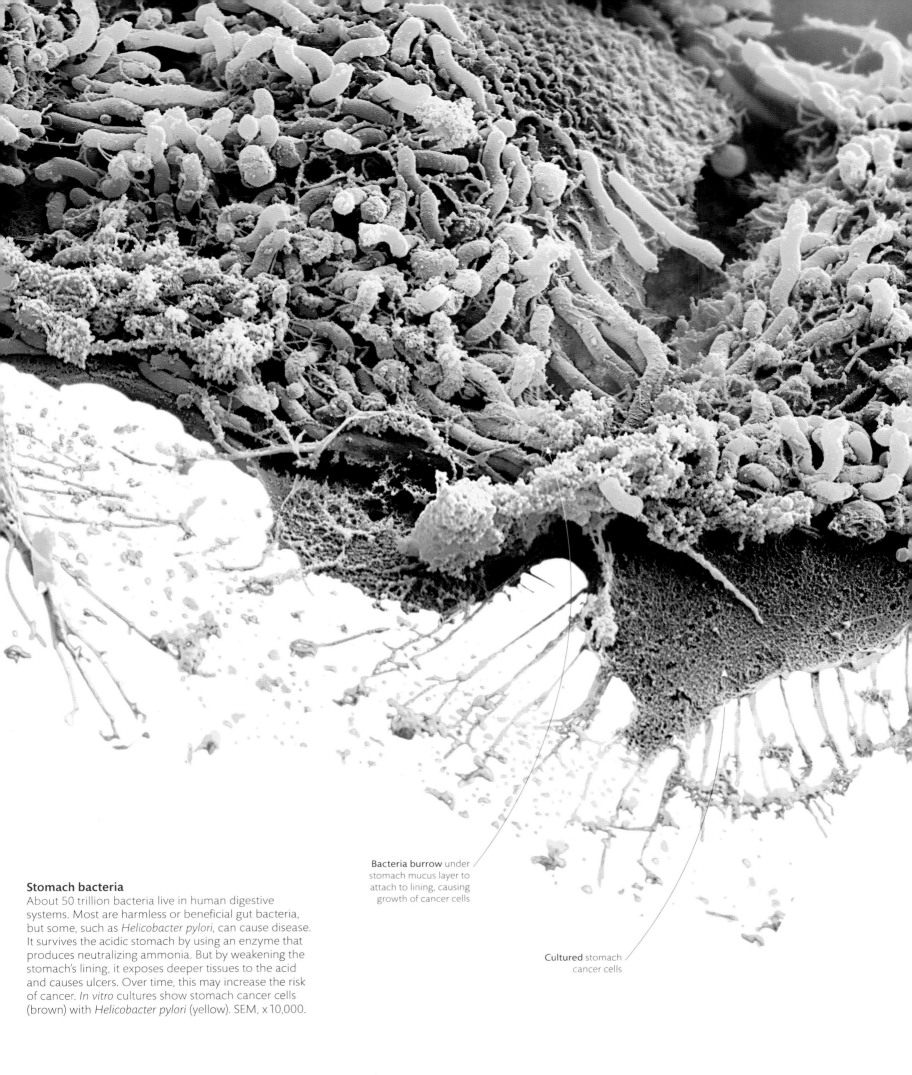

Bacteria burrow under stomach mucus layer to attach to lining, causing growth of cancer cells

Cultured stomach cancer cells

Stomach bacteria
About 50 trillion bacteria live in human digestive systems. Most are harmless or beneficial gut bacteria, but some, such as *Helicobacter pylori*, can cause disease. It survives the acidic stomach by using an enzyme that produces neutralizing ammonia. But by weakening the stomach's lining, it exposes deeper tissues to the acid and causes ulcers. Over time, this may increase the risk of cancer. *In vitro* cultures show stomach cancer cells (brown) with *Helicobacter pylori* (yellow). SEM, x10,000.

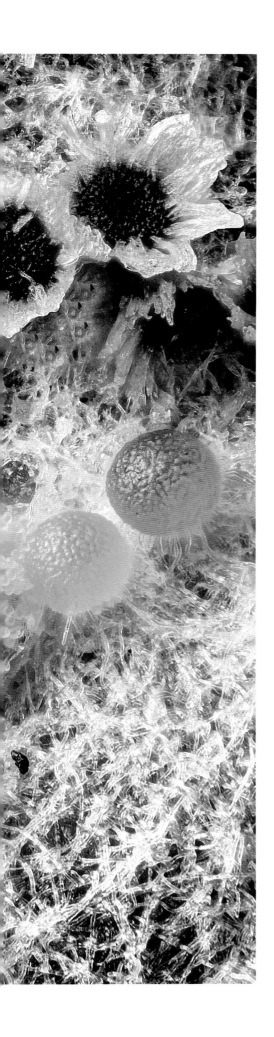

Toadstools are the spore-producing part of the fungus

Essential rot
Hyphae connected to these glistening inkcap (*Coprinellus micaceus*) toadstools are hidden in the rotting log. They rot the dead wood for food, releasing nutrients into the forest floor.

Fungal carpet
A white fuzz of hyphae from the powdery mildew fungus *Erysiphe adunca* covers the surface of a willow leaf. Needlelike side branches from the hyphae, known as haustoria, grow into the leaf and absorb nutrients. The carpet of hyphae restricts photosynthesis, damaging but not killing the leaf beneath. What look like fancy confectionery items are cleistothecia. Less than 0.3 mm across, they are the spore-producing structures of the fungus.

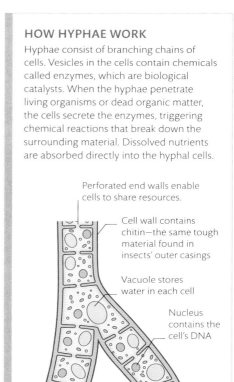

HOW HYPHAE WORK
Hyphae consist of branching chains of cells. Vesicles in the cells contain chemicals called enzymes, which are biological catalysts. When the hyphae penetrate living organisms or dead organic matter, the cells secrete the enzymes, triggering chemical reactions that break down the surrounding material. Dissolved nutrients are absorbed directly into the hyphal cells.

- Perforated end walls enable cells to share resources.
- Cell wall contains chitin—the same tough material found in insects' outer casings
- Vacuole stores water in each cell
- Nucleus contains the cell's DNA
- Vesicle contains enzymes
- Growing tip has a greater concentration of vesicles

STRUCTURE OF A HYPHA

absorbing food

Organisms that cannot produce their own food by photosynthesis must find other ways to obtain the energy they need to live, grow, and reproduce. Fungi do this via microscopic threads called hyphae. These threads link up to form a network known as a mycelium, which is the main body of a fungus. Collectively, the hyphae provide a large surface area for absorbing nutrients from their surroundings. Most fungi are decomposers—they feed on dead or decaying plant and animal matter. Some are parasitic, feeding off living things and sometimes penetrating deep into their tissues to do so.

Named after the Latin for "paintbrush" because of the appearance of their branched asexual structures, *Penicillium* species of fungi are ubiquitous in damp soils worldwide. These fungi are most commonly encountered as the molds that attack fruit, bread, and other foodstuffs, and as the blue veins and white edible coatings of some cheeses. They are also the natural source of penicillin, the world's first antibiotic.

spotlight *Penicillium*

Like many fungi, *Penicillium* species are significant decomposers, reducing dead plant and animal remains to organic fragments that enrich the soil. While detritivores ingest dead material and digest it internally, decomposer fungi such as *Penicillium* use external chemical reactions to break down food and absorb its nutrients into the threadlike hyphae that form the mycelium—the fungal body. As they feed, they also secrete chemicals that inhibit the growth of bacteria, which would otherwise compete for the same food sources. These chemicals have been harnessed by medicine for use as infection-fighting antibiotics, and are thought to have saved hundreds of millions of lives.

Penicillium fungi can reproduce vegetatively—with hyphae that break off being able to develop into new mycelia—and some species are capable of sexual reproduction, but they mostly rely on asexually produced spores to spread copies of themselves. The microscopic spores are dispersed by air and water. It is estimated that humans inhale between 1,000 and 10 billion *Penicillium* spores every day. Although typically harmless, the spores of some species can produce allergic reactions and may even trigger more serious conditions.

Droplets of exudate (in yellow), a source of penicillin, form on the surface of a *Penicillium* colony

penicillin forming on a colony

Spore production
Penicillium fungi produce asexual spores known as conidia (shown in yellow) on the tips of brushlike hyphal threads called conidiophores (the pink stalks). The spores often contain blue or green pigments, which is why the fungi are sometimes referred to as blue-green molds. SEM, x220.

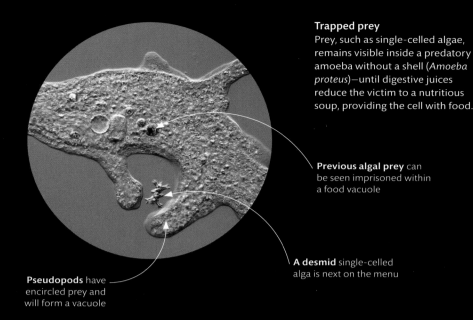

Trapped prey
Prey, such as single-celled algae, remains visible inside a predatory amoeba without a shell (*Amoeba proteus*)—until digestive juices reduce the victim to a nutritious soup, providing the cell with food.

Previous algal prey can be seen imprisoned within a food vacuole

A desmid single-celled alga is next on the menu

Pseudopods have encircled prey and will form a vacuole

Epipodium is the shell opening, through which the organism extends its pseudopods

engulfing prey

Like animals, many kinds of single-celled organisms survive by consuming organic food. A lot of nutrients, such as sugars, are soluble and can be absorbed by the organisms through their cell-surface membranes. Some microbes, however, can deal with more solid meals. These shape-shifting micro-predators probe outward with the thin jelly of their cytoplasm to form pseudopods, or "false feet," that ensnare smaller organisms such as bacteria, algae, and even the occasional tiny animal. The prey gets trapped inside a bubblelike food vacuole, where it is digested alive.

PHAGOCYTOSIS

Food particles are gathered by pseudopods in a process called phagocytosis, meaning "cell eating." The cell makes vesicles called lysosomes that contain digestive enzymes. As pseudopods join to create a food vacuole, the lysosomes fuse with the vacuole and release their contents. The liquefied products of digestion are then absorbed into the cell's cytoplasm.

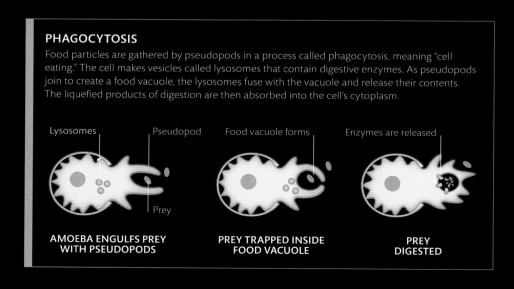

Lysosomes | Pseudopod | Food vacuole forms | Enzymes are released
Prey

AMOEBA ENGULFS PREY WITH PSEUDOPODS | **PREY TRAPPED INSIDE FOOD VACUOLE** | **PREY DIGESTED**

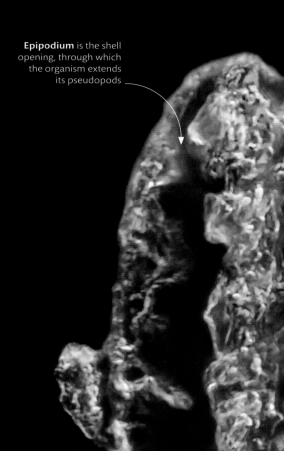

Micro-predator within
This microscopic "vase," seen at 2,000 times life size, is built from hundreds of sand grains and other glassy particles. It harbors a predatory shelled amoeba (*Difflugia sp.*). From the protection of its casing, the microbe extends its pseudopods through the flared opening to catch prey and also to collect more construction materials. The particles are cemented together with extraordinary intricacy using sticky secretions.

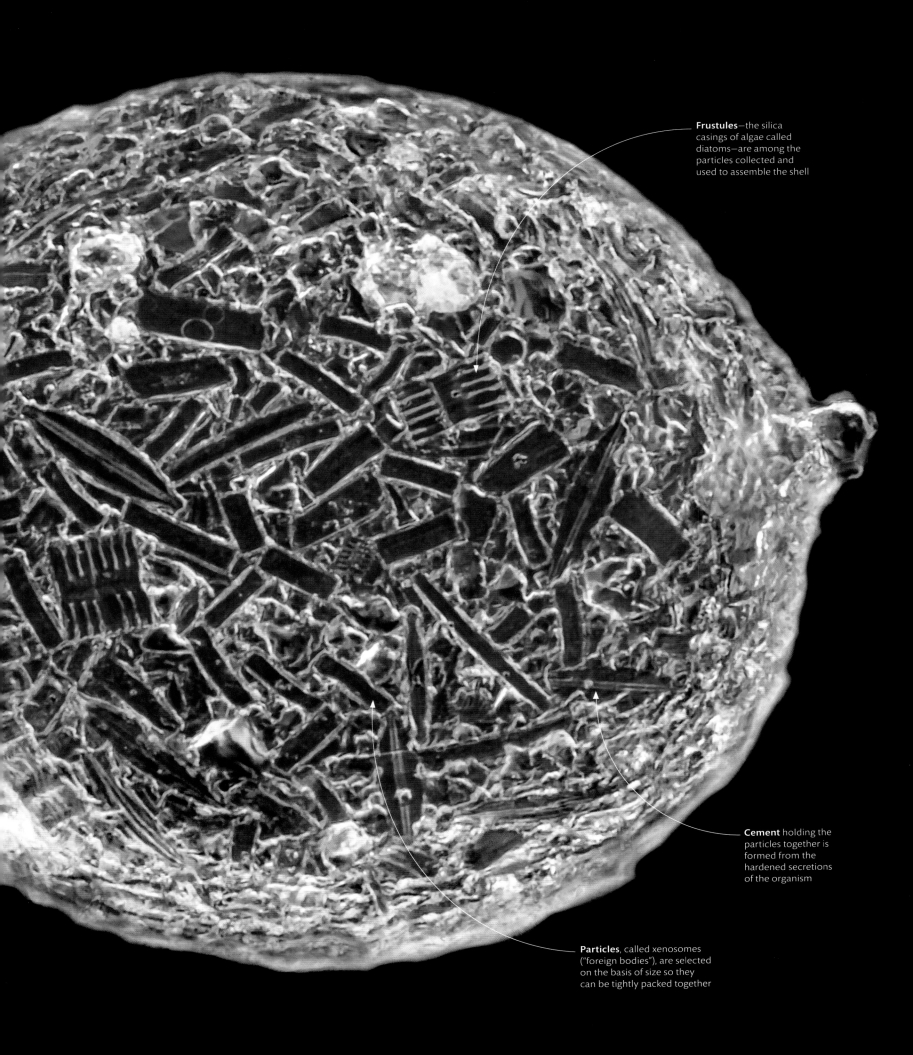

Frustules—the silica casings of algae called diatoms—are among the particles collected and used to assemble the shell

Cement holding the particles together is formed from the hardened secretions of the organism

Particles, called xenosomes ("foreign bodies"), are selected on the basis of size so they can be tightly packed together

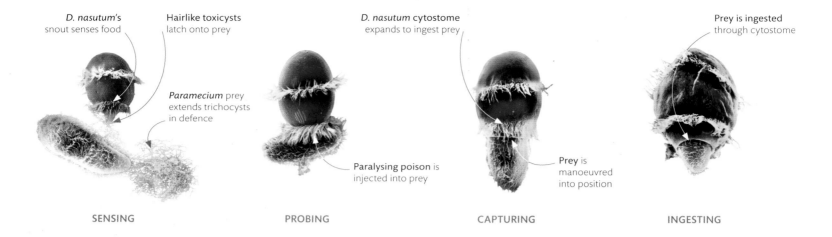

| SENSING | PROBING | CAPTURING | INGESTING |

microbial predator

While some microbes create their own food or scavenge for organic matter, others target prey to feed on. There are two types of predatory microbe: facultative predators that have a varied diet of prey and other organic matter, and obligate predators that hunt and feed exclusively on other microbes. One such obligate microbe is *Didinium nasutum*, a prodigious hunter that uses specialized hairlike structures, called toxicysts, to paralyze its prey. When prey is scarce, the microbe can lie dormant as a cyst for up to 10 years until prey returns.

Predation in action
Didinium nasutum posesses toxicysts (threadlike poisonous cells) that it ejects from an organelle in its proboscis when it senses food. *D. nasutum*, and related species, then maneuver their prey for ingestion so that the cytostome (mouth) can expand to engulf the prey completely prior to digestion.

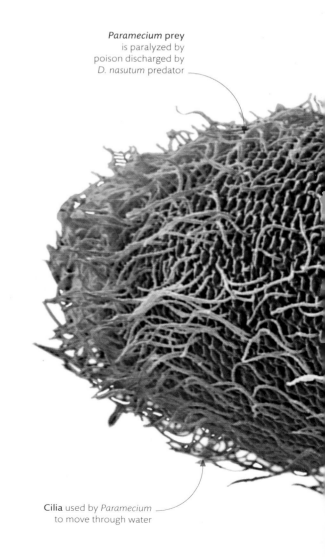

CONSUMING BACTERIUM
Even simple bacteria can be predators. *Bdellovibrio bacteriovorus*, a species of bacterium often found in the human gut, has been shown to eradicate a number of bacteria, including drug-resistant *Escherichia coli*, a major cause of food poisoning. This predatory bacterium enters its prey and kills it from the inside.

PREDATORY BACTERIUM ATTACKING PREY

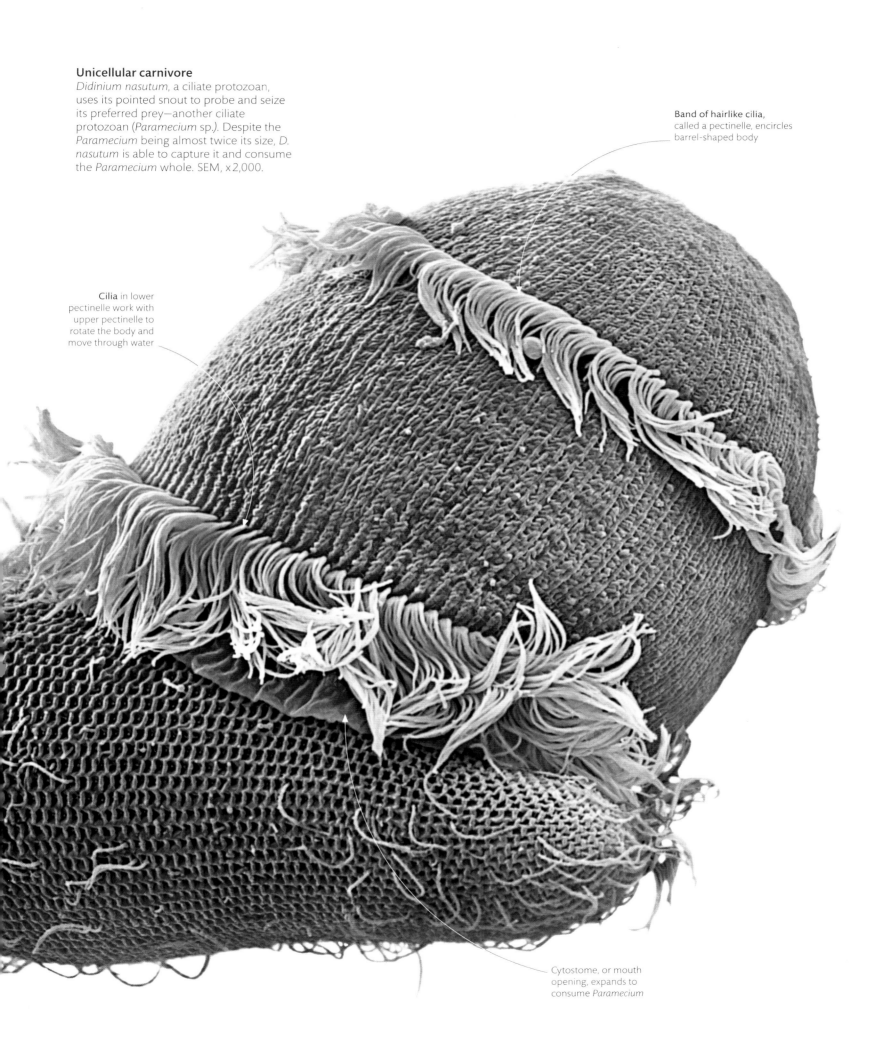

Unicellular carnivore
Didinium nasutum, a ciliate protozoan, uses its pointed snout to probe and seize its preferred prey—another ciliate protozoan (*Paramecium* sp.). Despite the *Paramecium* being almost twice its size, *D. nasutum* is able to capture it and consume the *Paramecium* whole. SEM, ×2,000.

Band of hairlike cilia, called a pectinelle, encircles barrel-shaped body

Cilia in lower pectinelle work with upper pectinelle to rotate the body and move through water

Cytostome, or mouth opening, expands to consume *Paramecium*

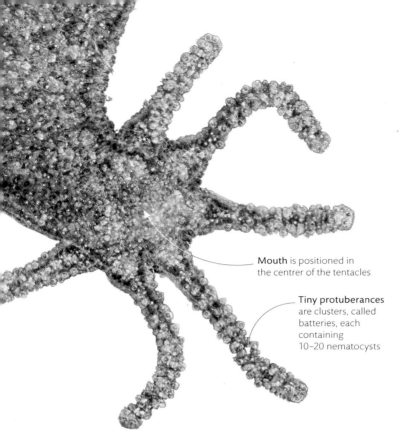

Mouth is positioned in the center of the tentacles

Tiny protuberances are clusters, called batteries, each containing 10–20 nematocysts

Coiled and ready to strike
The large nematocysts found in the folds of the mushroom polyp *Discosoma* are highly effective weapons, used primarily not for feeding, but for defence. The long, coiled tubule bears tiny spines along its length and is armed with barbs at its base. When stimulated, the tubule explodes out at high speed, generating pressure sufficient for the barbs to penetrate their target. The orange spheres are zooxanthellae—algae living in the coral's tissues. LM, x 1,300.

Miniature pond predator
Highly effective predators, hydra (*Hydra* spp.) are less than ½ in (15mm) long, but catch prey up to half their size. Movement near the tentacles causes nematocysts to discharge. The prey is pushed though the mouth and into the gastric cavity.

stinging cells

Hydroids, anemones, corals, and jellyfish possess unique cellular structures called cnidae, used mainly for piercing, poisoning, or entangling their prey. Cnidae are tiny capsules up to 0.1 mm long, with a hollow, coiled tubule extending from one end. When stimulated, the capsule opens, firing its tubule. A fired tubule is everted—turned inside out. It cannot be fired again and must be replaced. Cnidae develop within the body of the animal and then move to the ectoderm (outer layer), especially on the tentacles, where they are clustered into batteries.

TYPES OF CNIDAE
There are 28 known kinds of cnidae. Nematocysts inject stinging venom and usually have barbs to keep the threads embedded in their prey. Spirocysts entangle their prey with sticky hairs. Ptychocysts, found only in cerianthid tube anemones, are used for tube building and entangling prey.

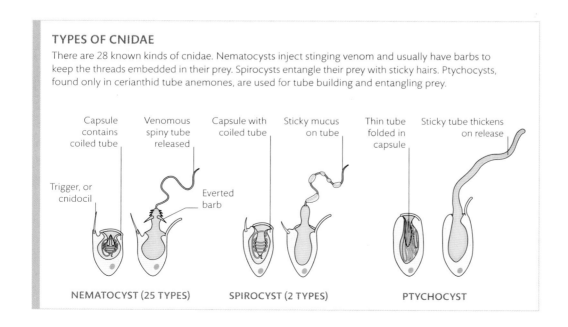

NEMATOCYST (25 TYPES) — Capsule contains coiled tube; Trigger, or cnidocil; Venomous spiny tube released; Everted barb

SPIROCYST (2 TYPES) — Capsule with coiled tube; Sticky mucus on tube

PTYCHOCYST — Thin tube folded in capsule; Sticky tube thickens on release

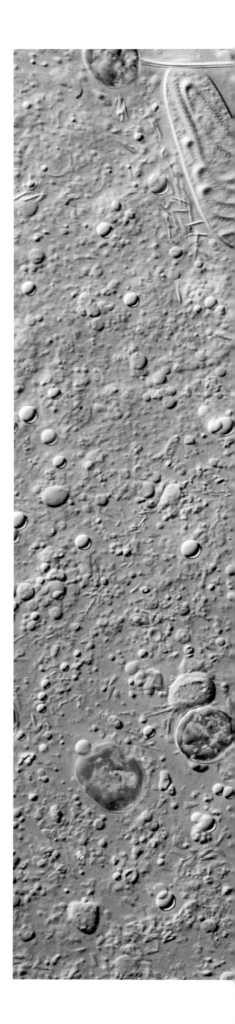

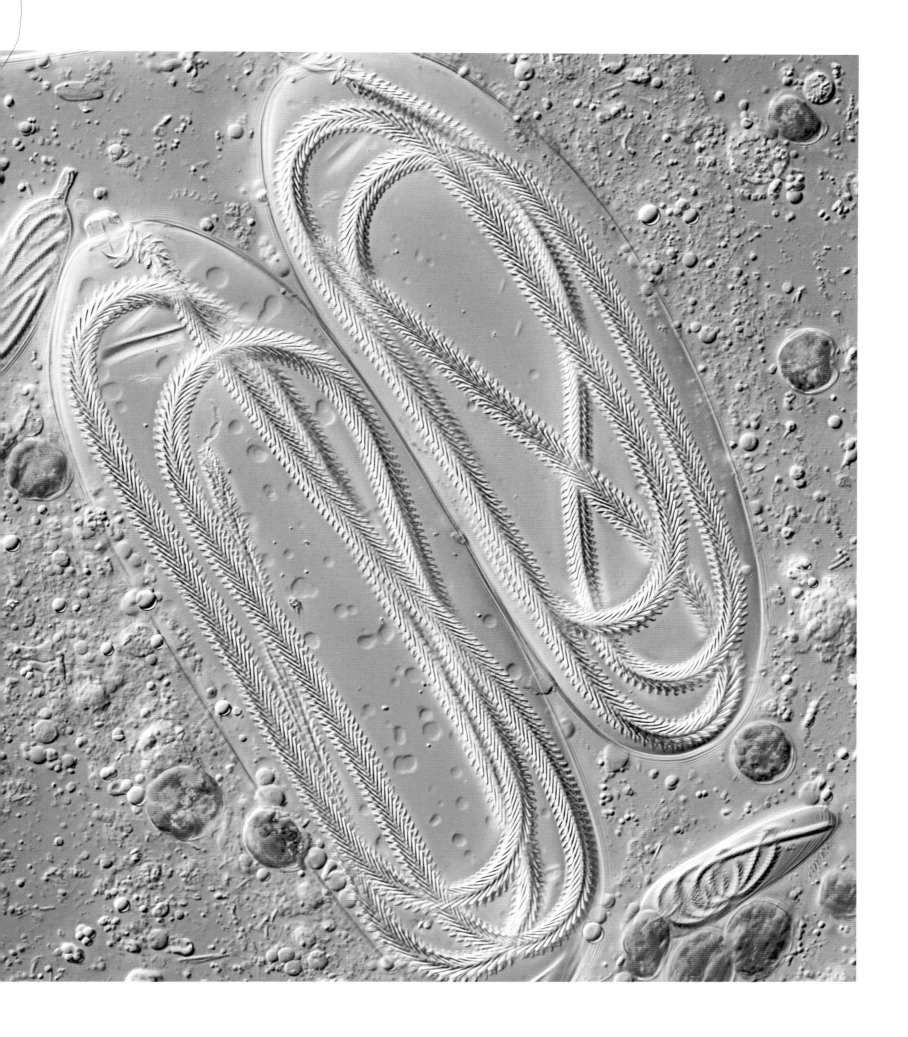

feeding on particles

From microorganisms to worms and whales, aquatic animals thrive on food particles such as plankton and organic detritus that float in the water. Most microorganisms that feed in this way are passive suspension feeders that depend on water flow to bring particles to their feeding structures. Active suspension feeders move through the water or create their own currents with a pumping or wafting mechanism. Most large, complex animals—and some microorganisms—that actively suspension-feed use a special net or mesh to filter particles.

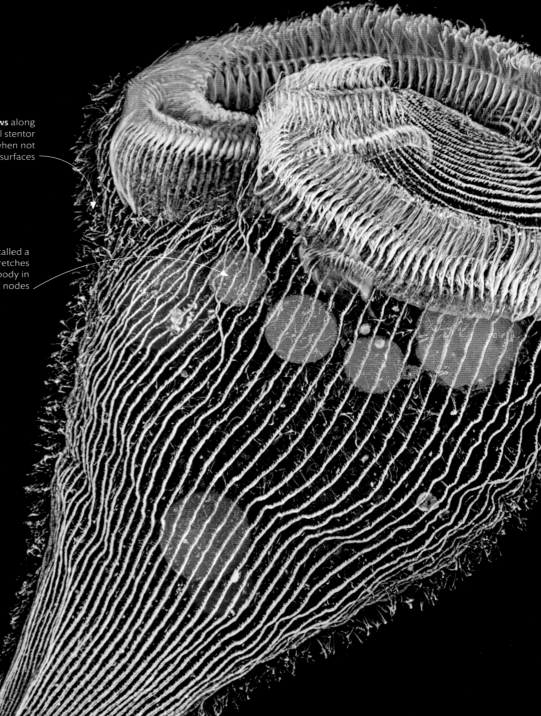

Broad stripes of pigment granules run along narrow furrows that carry microtubules for changing cell shape

Cilia aligned in rows along body propel stentor through water when not attached to surfaces

Large nucleus, called a macronucleus, stretches down length of body in interconnected nodes

Single cell, complex body
Stentors are trumpet-shaped microbes that have a nucleus and complex body, but lack features that would classify them as plants, animals, or fungi. They are large by single-celled standards—growing up to 0.1 in (3 mm) tall—and mostly live in fresh water, attached to vegetation or occasionally swimming. Stentors are ciliates and, when anchored, they suspension-feed by using oral cilia to waft water into mouthlike openings. SEM, x 100.

Oral cilia on membranellar band beat constantly, creating a vortex to capture food particles

About 250 plates, each formed from two or three rows of tightly-packed cilia, create a mouthlike opening called the membranellar band

Holdfast anchors stentor to plants and other surfaces

Each cell has a beating flagellum

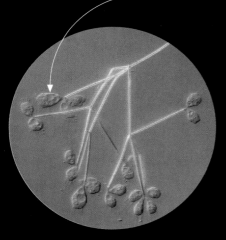

Feeding colony
Each cell of the colonial choanoflagellate *Codosiga umbellata* creates a feeding current using its flagellum. This pulls water toward a collar of protrusions (microvilli) where foods such as bacteria are trapped.

TYPES OF SUSPENSION FEEDERS

Both microscopic ciliates and larger barnacles are suspension feeders, but the way that food is captured is different. The barnacle is a filter-feeder—its leg apparatus sweeps through the water, orienting into the current, and traps particles in a mesh. By contrast, the ciliate creates a current but has no sieve or mesh, so engulfs particles one by one.

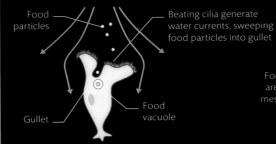

Food particles

Beating cilia generate water currents, sweeping food particles into gullet

Gullet

Food vacuole

SUSPENSION-FEEDING MICROSCOPIC CILIATE

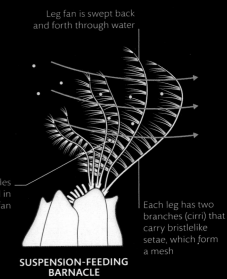

Leg fan is swept back and forth through water

Food particles are trapped in mesh of leg fan

Each leg has two branches (cirri) that carry bristlelike setae, which form a mesh

SUSPENSION-FEEDING BARNACLE

COMPONENT MOUTHPARTS

The complex mouthparts of all insects consist of five main components that develop from successive sections of the head. At the front is a liplike labrum, followed by paired mandibles and a tonguelike hypopharynx. There are two pairs of maxillae, including the labium, that work like accessory mouthparts. In most insects, the maxillae have palps at the end that are used to manipulate or taste the food.

Mandible with serrated edges used for biting and chewing

First maxilla good for grasping

Labrum (upper lip)

Second maxilla (labium or lower lip) used for biting

MOUTHPARTS OF A GRASSHOPPER

Mandible — *Labrum*

Hypopharynx—a tonguelike organ that releases saliva

First maxilla

Serrated edge of mandible

Maxillary palp

Labium

Second maxilla

Labial palp

DISARTICULATED MOUTHPARTS

Multi-jointed palps of first maxilla

Raptorial limb has serrated edges that grab prey

Mighty mandibles
The head of this Reddish ground beetle (*Carabus rutilans*) is armed with mouthparts that make it a formidable predator of other invertebrates such as slugs, snails, worms, and insects. With more than 40,000 species worldwide, the ground beetle family has thrived due to its biting mouthparts.

Grasping limbs
The Asian praying mantis (*Hierodula doveri*) catches its prey—here its mate—with its limbs. Its mandibles—much smaller than those of a ground beetle—are sharp enough to dismember its kill.

connecting mouthparts

An animal's mouth needs a special set of equipment to ingest food, whether it bites and chews solids or sucks up liquids. Insects have multiple mouthparts that they use to manipulate, eat, and taste food. Their mouthparts have evolved into different tools that deal with different foodstuffs—from vegetation or flesh to blood, sap, or waste—specialization that is adapted to diet and habitat. Insects that chew, such as plant-eating locusts or predatory ground beetles, demonstrate the clearest "ground plan" of these mouthparts, with their cutting mandibles and fingerlike palps.

Piercing

Many insects feed on nectar, plant sap, and even blood, by piercing. While their mouthparts all have the superficial resemblance of a hollow tube, they are constructed in different ways. True bugs (an order of insects including cicadas and aphids) have mouthparts formed into a piercing "beak," made of elongated mandibles, labium, and maxillae. The maxillae are formed into blades called stylets, used for cutting into plant or animal matter, depending on the species.

WESTERN DUSK-SINGING CICADA
Megatibicen resh

Noselike postclypeus contains muscles that suck up sap through beak

SHIELD BUG
Loxa viridis

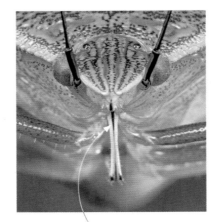

Long, narrow beak used for puncturing leaves for sap

LEATHER BUG
Coreus sp.

Beak punctures seeds and fruits

Chewing

The first insects to evolve about 400 million years ago had chewing mouthparts. They are also common in insect larvae, such as maggots and caterpillars, which use them even though their adult forms may not. Mouthparts used for chewing generally have a pair of sharp mandibles that bite laterally and open to the side. Other mouthparts, such as maxillae, may form limblike appendages called palps, which help to handle pieces of food.

COMMON WASP
Vespula vulgaris

Mandibles slice food into mouth-sized chunks

GIANT FOREST ANT
Camponotus gigas

Ant can carry heavy loads with its mandibles

MIGRATORY LOCUST
Locusta migratoria

Chewing mouthparts move from side to side in a sawing motion

Lapping and mopping

Most flies and all bees have mouthparts ideal for lapping up food. Muscomorphia flies, such as houseflies and bottle flies, lack mandibles, so regurgitate powerful digestive juices onto food to break it down. Rounded, spongelike tips of the labellum then soak up the resulting mix of liquid and solid food. Bees have a labium that extends into a brushlike tongue, called a glossa, for lapping nectar and honey.

HOUSEFLY
Musca domestica

Drops of liquid food can cling to hairy mouthparts

BOTTLE FLY
Calliphora vicina

Tip of labium expands to form spongy pad (labellum)

HOVERFLY
Helophilus sp.

Short proboscis collects nectar and pollen

Nectar feeder
Adult butterflies and almost all moths feed on liquids, mostly the sweet nectar provided by flowers, which they suck up using a long feeding tube, or proboscis. The proboscis of the painted lady buttefly (*Vanessa cardui*), right, is formed from highly elongated maxillae, which hook together to form a flexible cylinder. It is straightened for feeding and can be tucked away when the animal is in flight.

GREEN TIGER BEETLE
Cicindela campestris

Large, curved mandibles grasp prey that beetle drags into burrow

EUROPEAN HONEYBEE
Apis mellifera

Glossa (tongue) extends to lap nectar

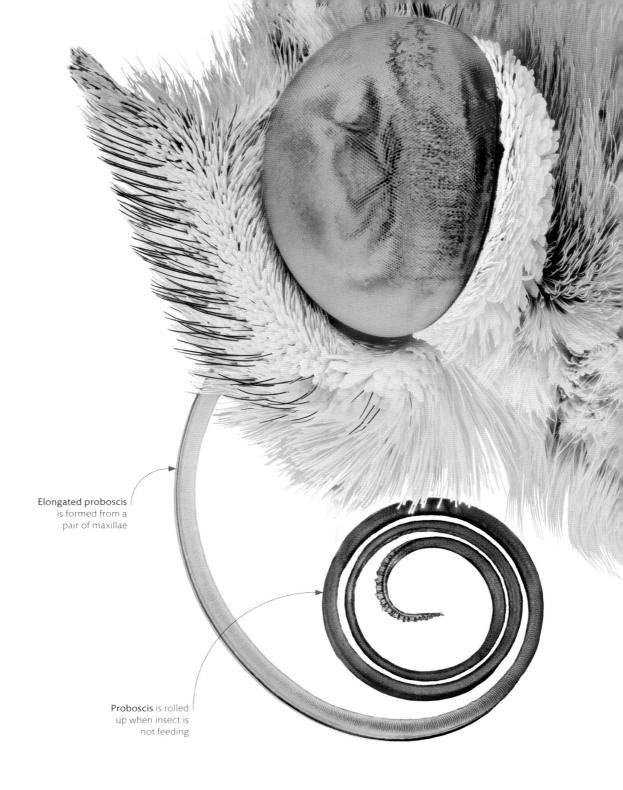

Elongated proboscis is formed from a pair of maxillae

Proboscis is rolled up when insect is not feeding

insect mouthparts

Despite the enormous diversity of forms and functions, insects' mouthparts are all assembled from one or more of the five anatomical units of the animal's head—the labrum (upper "lip"), mandible, maxilla, labium (lower "lip"), and tonguelike hypopharynx (see pp.46–47). Paired mandibles and maxillae may form limblike, jointed appendages that are used in feeding. Equipped with these mouthparts, insects can take in food ranging from blood or plant sap to wood or rotten flesh.

Small, simple eye with a single lens is capable of sensing light and dark and detecting movement

Protective spines point backward on the flea's body to offer minimal resistance as it crawls forward through fur

Body is covered in tough protective scales called sclerites; it can also be compressed from side to side, which helps the parasite slip between hairs without being caught or harmed by its host

Fingerlike palps, on either side of the needlelike mouthparts, carry sensory hairs that can detect the scent of a nearby blood-meal

blood sucker

Miniaturization brings benefits to a blood-sucker: a parasite can spend its life clinging to the body of a host, hidden beneath fur with easy access to unlimited food just below the skin's surface. Adult fleas have hard, compressed bodies and can jump to evade capture or move between hosts. Like other biting insects, they use mouthparts with cutting blades called stylets to puncture the skin and inject an anticlotting chemical in their saliva. This ensures that blood keeps flowing while they suck until their tiny bodies are satiated. Both sexes bite, and females use the protein-rich nourishment to produce hundreds of eggs in their lifetime.

Larva of cat flea (*Ctenocephalides felis*) is engorged with food

Second-hand blood
Feeding on detritus in their host's bedding or nest, flea larvae have a more solid diet than the parasitic adults. The larvae obtain protein by eating the feces of adult fleas, which contain undigested blood.

Feeding on felines
There are more than 2,500 species of fleas, and most—like the cat flea (*Ctenocephalides felis*), which mainly parasitizes cats and dogs—are more or less host-specific. They favor warm-bodied hosts: 94 percent attack mammals, and the rest live on birds. All are flightless, with slim, tough, scratch-resistant bodies and mouthparts that pierce like a needle. SEM, x300.

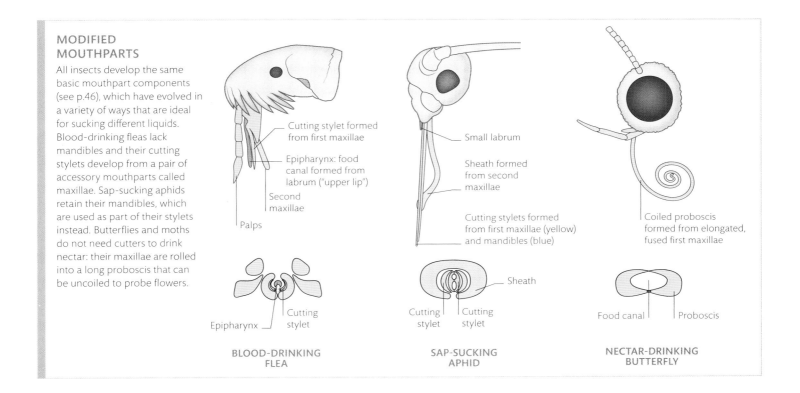

MODIFIED MOUTHPARTS

All insects develop the same basic mouthpart components (see p.46), which have evolved in a variety of ways that are ideal for sucking different liquids. Blood-drinking fleas lack mandibles and their cutting stylets develop from a pair of accessory mouthparts called maxillae. Sap-sucking aphids retain their mandibles, which are used as part of their stylets instead. Butterflies and moths do not need cutters to drink nectar: their maxillae are rolled into a long proboscis that can be uncoiled to probe flowers.

BLOOD-DRINKING FLEA — Cutting stylet formed from first maxillae; Epipharynx: food canal formed from labrum ("upper lip"); Second maxillae; Palps; Epipharynx; Cutting stylet

SAP-SUCKING APHID — Small labrum; Sheath formed from second maxillae; Cutting stylets formed from first maxillae (yellow) and mandibles (blue); Sheath; Cutting stylet; Cutting stylet

NECTAR-DRINKING BUTTERFLY — Coiled proboscis formed from elongated, fused first maxillae; Food canal; Proboscis

> By stroking aphids with their antennae, ants encourage them to release honeydew, which the ants drink

Insect-milking insect
Aphids secrete excess sugars from their anus as honeydew, which attracts sweet-loving ants. Many ants "milk" aphids for their sugar and even protect them from predators.

Wingless infestation
Like other aphids, common nettle aphids (*Microlophium carnosum*) find and colonize suitable growing plants by flight. Once established, they reproduce wingless forms that stay close to the sugar-rich veins of a shoot or leaf. Females reach a length of 1/6 in (4 mm) and give birth without mating (see p.243), rapidly infesting the plant with offspring.

drinking sap

Many insects draw sap from plants using mouthparts that work like a hypodermic needle. The sap flows just below the surface of a leaf or stem through microscopic pipes called phloem vessels (see pp.206–07). Aphids find these vessels by probing and tasting with their proboscis. Phloem sap, which transports food made in photosynthesis, is at least 50 percent sugar; the rest consists of amino acids and other vital nutrients. Aphids grow and reproduce quickly on their energy-rich diet, and they can soon smother shoots in such huge numbers that they can stunt the plant's growth.

SUGAR ON TAP
Sensors in an aphid's mouthparts tell the insect where to probe to locate phloem, which it pierces with the needlelike stylet housed in its proboscis (see p.51). Once a vessel is tapped, the pressure in the phloem is enough to push the sap into the aphid's gut without need for sucking.

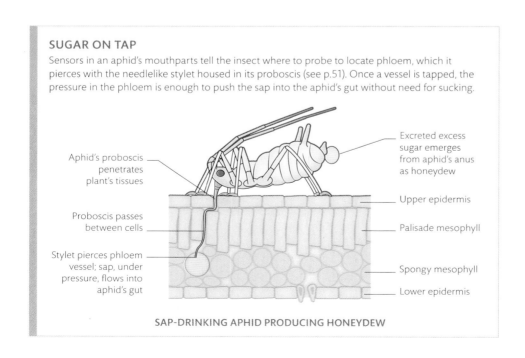

- Aphid's proboscis penetrates plant's tissues
- Proboscis passes between cells
- Stylet pierces phloem vessel; sap, under pressure, flows into aphid's gut
- Excreted excess sugar emerges from aphid's anus as honeydew
- Upper epidermis
- Palisade mesophyll
- Spongy mesophyll
- Lower epidermis

SAP-DRINKING APHID PRODUCING HONEYDEW

rasping food

When animals consume a meal, the food they take into their bodies must be digested into particles small enough to be absorbed into their cells. For a plant-eating snail, like many other animals, this process begins by mechanically crumbling the food in the mouth—but a snail achieves this without chewing. Instead, it uses a tonguelike structure called a radula that is coated with microscopic teeth. By lapping backward and forward, the radula scrapes the plant surface, dislodging pieces of vegetation that can be swallowed. Chemical processing by enzymes in the gut completes the digestion.

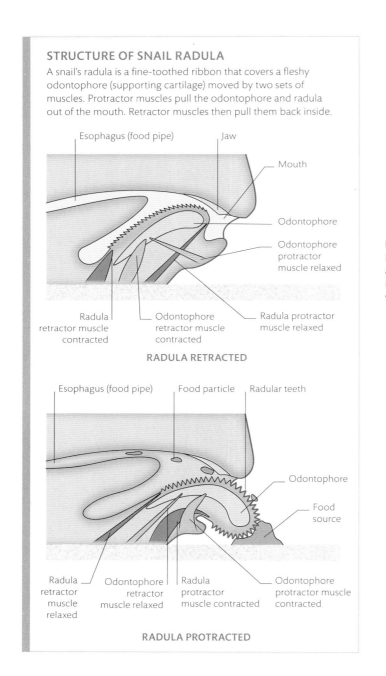

STRUCTURE OF SNAIL RADULA
A snail's radula is a fine-toothed ribbon that covers a fleshy odontophore (supporting cartilage) moved by two sets of muscles. Protractor muscles pull the odontophore and radula out of the mouth. Retractor muscles then pull them back inside.

RADULA RETRACTED

RADULA PROTRACTED

Shell, formed mostly of calcium carbonate, grows with the snail until the animal reaches adult size

Leaf eater
Despite its sluggishness, the Roman snail is a voracious herbivore and is most abundant in chalky habitats—where calcium-rich vegetation helps it to maintain its shell.

Rows of teeth are arranged with the points facing backward, which helps direct food particles further into the mouth

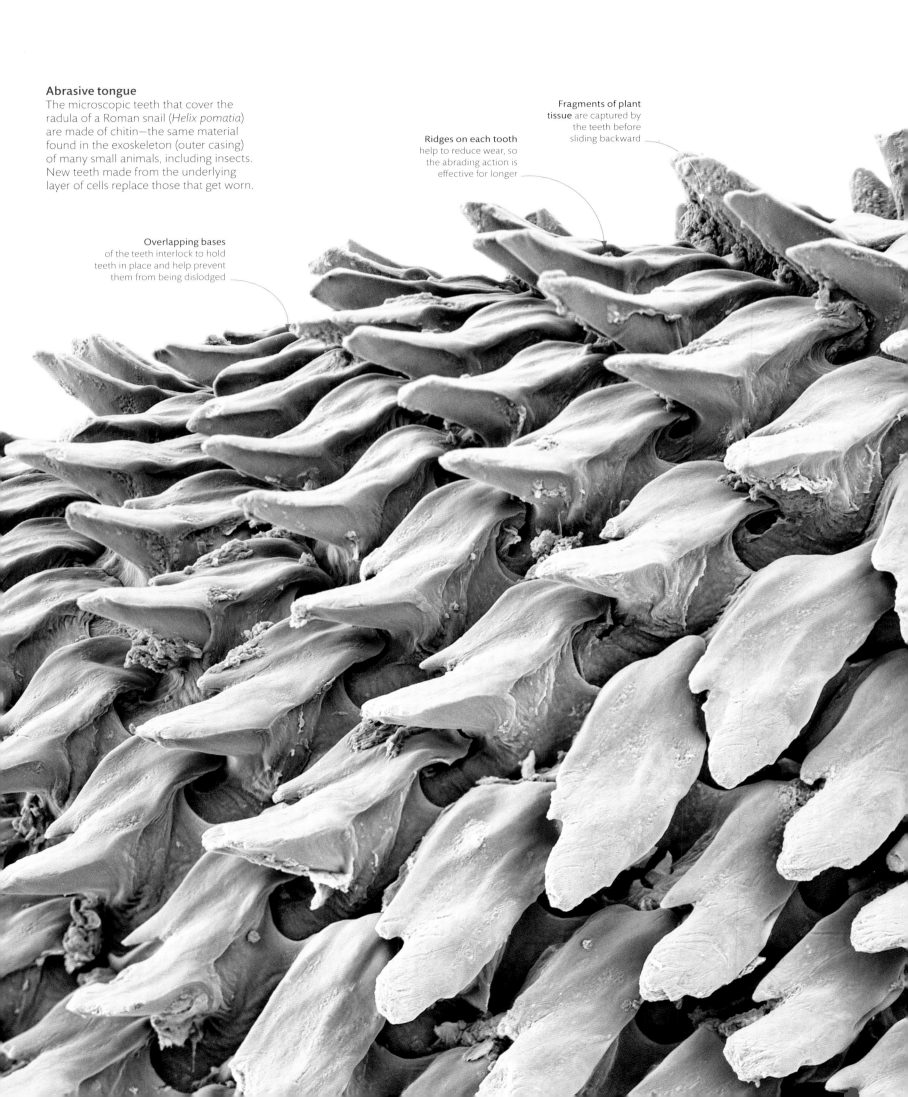

Abrasive tongue
The microscopic teeth that cover the radula of a Roman snail (*Helix pomatia*) are made of chitin—the same material found in the exoskeleton (outer casing) of many small animals, including insects. New teeth made from the underlying layer of cells replace those that get worn.

Fragments of plant tissue are captured by the teeth before sliding backward

Ridges on each tooth help to reduce wear, so the abrading action is effective for longer

Overlapping bases of the teeth interlock to hold teeth in place and help prevent them from being dislodged

venomous pincers

The hidden world of micro-life is occupied by hunters that can appear formidable up close. The largest pseudoscorpions are no bigger than a grain of rice, and most species are so tiny they are usually overlooked. But all are armed with pincers that grab prey, and most can inject paralyzing venom through their pincer tips before sucking the body juices of their victim, just like related spiders and true scorpions. Pseudoscorpions thrive in leaf litter, compost heaps, birds' nests, caves—and wherever there are plenty of springtails, mites, and other prey small enough to tackle. Some even live among the books of old libraries, where they hunt book lice that graze on binding paste.

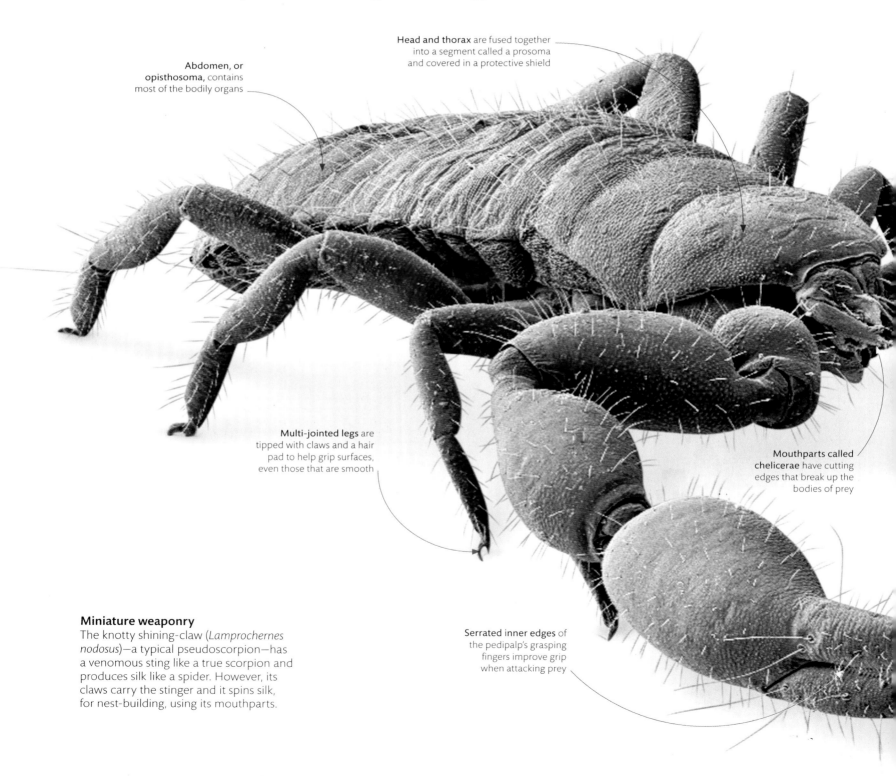

Head and thorax are fused together into a segment called a prosoma and covered in a protective shield

Abdomen, or opisthosoma, contains most of the bodily organs

Multi-jointed legs are tipped with claws and a hair pad to help grip surfaces, even those that are smooth

Mouthparts called chelicerae have cutting edges that break up the bodies of prey

Serrated inner edges of the pedipalp's grasping fingers improve grip when attacking prey

Miniature weaponry
The knotty shining-claw (*Lamprochernes nodosus*)—a typical pseudoscorpion—has a venomous sting like a true scorpion and produces silk like a spider. However, its claws carry the stinger and it spins silk, for nest-building, using its mouthparts.

Hitchhiking

Pseudoscorpions often use their pincers to hold onto a larger animal—such as a flying insect or another arachnid—so they are carried from place to place. These hitchhikers are invariably females that help disperse their offspring to new patches of habitat (see also pp.172–73).

Pseudoscorpion pincers are employed in hanging onto their hosts—sometimes for days

The host is a much bigger animal—a wasp (*Dolichomitus* sp.)

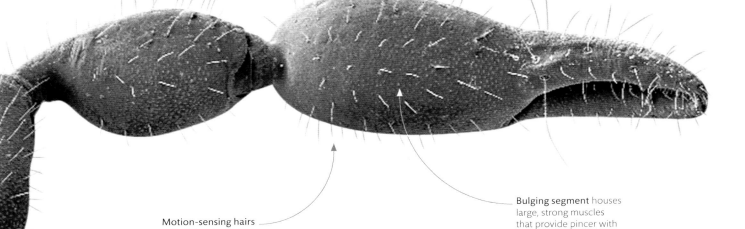

Motion-sensing hairs called chaetae detect the movement of nearby prey

Bulging segment houses large, strong muscles that provide pincer with closing force

STINGING FINGERS

The pincers of a pseudoscorpion are modified food-handling appendages called pedipalps. Like those of true scorpions, each pedipalp has two fingers—one fixed and another hinged for grasping. Unlike scorpions, which have venom in their tail, most pseudoscorpions have venom in their pedipalps that is dispensed through openings in the fingers. These are the pedipalps of three different species of pseudoscorpion, representing three different arrangements of venom glands.

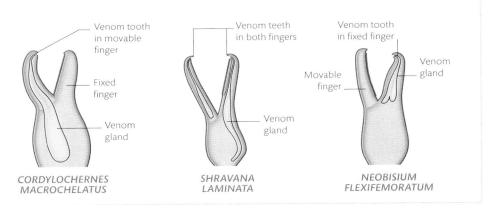

CORDYLOCHERNES MACROCHELATUS
Venom tooth in movable finger
Fixed finger
Venom gland

SHRAVANA LAMINATA
Venom teeth in both fingers
Venom gland

NEOBISIUM FLEXIFEMORATUM
Venom tooth in fixed finger
Movable finger
Venom gland

Sensory hairs, or setae, allow mites to feel or smell surroundings

Dorsal shield, a hard piece of exoskeleton, protects mite's unsegmented body

Chelicerae behave like jaws to grip prey and inject digestive juices

Pedipalp is a long, jointed, leglike appendage

predigesting prey

Predatory arachnids, including mites and spiders, feed either by sucking liquid directly from inside their prey or by regurgitating digestive juices—produced in the gut and their maxillary glands—that turn their prey into liquid. Some arachnids also break down food mechanically, using jawlike mouthparts called chelicerae. The liquid diet is ingested and filtered using hairs, or setae, that line their mouth cavities. The prey soup is sucked up using the heavily muscled stomach, which pumps the liquid into the oesophagus (throat). The chelicerae of many spiders are hollow, fanglike appendages that contain, or are connected to, venom glands. Venom is injected to paralyze a spider's prey, enabling them to predigest their victim.

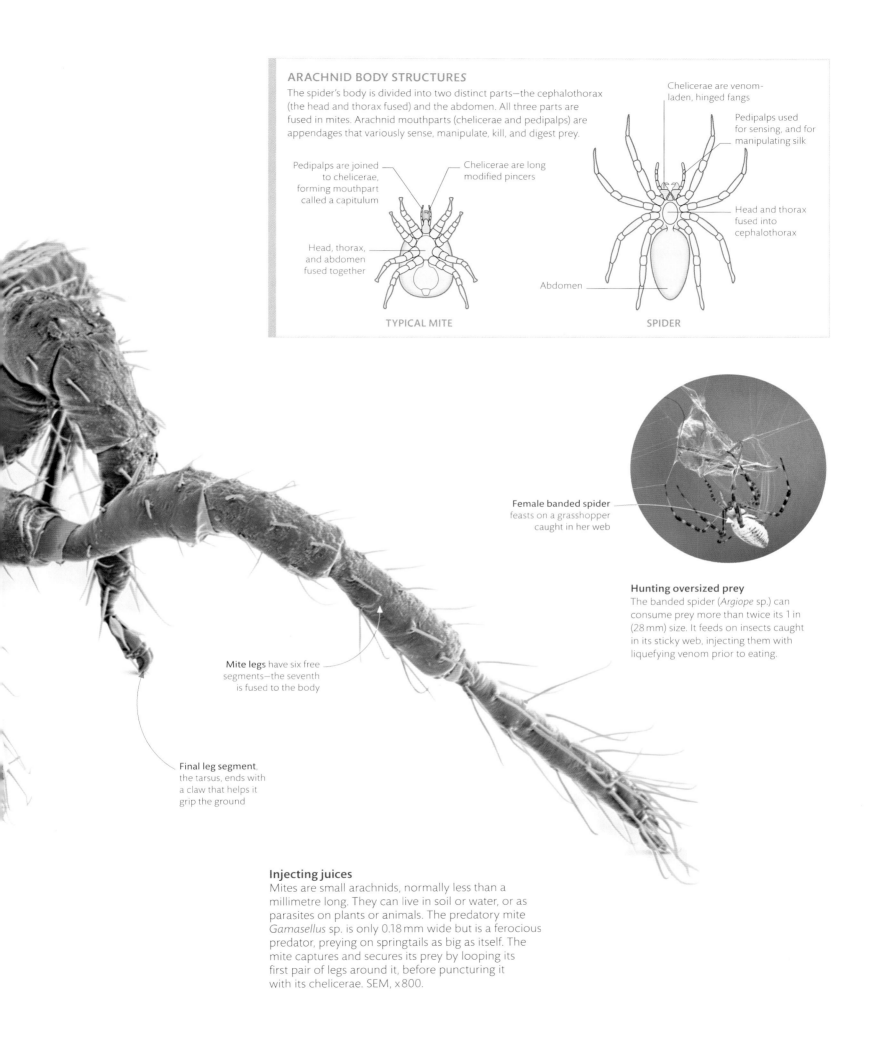

ARACHNID BODY STRUCTURES
The spider's body is divided into two distinct parts—the cephalothorax (the head and thorax fused) and the abdomen. All three parts are fused in mites. Arachnid mouthparts (chelicerae and pedipalps) are appendages that variously sense, manipulate, kill, and digest prey.

- Pedipalps are joined to chelicerae, forming mouthpart called a capitulum
- Chelicerae are long modified pincers
- Head, thorax, and abdomen fused together

TYPICAL MITE

- Chelicerae are venom-laden, hinged fangs
- Pedipalps used for sensing, and for manipulating silk
- Head and thorax fused into cephalothorax
- Abdomen

SPIDER

Female banded spider feasts on a grasshopper caught in her web

Hunting oversized prey
The banded spider (*Argiope* sp.) can consume prey more than twice its 1 in (28 mm) size. It feeds on insects caught in its sticky web, injecting them with liquefying venom prior to eating.

Mite legs have six free segments—the seventh is fused to the body

Final leg segment, the tarsus, ends with a claw that helps it grip the ground

Injecting juices
Mites are small arachnids, normally less than a millimetre long. They can live in soil or water, or as parasites on plants or animals. The predatory mite *Gamasellus* sp. is only 0.18 mm wide but is a ferocious predator, preying on springtails as big as itself. The mite captures and secures its prey by looping its first pair of legs around it, before puncturing it with its chelicerae. SEM, ×800.

MACROLIFE

ABSORBING NUTRIENTS

The human intestine absorbs simple substances—such as sugars, amino acids, and fatty acids—that are small enough to pass through the surface lining (epithelium) of the intestine. Sugar and amino acids, respectively the basic units of carbohydrates and proteins, pass quickly into blood vessels. A mesh of tiny blood vessels—capillaries—fills every villus. Less soluble fats are collected by the lymphatic system and fed into the bloodstream later.

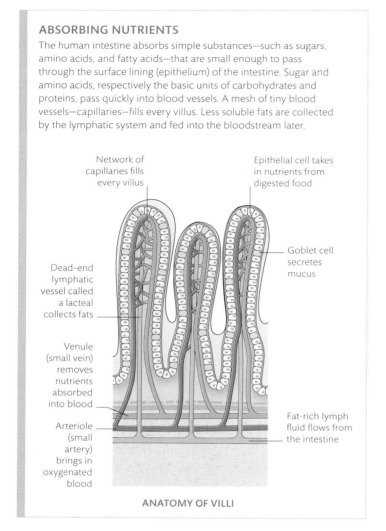

ANATOMY OF VILLI

- Network of capillaries fills every villus
- Epithelial cell takes in nutrients from digested food
- Dead-end lymphatic vessel called a lacteal collects fats
- Goblet cell secretes mucus
- Venule (small vein) removes nutrients absorbed into blood
- Arteriole (small artery) brings in oxygenated blood
- Fat-rich lymph fluid flows from the intestine

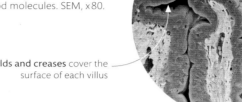

Added surface area
This cross-section of human intestinal wall reveals how the villi, seen here with a red lining of epithelial cells, dramatically enlarge the wall's surface area. This boosts the intestine's ability to absorb food molecules. SEM, x80.

Folds and creases cover the surface of each villus

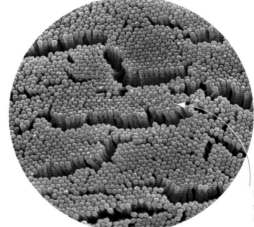

Microvilli together make a brushlike layer that forms the interface with the intestine's contents

Microvilli
The epithelial cells covering the surface of each villus have a highly convoluted upper membrane that forms another set of fingerlike projections, called microvilli. Just like the villi, the microvilli create a very large absorption surface. SEM, x8,900.

food into the bloodstream

All organisms that eat food must break down, or digest, food substances into small molecules that are easily absorbed into the organism's body tissues. Microscopic organisms can simply absorb nutrients across their body surface, but large animals need elaborate and sophisticated structures to absorb nutrients quickly. In humans, the main nutrient-absorbing surface is the small intestine, whose wall is formed into microscopic, fingerlike projections called villi. The villi are lined with cells that themselves are lined with more tiny fingerlike projections, called microvilli. Together, the villi and microvilli represent an extensive arena for absorption. The end of the small intestine, the 10-ft (3-m) section known as the ileum, contains villi that take in vitamins and any remaining nutrients.

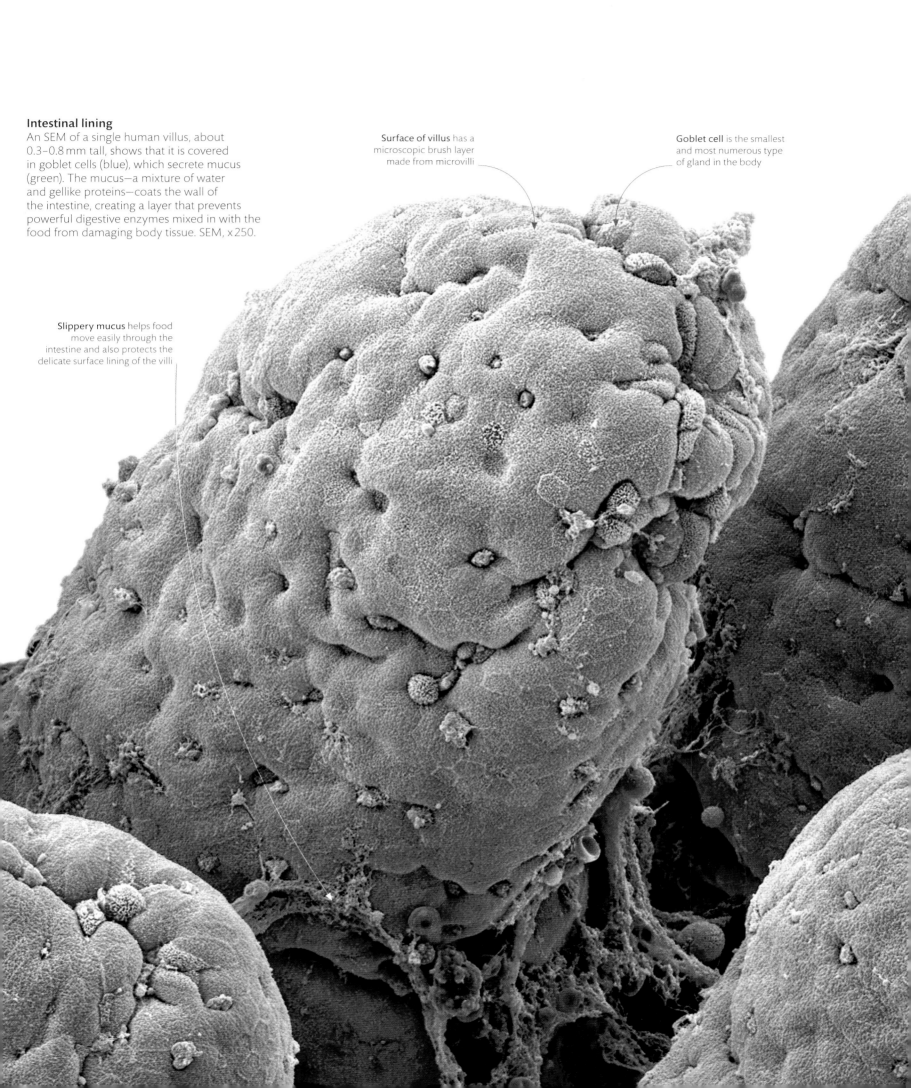

Intestinal lining
An SEM of a single human villus, about 0.3–0.8 mm tall, shows that it is covered in goblet cells (blue), which secrete mucus (green). The mucus—a mixture of water and gellike proteins—coats the wall of the intestine, creating a layer that prevents powerful digestive enzymes mixed in with the food from damaging body tissue. SEM, x250.

Surface of villus has a microscopic brush layer made from microvilli

Goblet cell is the smallest and most numerous type of gland in the body

Slippery mucus helps food move easily through the intestine and also protects the delicate surface lining of the villi

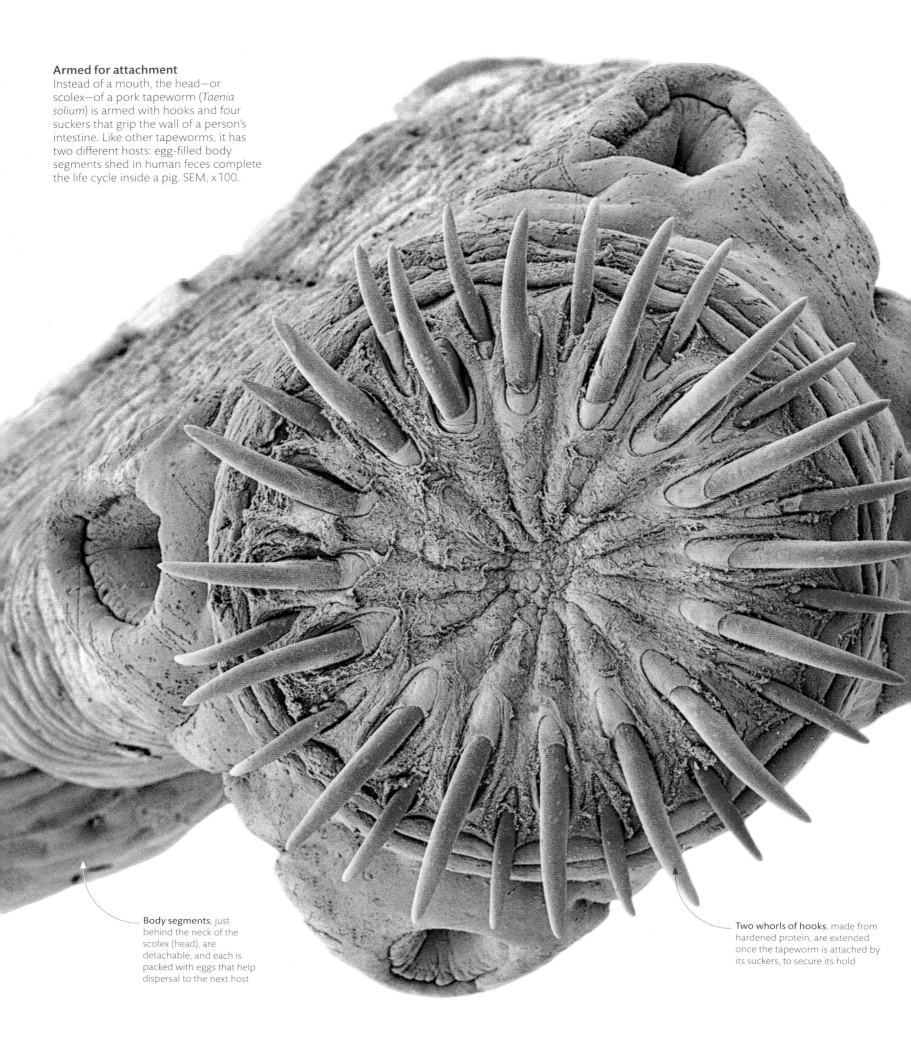

Armed for attachment
Instead of a mouth, the head—or scolex—of a pork tapeworm (*Taenia solium*) is armed with hooks and four suckers that grip the wall of a person's intestine. Like other tapeworms, it has two different hosts: egg-filled body segments shed in human feces complete the life cycle inside a pig. SEM, ×100.

Body segments, just behind the neck of the scolex (head), are detachable, and each is packed with eggs that help dispersal to the next host

Two whorls of hooks, made from hardened protein, are extended once the tapeworm is attached by its suckers, to secure its hold

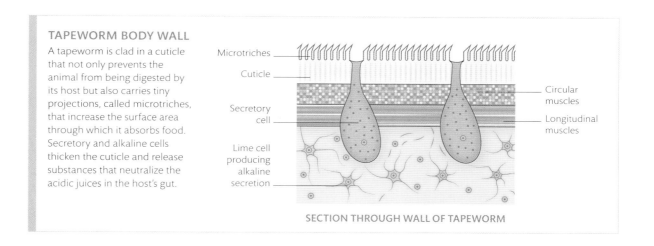

TAPEWORM BODY WALL

A tapeworm is clad in a cuticle that not only prevents the animal from being digested by its host but also carries tiny projections, called microtriches, that increase the surface area through which it absorbs food. Secretory and alkaline cells thicken the cuticle and release substances that neutralize the acidic juices in the host's gut.

SECTION THROUGH WALL OF TAPEWORM

living in the gut

A parasite gets its food from another organism while living on or inside its body. Deprived of nourishment, the host is weakened, but it is in the interest of the parasite for the host to stay alive—at least for as long as it takes for the parasite to reproduce. Tapeworms are masters of this unequal relationship. They live in the intestines of many animals, sufficiently far down the digestive tract to be bathed in pre-digested food. Lacking guts of their own, they need only the means to stay fastened to the gut wall and absorb what is around them. And once grown to maturity, they shed eggs in the host's waste—ready to be ingested by a new animal.

Tough surface of cuticle resists the corrosive action of the host's digestive juices

Suckers attach to the host's intestinal lining to prevent the tapeworm from being dislodged by the gut's waves of muscular peristalsis

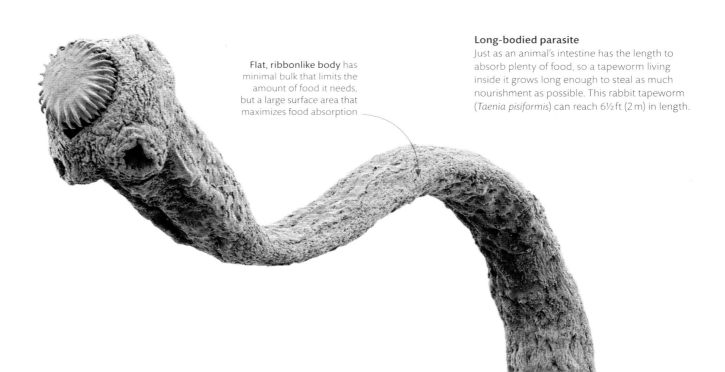

Flat, ribbonlike body has minimal bulk that limits the amount of food it needs, but a large surface area that maximizes food absorption

Long-bodied parasite
Just as an animal's intestine has the length to absorb plenty of food, so a tapeworm living inside it grows long enough to steal as much nourishment as possible. This rabbit tapeworm (*Taenia pisiformis*) can reach 6½ ft (2 m) in length.

Initial frame formed from thickest threads

Spider holds onto web with claws and bristles at end of its feet

Creating a trap
A four-spot orb weaver spiders (*Araneus quadratus*) uses its silk to build a netlike orb web for catching flying insects. Construction begins with a "bridge" that connects two supports, before the spider makes a pattern of spokes and spirals around a central point.

spinning webs

Silk is micro-life's ultimate fabric. It originates as a fluid inside silk-spinning animals and emerges as one of the strongest materials produced by any living thing. Spiders and insects such as caterpillars make it into cocoons, but spiders also use it to form traps for catching their prey. Silk is a protein and, like many proteins—such as insulin or digestive enzymes—it is made in glands and secreted in liquid form. But as it emerges from the glands, physical and chemical reactions cause it to become solid threads. Spiders can tease the silk into webs of extraordinary geometric complexity.

MAKING SILK
Glands in a spider's abdomen produce different kinds of silk. Some silk is substantial enough to protect eggs, other types are so delicate they transmit vibrations of prey and alert the spider, or are sticky to hold onto prey. But the silk is all processed in a similar way. The liquid protein stored in the silk gland is passed through ducts where secretions raise its acidity and the water is removed. These changes alter the chemical state of the protein and turn it into threads—much like boiling an egg turns it solid.

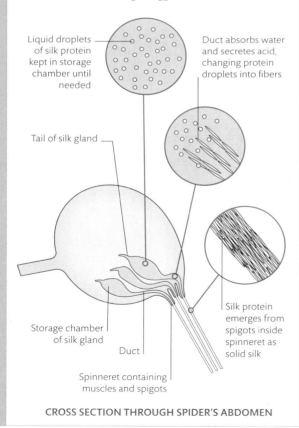

Liquid droplets of silk protein kept in storage chamber until needed

Duct absorbs water and secretes acid, changing protein droplets into fibers

Tail of silk gland

Storage chamber of silk gland

Duct

Silk protein emerges from spigots inside spinneret as solid silk

Spinneret containing muscles and spigots

CROSS SECTION THROUGH SPIDER'S ABDOMEN

Minuscule spigots inside a spinneret release silk strands that can be 20 times thinner than a human hair

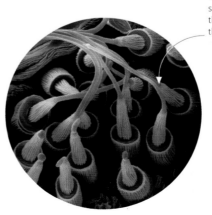

Releasing silk
Depending upon the species, a spider can have up to four pairs of spinnerets at the end of its abdomen. Inside each one, as in this spinneret of *Gasteracantha* sp., are muscles that control its position and that of a cluster of tiny holes, or spigots, at its tip that eject the silk threads.

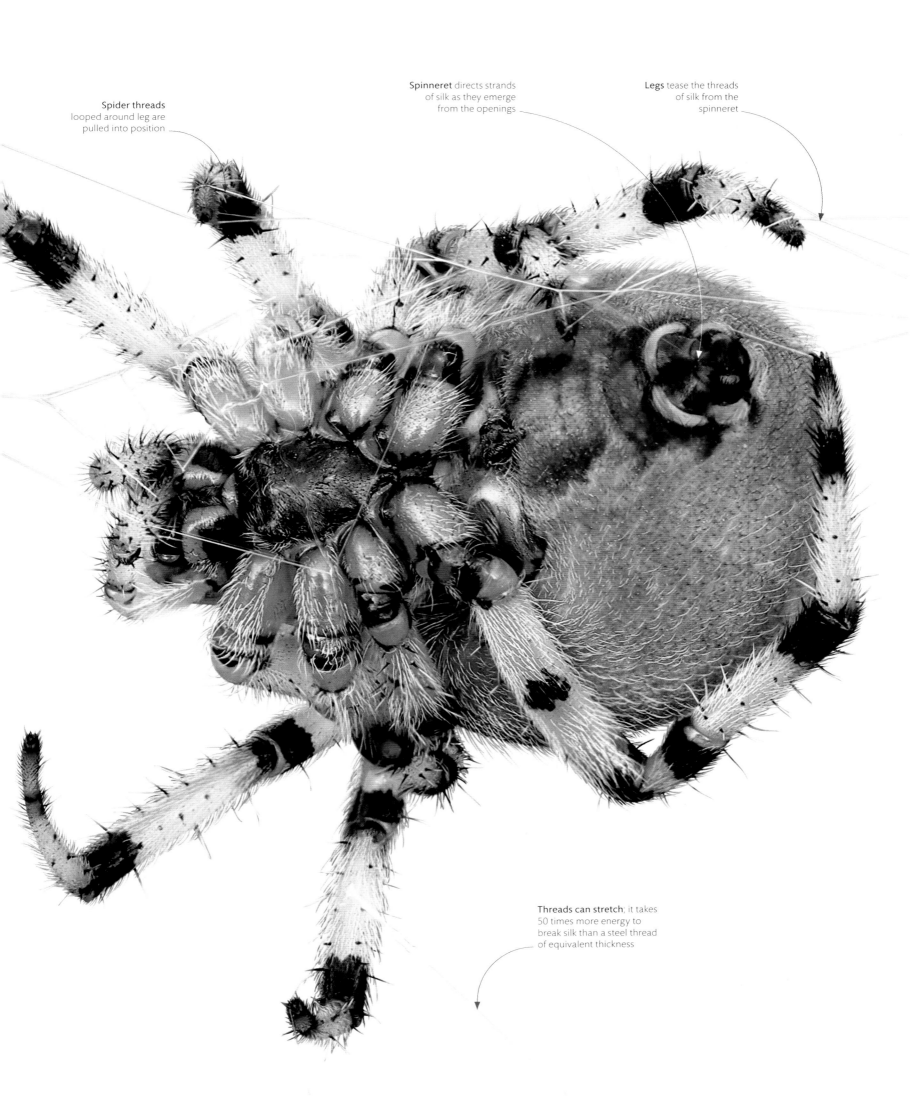

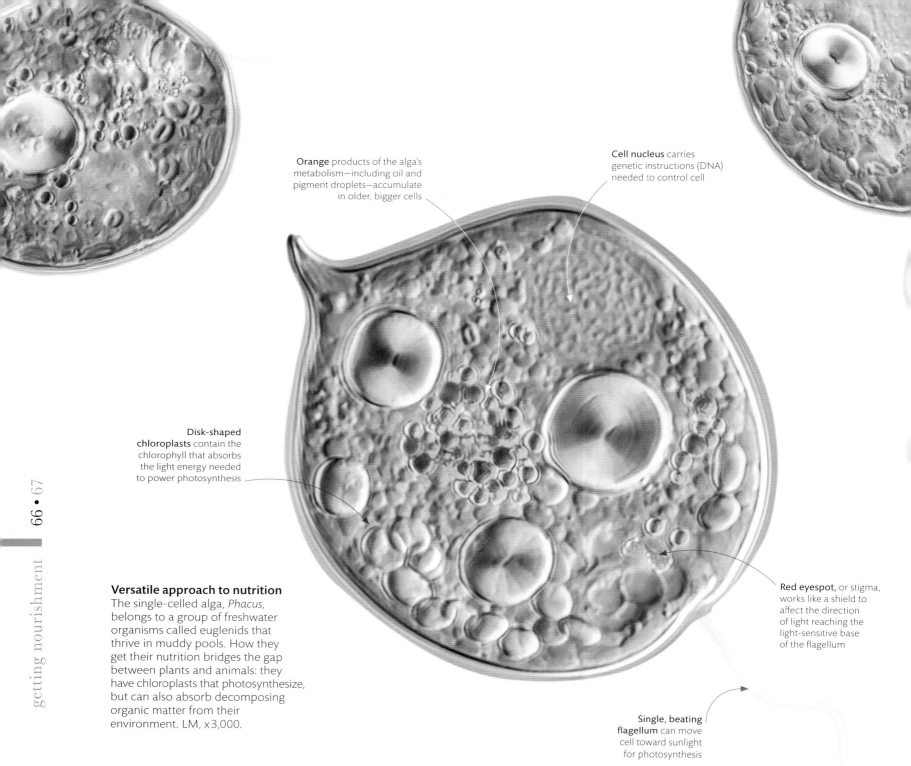

Orange products of the alga's metabolism—including oil and pigment droplets—accumulate in older, bigger cells

Cell nucleus carries genetic instructions (DNA) needed to control cell

Disk-shaped chloroplasts contain the chlorophyll that absorbs the light energy needed to power photosynthesis

Red eyespot, or stigma, works like a shield to affect the direction of light reaching the light-sensitive base of the flagellum

Single, beating flagellum can move cell toward sunlight for photosynthesis

Versatile approach to nutrition
The single-celled alga, *Phacus*, belongs to a group of freshwater organisms called euglenids that thrive in muddy pools. How they get their nutrition bridges the gap between plants and animals: they have chloroplasts that photosynthesize, but can also absorb decomposing organic matter from their environment. LM, x3,000.

neither animal nor plant

Photosynthesizing single-celled algae are important sources of organic food in all underwater food chains. Like plant cells, algal cells contain chloroplasts—chlorophyll-packed organelles that absorb light energy. But—like animals—many also move around, propelled by hairlike flagella that help them reach sunlit places (see pp.100–01). Some algae can absorb organic materials that supply vital nutrients, such as vitamins, which they are unable to make themselves from their surroundings. A few can even switch off photosynthesis when shaded, so they can obtain all their nutrition in this way.

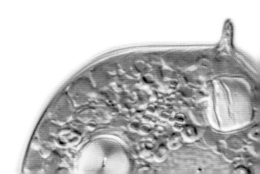

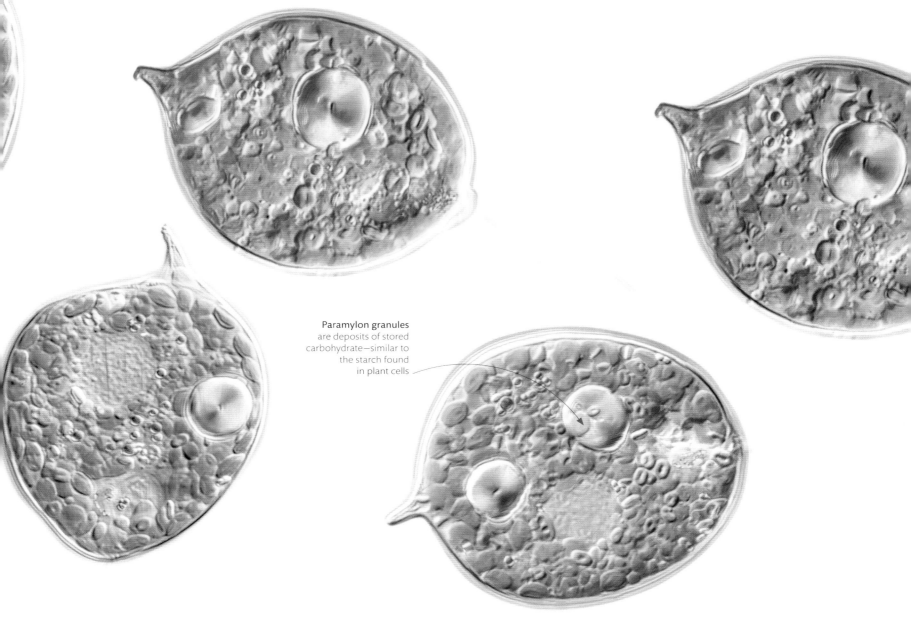

Paramylon granules are deposits of stored carbohydrate—similar to the starch found in plant cells

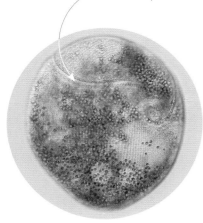

Spherical *Chlorella* algae perform a chloroplastlike service for *Stentor*, making sugars from carbon dioxide and passing them on to the protozoan

Exchanging resources
Stentor looks like an alga with chloroplasts, like *Phacus*, but it is an animallike protozoan. The green objects are single-celled algae, which have taken up residence in a mutually beneficial relationship called a symbiosis.

SWITCHING STRATEGIES

A few kinds of euglenid algae, such as *Euglena gracilis*, can actually switch between photosynthesizing and consuming foods, according to their circumstances. In bright light their chloroplasts enlarge and make practically all the food they need. However, if conditions become too dark for photosynthesis, their chloroplasts shrink and the cells rely on absorbing organic food from the surroundings instead.

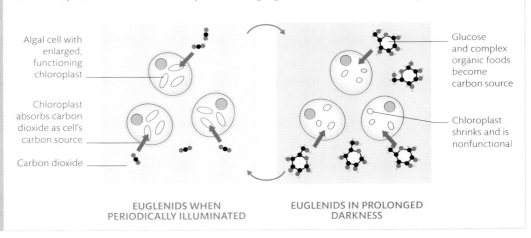

Algal cell with enlarged, functioning chloroplast

Chloroplast absorbs carbon dioxide as cell's carbon source

Carbon dioxide

Glucose and complex organic foods become carbon source

Chloroplast shrinks and is nonfunctional

EUGLENIDS WHEN PERIODICALLY ILLUMINATED

EUGLENIDS IN PROLONGED DARKNESS

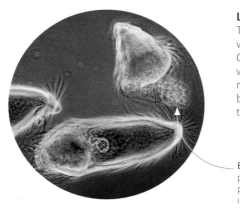

Living in wood
Termites are widespread in warm parts of the world. As in other species, wood digestion in *Coptotermes niger* of Central America begins with chewing mouthparts and ends with gut microbes. This not only provides nourishment, but also opens cavities in wood that become the nest space of a colony. SEM, x50.

Beating flagella help the protozoan swim toward particles of woody food in the termite's gut

Gut microbes
Trichonympha flagellates are gut-living protozoans that generate fatty acids, some of which are absorbed and used by the termite.

eating wood

The cellulose fibers in the walls of plant cells present a challenge for many herbivores—and the fiber in wood is even tougher. Wood contains lignin, which is resistant to digestion, but termites have adapted to eating wood by partnering with microbes. Termite guts harbor a diversity of microbes, which soften the wood with enzymes that turn cellulose into nourishment. Some gut bacteria extract nitrogen from the air, boosting the termites' nitrogen intake, while Archaea microbes process their neighbours' waste into methane, making termites a major producer of this greenhouse gas.

PROCESSING WOOD

A termite gut digests cellulose in wood, by reducing it to nutrients that can be absorbed, and leaving indigestible lignin as waste. Microbes in the gut aid this digestion. Some of these nitrogen-rich microbes are excreted, then consumed and digested in the midgut of other termites to augment their nitrogen-poor woody diet.

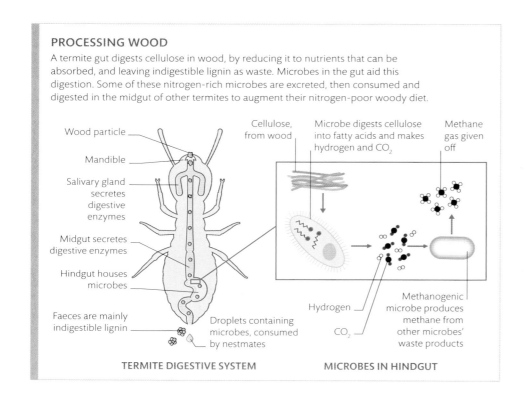

TERMITE DIGESTIVE SYSTEM — MICROBES IN HINDGUT

- Wood particle
- Mandible
- Salivary gland secretes digestive enzymes
- Midgut secretes digestive enzymes
- Hindgut houses microbes
- Faeces are mainly indigestible lignin
- Droplets containing microbes, consumed by nestmates
- Cellulose, from wood
- Microbe digests cellulose into fatty acids and makes hydrogen and CO_2
- Methane gas given off
- Hydrogen
- CO_2
- Methanogenic microbe produces methane from other microbes' waste products

powering the body

Living bodies contain chemical energy, and releasing that energy—and using it to power life processes—is the job of the chemical reactions of respiration. Some microbes respire without oxygen, but large organisms need oxygen, and many of the microscopically intricate structures in their bodies are involved with absorbing oxygen and distributing it to every cell.

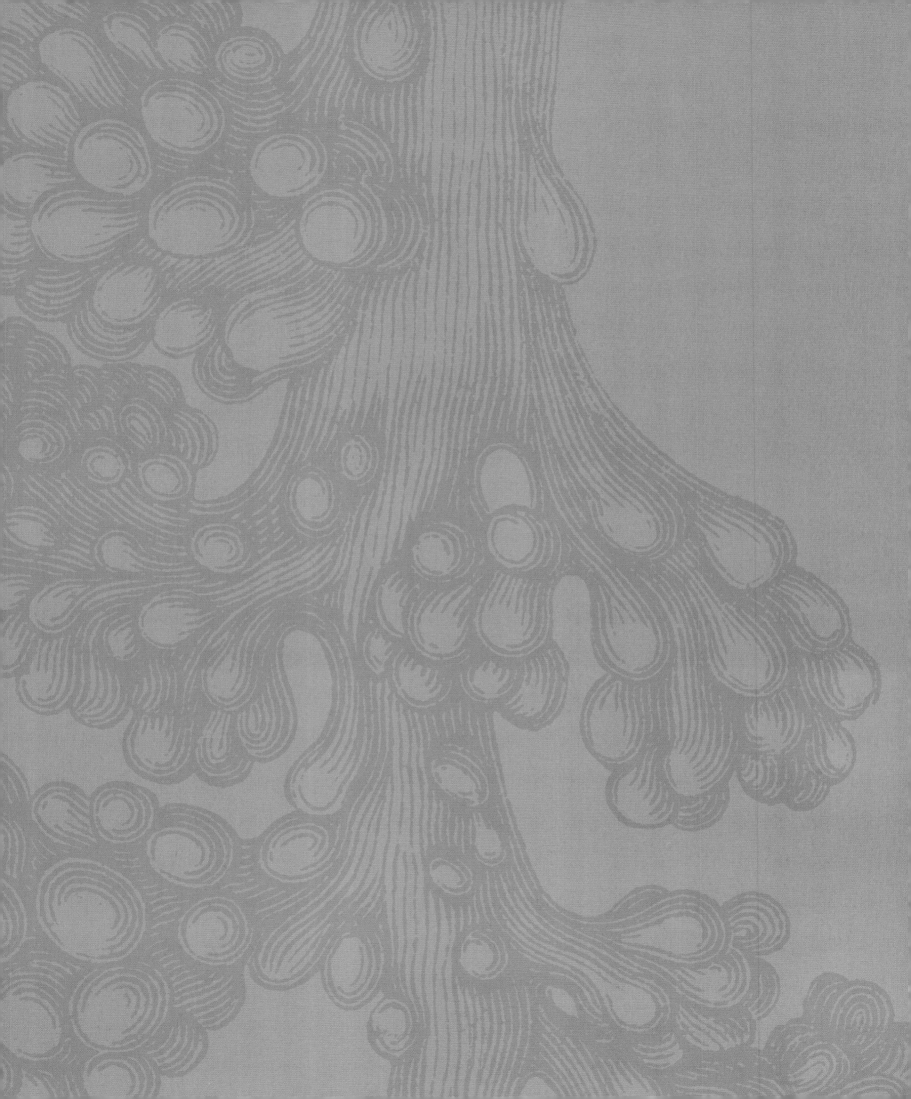

Flashing for mates
At night, male firefly beetles flash with light to attract females. The light they emit is "cold"—it entails little loss of heat—but it still consumes energy that comes from the food the fireflies eat.

energy release

All life uses energy. Even organisms that look inactive, such as plants, are busy inside their cells: cytoplasm is streaming and cells are dividing. Energy pumps nutrients into cells, generates warmth, and builds new material for growth and reproduction. Animals also need energy to fire electrical impulses, contract muscles, and—in some cases— emit light. The energy for doing this comes from food, either consumed or made by photosynthesis, and released by the chemical reactions of respiration inside all cells.

Light is produced by specialized light-emitting organs in the abdomen

Generating light
A firefly's light is produced by chemical reactions in a process called bioluminescence (see pp.118–19). The necessary chemicals are manufactured by the insect's metabolism.

LIFE'S ENERGY BUDGET

The organic molecules that make up the body of an organism contain chemical energy. Animals obtain these molecules from the food they eat, while plants manufacture them by photosynthesis using light energy. Calorific food molecules, such as sugars (carbohydrates) and fats (lipids), react in the chemical reactions of respiration to transfer their energy into molecules of ATP (adenosine triphosphate)—the main chemical energy "currency" of the cells. ATP is then broken down to release the energy that drives life's vital processes, such as growth and movement.

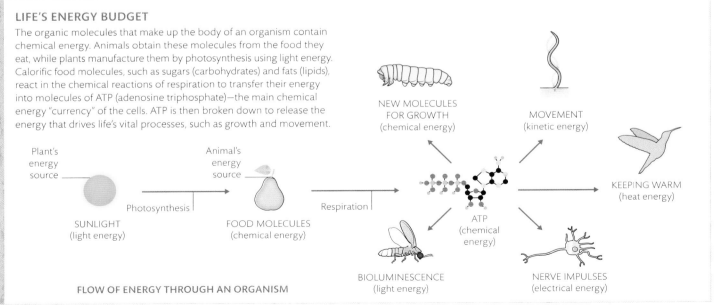

FLOW OF ENERGY THROUGH AN ORGANISM

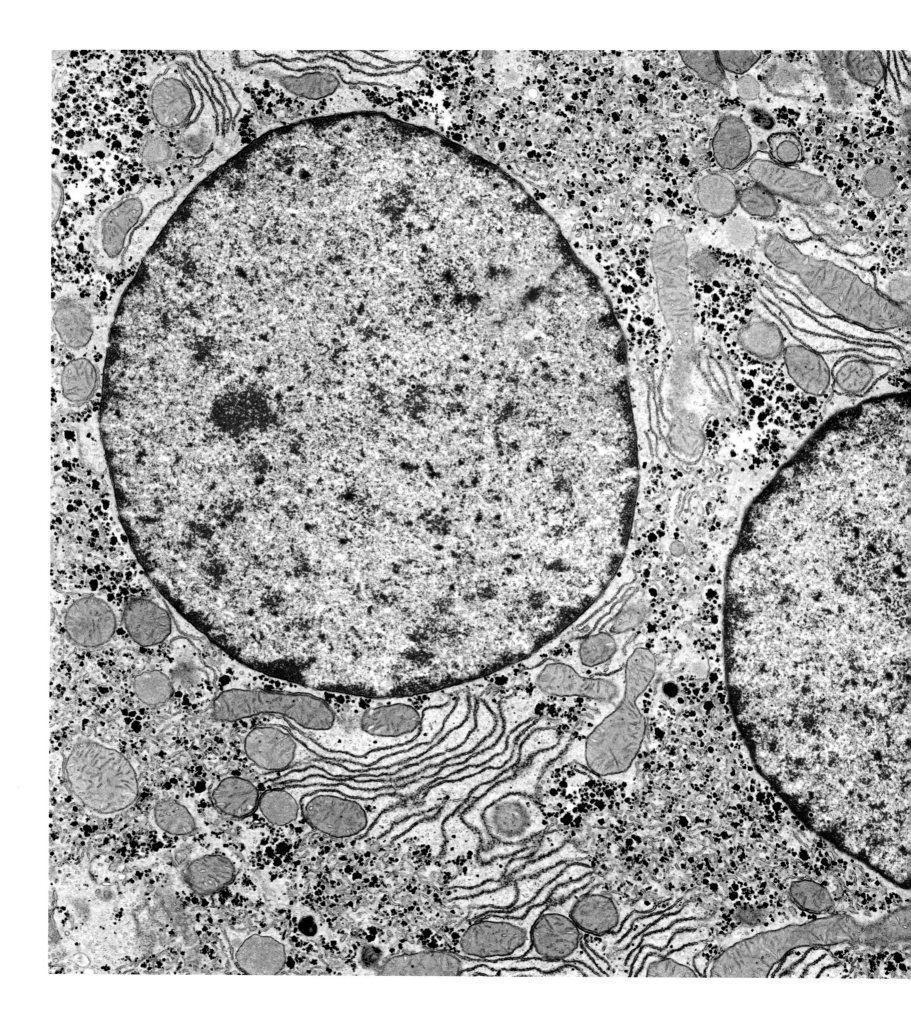

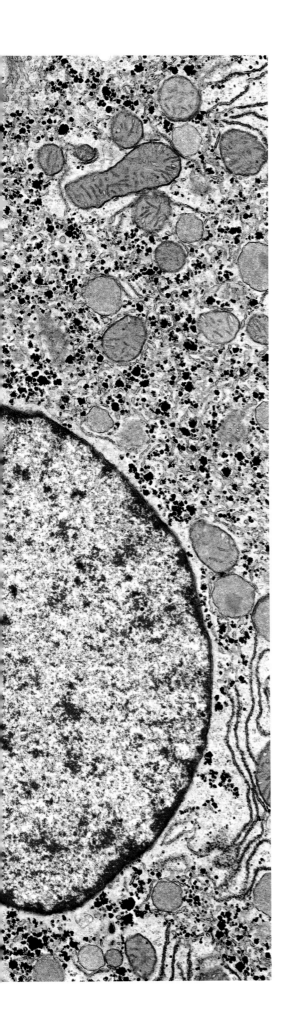

Busy cell
A human liver cell uses chemical reactions to store carbohydrates, detoxify harmful substances, and perform other vital processes. All this activity, driven by enzymes, demands energy from the cell's mitochondria, visible here as light blue structures. TEM, x 15,000.

Powering the heart
Heart cells have numerous mitochondria, which supply the energy that allows the heart to beat continuously. The mitochondria are especially well-packed with membranes called cristae.

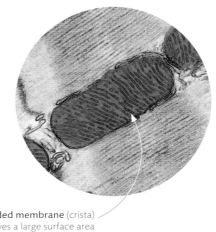

Folded membrane (crista) gives a large surface area for carrying enzymes

using oxygen

The chemical reactions of respiration split energy-rich molecules, notably glucose, into smaller molecules, releasing their stored energy. Turning all the carbon in glucose into carbon dioxide (CO_2) yields the most energy, but it requires oxygen. Cell structures called mitochondria use enzymes to bring the right molecules together and make the reactions happen. This "aerobic respiration" (respiration using oxygen) is a cell's main power source. Tissues that work hard, such as muscle or liver, have more mitochondria.

RELEASING ENERGY FROM GLUCOSE

Reactions that release energy involve the "oxidation" of fuel, whereby hydrogen is removed and oxygen is added. The fuel in cells is glucose, and oxidation splits it into carbon dioxide. Hydrogen is removed in steps, but most energy is generated on membranes inside the mitochondria called cristae. Here, hydrogen follows a cascade of oxygen-dependent reactions, releasing 90 percent of the energy the cell gets from glucose.

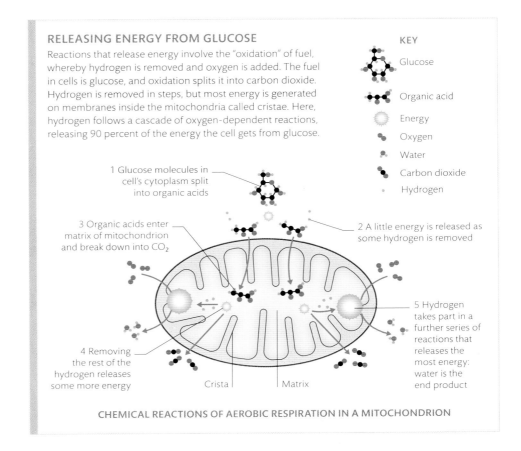

KEY
- Glucose
- Organic acid
- Energy
- Oxygen
- Water
- Carbon dioxide
- Hydrogen

1 Glucose molecules in cell's cytoplasm split into organic acids

2 A little energy is released as some hydrogen is removed

3 Organic acids enter matrix of mitochondrion and break down into CO_2

4 Removing the rest of the hydrogen releases some more energy

5 Hydrogen takes part in a further series of reactions that releases the most energy: water is the end product

Crista | Matrix

CHEMICAL REACTIONS OF AEROBIC RESPIRATION IN A MITOCHONDRION

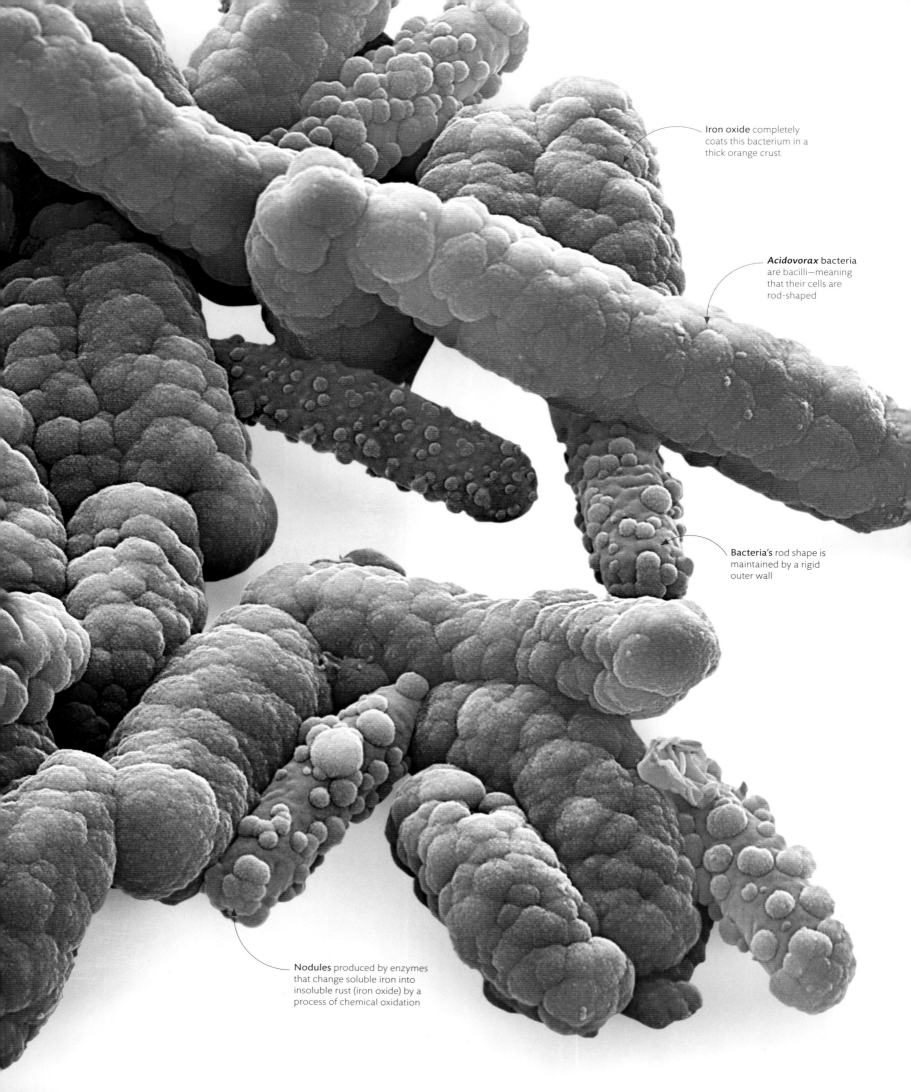

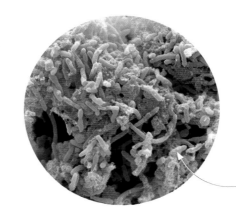

Using uranium
The bacterium *Geobacter metallireducens* uses dissolved metals—including iron, manganese, and uranium—to oxidize the organic substances that it feeds on, and in doing so, produces solid metal as a byproduct. Its tolerance to radioactivity means it can be used to separate dangerous uranium waste from drinking water.

Bacteria (here colored green) colonize the surface of minerals rich in uranium

Nodules of iron oxide beginning to appear on this bacterium

mineral energy

Of the 10,000 known species of bacteria, most look similar under the microscope. But this belies an astonishing diversity in what they are capable of doing. Bacteria use chemical processes that are impossible in other living things. Many, like plants, make food from carbon dioxide, but some of them use minerals—rather than light— as an energy source. Others use minerals as a substitute for oxygen, helping them thrive in oxygen-poor habitats, but still extract energy from food. These chemical tricks are also making bacteria useful in the modern world—as a way to clear up pollutants.

Energy from rusting
Acidovorax bacteria gather a coating of rusted iron due to their unusual respiration. They live in poorly aerated soil and use nitrate—the same nitrogen-containing mineral absorbed by plants—instead of oxygen. They collect dissolved iron and use the nitrate, not oxygen, to turn it to rust (iron oxide)—generating usable energy in the process. SEM, x 50,000.

DETOXIFYING INDUSTRIAL WASTE
Harmful elements—such as uranium or arsenic—are poisonous because they dissolve in water, and so can circulate in bodies of animals and plants. But some bacteria convert these elements into solid forms that can be separated. These bacteria can be used in bioremediation—whereby organisms are used to detoxify contaminated environments.

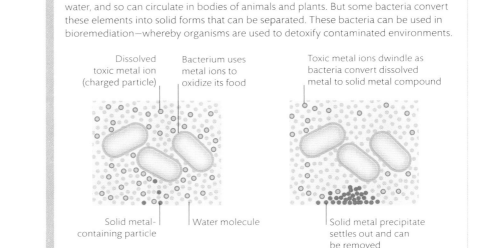

Dissolved toxic metal ion (charged particle)

Bacterium uses metal ions to oxidize its food

Toxic metal ions dwindle as bacteria convert dissolved metal to solid metal compound

Solid metal-containing particle

Water molecule

Solid metal precipitate settles out and can be removed

DECONTAMINATING WATER

THE BREWING PROCESS

Yeast grows best when it respires aerobically. In both aerobic and anaerobic respiration, enzymes in the yeast's cytoplasm break down sugar into organic acid, releasing a little energy. When oxygen is present, the mitochondria—organelles containing enzymes—break down the organic acid further to release copious energy. However, without oxygen, yeast can bypass the aerobic stage through fermentation, converting the organic acid into ethanol, released as waste.

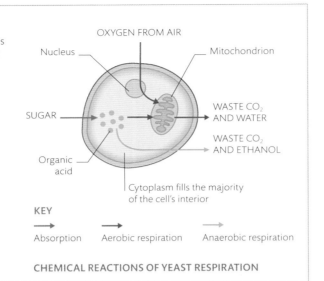

CHEMICAL REACTIONS OF YEAST RESPIRATION

YEAST BUDDING

Whether respiring aerobically or fermenting, yeast reproduces by budding. The nucleus divides and an outgrowth of the parent cell receives a nucleus to create a smaller daughter cell. When well nourished and oxygenated, the yeast has energy to produce chains of cells.

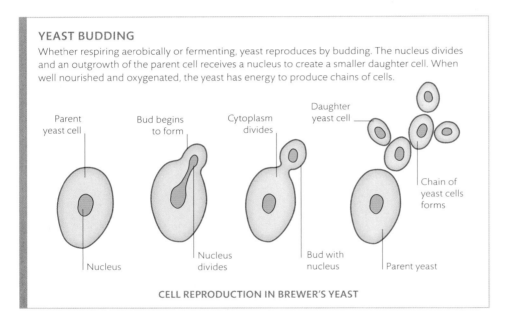

CELL REPRODUCTION IN BREWER'S YEAST

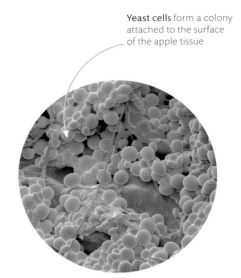

Sweet microhabitat
Yeasts thrive wherever there is a source of sugar for fueling their metabolism—such as here, in the tunnels made by a codling moth caterpillar inside a sweet apple.

fermentation energy

While respiration using oxygen yields the most energy (see pp.74–75), many organisms, including yeasts, can power their cells when oxygen is limited or even absent. The glucose fuel is broken down in fewer steps—not completely into carbon dioxide—so much of the energy remains in the products. For yeast, this product is alcohol. Anaerobic respiration—without oxygen—is also called fermentation and releases a fraction of the energy possible from aerobic respiration. It tends to be a temporary strategy and is mainly exploited by organisms—such as microbes—with low energy demands. But it can also supply emergency energy in aerobic animals working at peak capacity.

Useful microbe
The fermentation effect of brewer's yeast (*Saccharomyces cerevisiae*) is used in winemaking, brewing, and baking. As the cells multiply by budding, their respiration generates waste products that are used in food and drink production: ethanol for making alcoholic drinks and carbon dioxide gas for making bread rise. SEM, x 10,000.

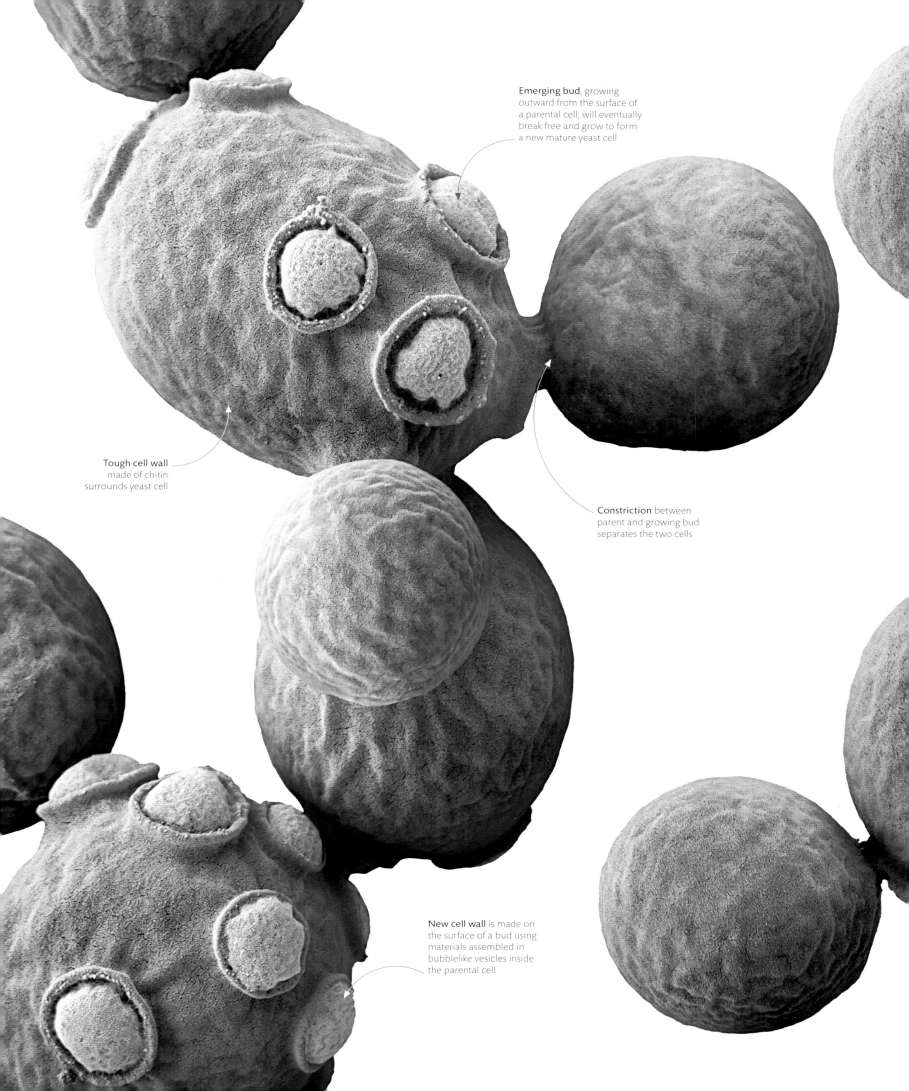

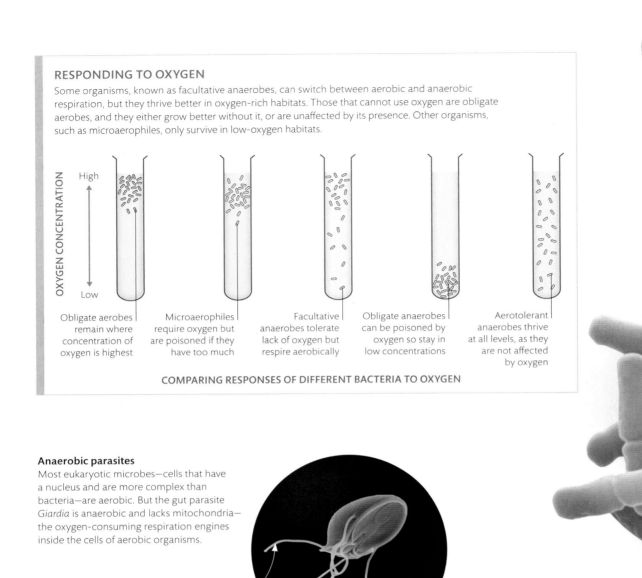

RESPONDING TO OXYGEN

Some organisms, known as facultative anaerobes, can switch between aerobic and anaerobic respiration, but they thrive better in oxygen-rich habitats. Those that cannot use oxygen are obligate aerobes, and they either grow better without it, or are unaffected by its presence. Other organisms, such as microaerophiles, only survive in low-oxygen habitats.

COMPARING RESPONSES OF DIFFERENT BACTERIA TO OXYGEN

- Obligate aerobes remain where concentration of oxygen is highest
- Microaerophiles require oxygen but are poisoned if they have too much
- Facultative anaerobes tolerate lack of oxygen but respire aerobically
- Obligate anaerobes can be poisoned by oxygen so stay in low concentrations
- Aerotolerant anaerobes thrive at all levels, as they are not affected by oxygen

Anaerobic parasites
Most eukaryotic microbes—cells that have a nucleus and are more complex than bacteria—are aerobic. But the gut parasite *Giardia* is anaerobic and lacks mitochondria—the oxygen-consuming respiration engines inside the cells of aerobic organisms.

Giardia swims in the gut by beating its threadlike flagella

Chains of bacteria form as a result of cell division

poisoned by oxygen

Most life on Earth has cells that use oxygen and enjoy the efficient release of energy via aerobic respiration. But where oxygen levels fall far below a normal atmospheric concentration of about 20 percent, organisms need to respire without it via a process called anaerobic respiration. Some microbes, such as archaeans, are descended from ancestors that lived billions of years ago—before the evolution of oxygen-generating photosynthesis, when Earth's atmosphere entirely lacked oxygen. They have remained in anaerobic habitats such as stagnant mud or the ocean floor ever since. Other oxygen-shunning microbes have evolved to occupy the digestive tracts of animals.

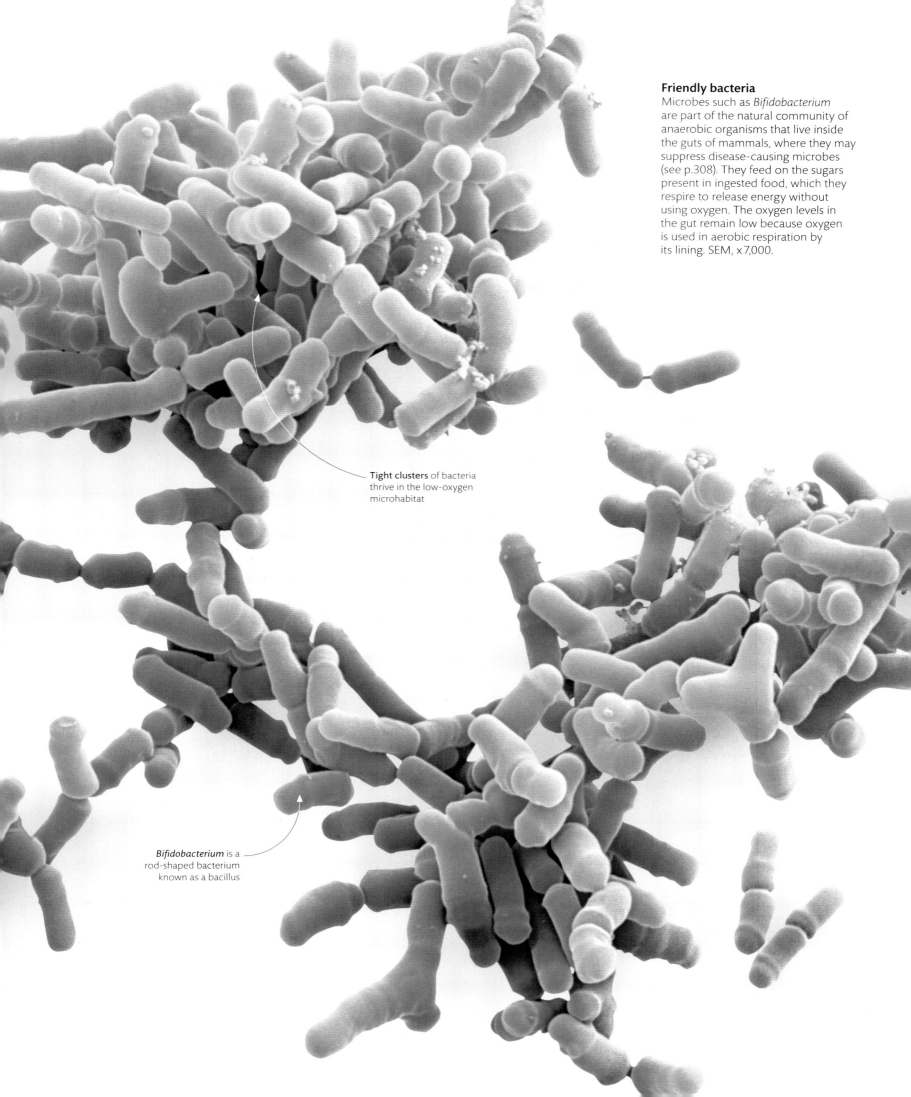

Friendly bacteria
Microbes such as *Bifidobacterium* are part of the natural community of anaerobic organisms that live inside the guts of mammals, where they may suppress disease-causing microbes (see p.308). They feed on the sugars present in ingested food, which they respire to release energy without using oxygen. The oxygen levels in the gut remain low because oxygen is used in aerobic respiration by its lining. SEM, x7,000.

Tight clusters of bacteria thrive in the low-oxygen microhabitat

Bifidobacterium is a rod-shaped bacterium known as a bacillus

gas exchange

A living animal is constantly using oxygen and producing carbon dioxide (see pp.74–75), creating differences in gas concentration between inside and out. Oxygen diffuses in and carbon dioxide diffuses out, flowing from areas of high concentration to low, and this diffusion can supply a tiny organism's oxygen needs. But diffusion is not fast enough above a certain body size, so aquatic animals use gills—finely divided body parts that make diffusion efficient, with a large surface area and thin walls filled with tiny blood vessels. As blood flows through the gills, and water across them, a sharp contrast in gas concentrations is maintained, leading to fast gas exchange.

RESPIRATORY SURFACES

Aquatic animals rely on oxygen dissolved in water. They need extensive respiratory surfaces, because oxygen is less plentiful than in the air. Small organisms can use their entire body surface, but larger ones need gills or similar external structures. In contrast, air is oxygen-rich, but dry, and provides no support. Air-breathing animals therefore have internal respiratory organs, which minimize water loss and stay moist and well supported.

BODY SURFACE IN WATER — Gases exchange across body surface

BODY SURFACE IN AIR — Water vapor lost through skin

GILLS IN WATER — Gills supported by water

LUNGS AND TRACHEAE IN AIR — Air passages in lungs (or tracheae—breathing tubes—in insects)

KEY → Oxygen → CO_2 → Water vapor

Protective tube contains worm's thorax and abdomen

Retractable breathing apparatus
Fanworms get their name from the crown of tentacles protruding from their protective tube, which they build with calcium carbonate or mucus. They withdraw the tentacles rapidly at any sign of danger.

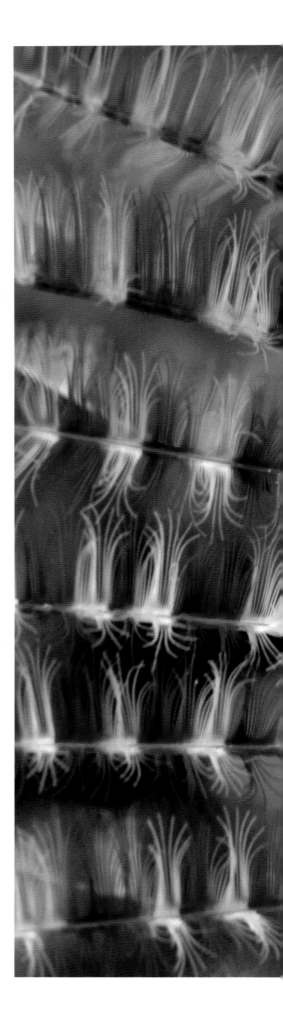

Feathery branches
Gills are formed from a range of body parts in different animals. The gills of *Sabellastarte magnifica*, the feather duster worm, seen here at 10 times life size, are its featherlike radioles (ciliated tentacles). On each radiole, many small appendages called pinnules can be seen. These are filled with blood and covered with beating cilia, which draw water through the fan for feeding and respiration.

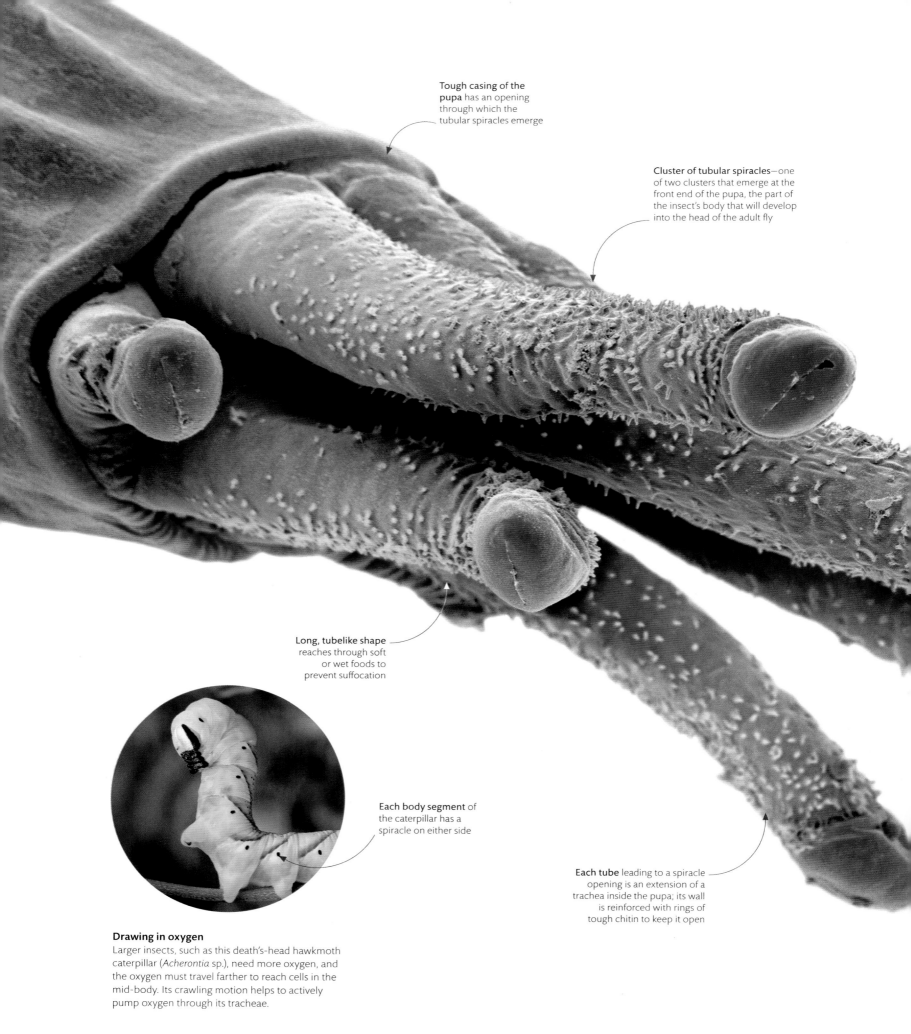

Tubular spiracles
The slitlike breathing holes, or spiracles, on the pupa of a fruit fly (*Drosophila melanogaster*) are borne on the ends of tubes. Fruit fly larvae feed among the sweet fluid of rotting fruit. If a larva cannot find a dry place to pupate, tubular spiracles prevent the insect's tracheal system from being suffocated by the sticky food, enabling it to reach oxygen-rich air and keep breathing in its runny surroundings. SEM, x 1,800.

Oxygenating a tiny body
In most adult flying insects, such as this fruit fly, the tracheal system opens through a row of spiracles along each side of the body. The tiny insect—just 1/6 in (4 mm) long—is small enough for oxygen to reach cells by diffusion, and does not need to actively draw it in.

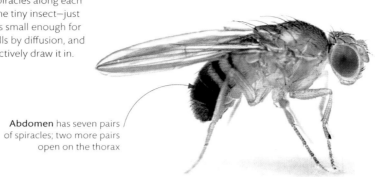

Abdomen has seven pairs of spiracles; two more pairs open on the thorax

Slitlike spiracle opening helps prevent the breathing tube from being clogged by detritus

breathing tubes

In many animals, oxygen is collected by lungs or gills and supplied to cells via circulating blood. Insects, however, have evolved a system of breathing that bypasses blood. A network of air-filled pipes called tracheae carries oxygen directly from holes, or spiracles, at the body's surface to respiring tissues. The pipes branch so finely that they deliver oxygen right into individual cells. This breathing system requires minimal energy input, and the tiny air holes lose little water. But the traveling distance for seeping oxygen must be short to be effective—which is why most insects are tiny.

TRACHEA AND TRACHEOLES
The tracheae of insects are kept open by reinforcements of hard chitin—the same material that forms their exoskeleton. But the tiniest, deepest tubes of their tracheal system, called tracheoles, are just lined with thin cell membrane, which enables oxygen to diffuse into surrounding cells.

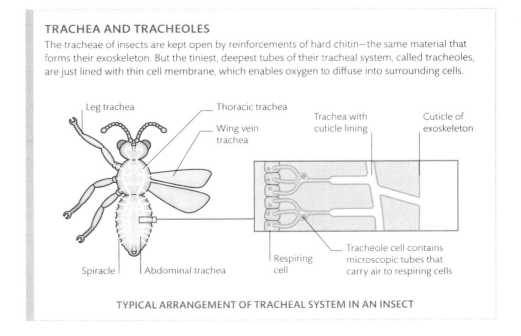

TYPICAL ARRANGEMENT OF TRACHEAL SYSTEM IN AN INSECT

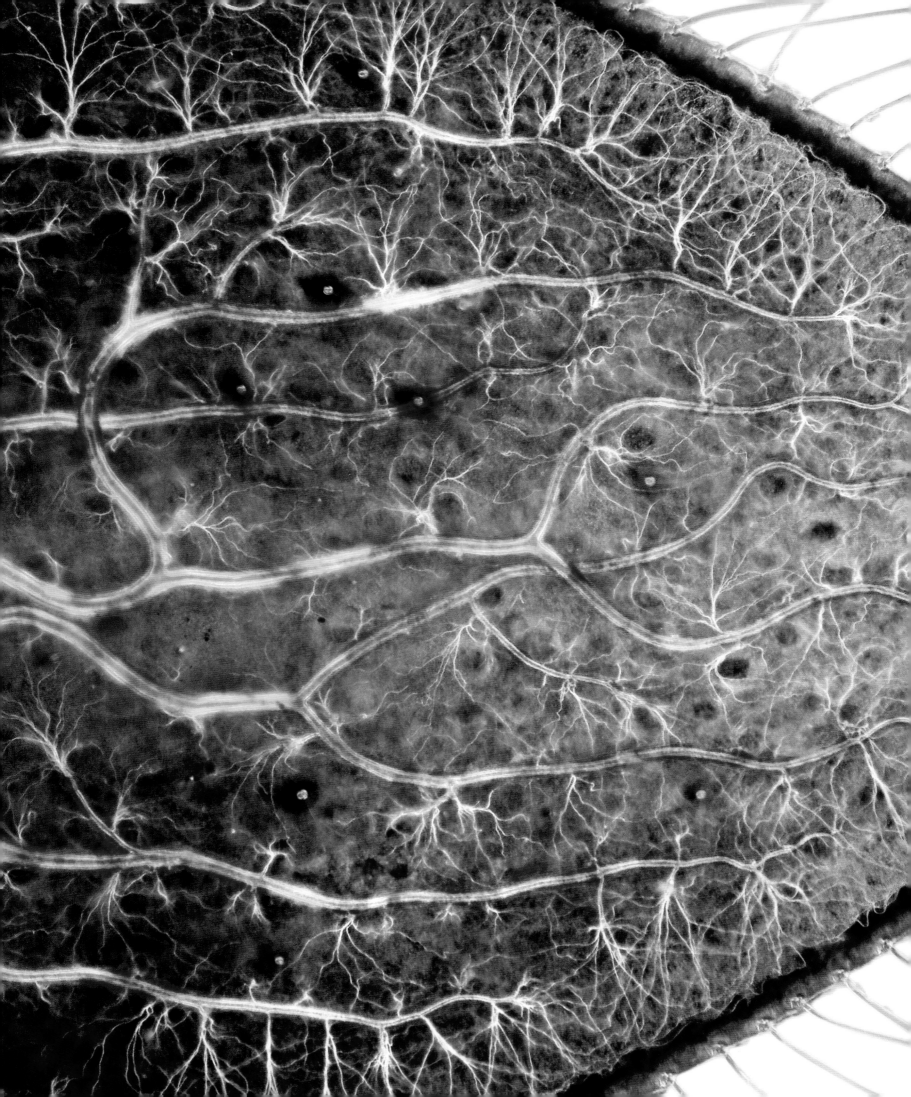

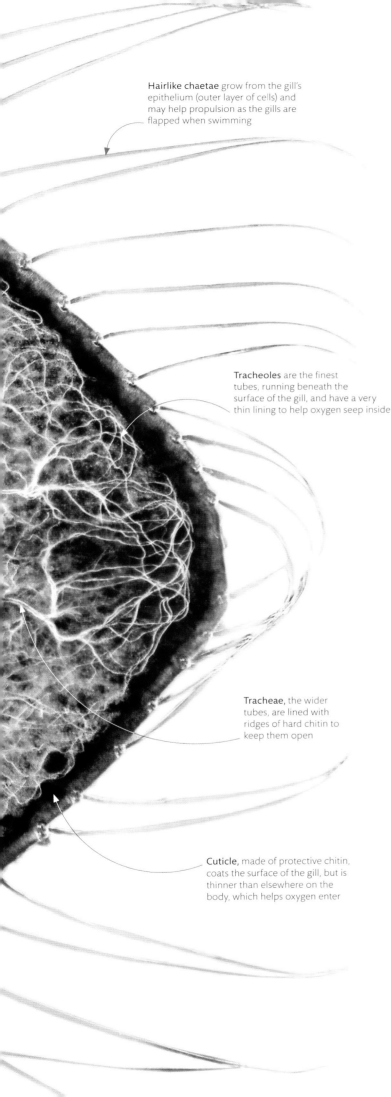

Hairlike chaetae grow from the gill's epithelium (outer layer of cells) and may help propulsion as the gills are flapped when swimming

Tracheoles are the finest tubes, running beneath the surface of the gill, and have a very thin lining to help oxygen seep inside

Tracheae, the wider tubes, are lined with ridges of hard chitin to keep them open

Cuticle, made of protective chitin, coats the surface of the gill, but is thinner than elsewhere on the body, which helps oxygen enter

the gills of insects

Insects have a network of air-filled tubes, called tracheae, which deliver oxygen into the body from the air. But some insects have evolved an aquatic lifestyle—many, including damselflies, start as underwater nymphs, and their tracheae branch finely in flaps on their body that work as gills. Instead of having breathing pores (spiracles, see pp.84–85), they rely on this mesh of extra-fine tracheae. Oxygen dissolved in the water seeps into these tubes across their thin gill wall. As long as the habitat is sufficiently oxygenated, oxygen continually diffuses into the body to replace the oxygen used by the insect's tissues.

Three gill filaments attached, like a tail, to the rear of the abdomen

Aquatic nymph
Damselflies are acrobatic fliers as adults but spend most of their life as nymphs, crawling in ponds and streams. Their gills work as swimming paddles as well as oxygen collectors.

Tracheal gill
The gills of most animals contain blood vessels, but a tracheal gill of a damselfly has silvery, gas-filled tracheae instead. These tubes are blind-ending and run so close to the gill surface that oxygen can easily pass into them. LM, x550.

AIR SUPPLY

Many underwater insects rely on an air-breathing tracheal system, which means they must carry an air supply. This must be replenished from time to time at the surface. Diving beetles carry a bubble beneath their wing cases, releasing it when the oxygen is exhausted, whereas many aquatic bugs have a film of air trapped by hairs on their body surface. Others, such as rat-tailed maggots, can stay submerged by breathing through siphons connected to the surface.

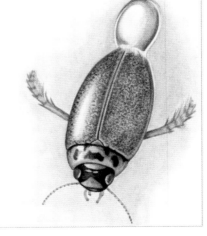

DIVING BEETLE

MACROLIFE

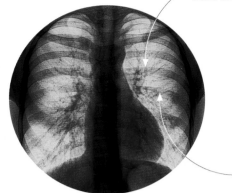

Veins and arteries carrying blood connect lungs and heart

Network of air sacs
Millions of narrow air passages, called bronchioles, terminate at clusters of thin-walled air sacs called alveoli, visible as spherical spaces in this image of a section of lung tissue. Red blood cells showing through breaks in the alveolar walls indicate the presence of a net of tiny blood vessels, called capillaries, around each alveolus. Red blood cells absorb oxygen for circulation to the body's cells. SEM, x1,860.

Ribcage surrounds and protects lungs

Human lungs
Like all mammals, humans have two lungs. Air enters the lungs through two tubes, called bronchi. A dedicated circulatory system connects the lungs to the heart, which then pumps oxygenated blood around the body.

mammal lungs

All animals need a supply of oxygen for releasing energy from their food. The larger air-breathing vertebrates, such as mammals, obtain it using lungs. They actively draw air containing oxygen into the lungs and expel waste carbon dioxide with breathing movements. The complex structure of tubes and air sacs in the lungs also maximizes the surface area through which oxygen can be acquired. As a result, animals with lungs can grow many times larger than those without them, such as insects.

EXCHANGING GAS

Gas exchange is the process by which oxygen is taken up by the blood, and carbon dioxide (a waste product of respiration) is removed from the blood; air entering the air sac (alveolus) is about 21 percent oxygen, and air leaving it about 16 percent. The two gases move in opposite directions by diffusion across the thin walls of the air sacs and capillaries that surround them, traveling from where they are highly concentrated to where they are not.

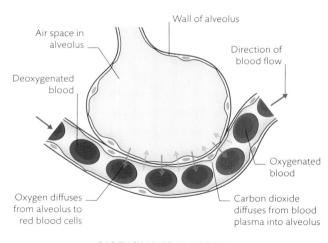

GAS EXCHANGE IN ALVEOLI

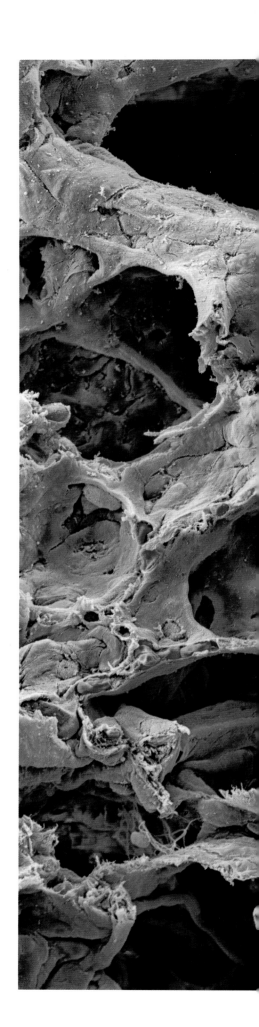

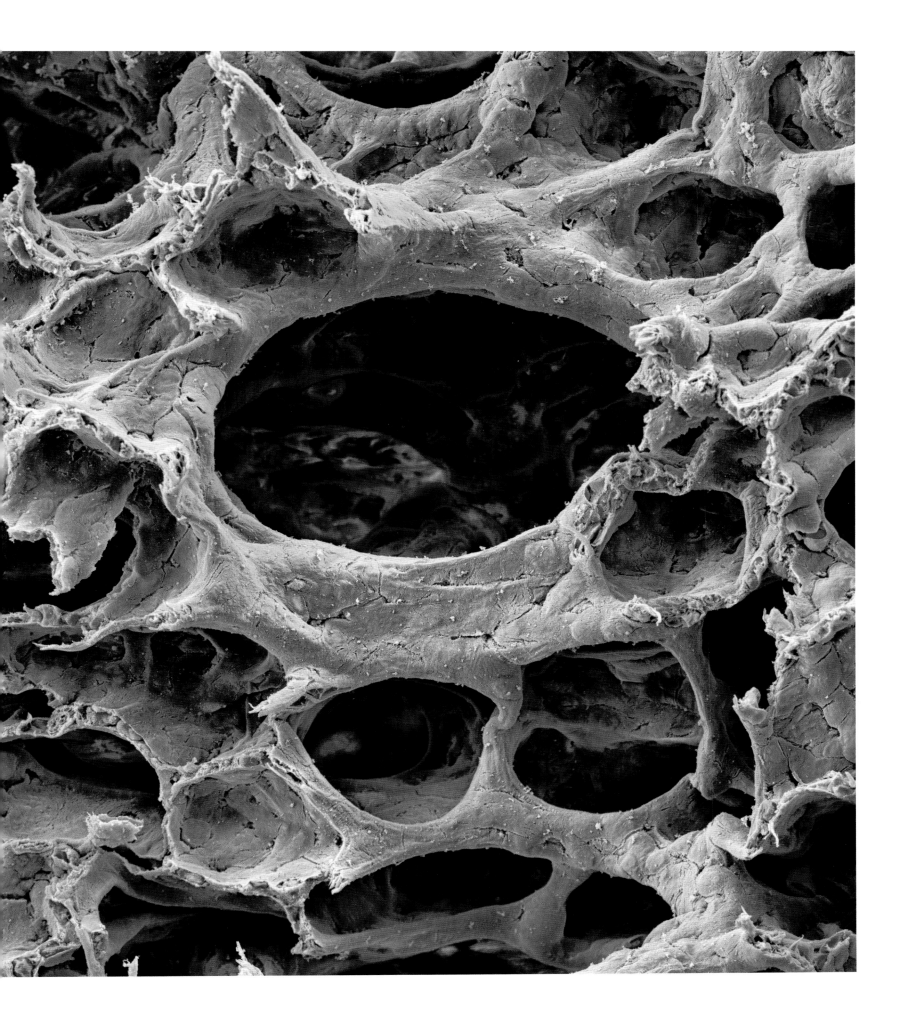

carrying oxygen

Oxygen is vital for respiration, but it dissolves less well than sugar and carbon dioxide in water. This is not a problem for insects, which pass oxygen to cells directly from the air via tracheal breathing tubes. But animals that transport oxygen in circulating, liquid blood need a carrier to pick up oxygen in lungs or gills and offload it at respiring cells. Vertebrates package their oxygen carrier, hemoglobin, in red blood cells. More than 80 percent of cells in a human body are red blood cells—an indication of just how important they are.

Oxygen-carrying disks
A sample of human blood contains red cells that carry oxygen and white cells that help fight infection, but red cells typically outnumber white ones by as much as 1,000 to 1. Each red cell is a disk that lacks a nucleus in order to pack as much hemoglobin as possible. SEM, x 2,800.

BINDING AND UNBINDING

Red blood cells are packed with hemoglobin and not much else. Those of mammals even lose their nucleus, with all their DNA, and devote themselves to hemoglobin. A hemoglobin molecule consists of four protein chains, each carrying an iron-containing component called heme. Oxygen binds to heme where oxygen levels are high—in lungs or gills—and lets go where oxygen levels are low—in respiring tissues.

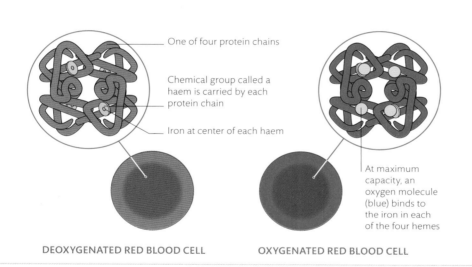

- One of four protein chains
- Chemical group called a haem is carried by each protein chain
- Iron at center of each haem
- At maximum capacity, an oxygen molecule (blue) binds to the iron in each of the four hemes

DEOXYGENATED RED BLOOD CELL   OXYGENATED RED BLOOD CELL

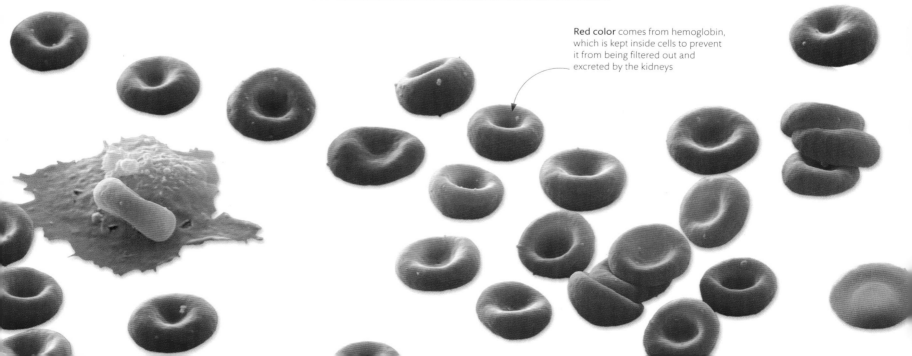

Red color comes from hemoglobin, which is kept inside cells to prevent it from being filtered out and excreted by the kidneys

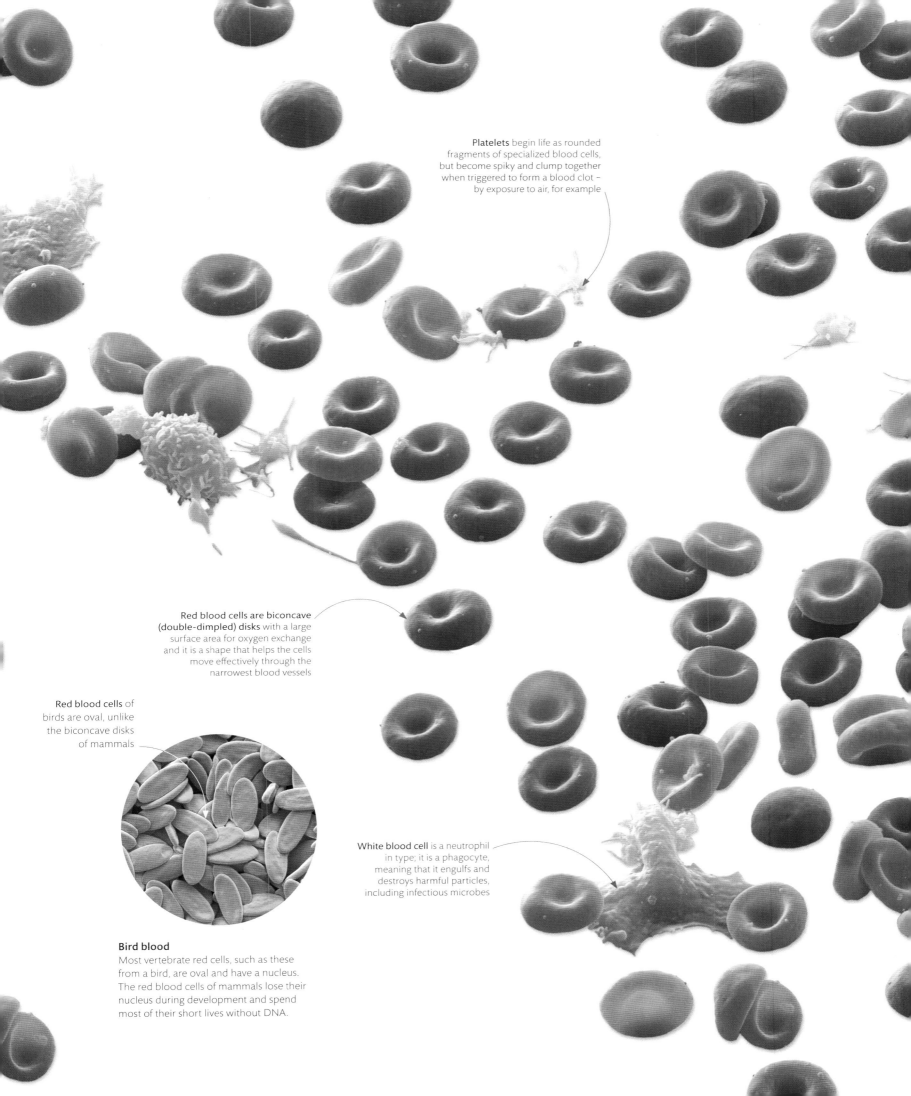

Platelets begin life as rounded fragments of specialized blood cells, but become spiky and clump together when triggered to form a blood clot – by exposure to air, for example

Red blood cells are biconcave (double-dimpled) disks with a large surface area for oxygen exchange and it is a shape that helps the cells move effectively through the narrowest blood vessels

Red blood cells of birds are oval, unlike the biconcave disks of mammals

White blood cell is a neutrophil in type; it is a phagocyte, meaning that it engulfs and destroys harmful particles, including infectious microbes

Bird blood
Most vertebrate red cells, such as these from a bird, are oval and have a nucleus. The red blood cells of mammals lose their nucleus during development and spend most of their short lives without DNA.

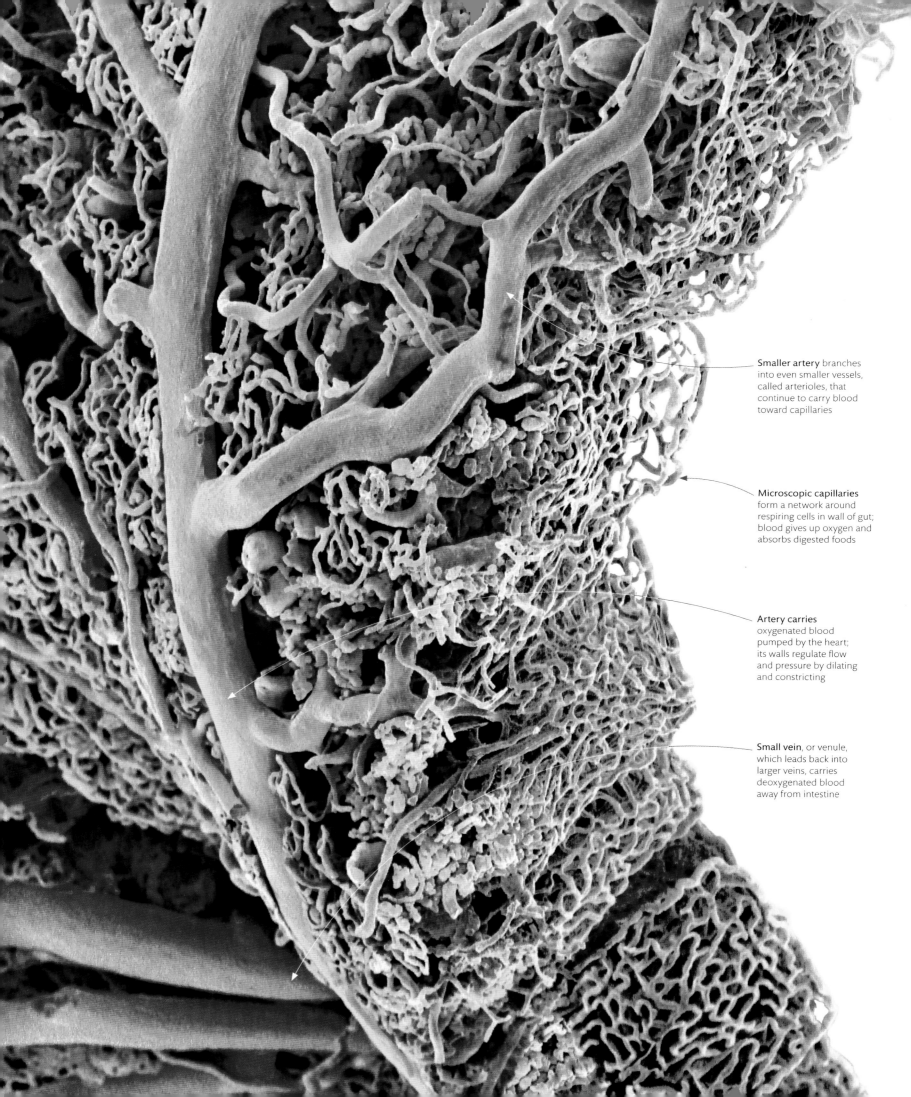

Smaller artery branches into even smaller vessels, called arterioles, that continue to carry blood toward capillaries

Microscopic capillaries form a network around respiring cells in wall of gut; blood gives up oxygen and absorbs digested foods

Artery carries oxygenated blood pumped by the heart; its walls regulate flow and pressure by dilating and constricting

Small vein, or venule, which leads back into larger veins, carries deoxygenated blood away from intestine

MACROLIFE

Supplying the body systems
Blood vessels supplying the small intestine carry blood that will absorb digested food, and then circulate it to feed cells all over the body. The vessels divide like tree branches, leading to a network of capillaries where blood absorbs food and delivers oxygen. SEM, x200.

Oxygen transfer
The thin walls of the microscopic capillaries appear almost transparent. Red blood cells pass along them in single file, which brings the oxygen they carry so close to the walls that it can seep out to the respiring tissues.

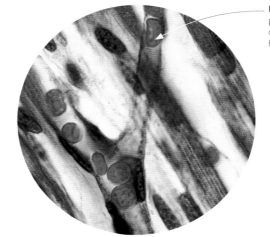

Red blood cells are packed with red, oxygen-carrying hemoglobin

circulatory system

All the cells of an animal's body need oxygen and food to stay alive, and most need a blood transport system to carry both to where they are needed. Circulating blood carries oxygen and food from the organs that acquire them, then delivers them around the body. A muscular pump, the heart, sustains the flow. The same system can transport waste products away from the cells for excretion. In many invertebrates, blood bathes the body cells in an open chamber, but in vertebrates the blood remains in a network of vessels, leading to microscopic tubes that reach between the cells.

TYPES OF BLOOD VESSELS
Blood is pumped from the heart into arteries, which have thicker walls that sustain the high pressure, and thick muscle that can constrict and control flow. Thinner-walled veins, with one-way valves that prevent backflow, return blood to the heart. Between the arteries and veins are microscopic capillaries, in which the exchange of materials with surrounding cells takes place.

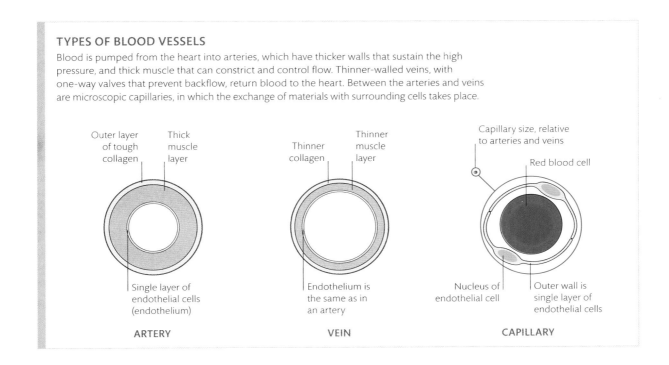

ARTERY — Outer layer of tough collagen; Thick muscle layer; Single layer of endothelial cells (endothelium)

VEIN — Thinner collagen; Thinner muscle layer; Endothelium is the same as in an artery

CAPILLARY — Capillary size, relative to arteries and veins; Red blood cell; Nucleus of endothelial cell; Outer wall is single layer of endothelial cells

powering the body

MACROLIFE

TRANSPIRATION

When stomata open to take in carbon dioxide, some water vapor escapes, or transpires, from the leaf into the air. The loss of water produces negative pressure that draws more water into the leaf from the plant's vascular system. The combined transpiration by all of a plant's leaves creates enough force to pull water up from the roots. Some is transpired and some is used for photosynthesis in the mesophyll (inner tissue).

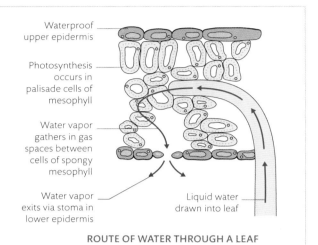

ROUTE OF WATER THROUGH A LEAF

LILY OF THE VALLEY

Row of bell-shaped flowers with a sweet scent

Upper epidermis of leaf has few pores and a thick, waxy cuticle

Feeding flowers
Lily of the valley is a moisture-loving plant. Most stomata occur on the underside of the leaf. This could be to reduce water loss, or because pathogens, such as fungal spores, can enter the leaf more easily from above.

Active opening
Each green-ringed aperture on this lily of the valley (*Convallaria majalis*) leaf is a stoma, formed by two guard cells. Light-sensitive phototropin pigments in these cells trigger the stoma to open. The cells swell with water and bow outwards, since their outer walls are more elastic, creating an air gap a few micrometers wide. Unlike other surface cells, guard cells have green chloroplasts, which provide energy for the opening mechanism. When insufficient light is detected, the guard cells lose water, shrink, and the opening closes.

leaf pores

As the main site on a plant for photosynthesis, leaves need a good supply of two raw ingredients: water, delivered from the roots, and carbon dioxide gas, collected from the air. Carbon dioxide enters the leaf through pores called stomata (singular: stoma). The stomata are not simple holes—they respond to the environment. In most plants, they open when there is light for photosynthesis and close when it is too dark—or too dry—for the process to continue. As well as allowing carbon dioxide in, stomata release unwanted oxygen gas, which, like sugar, is a product of photosynthesis.

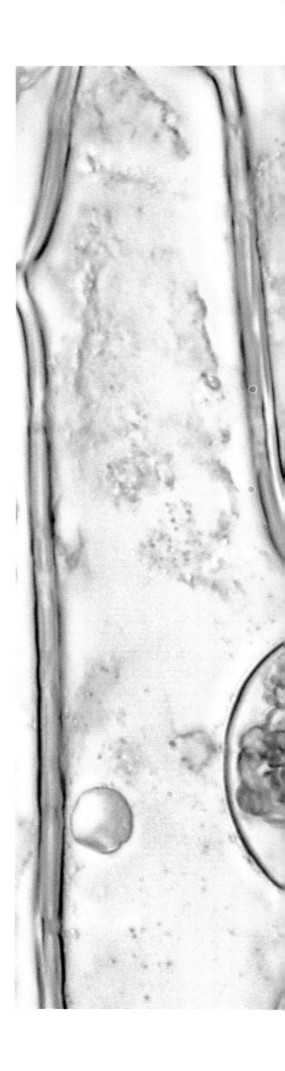

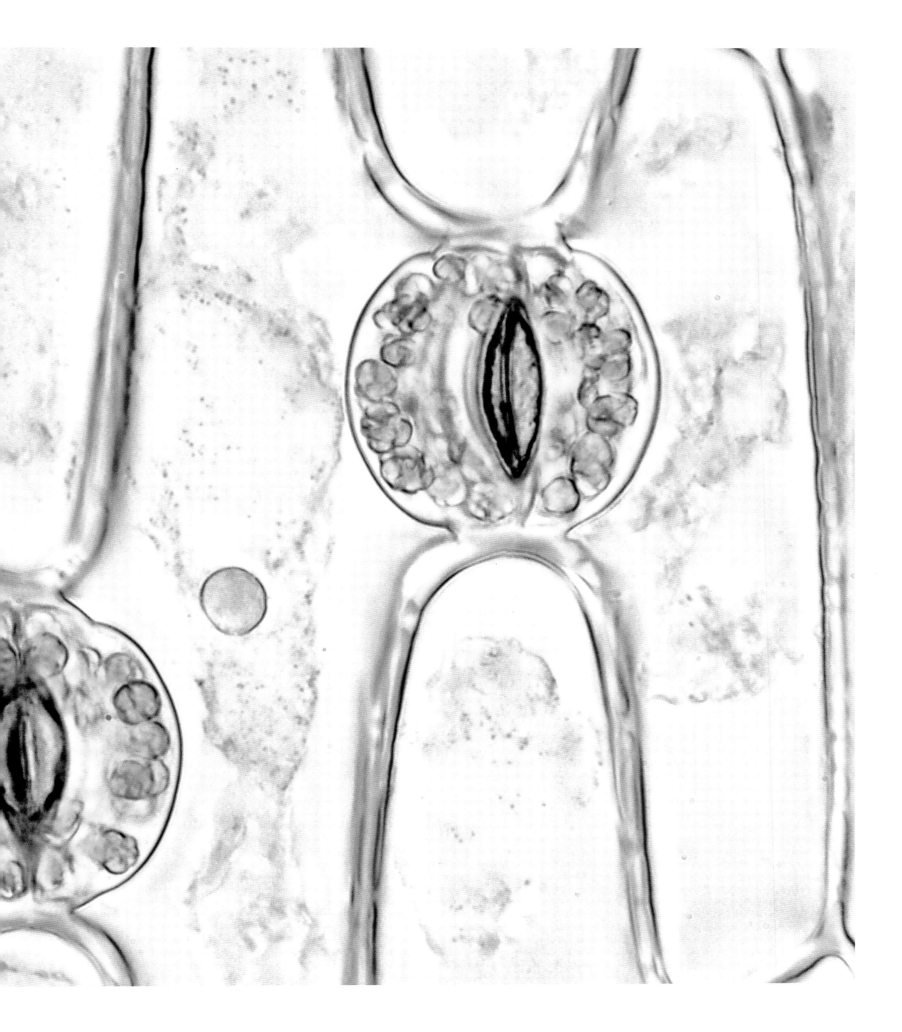

keeping warm

It is not easy for a warm-blooded animal to be small. Birds and mammals generate body heat from their internal metabolism—rather than relying on obtaining warmth from their surroundings, as most other animals do. There are pros and cons to being warm-blooded: a body can keep active, even in places that are cold, but the smallest birds and mammals pay a hefty price in fuel consumption. The tiniest bodies lose the biggest proportion of body heat and thus chill more easily in the cold. As a result, hummingbirds and shrews—which are smaller than the biggest insects and spiders—need a constant supply of high-energy food just to survive.

Long, pointed bill reaches deep into tubular flowers as the bird hovers and laps up nectar with its long tongue

Shrew's furry skin traps body heat, as in all mammals

Tiny mammal
A Eurasian pygmy shrew (*Sorex minutus*) weighs the same as a penny. To power its tiny body it must eat more than its own mass in invertebrates every 24 hours.

Nectar feeder
A hummingbird, such as a green-breasted mango (*Anthracothorax prevostii*), survives on a diet dominated by high-energy nectar. Hummingbirds will aggressively defend a patch of habitat with a good supply of flowers, since access to fuel can be a matter of life or death. The flight of a hummingbird involves beating the wings many times each second. The muscle activity involved is costly in energy terms, but it is offset by an ability to hover, practically motionless, in front of a flower for efficient refueling.

THE PERILS OF BEING SMALL
While the amount of heat lost from a warm body depends on its surface area, the heat generated by the body's metabolism depends on its volume. The ratio between the two—most easily calculated for a cube—increases as the body gets smaller, so small bodies lose heat faster, proportionally.

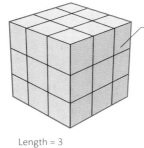

Small surface area : volume ratio results in little loss of generated heat

Large surface area : volume ratio results in rapid heat loss

Length = 3
Surface area = $3^2 \times 6 = 54$
Volume = $3^3 = 27$

SURFACE : VOLUME RATIO = 2

Length = 2
Surface area = $2^2 \times 6 = 24$
Volume = $2^3 = 8$

SURFACE : VOLUME RATIO = 3

Length = 1
Surface area = $1^2 \times 6 = 6$
Volume = $1^3 = 1$

SURFACE : VOLUME RATIO = 6

sensing and responding

Even the simplest microbes sense their environment—sensing is one of the core attributes of life. Life-forms use their sensory information to take action, moving toward food, light, or mates, or away from danger. Chemical senses are perhaps the simplest, but the other fundamental senses—touch and light—are also almost universal.

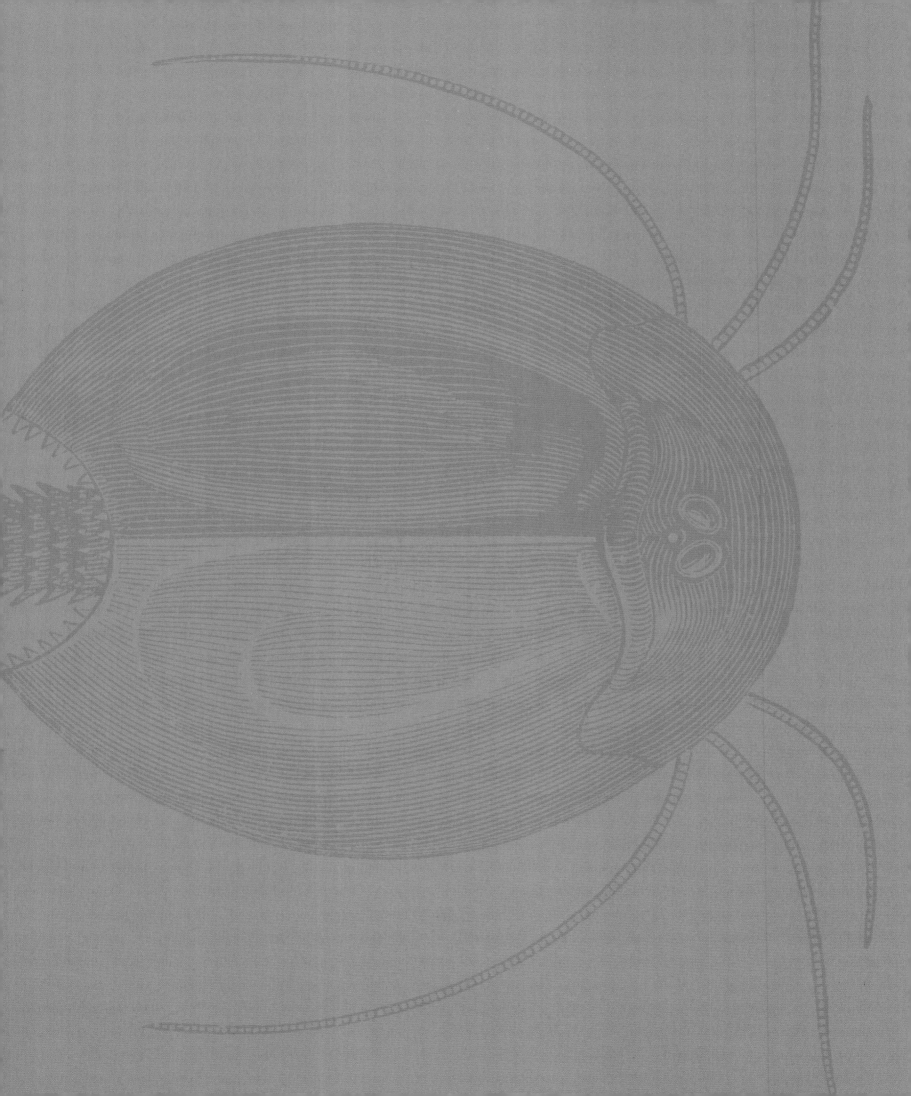

sensing the environment

The ability to sense something and then respond is one of the characteristics that sets living things apart from the nonliving world. Sense improves survival, whether it is to help track food or light for energy, find a mate, or avoid danger. Animals and plants have sense organs made of specialized cells in different parts of their bodies, but microbes can sense within a single cell. Single-celled algae use light to power their photosynthesis so they can make food, and many have beating whips, called flagella, that propel them. By producing a light-absorbing pigment, or photoreceptor, which activates their flagella, they can link sense to response and swim toward the light.

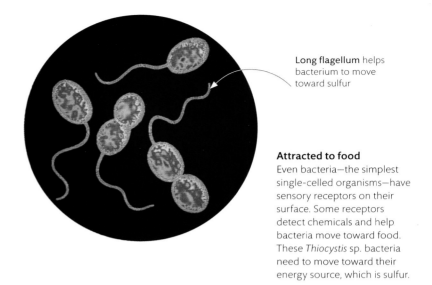

Long flagellum helps bacterium to move toward sulfur

Attracted to food
Even bacteria—the simplest single-celled organisms—have sensory receptors on their surface. Some receptors detect chemicals and help bacteria move toward food. These *Thiocystis* sp. bacteria need to move toward their energy source, which is sulfur.

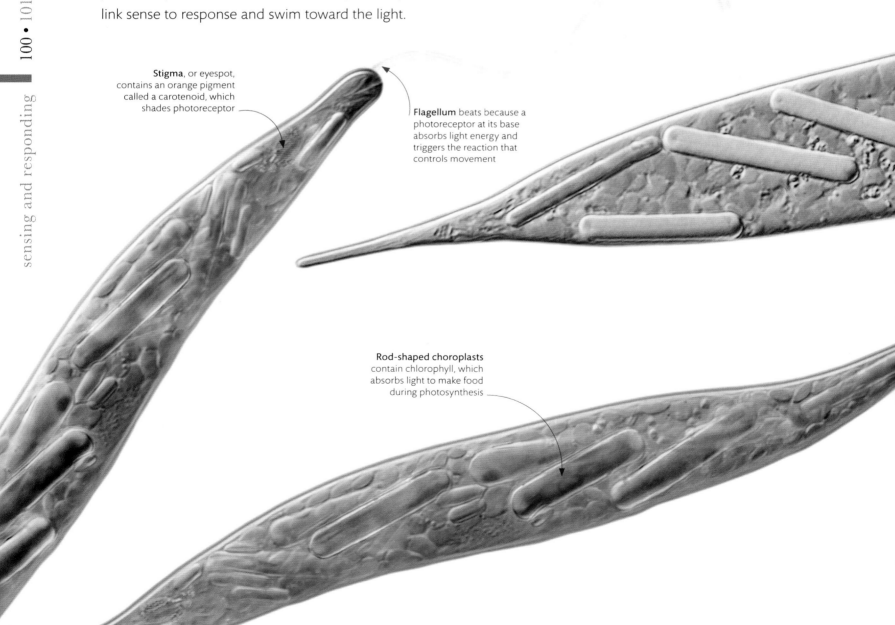

Stigma, or eyespot, contains an orange pigment called a carotenoid, which shades photoreceptor

Flagellum beats because a photoreceptor at its base absorbs light energy and triggers the reaction that controls movement

Rod-shaped choroplasts contain chlorophyll, which absorbs light to make food during photosynthesis

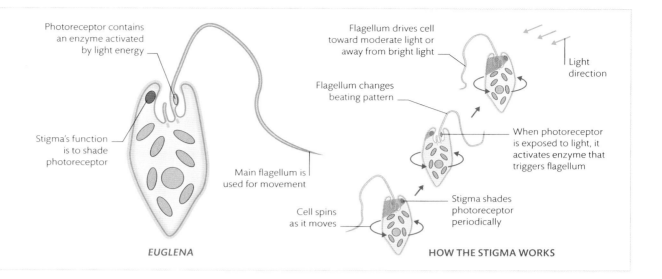

SENSING AND RESPONDING TO LIGHT

The beating of the main flagellum of each *Euglena* sp. cell is driven by chemical reactions triggered by the light-activated photoreceptor. It works by a system of sliding cables that cause bending. The cell spins as it moves, and a red stigma or "eyespot" near the base of the flagellum casts a shadow on the photoreceptor once in every spin, giving *Euglena* sp. frequent updates on light direction. This may help to orientate the cell in the direction of the light.

- Photoreceptor contains an enzyme activated by light energy
- Stigma's function is to shade photoreceptor
- Main flagellum is used for movement
- Flagellum drives cell toward moderate light or away from bright light
- Light direction
- Flagellum changes beating pattern
- When photoreceptor is exposed to light, it activates enzyme that triggers flagellum
- Cell spins as it moves
- Stigma shades photoreceptor periodically

EUGLENA

HOW THE STIGMA WORKS

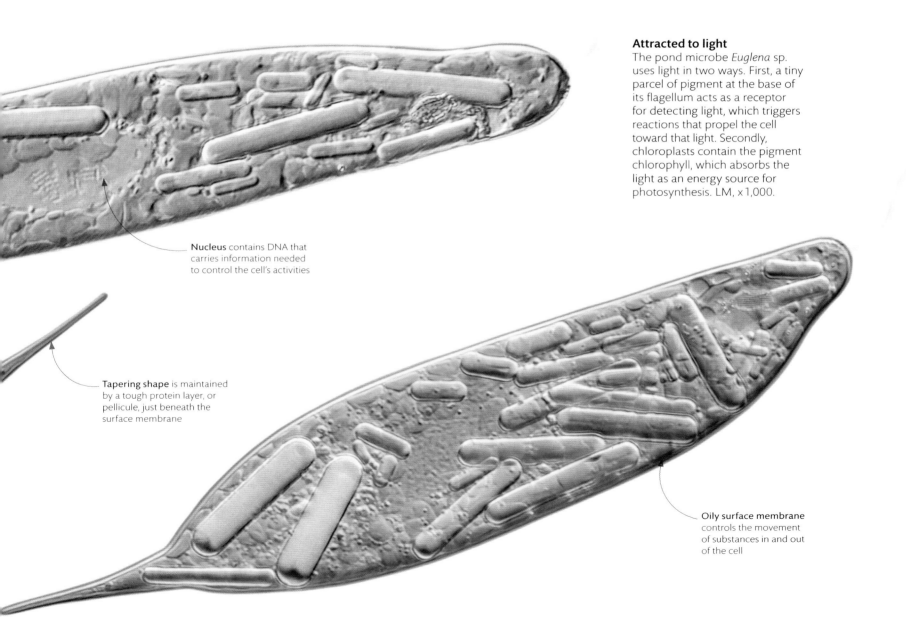

Attracted to light

The pond microbe *Euglena* sp. uses light in two ways. First, a tiny parcel of pigment at the base of its flagellum acts as a receptor for detecting light, which triggers reactions that propel the cell toward that light. Secondly, chloroplasts contain the pigment chlorophyll, which absorbs the light as an energy source for photosynthesis. LM, x 1,000.

- **Nucleus** contains DNA that carries information needed to control the cell's activities
- **Tapering shape** is maintained by a tough protein layer, or pellicle, just beneath the surface membrane
- **Oily surface membrane** controls the movement of substances in and out of the cell

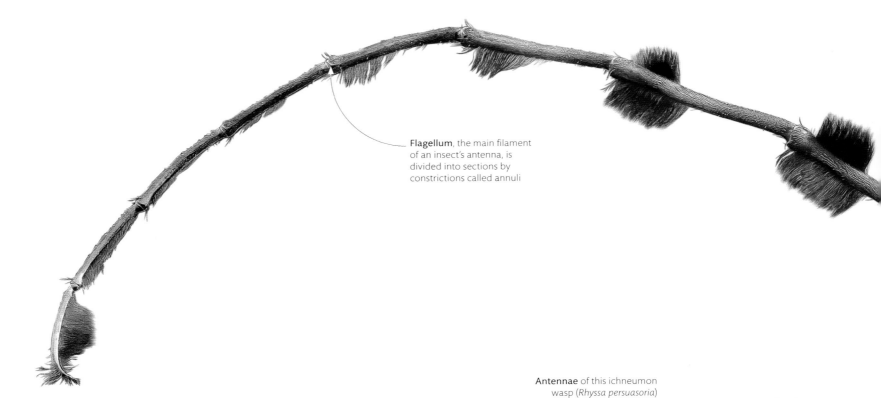

Flagellum, the main filament of an insect's antenna, is divided into sections by constrictions called annuli

Antennae of this ichneumon wasp (*Rhyssa persuasoria*) detect the vibrations and scents of wood-boring grubs under tree bark

Long ovipositor (egg-laying tube) drills into bark to lay eggs on larvae inside tree trunks

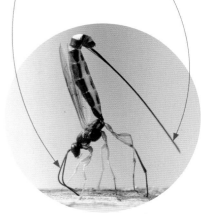

antennae

Animals with antennae that stretch far beyond the head are well equipped to sense the world around them. Antennae are found in arthropods: crustaceans typically possess two pairs, while insects, millipedes, and centipedes have one pair; only arachnids lack them altogether. The prime function of antennae is sensory, but they sometimes have other uses too, such as the swimming antennae of water fleas. Insect antennae have motion sensors at their base that help flight control, as well as long filaments packed with a battery of sensors that detect chemicals, pressure, and heat.

Sensing food
Female ichneumon wasps use their antennae—bigger than those of males—to find a host, such as a grub or caterpillar, on which to lay eggs. The eggs hatch into larvae that eat the host alive.

LONGHORN DIVERSITY
The antennae of longhorn beetles are typically at least two-thirds the combined length of their head and body. There are more than 30,000 species of longhorn beetle, making the group one of the largest of all beetle families, and its diversity is especially high in tropical regions. Males sometimes have longer antennae than females and may court their mates using touch. Like those of other insects, longhorn antennae are worked by muscles at their base and can be reflexed backward.

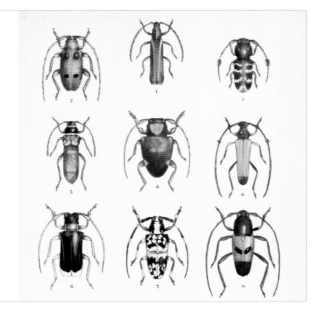

ILLUSTRATION OF LONGHORN BEETLES, C.1880–1890

Multipurpose sensing
Antennae are often described as "feelers," but their sensory adaptations go far beyond just touching their surroundings. The long antennae of a flat-faced longhorn beetle (*Thysia wallichii*) can smell and taste, as well as touch. By responding to multiple stimuli, they guide the beetle to mates and food plants.

sensing taste

Smell and taste both involve sensors that detect substances and fire off electrical impulses to the brain. Whereas smell picks up on particles drifting through the air, taste works by direct contact with liquids or solids. Insects, which lack the kind of fleshy tongue found in vertebrates, have sensilla (sensory organs) adapted for taste on their own unique mouthparts. Other sensilla are found elsewhere, including on antennae or feet. The sensors on the proboscis of a nectar-drinking moth or butterfly are triggered by sugar.

Sugar sensors
Peglike taste sensilla (blue) are clustered near the coiled tip of a moth's proboscis, shown right. Other sensilla lie inside the tube. Contact with sweet nectar triggers a feeding reflex, whereby nerve impulses make the proboscis fill with blood. The pressure causes the proboscis to uncoil so that the moth can drink. SEM, x120.

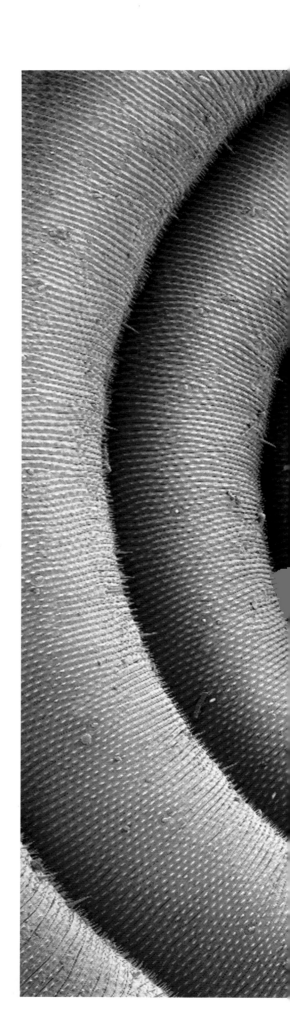

Long proboscis coils up when not in use, to prevent it from being damaged as the insect moves from place to place

Sensory mouthparts
A butterfly's proboscis can delve deep into a flower to reach the nectaries—the glands that secrete nectar. The proboscis has receptors for touch as well as taste, allowing it to feel its way to its target.

TASTE SENSILLUM

An insect's taste sensillum is a microscopic cone that contains a cluster of sensory nerve cells. The fibers of these cells carry molecular receptors that lock on to molecules seeping into the pore at the tip of the sensillum. The fibers have different-shaped receptors that fit sugars, salt, and various other detectable substances.

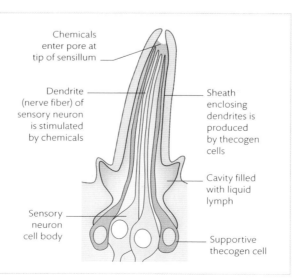

CELLS IN A TASTE SENSILLUM

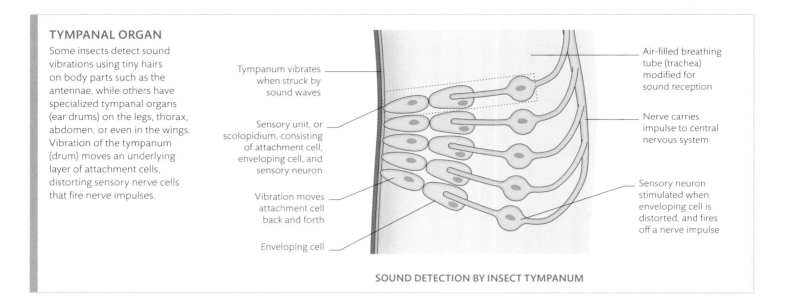

TYMPANAL ORGAN

Some insects detect sound vibrations using tiny hairs on body parts such as the antennae, while others have specialized tympanal organs (ear drums) on the legs, thorax, abdomen, or even in the wings. Vibration of the tympanum (drum) moves an underlying layer of attachment cells, distorting sensory nerve cells that fire nerve impulses.

- Tympanum vibrates when struck by sound waves
- Sensory unit, or scolopidium, consisting of attachment cell, enveloping cell, and sensory neuron
- Vibration moves attachment cell back and forth
- Enveloping cell
- Air-filled breathing tube (trachea) modified for sound reception
- Nerve carries impulse to central nervous system
- Sensory neuron stimulated when enveloping cell is distorted, and fires off a nerve impulse

SOUND DETECTION BY INSECT TYMPANUM

hearing sound

Sound—the waves of pressure that travel through air, water, or the ground—creates movement, which many animals can pick up with simple sensory hairs. Animals with specialized hearing organs, or ears, are especially sensitive to sound. Moths can tune in to the echolocating sounds of predatory bats, while insects that communicate by sound have especially well-developed ears. In crickets, for example, a patch of cuticle stretched over an air sac on the leg works like a drum, helping a female pick up the vibrations of a male's chirping "love song."

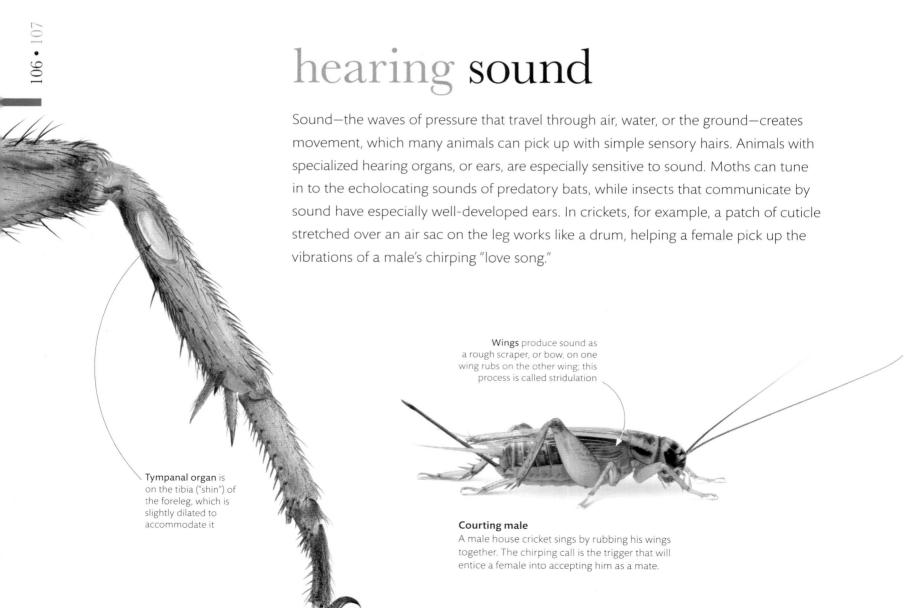

Tympanal organ is on the tibia ("shin") of the foreleg, which is slightly dilated to accommodate it

Wings produce sound as a rough scraper, or bow, on one wing rubs on the other wing; this process is called stridulation

Courting male
A male house cricket sings by rubbing his wings together. The chirping call is the trigger that will entice a female into accepting him as a mate.

Listening with legs
A house cricket (*Acheta domesticus*) has eardrums, called tympanal organs, on its two forelegs. They detect the pitch and loudness of a sound—the frequency and size of the vibrations in the air—and send the information to the insect's brain.

Tympanum cuticle lacks the protective thickening of the rest of the exoskeleton, so it is free to vibrate

Stiff, spinelike setae may help protect the thin cuticle underneath

Fibers of elastic protein called resilin help the cuticle to stretch when vibrating

Reinforced cuticle of the surrounding exoskeleton is continuous with the cuticle of the tympanum

MACROLIFE

sensory cells in the ear

The mammalian ear is an incredibly sensitive organ with a unique amplification apparatus—three tiny bones, called ossicles. Airborne sound makes the eardrum vibrate, and the ossicles transmit these vibrations into the fluid-filled inner ear. At the heart of the inner ear are hair cells, simple movement receptors that are embedded in a membrane. The membrane changes in stiffness along its coiled length, so it is selectively sensitive to sounds of different pitches. The hair cells can therefore convert detailed sound information into nerve signals.

Highly sensitive hair cells
This image shows the V-shaped hairlike tufts in a guinea pig's (*Cavia porcellus*) organ of Corti. Each set of "hairs" protrudes from one sensory cell, and their tiny movements—caused by sound waves—are translated into electrical signals that are sent to the brain. SEM, x10,600

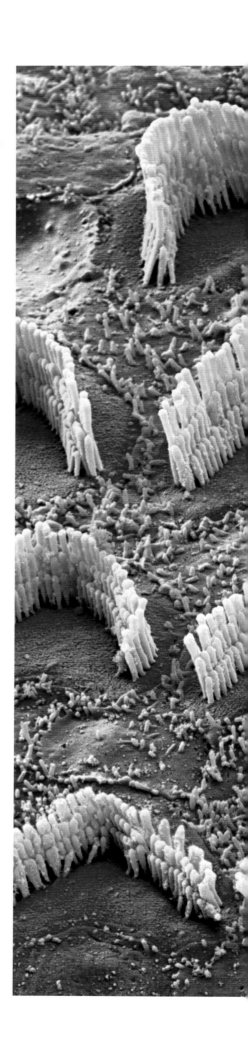

HOW THE COCHLEA WORKS
The coil-shaped cochlea receives sound as a vibration from the tiny bones of the middle ear. This sound ripples through fluid inside the cochlea, moving the sensory cells in the organ of Corti. The brain distinguishes pitch by receiving signals from hair cells at different points in the coil.

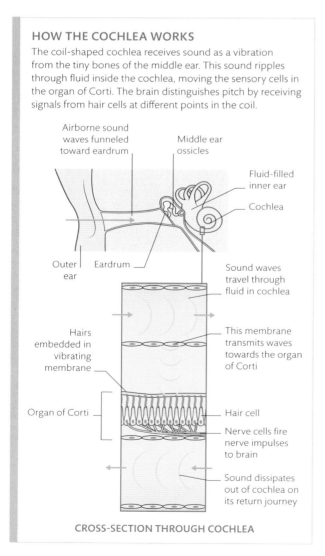

CROSS-SECTION THROUGH COCHLEA

Organ of Corti, with its long tubes of membranes, runs the full length of the cochlea coil

COCHLEA

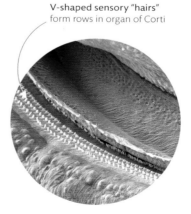

V-shaped sensory "hairs" form rows in organ of Corti

Rows of cells
The organ of Corti is composed of a row of inner "hairs" and three outer rows. These rows are separated by supporting cells and bathed in watery cochlear fluid.

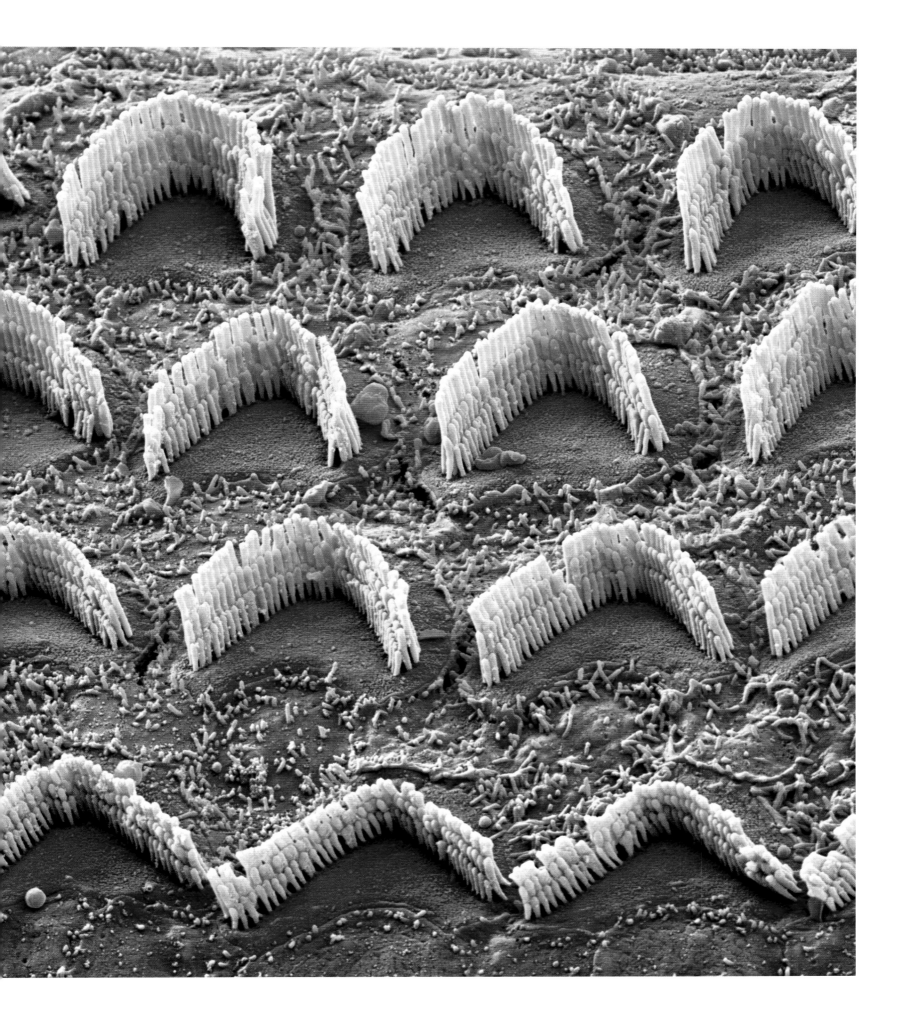

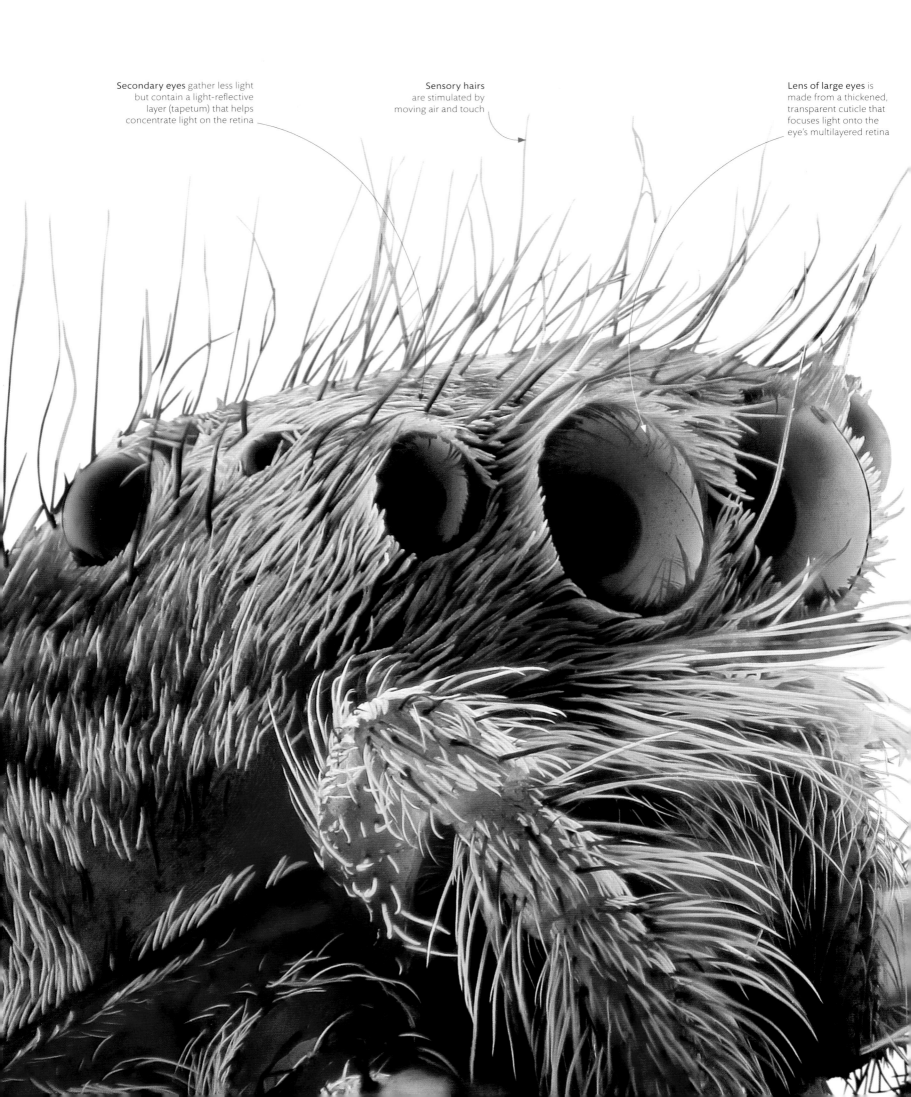

Secondary eyes gather less light but contain a light-reflective layer (tapetum) that helps concentrate light on the retina

Sensory hairs are stimulated by moving air and touch

Lens of large eyes is made from a thickened, transparent cuticle that focuses light onto the eye's multilayered retina

Two smaller, forward-facing eyes provide a wide field of vision

Trapped prey
A fast reaction and a venomous bite mean that even a stinging wasp is no match for a jumping spider. The venom kills the prey quickly so the spider can feed.

Eyes of a hunter
Like most spiders, the jumping spider has eight complex eyes. But it is its two larger, forward-facing eyes that sets this spider apart as an ambush hunter. The big eyes have larger lenses and more photoreceptors, so they are not only better at collecting light, but also produce a clearer image.

ASSESSING JUMPING DISTANCE
Vertebrates with forward-facing eyes judge distance by binocular vision—each eye sends a slightly different image to the brain, which the brain combines to give a sense of three dimensions. Jumping spiders use a multilayered retina to do this. While the deepest layers focus green light, others sense fuzzy ultraviolet (UV) light—the closer the spider is to the target, the fuzzier this gets. The spider judges distance by assessing the amount of UV blur compared with focused green.

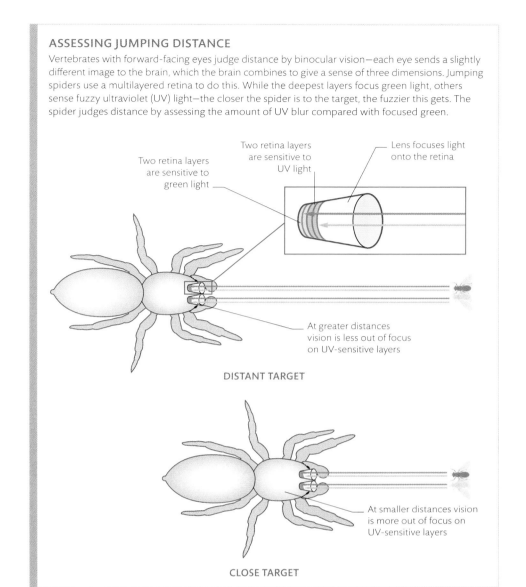

DISTANT TARGET

- Two retina layers are sensitive to green light
- Two retina layers are sensitive to UV light
- Lens focuses light onto the retina
- At greater distances vision is less out of focus on UV-sensitive layers
- At smaller distances vision is more out of focus on UV-sensitive layers

CLOSE TARGET

judging distance

The complex eyes of animals take them beyond just sensing light and dark to generating images of the world around them—they enable them to see shapes, movement, and color, and even sense depth and distance. The advantages of good vision are enormous. A jumping spider hunts its prey instead of trapping it in a web. Its big complex eyes help it see a nearby insect as a meal and provides its brain with all the information it needs for a successful ambush. Each eye has a lens that focuses the light and a retina made up of light-sensitive cells, or photoreceptors, that are wired into its nervous system, so the spider can not only assess whether its prey is a safe target of manageable size, but also how far it has to jump to make a catch.

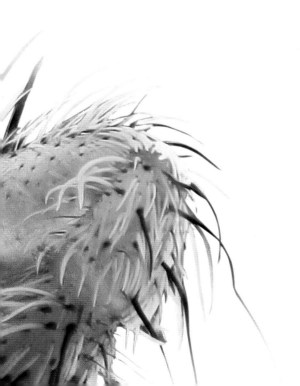

Pigmented areas may aid vision when switching between light and shade in sun-dappled habitats

Antenna contains sensory cells that detect air movement; combined with vision, this information is used to help control flight

Colored eyes
The violet stripes on the eyes of a black soldier fly (*Hermetia illucens*) are caused by pigments that screen different parts of the eye.

compound eyes

The eyes of many animals—including spiders, snails, and vertebrates—contain a single lens. A single, big lens focused on more light-sensitive cells can produce a more detailed image. Insects have compound eyes made up of lots of lenses focused on fewer cells. Each lens is tiny and resolution is poor: even the best dragonfly eyes see ten times less detail than a human eye. However, hundreds of lenses spread over a compound eye give a wide field of view. The slightest movement across them triggers one lens after another, making insect eyes ultrasensitive motion-detectors.

LIGHT AND DARK ADAPTATION
Compound eyes with the best resolution have facets, or ommatidia, enveloped by sleeves of pigment that guide light to nerve cells without interfering with adjacent facets. These eyes work best in bright light. Insects that fly at dusk or at night have facets with less pigment, so that each sensory cell benefits from light gathered by multiple facets. While this is an efficient use of the weak light available, it gives poorer resolution.

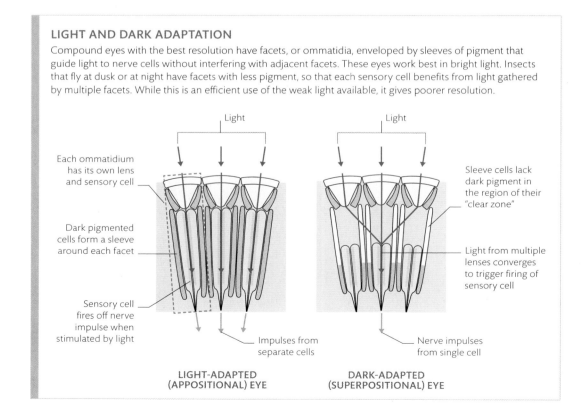

Each ommatidium has its own lens and sensory cell

Dark pigmented cells form a sleeve around each facet

Sensory cell fires off nerve impulse when stimulated by light

Impulses from separate cells

Sleeve cells lack dark pigment in the region of their "clear zone"

Light from multiple lenses converges to trigger firing of sensory cell

Nerve impulses from single cell

LIGHT-ADAPTED (APPOSITIONAL) EYE

DARK-ADAPTED (SUPERPOSITIONAL) EYE

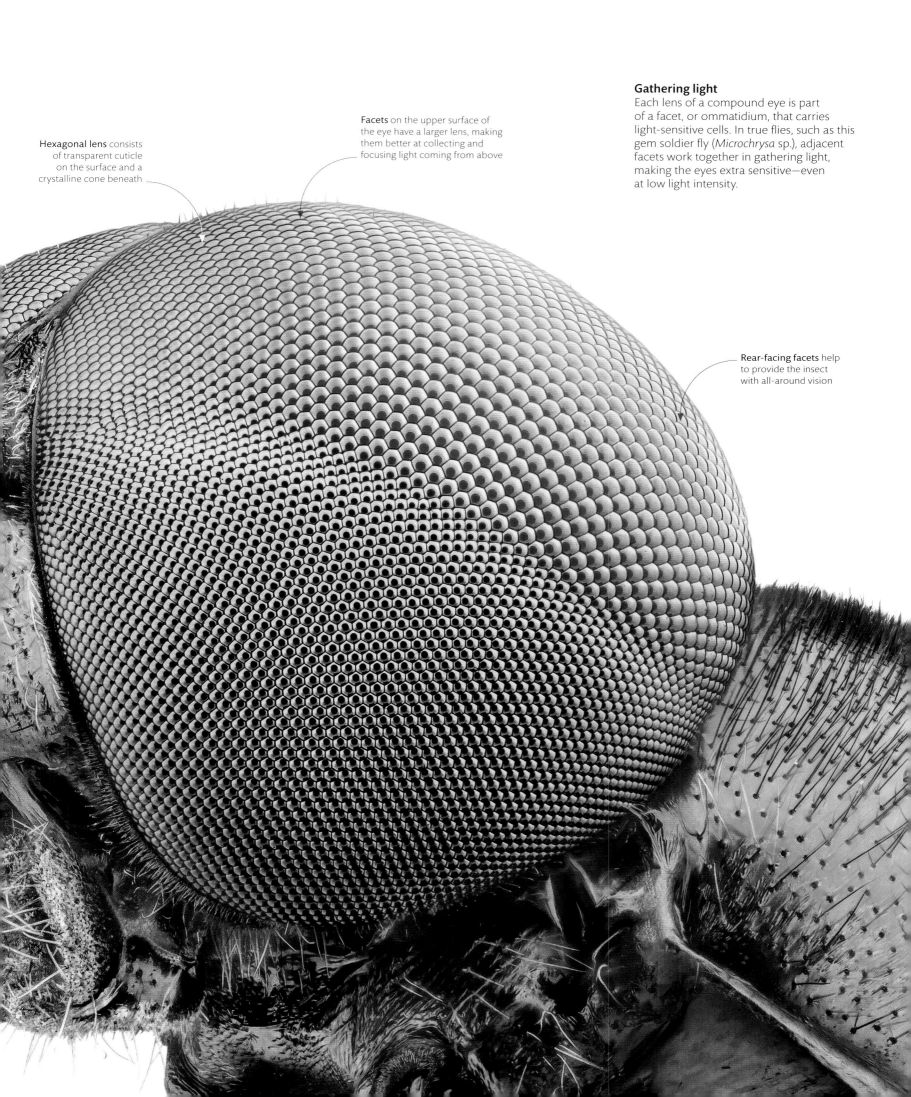

Hexagonal lens consists of transparent cuticle on the surface and a crystalline cone beneath

Facets on the upper surface of the eye have a larger lens, making them better at collecting and focusing light coming from above

Rear-facing facets help to provide the insect with all-around vision

Gathering light
Each lens of a compound eye is part of a facet, or ommatidium, that carries light-sensitive cells. In true flies, such as this gem soldier fly (*Microchrysa* sp.), adjacent facets work together in gathering light, making the eyes extra sensitive—even at low light intensity.

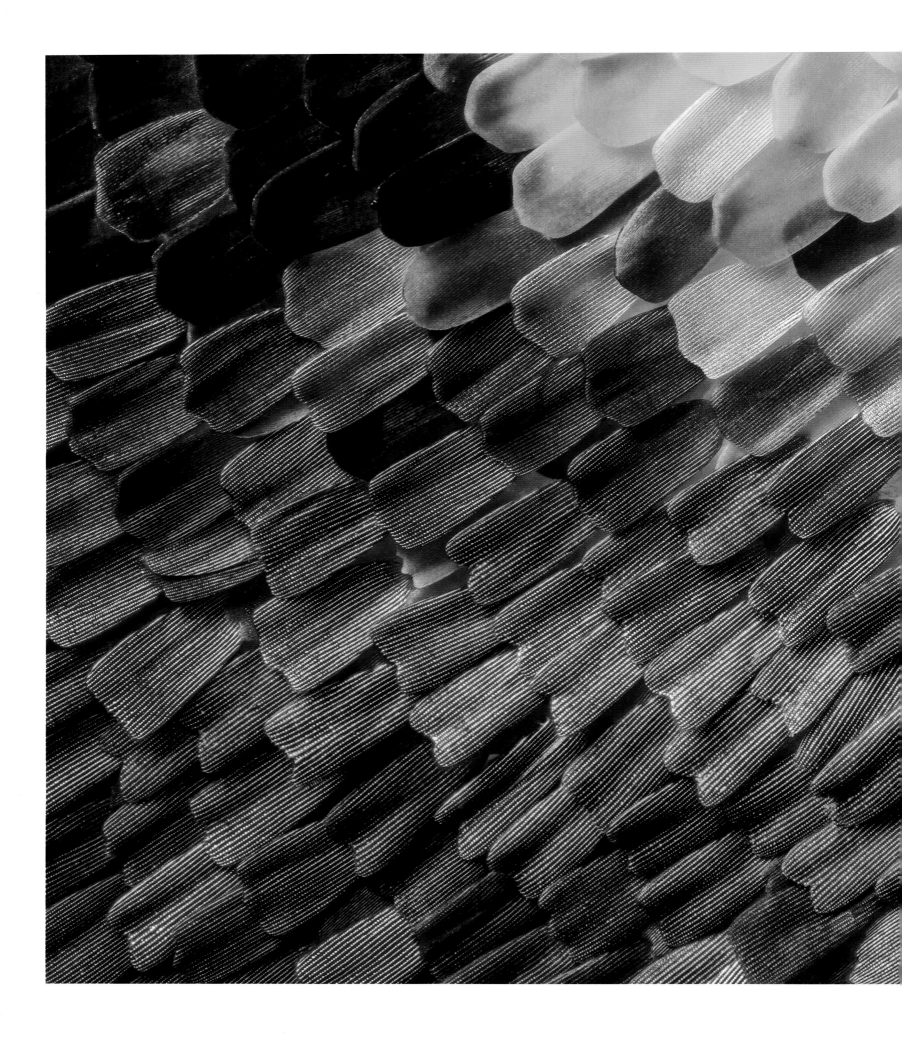

Multi-colored wing
Blues and yellows come together in the wing scales of an Asian day-flying moth (*Eterusia repleta*), producing green where their effects overlap. In moths and butterflies, blue and yellow can both arise from scales scattering light, but yellow may also be due to pigments called papiliochromes.

producing color

Life uses color to display, conceal, attract, or repel, and this involves tricks of the light at a microscopic level. Eyes see colors because they detect different wavelengths of light, from shortwave blue to long-wave red. The colorful wings of a butterfly or moth are clad in scales that affect how wavelengths bounce off them. Some colors, such as reds and blacks, are produced by pigments in the scales; others, such as blues, result from the way the surface of the scales scatters light—just as raindrops scatter light to make a rainbow.

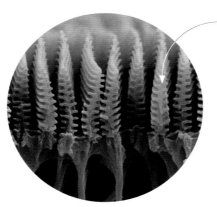

Each microscopic ridge makes blue wavelengths superimpose, intensifying the blue color

Scale ridges
The scales of a morpho butterfly (*Morpho* sp.) have vertical ridges that scatter and reinforce blue wavelengths as they reflect back to the observer.

COLORS FROM CHEMICALS AND TEXTURES
Pigments absorb and reflect specific wavelengths, according their chemical makeup. Structural color occurs because microscopic irregularities in surface textures refract (bend) different wavelengths to different degrees, so surfaces appear colored even though they lack pigments.

Red pigment reflects red wavelengths and absorbs the others

Black pigment absorbs all wavelengths and reflects none

White pigment reflects all wavelengths and absorbs none

Yellow pigment reflects yellow wavelengths and absorbs the others

PIGMENTARY COLOR

Vertical ridges scatter and strengthen blue wavelengths, making the surface look blue, as in morpho butterflies

Granules scatter all wavelengths, making the surface look white, as in cabbage white butterflies

STRUCTURAL COLOR

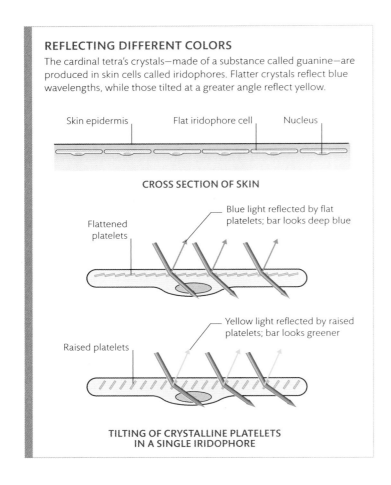

REFLECTING DIFFERENT COLORS

The cardinal tetra's crystals—made of a substance called guanine—are produced in skin cells called iridophores. Flatter crystals reflect blue wavelengths, while those tilted at a greater angle reflect yellow.

CROSS SECTION OF SKIN

Skin epidermis — Flat iridophore cell — Nucleus

TILTING OF CRYSTALLINE PLATELETS IN A SINGLE IRIDOPHORE

Flattened platelets — Blue light reflected by flat platelets; bar looks deep blue

Raised platelets — Yellow light reflected by raised platelets; bar looks greener

Trick of the light

A cardinal tetra (*Paracheirodon axelrodi*) is a tiny shoaling fish of the Amazon Basin. Under magnification, the skin has glittering flecks of blue and yellow produced by crystals that tilt at different angles. At a distance, the colors combine to make the iridescent blue stripe appear greener.

Dark spots are caused by granules of the pigment melanin, produced by skin cells called melanophores

Red band intensifies when the fish expands microscopic pigment sacs

Multicolored species

The Amazon has such a huge diversity of fish species that many use colors to identify their own kind. The cardinal tetra combines its iridescent blue with a stripe of red pigment.

iridescence

The multicolored, metallic hues of animals such as hummingbirds and beetles depend on the angle from which they are viewed. This property—iridescence—is caused by microscopic structures in the cuticle or feathers reflecting different wavelengths of light. As a result, the shimmering colors change when the animals shift their position. Some animals can control this optical effect. The iridescent stripe of a cardinal tetra changes from blue to green—even when the viewing angle stays the same. The fish achieves this optical trick by altering the tilt of light-reflecting crystals just below the skin's surface.

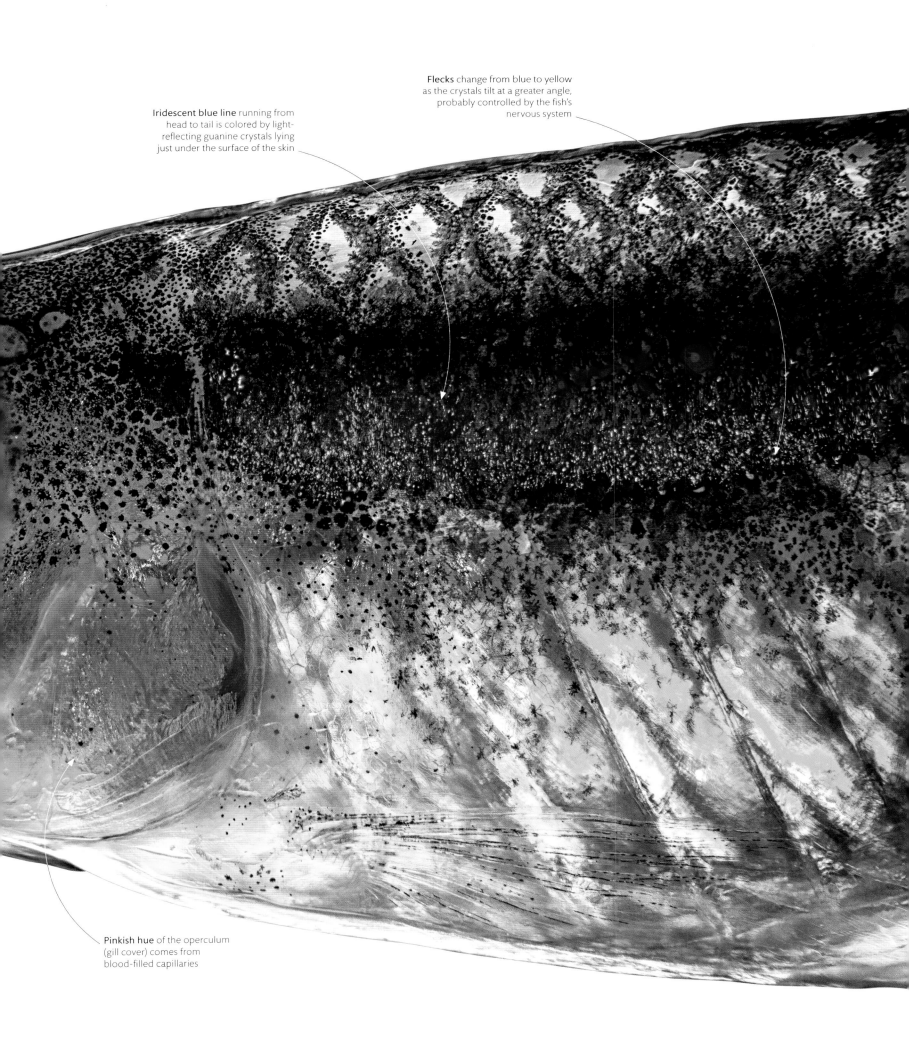

Iridescent blue line running from head to tail is colored by light-reflecting guanine crystals lying just under the surface of the skin

Flecks change from blue to yellow as the crystals tilt at a greater angle, probably controlled by the fish's nervous system

Pinkish hue of the operculum (gill cover) comes from blood-filled capillaries

Tiny vesicles called scintillons contain light-producing chemicals

Chloroplast expands during daylight to maximize absorption of light for photosynthesis; it contracts at night to improve the glow of the light-producing scintillons

Strands of cytoplasm hold the chloroplast in the center of the cell

Using light

Each cell of the ocean dinoflagellate *Pyrocystis fusiformis* has a prominent yellow chloroplast that absorbs the energy of sunlight and uses it to make food by photosynthesis. When disturbed by moving water, bioluminescent vesicles in the dinoflagellates emit a blue glow, changing some of the chemical energy in the food back into light. LM. ×100

Rigid cellulose cell wall envelopes the dinoflagellate and helps maintain its tapered, spindle shape

Blue light is produced by the plankton as they are buffeted by waves breaking over the shallows

Glowing tide When dinoflagellates and other bioluminescent plankton bloom in great numbers, they sometimes make inshore waters glow as the churning surf stimulates their cells to give off light.

producing light

Some organisms produce light for the same reasons that others reflect light: to confuse or send signals to animals that can see them. In the darkness of the night or the deep sea, a sudden flash might attract a mate or deter a predator. The light is produced by a chemical reaction inside living cells. This process, called bioluminescence, is triggered by a stimulus. Some dinoflagellates, members of ocean-dwelling phytoplankton (see pp.288–89), produce a bioluminescent glow when the sea around them is agitated—for example, by wave action or the movements of plankton-eating animals.

LIGHT-PRODUCING CELLS

Scintillons are vesicles in dinoflagellates that contain a mixture of a light-producer, luciferin, and an enzyme, luciferase. When the cell's surface is physically stressed by water pressure, the luciferase affects the luciferin in the scintillons so it releases light.

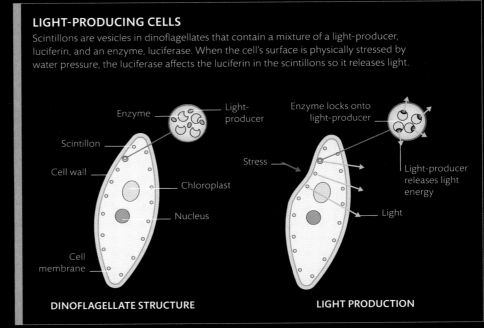

DINOFLAGELLATE STRUCTURE — Enzyme, Light-producer, Scintillon, Cell wall, Chloroplast, Nucleus, Cell membrane

LIGHT PRODUCTION — Enzyme locks onto light-producer, Stress, Light-producer releases light energy, Light

color changers

A wide range of animals have chromatophores in their skin—tiny, complex organs that can bring about color change. Those found in the skin of cephalopods—squid, octopuses, and cuttlefish—are the most responsive of all, because they are under the direct control of the central nervous system, enabling them to change color faster than any other animal, even chameleons. The neurons that innervate the chromatophores are in specific lobes of their brain—the common octopus (*Octopus vulgaris*) has more than half a million of them. The ability of cephalopods to change their appearance almost instantaneously is key to some escape behaviors and is also important in hunting and for camouflage and signaling.

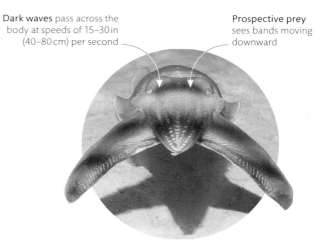

Dark waves pass across the body at speeds of 15–30 in (40–80 cm) per second

Prospective prey sees bands moving downward

Changing tactics
Broadclub cuttlefish (*Sepia latimanus*) approach their prey in camouflage, but once within a range of 20–40 in (50–100 cm) they switch to a passing wave display that tricks and hypnotizes their prey.

Conspicuous octopus
The greater argonaut octopus (*Argonauta argo*) lives near the surface of tropical and subtropical oceans. Predators abound in these waters, so it uses chromatophores as well as countershading to make itself less conspicuous. Larger than the males, females have a shell used as a brood chamber and to trap surface air for buoyancy control. This female is about 1½ in (4 cm) long.

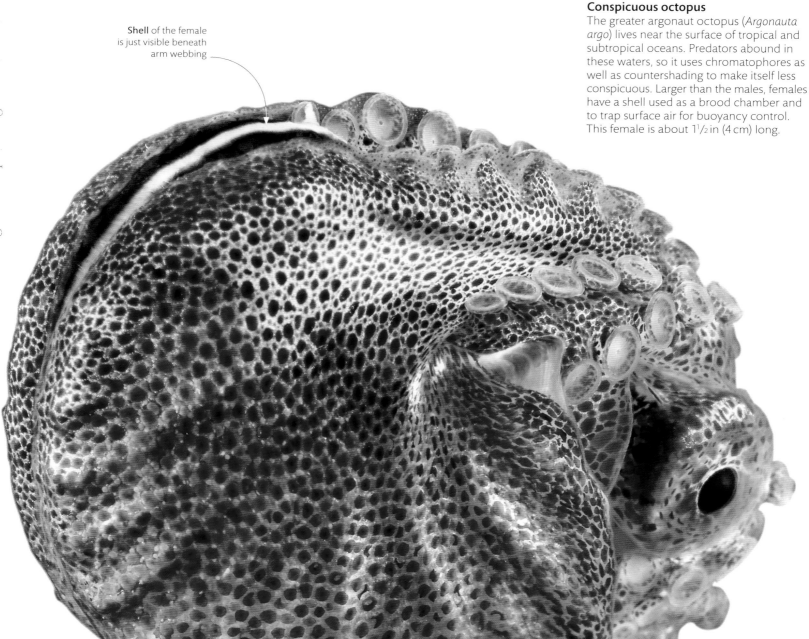

Shell of the female is just visible beneath arm webbing

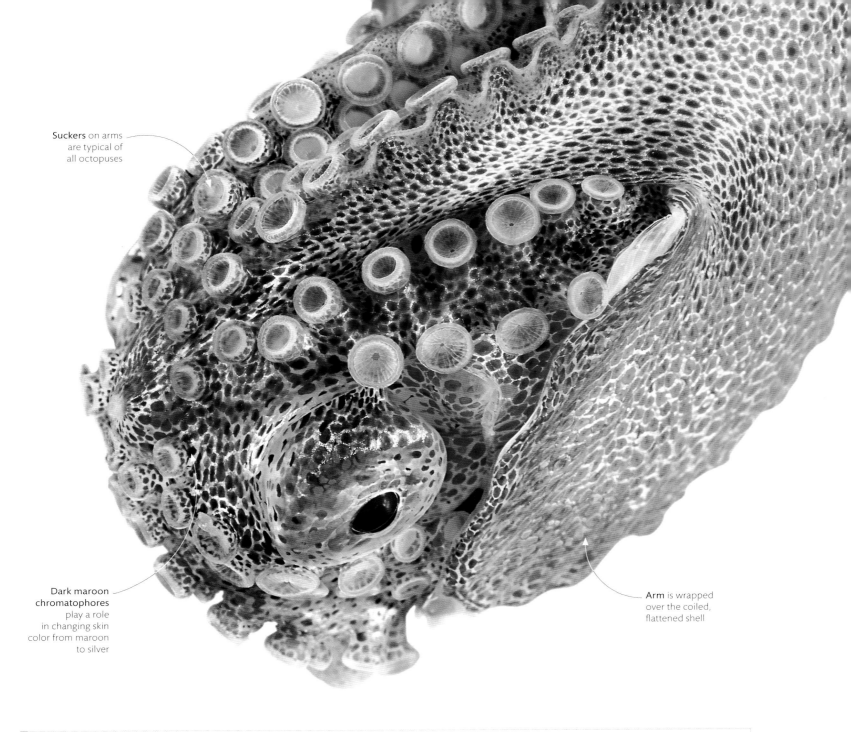

Suckers on arms are typical of all octopuses

Dark maroon chromatophores play a role in changing skin color from maroon to silver

Arm is wrapped over the coiled, flattened shell

HOW CEPHALOPOD CHROMOTOPHORES CHANGE SKIN COLOR

There are thousands of chromatophores in the skin of every cephalopod. Every chromatophore has an elastic, pigment-filled sac at its center, attached to the outer edges by sets of radial muscles, each with its own nerve supply. If the muscles are relaxed, the sac remains at the center. When stimulated, the muscles contract, expanding the sac so the pigment granules spread out.

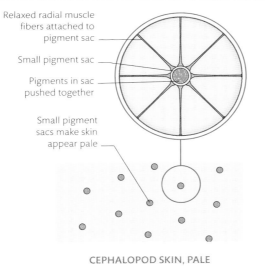

Relaxed radial muscle fibers attached to pigment sac

Small pigment sac

Pigments in sac pushed together

Small pigment sacs make skin appear pale

CEPHALOPOD SKIN, PALE

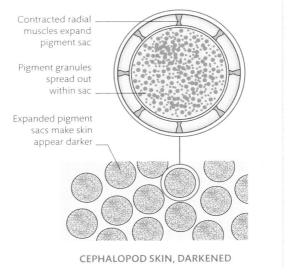

Contracted radial muscles expand pigment sac

Pigment granules spread out within sac

Expanded pigment sacs make skin appear darker

CEPHALOPOD SKIN, DARKENED

MACROLIFE

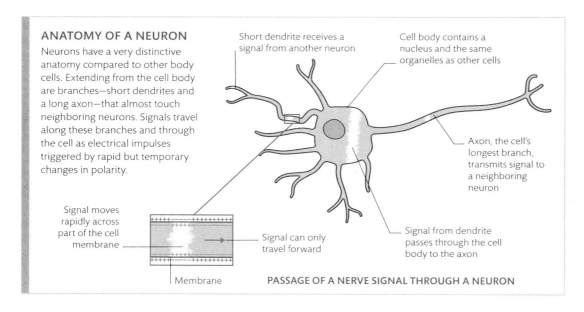

ANATOMY OF A NEURON
Neurons have a very distinctive anatomy compared to other body cells. Extending from the cell body are branches—short dendrites and a long axon—that almost touch neighboring neurons. Signals travel along these branches and through the cell as electrical impulses triggered by rapid but temporary changes in polarity.

- Short dendrite receives a signal from another neuron
- Cell body contains a nucleus and the same organelles as other cells
- Axon, the cell's longest branch, transmits signal to a neighboring neuron
- Signal from dendrite passes through the cell body to the axon
- Signal moves rapidly across part of the cell membrane
- Signal can only travel forward
- Membrane

PASSAGE OF A NERVE SIGNAL THROUGH A NEURON

Gray matter of the cerebral cortex—the brain's surface layer—is highly folded, giving a greater surface area that can accommodate more neurons

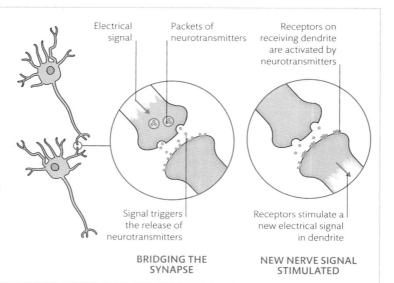

COMMUNICATION BETWEEN NEURONS
Axons do not form a physical connection with the dendrites of neighboring cells. Instead, there is a tiny gap, a mere 30 nanometers wide, called a synapse. An electrical impulse arriving at the end of the axon stimulates the release of chemical neurotransmitters, which diffuse across the gap. The chemical binds to receptors on the dendrite, which stimulates a new electrical impulse. Neurotransmitters either excite neighboring cells into sending signals or inhibit their transmission.

- Electrical signal
- Packets of neurotransmitters
- Receptors on receiving dendrite are activated by neurotransmitters
- Signal triggers the release of neurotransmitters
- Receptors stimulate a new electrical signal in dendrite

BRIDGING THE SYNAPSE **NEW NERVE SIGNAL STIMULATED**

Gray or white?
The tissue of the human brain's outer regions, where neuron cell bodies are densely packed, is called gray matter. The cell bodies are gray because they lack the white, fatty insulation of axons (white matter). The axons of these cells extend inward from gray matter to white.

Giant brain cells
Purkinje cells (green in this false-color image) are giant neurons found in the gray matter of the cerebellum—the part of the human brain where neural networks manage learned patterns of movement, such as walking. The human body contains 86 billion neurons, and the majority of these are in the brain and spinal cord. LM ×570.

nerve cells

All animals with sufficient size, complexity, and speed of movement need a signaling system that is fast enough to coordinate their bodies and react to changes both outside and inside their bodies. Due to the need for quick responses, it must be faster than the chemical signaling systems of microbes and plants. For nearly all animals—all those that are at least as complex as a jellyfish, unlike sponges—that system is an electrical network called the nervous system. It is made up of individual units called neurons, or nerve cells.

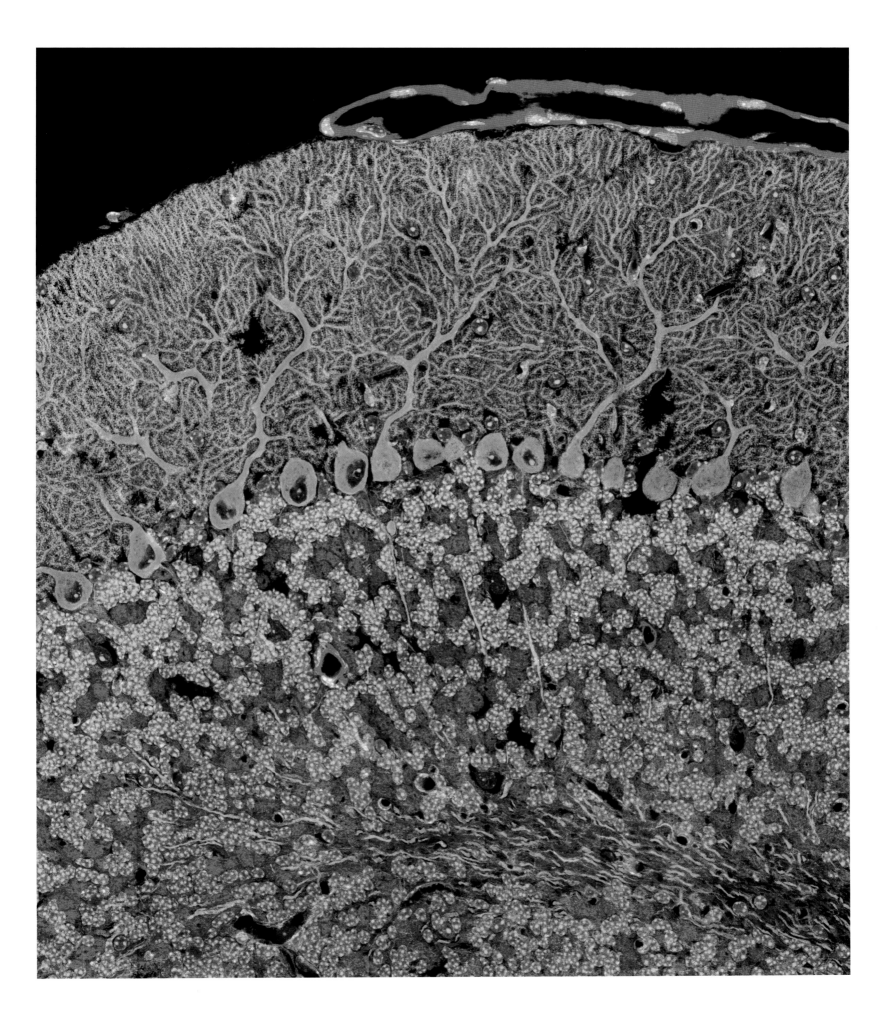

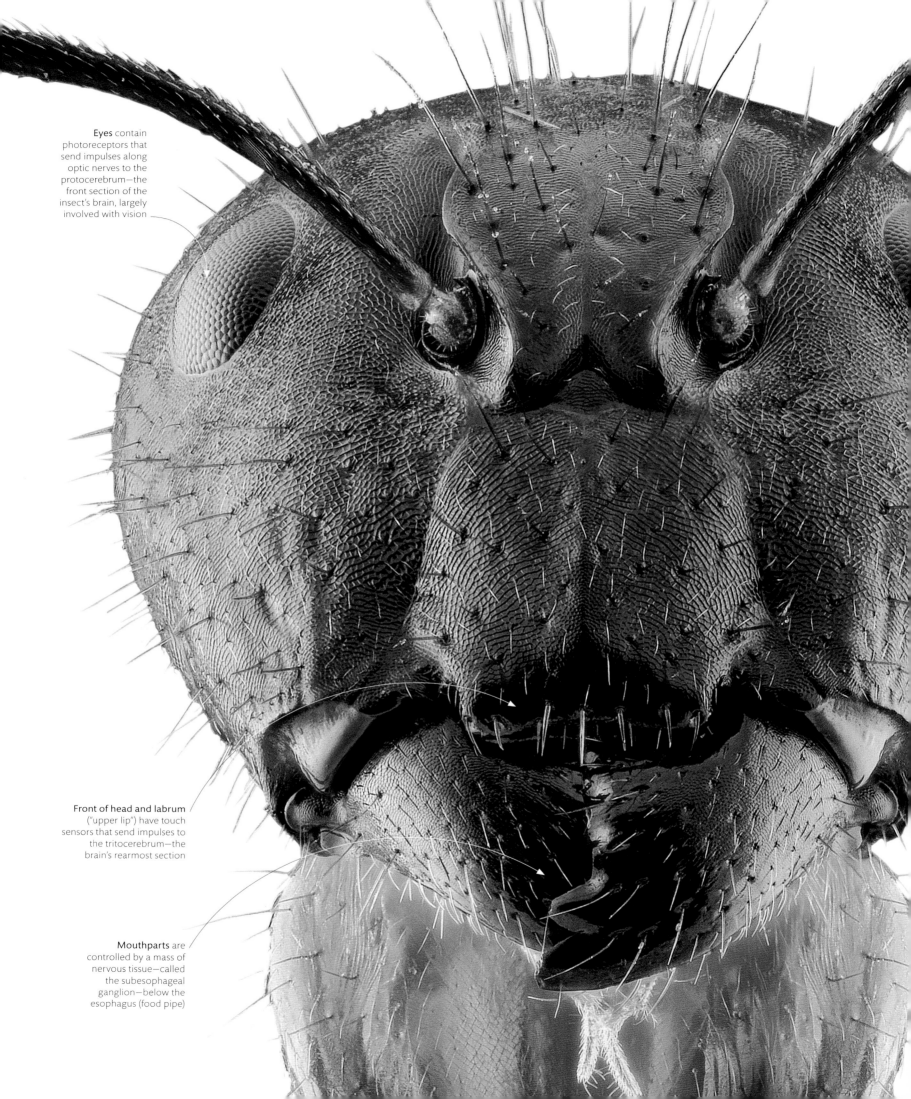

Eyes contain photoreceptors that send impulses along optic nerves to the protocerebrum—the front section of the insect's brain, largely involved with vision

Front of head and labrum ("upper lip") have touch sensors that send impulses to the tritocerebrum—the brain's rearmost section

Mouthparts are controlled by a mass of nervous tissue—called the subesophageal ganglion—below the esophagus (food pipe)

Antennae carry touch and odor receptors that send impulses along antennal nerves to the deuterocerebrum—the brain's midsection

A head for behavior
Even though much of the head of this banded sugar ant (*Camponotus consobrinus*) is made up of muscles that work powerful jaws, its brain is still crucial for controlling stereotypical and hardwired behavior. Three sections of the ant's brain—the protocerebrum, deuterocerebrum, and tritocerebrum—receive information from different sensors on the head.

Facing forward
A cephalized animal has symmetrical right and left halves, distinct front and back ends to its body, and it gathers sensory data as it moves forward. Some animals, such as sponges, corals, and jellyfish, do not have a left or right side, nor a front or back, nor a head.

Thorax carries a nerve cord leading from the head through the abdomen

coordinating behavior

Developing a front end (with a left and right side, and a back end) was a major milestone in animal evolution. Since this front end would most likely be the first part of the animal to encounter new sensory stimuli, its sense organs clustered there. To deal with the sensory information pouring in from these organs, its central nervous system (see pp.122–23) became concentrated in that front end. This was the process of cephalization—the evolution of a head with a brain encased inside. The brain is a hub for receiving information and dispensing signals that direct the body.

NERVOUS SYSTEM

The nervous system of an animal with a head is organized into the central nervous system—consisting of the brain and nerve cords—and the peripheral nervous system, which is made up of nerves carrying impulses from sensors and to muscles. The central nervous system of most invertebrates—including insects such as grasshoppers—has a ventral nerve cord running through the underside of the body. In contrast, vertebrates have a dorsal spinal cord running through the backbone.

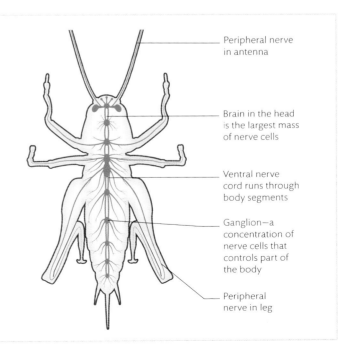

- Peripheral nerve in antenna
- Brain in the head is the largest mass of nerve cells
- Ventral nerve cord runs through body segments
- Ganglion—a concentration of nerve cells that controls part of the body
- Peripheral nerve in leg

NERVOUS SYSTEM OF A GRASSHOPPER

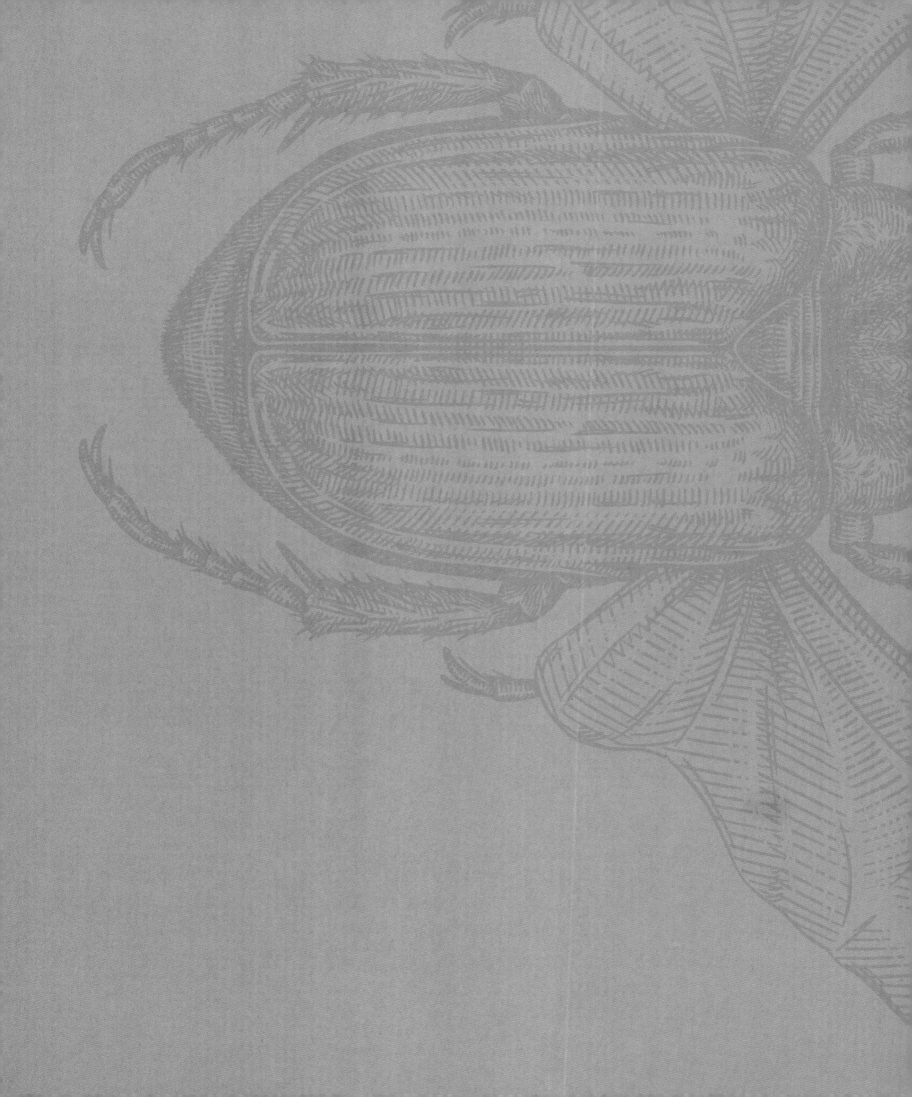

moving

Movement is one of life's fundamental characteristics. Some small organisms run and jump on jointed legs, or flap tiny aerofoils to stay aloft. Even bacteria can rotate cellular machinery that powers them forward. And in every organism, cells are moving on the inside, so they can change shape or move substances in and out.

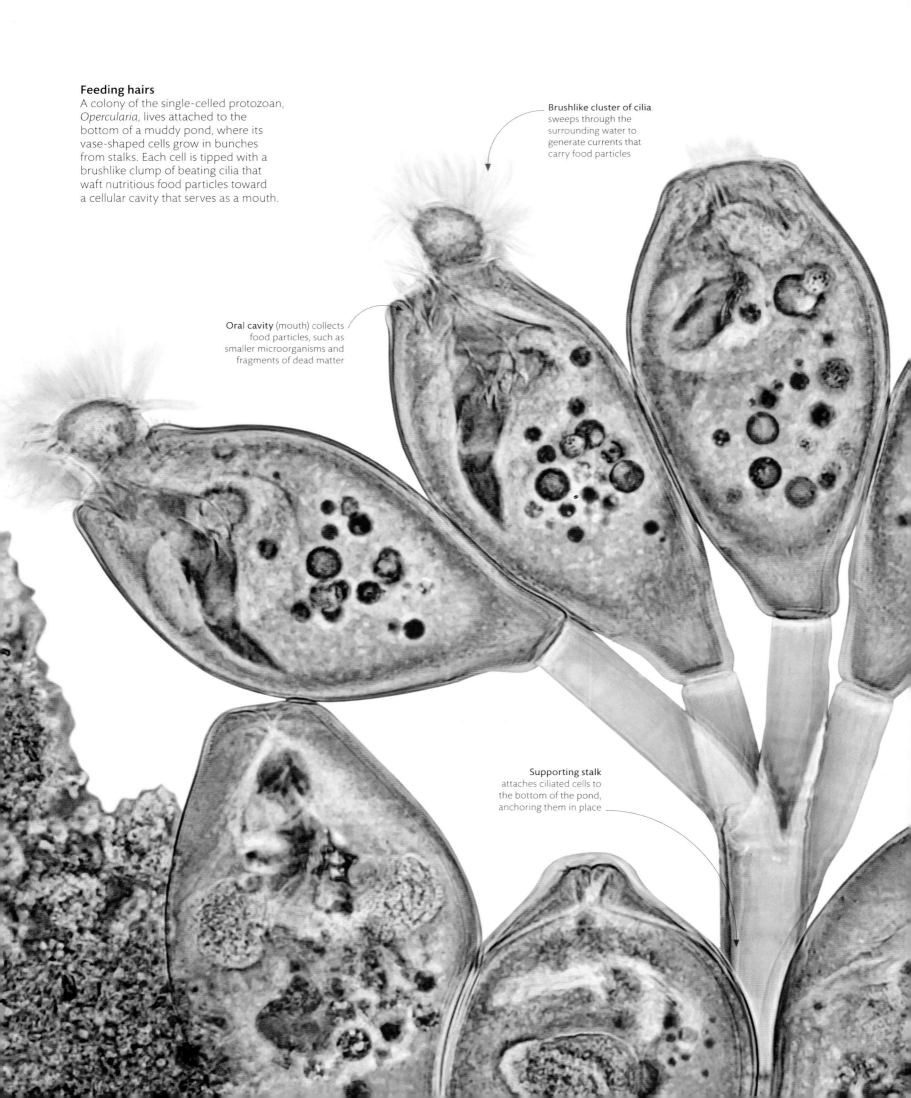

Feeding hairs
A colony of the single-celled protozoan, *Opercularia*, lives attached to the bottom of a muddy pond, where its vase-shaped cells grow in bunches from stalks. Each cell is tipped with a brushlike clump of beating cilia that waft nutritious food particles toward a cellular cavity that serves as a mouth.

Brushlike cluster of cilia sweeps through the surrounding water to generate currents that carry food particles

Oral cavity (mouth) collects food particles, such as smaller microorganisms and fragments of dead matter

Supporting stalk attaches ciliated cells to the bottom of the pond, anchoring them in place

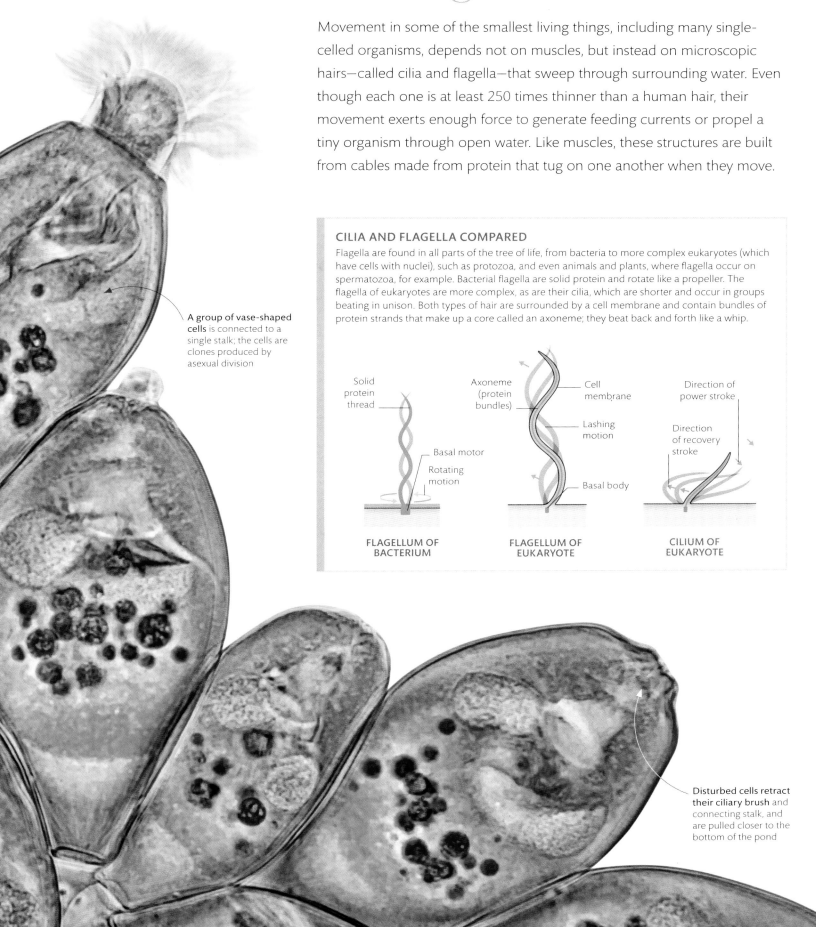

beating hairs

Movement in some of the smallest living things, including many single-celled organisms, depends not on muscles, but instead on microscopic hairs—called cilia and flagella—that sweep through surrounding water. Even though each one is at least 250 times thinner than a human hair, their movement exerts enough force to generate feeding currents or propel a tiny organism through open water. Like muscles, these structures are built from cables made from protein that tug on one another when they move.

A group of vase-shaped cells is connected to a single stalk; the cells are clones produced by asexual division

CILIA AND FLAGELLA COMPARED
Flagella are found in all parts of the tree of life, from bacteria to more complex eukaryotes (which have cells with nuclei), such as protozoa, and even animals and plants, where flagella occur on spermatozoa, for example. Bacterial flagella are solid protein and rotate like a propeller. The flagella of eukaryotes are more complex, as are their cilia, which are shorter and occur in groups beating in unison. Both types of hair are surrounded by a cell membrane and contain bundles of protein strands that make up a core called an axoneme; they beat back and forth like a whip.

FLAGELLUM OF BACTERIUM — Solid protein thread, Basal motor, Rotating motion

FLAGELLUM OF EUKARYOTE — Axoneme (protein bundles), Cell membrane, Lashing motion, Basal body

CILIUM OF EUKARYOTE — Direction of power stroke, Direction of recovery stroke

Disturbed cells retract their ciliary brush and connecting stalk, and are pulled closer to the bottom of the pond

When these single-celled organisms, which mostly live in freshwater habitats, were seen by the microbiologist Louis Joblot in 1718, he named them "slipper animalcules." The modern name came a few decades later, and means "elongated" or "oblong." Nevertheless, the resemblance to a comfortable slip-on shoe is still there, enhanced by the oral groove, or vestibulum, which is present near the widest part of the cell.

spotlight paramecium

The vestibulum is the entry point for food, such as bacteria and yeasts. At up to 300 micrometers in length, a paramecium is hundreds of times bigger than its food, which is swept up in a current of water created by the coordinated wafting of cilia on the paramecium's surface. As well as an entry point, the cell also has an exit, called the anal pore. This is less obvious than the oral opening but is located in the "heel" area of the cell.

The paramecium body is contained within a stiffened outer layer called a pellicle. The stiffness of the pellicle enables the cell to move using a whiplash motion of the cilia. During the fast downstroke, the cilia are stiffened to push against the water, but they then soften as they flex back to the starting position for the next downstroke.

A paramecium has two nuclei. The larger "macronucleus" contains multiple sets of its genes and is involved in the day-to-day control of the cell. The smaller "micronucleus" is involved in reproduction, which occurs by binary fission—a simple cell division resulting in two daughter cells. There is also occasional conjugation, where individual paramecia swap half their genetic material.

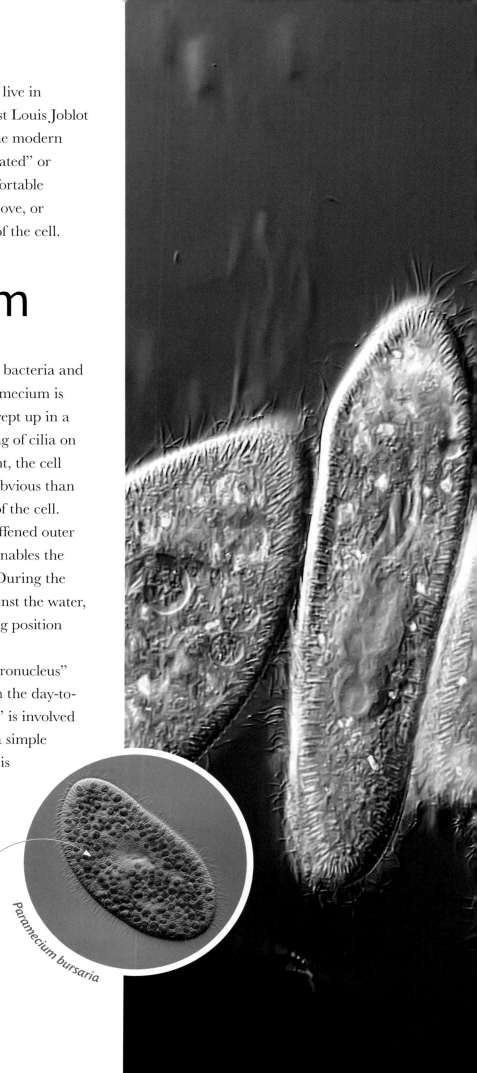

Green algae alive inside the paramecium, providing sustenance in return for protection

Paramecium bursaria

Food supplies
Vacuoles store ingested food inside the cells of the *Paramecium caudatum*, and they shrink as the contents are digested. Some species of *Paramecium* supplement their food intake with nutrients provided by photosynthetic endosymbionts—algae that live inside the cell. LM, x950.

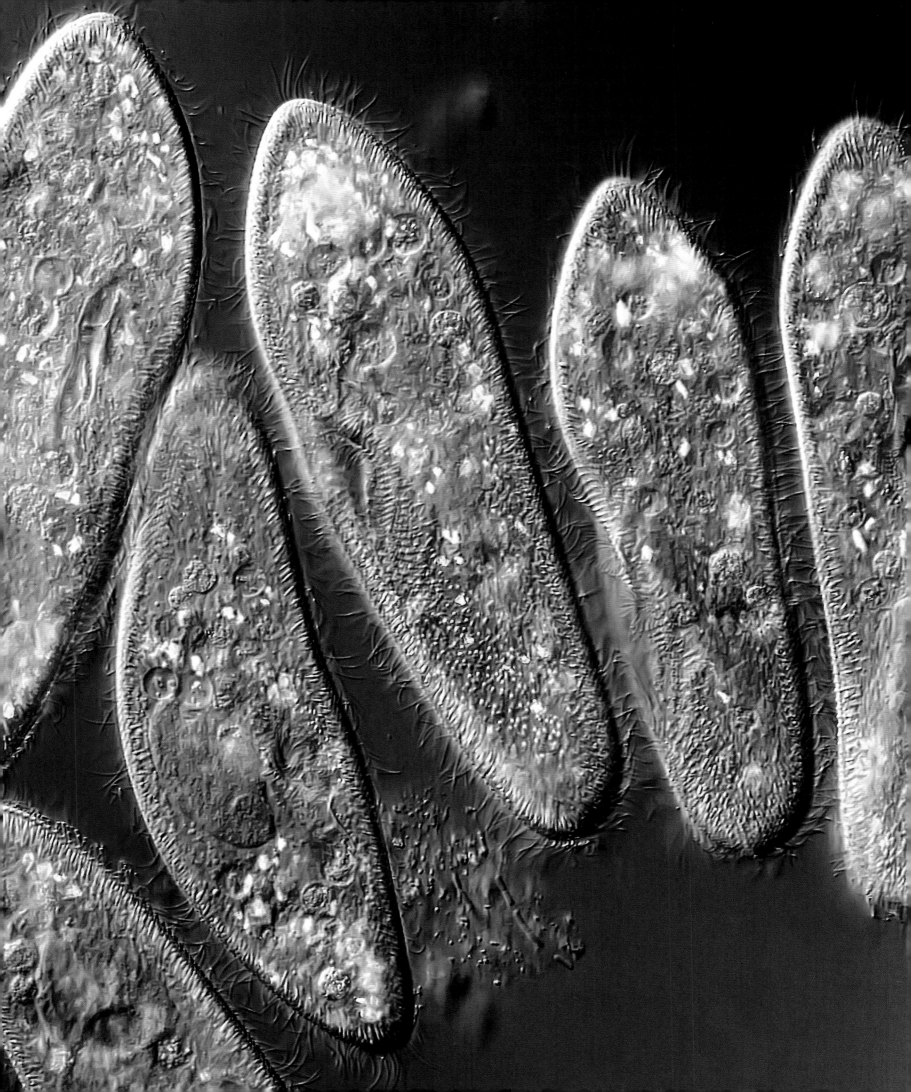

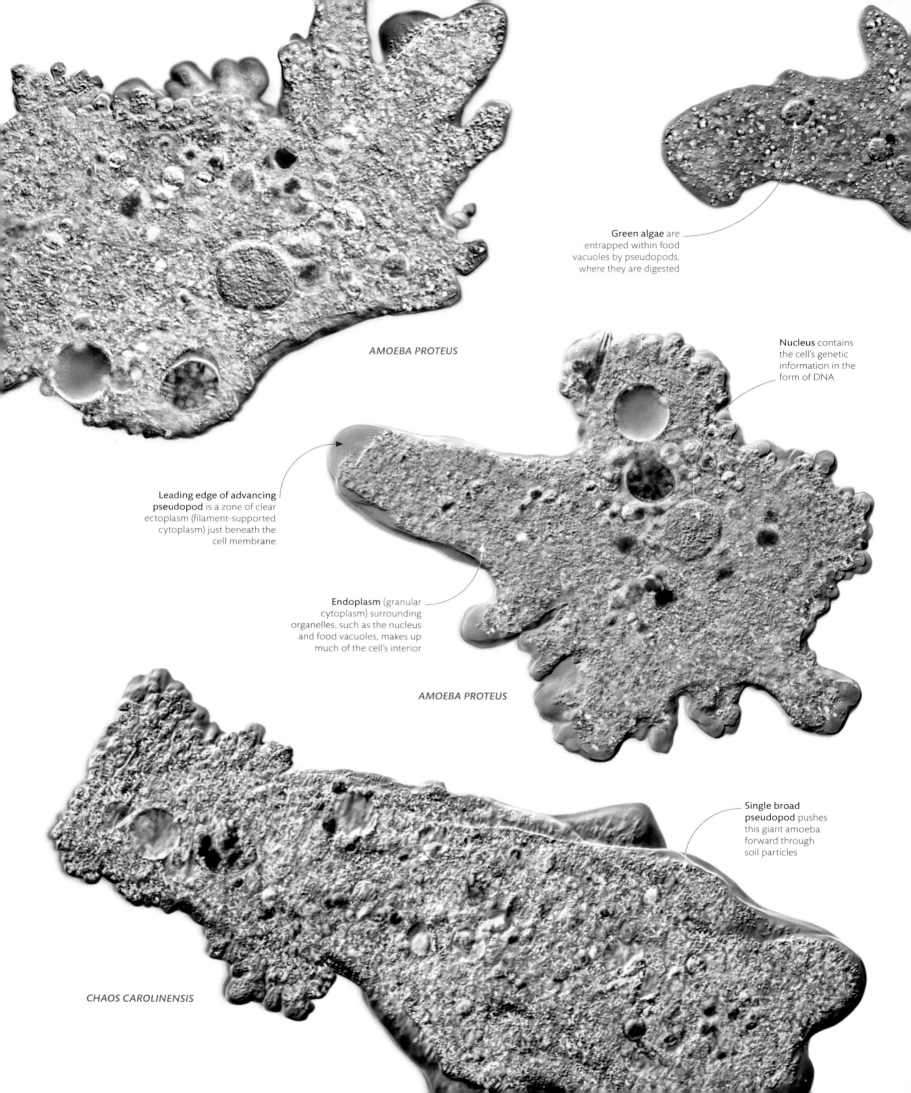

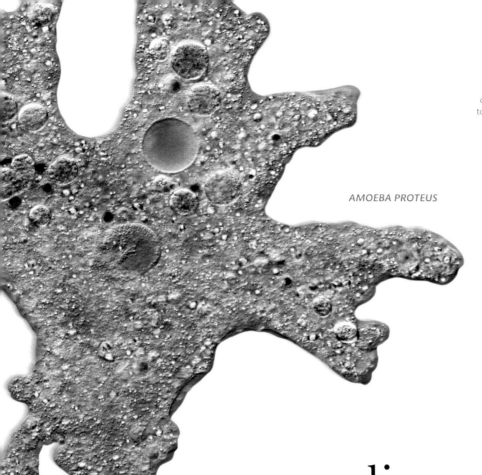

AMOEBA PROTEUS

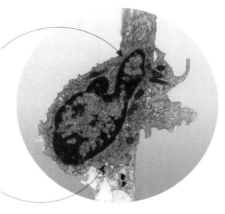

White blood cell uses cytoplasmic streaming to extend a pseudopod through pore

Wall of blood capillary has pores large enough to squeeze through

Fighting infection
White blood cells in an animal's circulation have many similarities with amoebas. They squeeze through pores in blood capillary walls to reach a site of infection, where they engulf alien particles with their pseudopods.

crawling cells

A single cell has few options for getting from place to place. Hairlike flagella or cilia (see pp.128–29) can propel a microbe or swimming sperm, but cells lacking such adaptations must change shape to make progress. All cells are filled with a thin jelly, or cytoplasm, reinforced by a scaffold of protein filaments. Amoebas and other crawling cells can assemble these filaments to stretch their cytoplasm into pseudopods ("false feet") that pull the cell along. Progress is slow: it would take a week for a predatory amoeba to cross this page, a distance a sperm could cover in 10 minutes. But in the micro-world, it is fast enough to catch prey that are even slower and tinier.

False feet
Amoeba species vary in the way they produce pseudopods. The common amoeba (*Amoeba proteus*, upper three cells), produces several pseudopods at once, enabling it to stretch and fan outward to ensnare prey such as bacteria and algae. The giant amoeba (*Chaos carolinensis*, bottom) produces a single, thick pseudopod and travels with sluglike movements. LM, x300.

FORMING PSEUDOPODS
Pseudopods move by a process called cytoplasmic streaming. This relies on the same protein, actin, that animals use to contract their muscles. At the front of the pseudopod, actin filaments are assembled from subunits. This creates a "cap" of stiffer ectoplasm that pushes forward and then streams back along the sides to the rear of the cell. There, the actin is broken down to recycle the subunits back into the watery endoplasm.

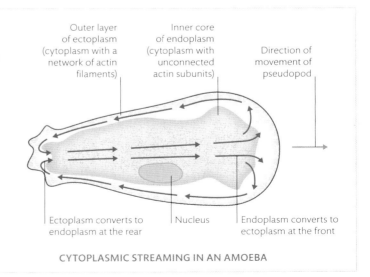

CYTOPLASMIC STREAMING IN AN AMOEBA

ROTIFER DIVERSITY

More than 2,000 described species of rotifer live in freshwater and marine habitats, and also in moist microhabitats—such as moss or lichens—on land. The smallest rotifer is shorter than a human sperm; the largest is nearly as big as a grain of sand. Most rotifers use their ciliated crowns to collect food particles, but others catch prey with funnel-shaped arrangements of spines and bristles. A few rotifers even have extendable biting jaws.

ENGRAVING OF ROTIFERS BY ERNST HAECKEL (1904)

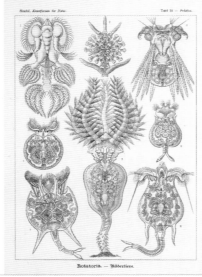

Mastax is a wide chamber of the gut lined with toothlike ridges called trophi that help to pulverize the ingested algae

The corona's ring of beating cilia surrounds a tiny mouth, which ingests algae collected by the currents

Narrow throat leading from the mouth is lined with beating cilia that sweep algae deeper into the rotifer's gut

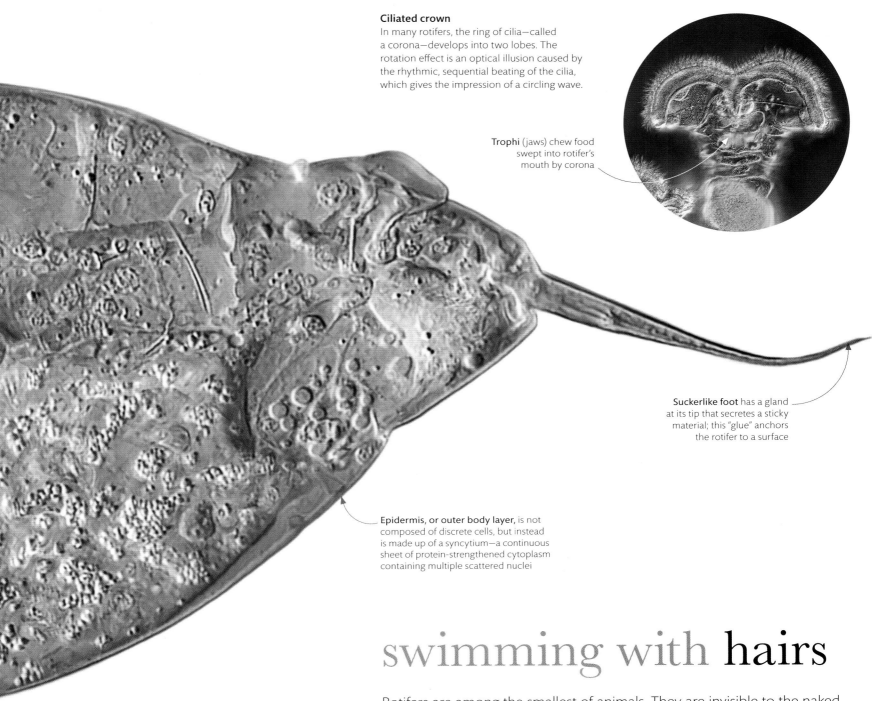

Ciliated crown
In many rotifers, the ring of cilia—called a corona—develops into two lobes. The rotation effect is an optical illusion caused by the rhythmic, sequential beating of the cilia, which gives the impression of a circling wave.

Trophi (jaws) chew food swept into rotifer's mouth by corona

Suckerlike foot has a gland at its tip that secretes a sticky material; this "glue" anchors the rotifer to a surface

Epidermis, or outer body layer, is not composed of discrete cells, but instead is made up of a syncytium—a continuous sheet of protein-strengthened cytoplasm containing multiple scattered nuclei

Algal grazer
The semi-digested algal food of this monogonont pond rotifer is clearly visible through its transparent, unpigmented body wall. The algae—wafted by beating cilia into the tiny mouth—collect in a chamber called a mastax, where muscular walls help to grind them into a pulp. LM, x 2,000.

swimming with hairs

Rotifers are among the smallest of animals. They are invisible to the naked eye, and—despite having bodies made up of many cells—can be even tinier than many single-celled microbes. Thousands can live in a cupful of nutrient-rich pond water, where their translucent, worm-shaped bodies glide between weeds and sediment. Their head end carries one or more tracts of beating hairlike cilia, whose coordinated movements look so much like rotating disks that early microscopists called rotifers "wheel animalcules." The tail end has a suckerlike "foot," which clings to objects while the rotifer extends its body to probe the surroundings. Most rotifers move by swimming with their cilia or creeping along like an inchworm, but some live permanently attached by their foot.

Base of the tentacles enclosed in sheaths, which contain the muscles that control their movement

Tentacles are lined with cells, called colloblasts, which discharge sticky threads that ensnare planktonic prey

Wave of muscular contraction passes along body, undulating it like a fish

Row of ctenes runs along each edge of the animal

combs of cilia

Many microscopic organisms swim using tiny beating hairs called flagella, or brushlike clusters of shorter cilia. For larger animals, flagella and cilia alone usually fail to generate enough thrust to move their heavier bodies, so muscle power is needed too. But one group of ocean jellies—called ctenophores—rely on cilia, even though some can grow as big as a small fish. They can do this because the cilia are fused together in comblike arrays, or ctenes, that push against the water with enough force to propel their bodies through the ocean.

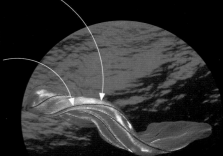

Muscle power
The largest ctenophores, such as this Venus's girdle (*Cestum veneris*), can be 3 ft (1 m) long. Too big for propulsion only by cilia, they also use muscles to undulate though the water.

Sea gooseberry

The common Atlantic ctenophore (*Pleurobrachia pileus*), also known as a sea goosebery, has a spherical body 0.4–1 in- (1–2.5 cm-) long, and relies on eight rows of ctenes, which are "combs" of fused cilia, to swim up and down the water column. These animals use their long, muscle-powered tentacles to prey on and catch floating organisms, known as plankton (see pp.286–87).

Iridescence occurs when beating combs reflect light of different wavelengths; this may attract prey or repel larger predators

Combs of fused cilia sweep downward, pushing against the water, driving the sea gooseberry upward

Anus ejects particles of undigested food

Long tentacles can be withdrawn into a sheath when not deployed to capture prey

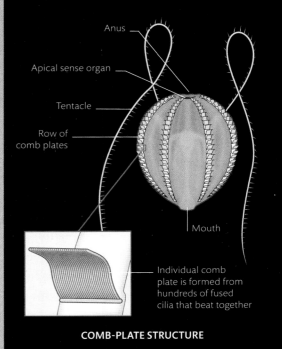

SWIMMING COMBS

Every ctenophore (comb jelly) has eight rows of ctenes (also known as comb plates) that run from its mouth to its anus. An apical sense organ near the anus not only balances the body in the water, but also controls the beating ctenes via nerve fibers that run along each row.

- Anus
- Apical sense organ
- Tentacle
- Row of comb plates
- Mouth
- Individual comb plate is formed from hundreds of fused cilia that beat together

COMB-PLATE STRUCTURE

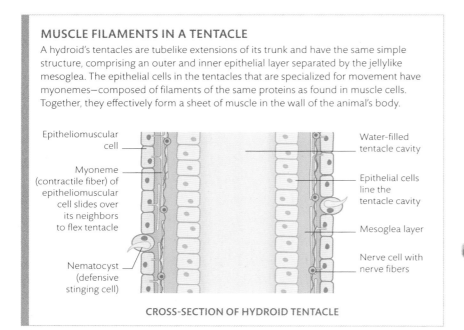

MUSCLE FILAMENTS IN A TENTACLE
A hydroid's tentacles are tubelike extensions of its trunk and have the same simple structure, comprising an outer and inner epithelial layer separated by the jellylike mesoglea. The epithelial cells in the tentacles that are specialized for movement have myonemes—composed of filaments of the same proteins as found in muscle cells. Together, they effectively form a sheet of muscle in the wall of the animal's body.

- Epitheliomuscular cell
- Myoneme (contractile fiber) of epitheliomuscular cell slides over its neighbors to flex tentacle
- Nematocyst (defensive stinging cell)
- Water-filled tentacle cavity
- Epithelial cells line the tentacle cavity
- Mesoglea layer
- Nerve cell with nerve fibers

CROSS-SECTION OF HYDROID TENTACLE

Tentacle movement
The hydroid *Ectopleura larynx* is up to 2 in (5 cm) tall and lives attached to rocks, shells, and seaweed in strong tidal streams. Pink or reddish in real life, it has an outer and inner ring of tentacles between which are clusters of reproductive buds called gonophores. The tentacles are extremely mobile, waving, bending, and contracting as they search for food and pass it to the mouth. SEM, x70.

simple muscles

While jellyfish use striated muscle (see pp.140–41) for swimming, many of their cnidarian relatives—sea anemones, corals, and hydroids—lack this, and in fact have no dedicated muscle cells. They do have contractile fibers, however, called myonemes, in specialized epithelial (skin) cells. Some of the myonemes run longitudinally while others form rings around the body. When they contract, they work against fluid-filled spaces, which act as a hydrostatic skeleton. With these two sets of muscle fibers, a cnidarian can lengthen or shorten its body and flex it in all directions.

Outer tentacle, known as an aboral tentacle

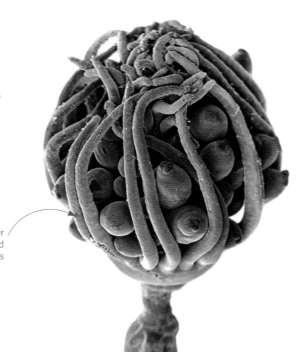

Defensive posture
The muscular system of *Ectopleura* is integrated with its sensory and nervous systems, so it can respond rapidly to danger. Here, its contracting epitheliomuscular cells have folded in both the oral and aboral tentacles to protect the gonophores.

Folded tentacles cover the mouth and reproductive structures

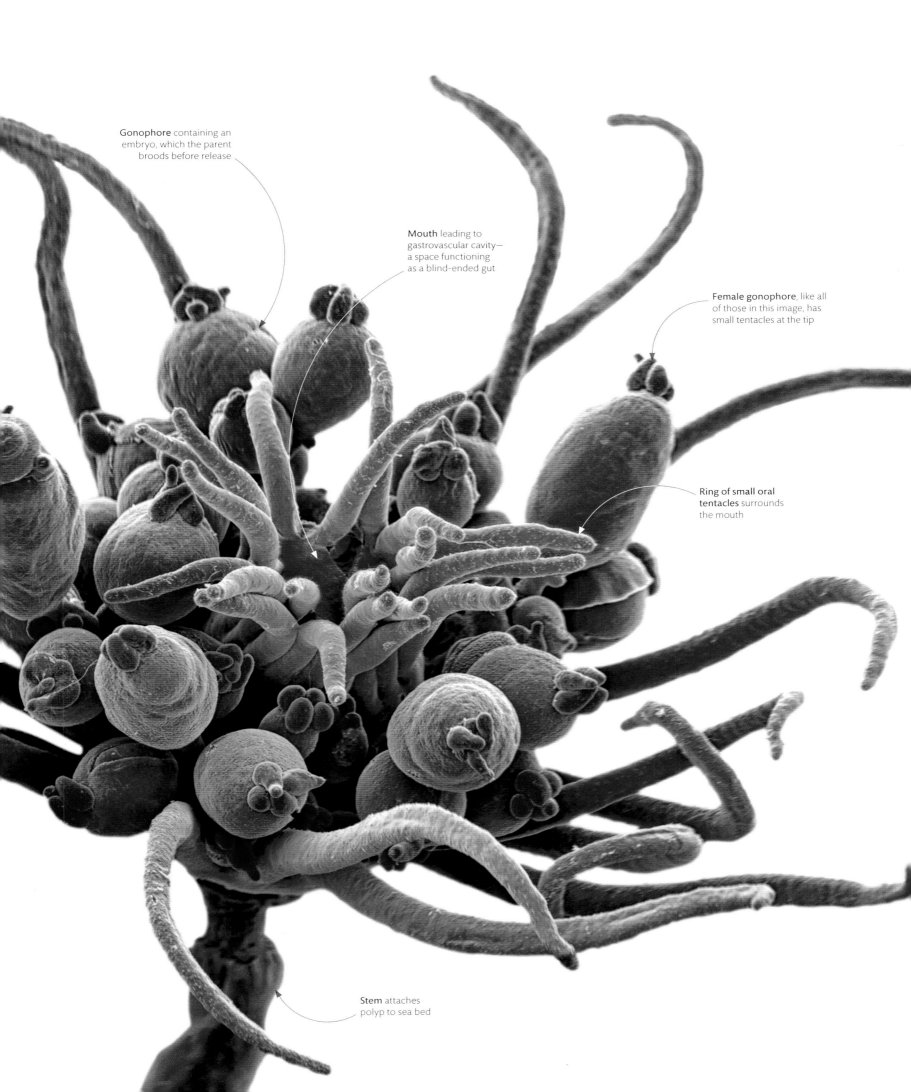

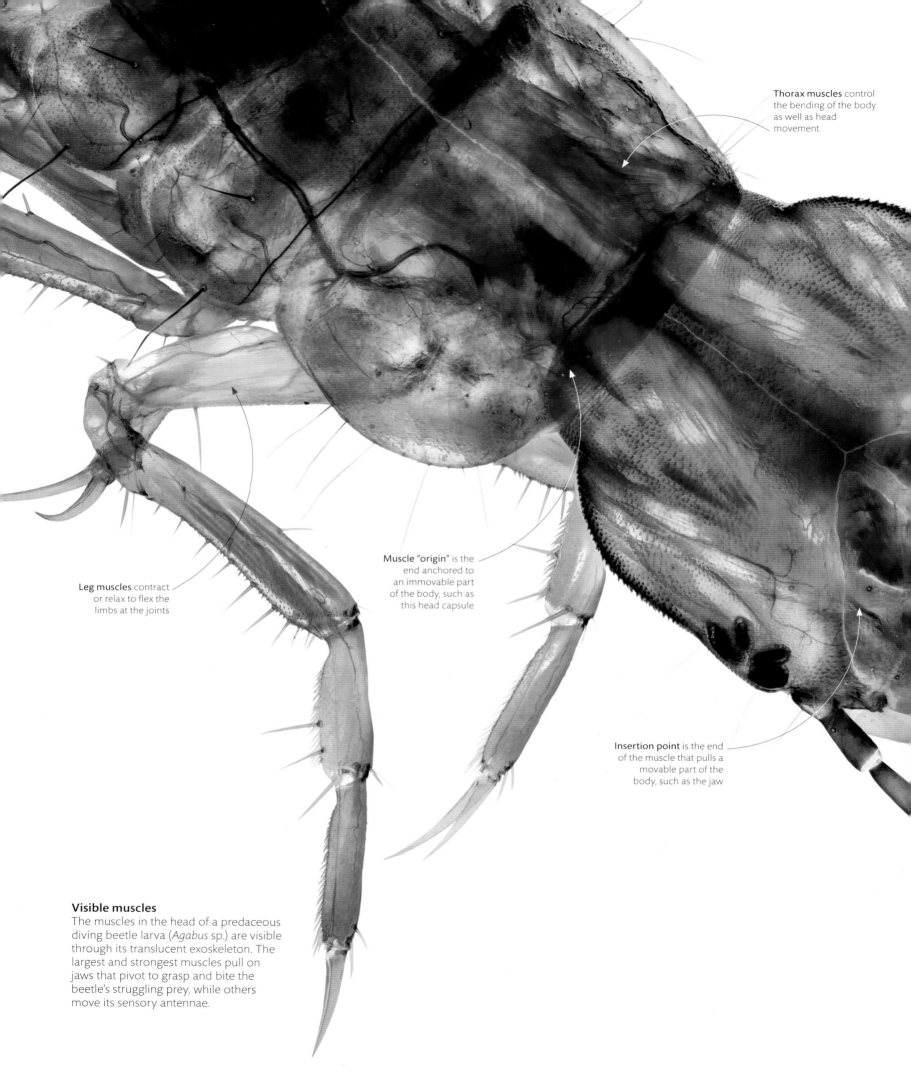

Thorax muscles control the bending of the body as well as head movement

Muscle "origin" is the end anchored to an immovable part of the body, such as this head capsule

Leg muscles contract or relax to flex the limbs at the joints

Insertion point is the end of the muscle that pulls a movable part of the body, such as the jaw

Visible muscles
The muscles in the head of a predaceous diving beetle larva (*Agabus* sp.) are visible through its translucent exoskeleton. The largest and strongest muscles pull on jaws that pivot to grasp and bite the beetle's struggling prey, while others move its sensory antennae.

contracting muscles

Muscles, and the fast movements they generate, set animals apart from other kinds of organisms. Muscles are packed with protein filaments that work to contract, or shorten, their long cells, using energy from respiration. Like nerves, muscles carry electrical impulses, and an electrical trigger—typically from the nervous system—stimulates their contraction. The muscles in the walls of vital organs squeeze down to move food, blood, or waste along them. Skeletal muscles have their ends attached to parts of the skeleton, and pull on different parts of the body to move them.

Striped appearance of muscles results from alternating stacks of thick and thin protein filaments within the cells

Flight muscles
The power of muscles enables extraordinary feats of locomotion. The wing muscles of the hummingbird hawkmoth (*Macroglossum stellatarum*) help it hover in midair while drinking nectar.

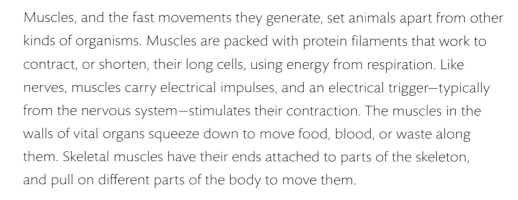

Wing muscles are packed into the thorax, or mid section, of the moth's body

Jaws are part of the exoskeleton but swivel around a flexible joint at the base

SLIDING FILAMENTS IN MUSCLES

The protein filaments (actin and myosin) inside muscle cells are stacked in overlapping bundles, giving skeletal muscle a banded appearance. When stimulated by an electrical impulse, the filaments use chemical energy to slide over one another, pulling the ends of the muscle cells together.

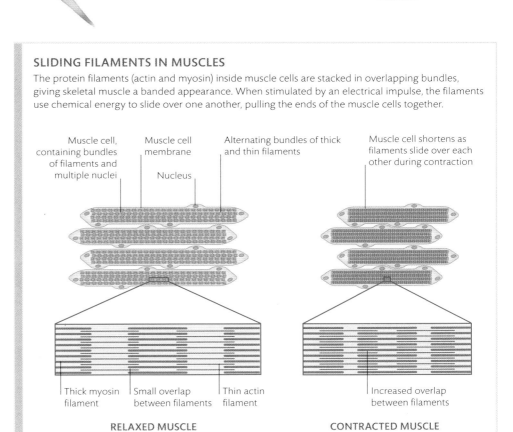

Muscle cell, containing bundles of filaments and multiple nuclei

Muscle cell membrane

Nucleus

Alternating bundles of thick and thin filaments

Muscle cell shortens as filaments slide over each other during contraction

Thick myosin filament | Small overlap between filaments | Thin actin filament

Increased overlap between filaments

RELAXED MUSCLE | **CONTRACTED MUSCLE**

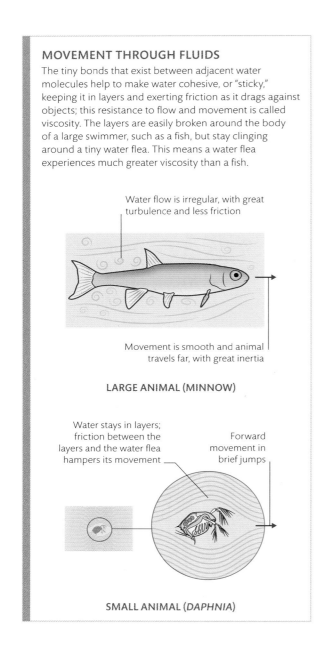

MOVEMENT THROUGH FLUIDS

The tiny bonds that exist between adjacent water molecules help to make water cohesive, or "sticky," keeping it in layers and exerting friction as it drags against objects; this resistance to flow and movement is called viscosity. The layers are easily broken around the body of a large swimmer, such as a fish, but stay clinging around a tiny water flea. This means a water flea experiences much greater viscosity than a fish.

Water flow is irregular, with great turbulence and less friction

Movement is smooth and animal travels far, with great inertia

LARGE ANIMAL (MINNOW)

Water stays in layers; friction between the layers and the water flea hampers its movement

Forward movement in brief jumps

SMALL ANIMAL (DAPHNIA)

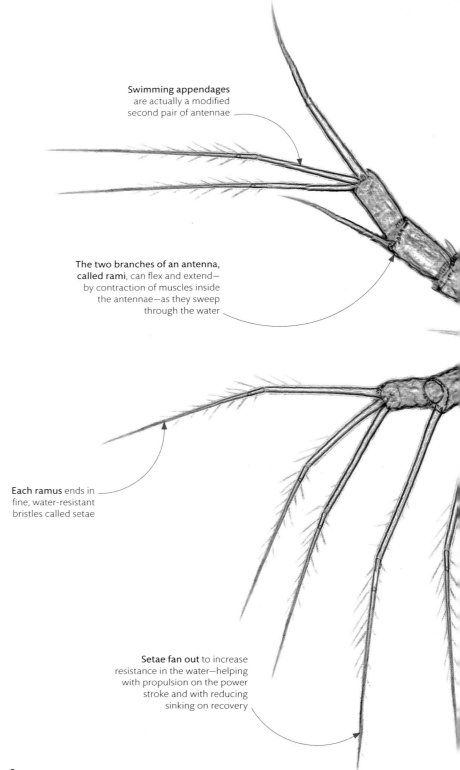

Swimming appendages are actually a modified second pair of antennae

The two branches of an antenna, called rami, can flex and extend—by contraction of muscles inside the antennae—as they sweep through the water

Each ramus ends in fine, water-resistant bristles called setae

Setae fan out to increase resistance in the water—helping with propulsion on the power stroke and with reducing sinking on recovery

overcoming friction

For any animal that swims, the thrust of its movement needs to overcome the friction between its body and the water. This is easier for large animals with powerful muscles and streamlined bodies. But for tiny animals, such as a water flea, friction is significant. And at the water flea's scale (about 0.04 in, or 1 mm, long), the bonds between water molecules are noticeable, so a water flea's experience is like a human swimming through molasses. This means a water flea expends more energy than a human in swimming—and water's stickiness quickly brings it to a halt with each effort, making its movements characteristically jerky.

Propulsive antennae
Like many other crustaceans, water fleas (Daphnia sp.) have branched antennae. But while most animals use their antennae for sensing, those of water fleas are vast and are used as swimming paddles. Strong muscles sweep the antennae through the water, pushing back and fanning outward on the power stroke.

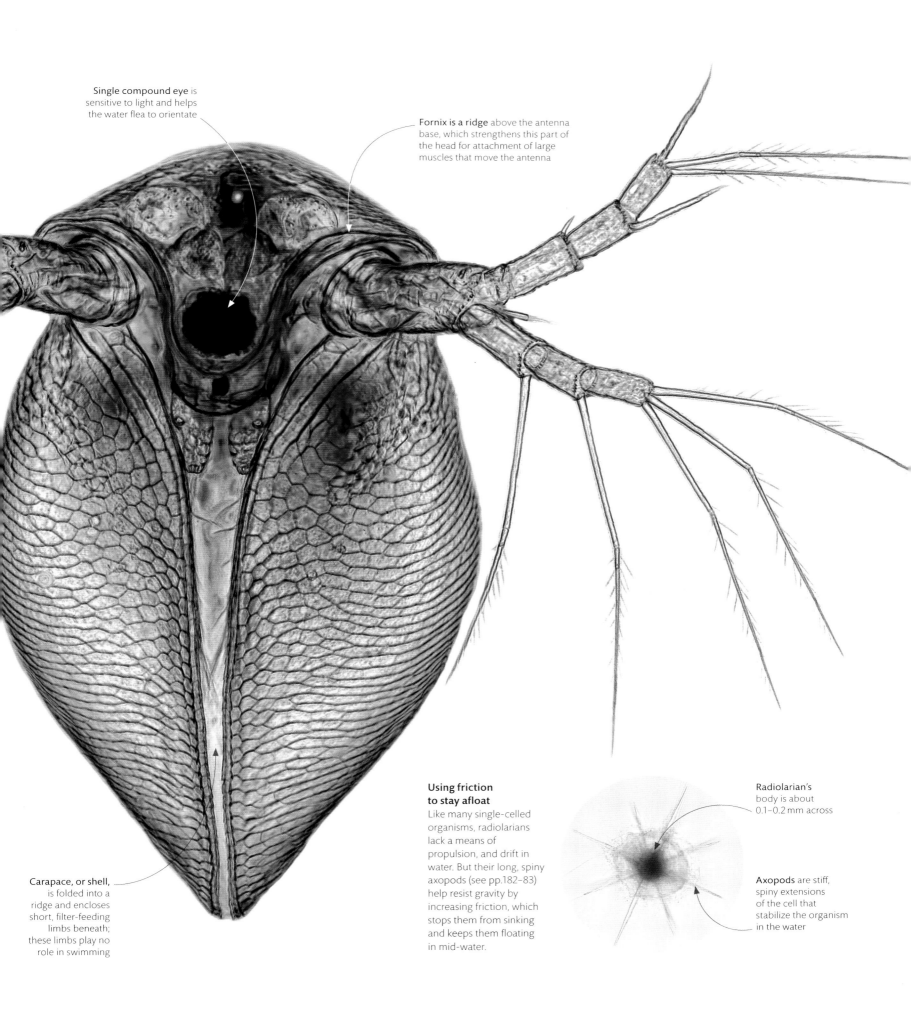

Single compound eye is sensitive to light and helps the water flea to orientate

Fornix is a ridge above the antenna base, which strengthens this part of the head for attachment of large muscles that move the antenna

Carapace, or shell, is folded into a ridge and encloses short, filter-feeding limbs beneath; these limbs play no role in swimming

Using friction to stay afloat
Like many single-celled organisms, radiolarians lack a means of propulsion, and drift in water. But their long, spiny axopods (see pp.182–83) help resist gravity by increasing friction, which stops them from sinking and keeps them floating in mid-water.

Radiolarian's body is about 0.1–0.2 mm across

Axopods are stiff, spiny extensions of the cell that stabilize the organism in the water

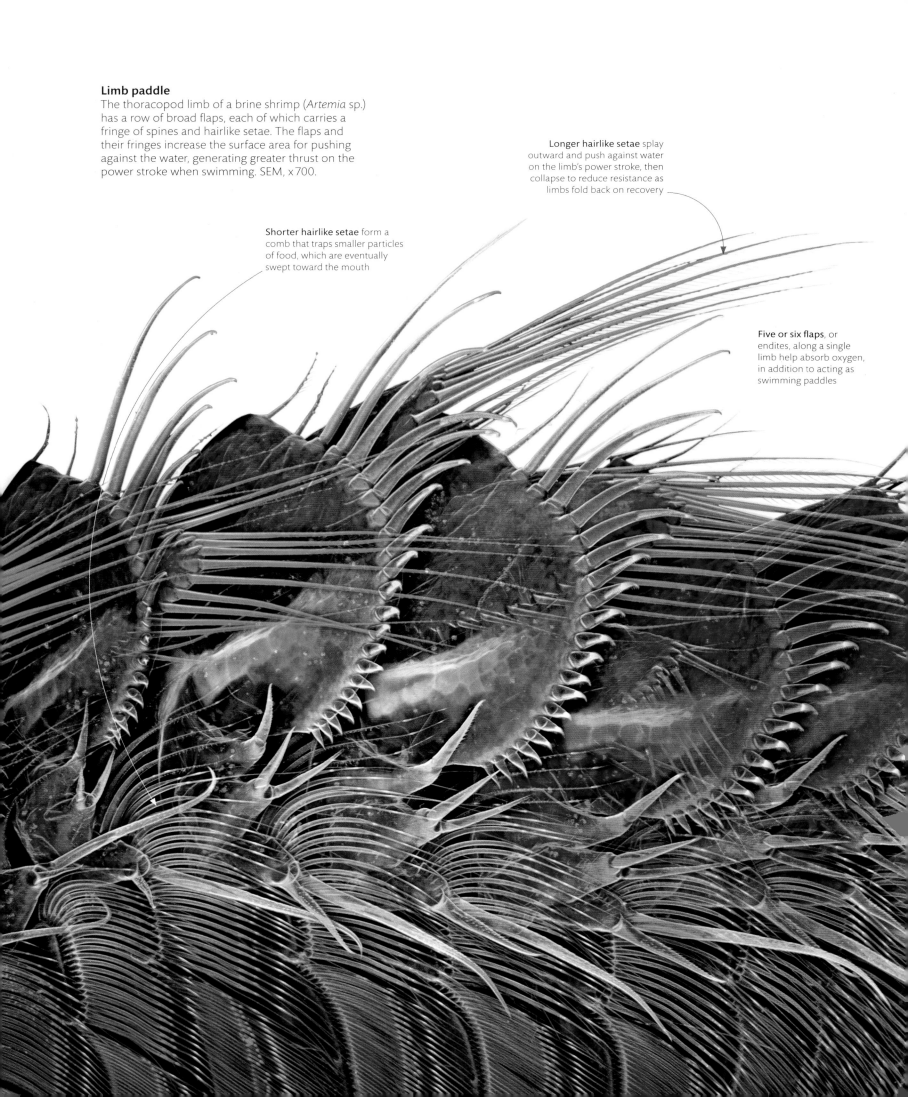

Limb paddle
The thoracopod limb of a brine shrimp (*Artemia* sp.) has a row of broad flaps, each of which carries a fringe of spines and hairlike setae. The flaps and their fringes increase the surface area for pushing against the water, generating greater thrust on the power stroke when swimming. SEM, x700.

Shorter hairlike setae form a comb that traps smaller particles of food, which are eventually swept toward the mouth

Longer hairlike setae splay outward and push against water on the limb's power stroke, then collapse to reduce resistance as limbs fold back on recovery

Five or six flaps, or endites, along a single limb help absorb oxygen, in addition to acting as swimming paddles

smooth rowing

The swimming actions of animals that live in mid-water must be strong enough to generate the thrust needed to move them forward. For small crustaceans such as brine shrimp, thrust comes from broad, leafy limbs that push against the water on the power stroke, then fold back on their hinged joints on the recovery. Brine shrimp swim on their backs and have 11 pairs of legs that work together in a rowing action, with a wave of coordinated movements that ripples down the body from back to front. The effect is to make the shrimp glide smoothly through the water.

LIMB FUNCTIONS

Brine shrimp belong to a group of crustaceans called branchiopods, meaning "gill-legged"—a reference to their multipurpose limbs. As well as being used for swimming, the limbs absorb oxygen. Their rowing action also draws a stream of water along the belly of the shrimp, where food particles get trapped between the limbs' flaplike extensions.

- Cercopods
- Thorax
- Adults have two compound eyes
- Abdomen
- Thoracopod has flaplike appendages used for swimming

ADULT BRINE SHRIMP

Setae thickened as spines trap larger food particles

Single eyespot (red) detects the presence and direction of light

Young shrimp
Like most other crustaceans, brine shrimp eggs hatch into a larva called a nauplius. It has three pairs of swimming appendages on the head that will form antennae and mouthparts in the adult.

Feathery setae near the ends of the limbs act like accessory paddles and provide extra thrust

controlling buoyancy

Organisms in mid-water need to be neutrally buoyant, meaning they float without sinking to the bottom or rising to the surface. To stay suspended, their overall density must match that of the water, even though their body parts vary, from heavy muscles in some organisms, to lighter oil. Insects—even aquatic ones—breathe with a network of air-filled tracheal tubes, which buoys them upward. But the underwater larvae of phantom midges employ adjustable sacs in their tracheal system, which they can expand or constrict to control their level in the water column.

Buoyant micro-predator
The head of a phantom midge larva (*Chaoborus* sp.) carries a complex set of equipment for life as a fierce mid-water predator, including grasping antennae and mandibles used for catching mosquito larvae and water fleas. Behind the head are two pigment-dappled air sacs. Another pair of air sacs is located near the tail. LM, x40.

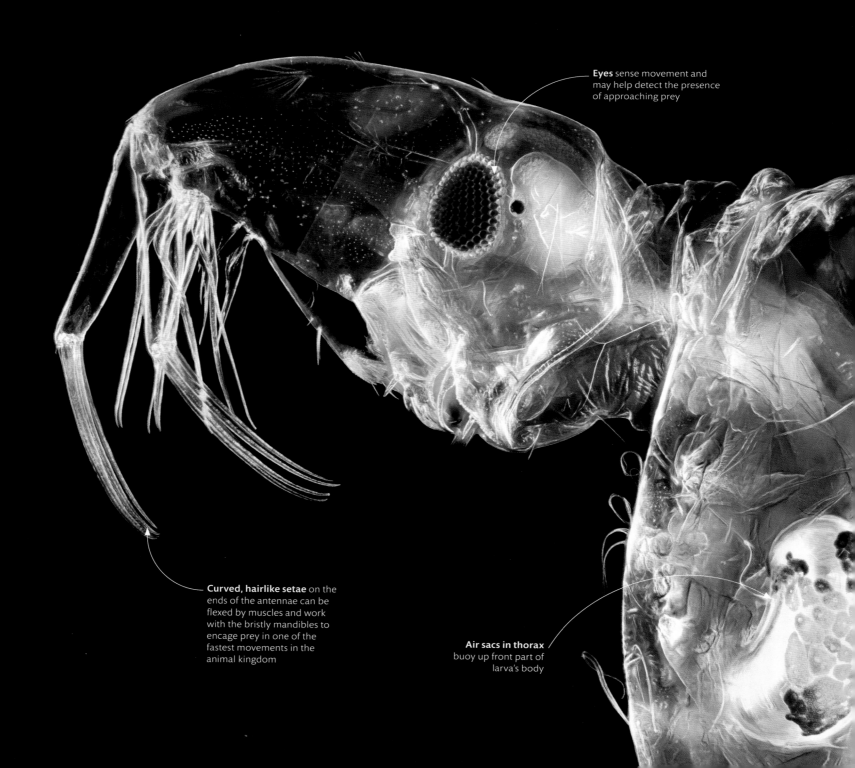

Eyes sense movement and may help detect the presence of approaching prey

Curved, hairlike setae on the ends of the antennae can be flexed by muscles and work with the bristly mandibles to encage prey in one of the fastest movements in the animal kingdom

Air sacs in thorax buoy up front part of larva's body

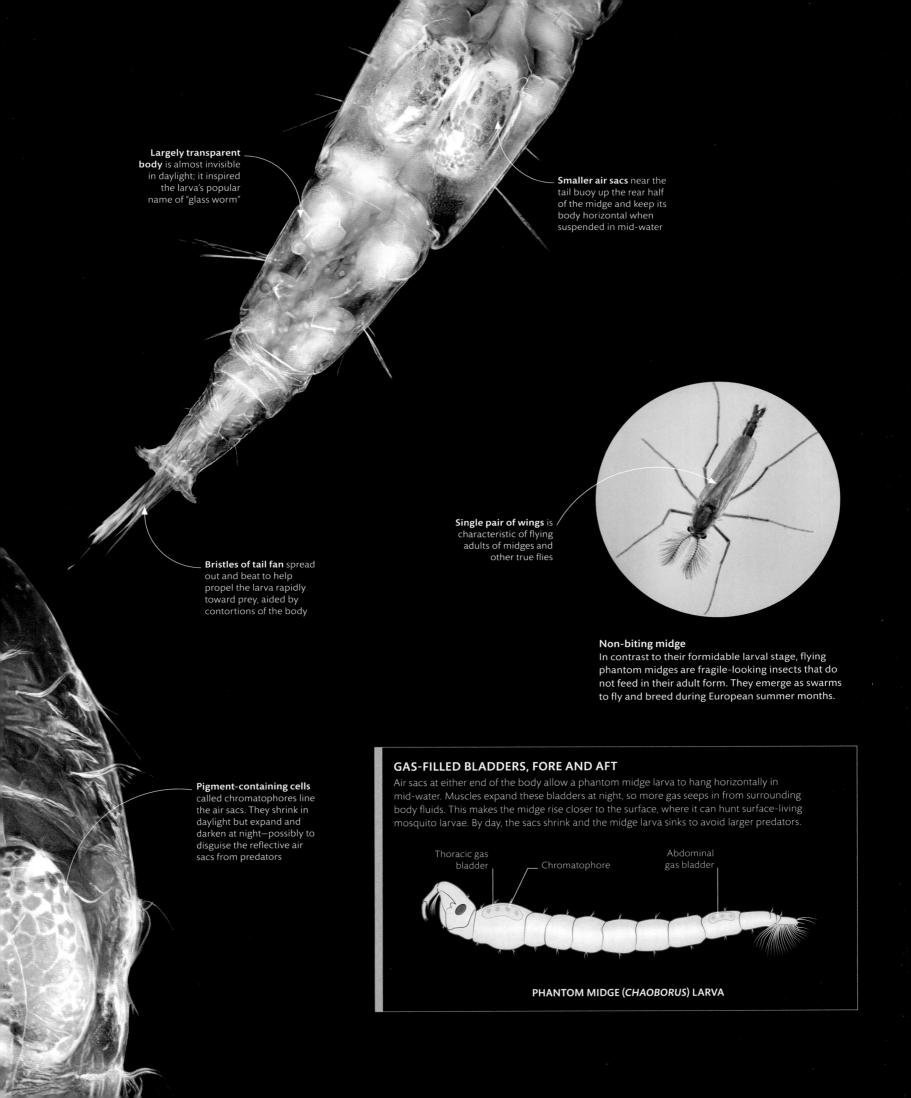

Largely transparent body is almost invisible in daylight; it inspired the larva's popular name of "glass worm"

Smaller air sacs near the tail buoy up the rear half of the midge and keep its body horizontal when suspended in mid-water

Bristles of tail fan spread out and beat to help propel the larva rapidly toward prey, aided by contortions of the body

Single pair of wings is characteristic of flying adults of midges and other true flies

Non-biting midge
In contrast to their formidable larval stage, flying phantom midges are fragile-looking insects that do not feed in their adult form. They emerge as swarms to fly and breed during European summer months.

Pigment-containing cells called chromatophores line the air sacs. They shrink in daylight but expand and darken at night—possibly to disguise the reflective air sacs from predators

GAS-FILLED BLADDERS, FORE AND AFT

Air sacs at either end of the body allow a phantom midge larva to hang horizontally in mid-water. Muscles expand these bladders at night, so more gas seeps in from surrounding body fluids. This makes the midge rise closer to the surface, where it can hunt surface-living mosquito larvae. By day, the sacs shrink and the midge larva sinks to avoid larger predators.

Thoracic gas bladder — Chromatophore — Abdominal gas bladder

PHANTOM MIDGE (*CHAOBORUS*) LARVA

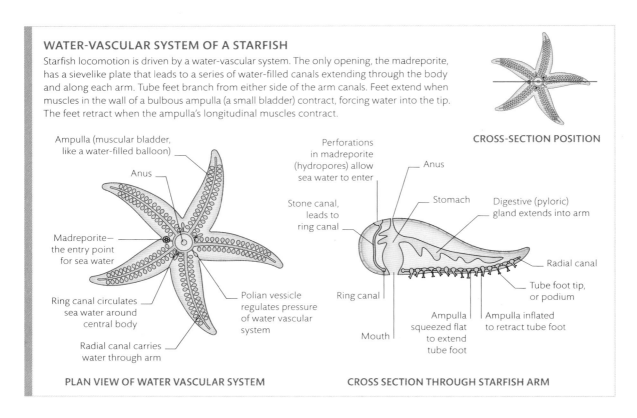

WATER-VASCULAR SYSTEM OF A STARFISH

Starfish locomotion is driven by a water-vascular system. The only opening, the madreporite, has a sievelike plate that leads to a series of water-filled canals extending through the body and along each arm. Tube feet branch from either side of the arm canals. Feet extend when muscles in the wall of a bulbous ampulla (a small bladder) contract, forcing water into the tip. The feet retract when the ampulla's longitudinal muscles contract.

tube feet

Echinoderms—starfish, urchins, brittle stars, and sea cucumbers—move with speed and agility with the aid of unique tube feet. These small, unjointed structures are operated by hydrostatic pressure created by an internal water vascular system (see above) and adapted for locomotion, attachment, and food manipulation. Each tube foot moves independently, lifting up and extending forward, then planting itself on the seabed and pushing back. Movements are achieved by coordination of opposing muscles in the walls of the tube feet and strategically placed flap valves that ensure water flows in the desired direction though the system. Most species also have an adhesive suction pad on the end of each foot.

Tube feet in action
This side view of an arm of a seven-armed starfish, *Luidia ciliaris*, shows the tube feet. The feet are different lengths and move independently, under the control of the central nervous system.

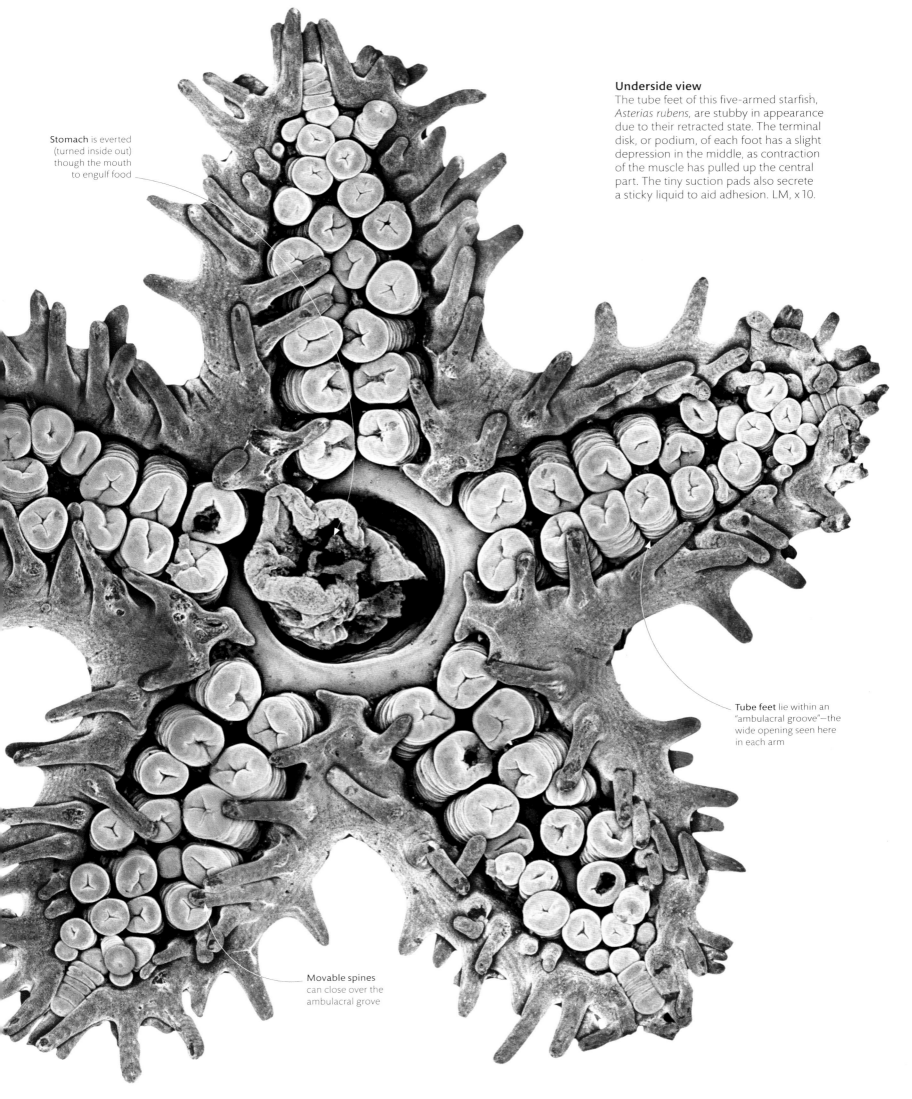

Underside view
The tube feet of this five-armed starfish, *Asterias rubens*, are stubby in appearance due to their retracted state. The terminal disk, or podium, of each foot has a slight depression in the middle, as contraction of the muscle has pulled up the central part. The tiny suction pads also secrete a sticky liquid to aid adhesion. LM, x10.

Stomach is everted (turned inside out) though the mouth to engulf food

Tube feet lie within an "ambulacral groove"—the wide opening seen here in each arm

Movable spines can close over the ambulacral grove

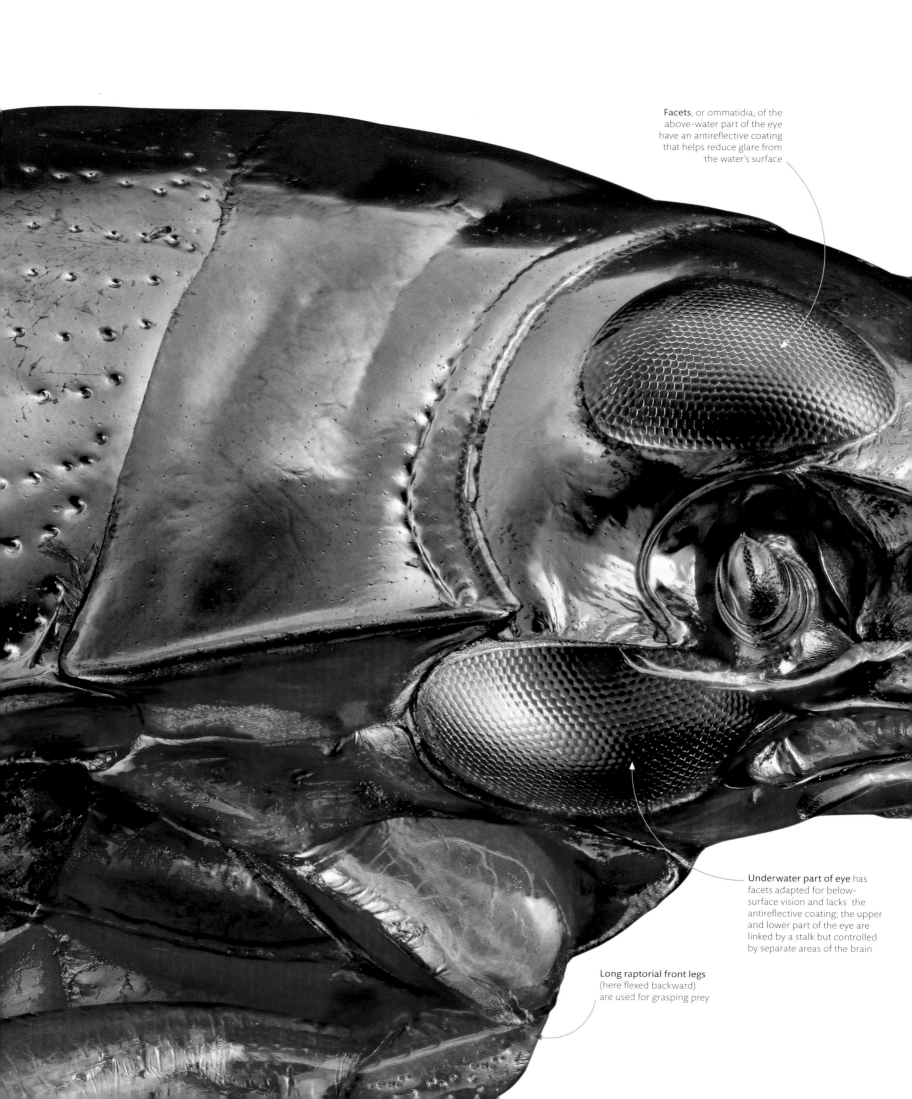

Facets, or ommatidia, of the above-water part of the eye have an antireflective coating that helps reduce glare from the water's surface

Underwater part of eye has facets adapted for below-surface vision and lacks the antireflective coating; the upper and lower part of the eye are linked by a stalk but controlled by separate areas of the brain

Long raptorial front legs (here flexed backward) are used for grasping prey

living at the surface

The surface of a pond is governed by unseen forces. Water molecules bond to others all around them, but those at the surface cannot do so with the air above. To compensate, they bond more strongly with their neighbors, creating a thin film of surface tension. Many animals living above the surface, such as pondskaters, use this film as a platform for walking on water; others, like mosquito larvae, hang from it by their breathing tubes. Whirligig beetles straddle the middle, half submerged, using split-level eyes to see above and below the surface simultaneously.

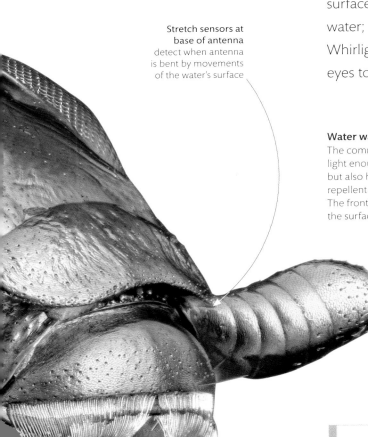

Surface opportunist
A whirligig beetle (*Gyrinus* sp.) has a water-repellent body only 1/5–1/4 in (5–7 mm) long. It preys on insects that are not so well adapted and get trapped in a pond's surface tension. As the beetle turns in endless circles, paddling with its hind legs, its motion-sensing antennae and eccentric vision alert it to the struggles of nearby prey.

Stretch sensors at base of antenna detect when antenna is bent by movements of the water's surface

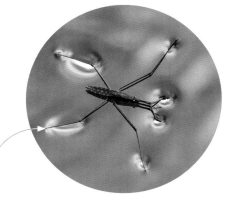

Water walking
The common pondskater (*Gerris lacustris*) is not only light enough to be supported by surface tension, but also has long legs that are tipped with water-repellent hairs that prevent its feet from getting wet. The front feet bear sensors that pick up ripples in the surface film caused by prey movement.

Water-repelling feet make dimples on the water's surface

SURFACE COMMUNITY

Organisms that live at a pond's surface either float because of their buoyancy or stand, like a pondskater, on the surface film. Collectively, they make up a community called the neuston. Whirligig beetles are buoyed up by air under their wing casings, and mosquito larvae by air-filled breathing tubes, or siphons, that break through the surface film, but are still held by it. Many microbes rely on buoyant oil droplets in their bodies.

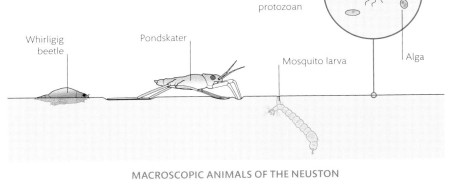

MICROSCOPIC ORGANISMS OF THE NEUSTON

Cyanobacterium
Protozoan
Colonial protozoan
Alga

Whirligig beetle
Pondskater
Mosquito larva

MACROSCOPIC ANIMALS OF THE NEUSTON

Backswimmers are freshwater insects named for their habit of floating upside down beneath the surface of the water, using the surface film to provide a foothold. Backswimmers are predators that lurk in this position, waiting to snatch prey that fall into the water. The insects also hunt more actively in deeper water, using their long, fringed back legs as a pair of oars. Growing to a maximum length of $1/2$ in (14 mm), the common backswimmer

spotlight backswimmers

(*Notonecta glauca*) preys on insects, tadpoles, and small fish. Backswimmers, like other true bugs (see p.48), have piercing mouthparts. They use them to inject a paralyzing toxin into prey before sucking out its body fluids.

Despite spending much of their time under water, backswimmers breathe air. They emerge every so often to restore their oxygen supply or make a short flight—and perhaps find a new hunting ground. When ready to submerge again they trap a layer of air bubbles on their bristled body and under the wings, providing an oxygen supply that can last more than 100 days under water. The bubbles also make a backswimmer highly buoyant. To avoid floating to the surface, it must hold on to vegetation nearer the bottom of the pond.

Backswimmers find prey mostly by sight and by detecting the ripples they cause in the water surface film. Their compound eyes are very large and can form images both in water and in air, and work well in daylight and at night. The common backswimmer's dark underside and pale back make it harder to see from above and below when it is on its back, lurking just beneath the surface.

Two pairs of legs grip surface film of water

common backswimmer

Freshwater predator
A common backswimmer (*Notonecta glauca*) has caught a fly that has landed on the water and become too wet to take off. The hunter uses the tips of its forelegs to feel for ripples in the surface film, and is able to discern those caused by struggling prey and other aquatic insects.

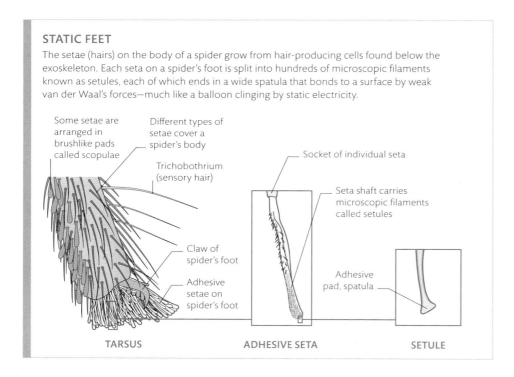

STATIC FEET
The setae (hairs) on the body of a spider grow from hair-producing cells found below the exoskeleton. Each seta on a spider's foot is split into hundreds of microscopic filaments known as setules, each of which ends in a wide spatula that bonds to a surface by weak van der Waal's forces—much like a balloon clinging by static electricity.

Some setae are arranged in brushlike pads called scopulae
Different types of setae cover a spider's body
Trichobothrium (sensory hair)
Claw of spider's foot
Adhesive setae on spider's foot
Socket of individual seta
Seta shaft carries microscopic filaments called setules
Adhesive pad, spatula

TARSUS — ADHESIVE SETA — SETULE

Multipurpose foot
The foot of an ant-mimic jumping spider (*Myrmarachne formicaria*) carries a complex array of hooks and hairs. Some of the hairs collect tactile information and even feel movements in the air. But the pad at the tip—with its feathery texture—enables the animal to cling to slopes of its grassy bank habitat. SEM, x 1,000.

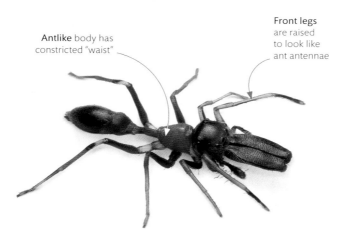

Ant-mimic jumping spider
Many predators avoid ants, as their acid-squirting defenses are unpalatable. This spider enjoys the obvious benefits by raising its front two legs and running on the other six.

Antlike body has constricted "waist"

Front legs are raised to look like ant antennae

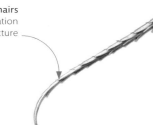

Long sensory hairs provide information about surface texture

clinging feet

The forces of nature at work in the micro-world mean that small animals can achieve feats that are impossible for larger ones. Spiders and insects are able to climb smooth surfaces because of tiny electrostatic attractions, called van der Waals forces, that take place between their feet and the surface. By themselves these forces are weak, but combined with the special adaptation of a spider's foot, the strength is multiplied many times over. Each foot has brushes of hairs, each split into tinier filaments with microscopic tips that can cling to a surface. While a spider may appear to use eight legs, in fact it has hundreds of points of attachment—more than enough to support its weight and enable it to run up vertical walls.

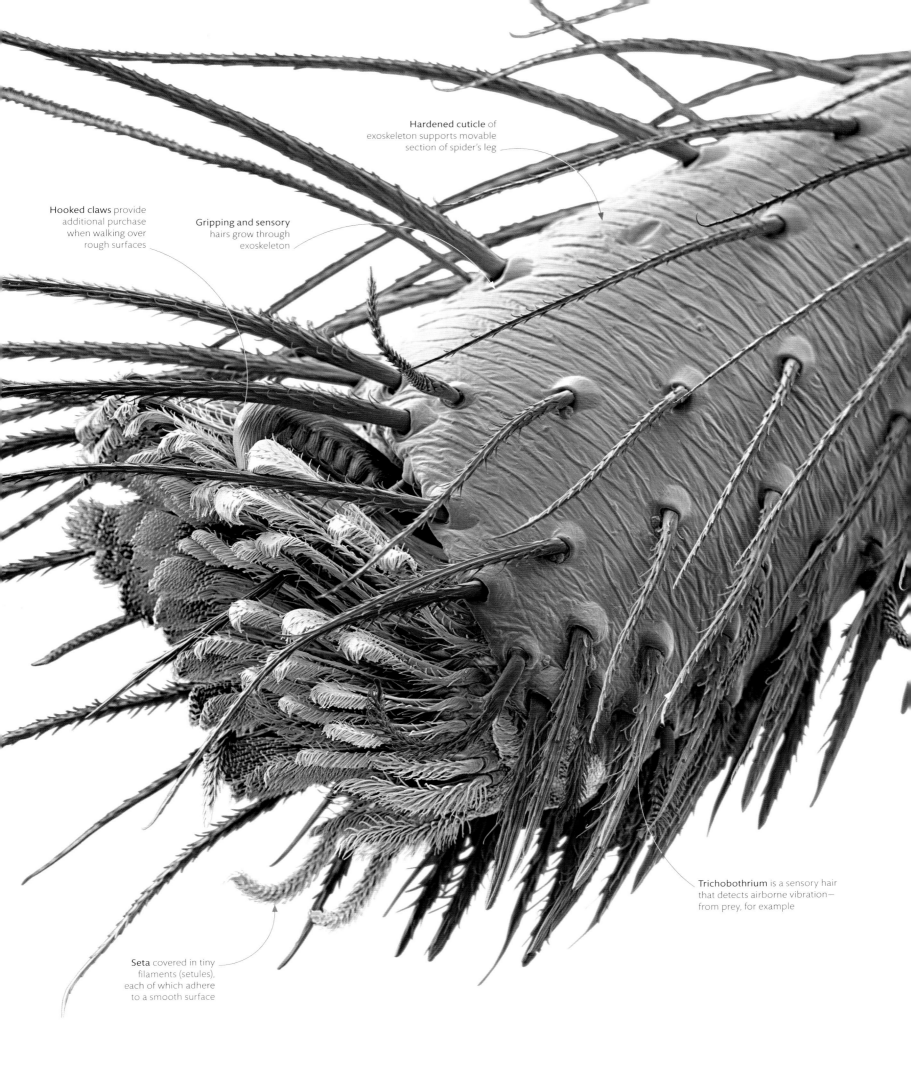

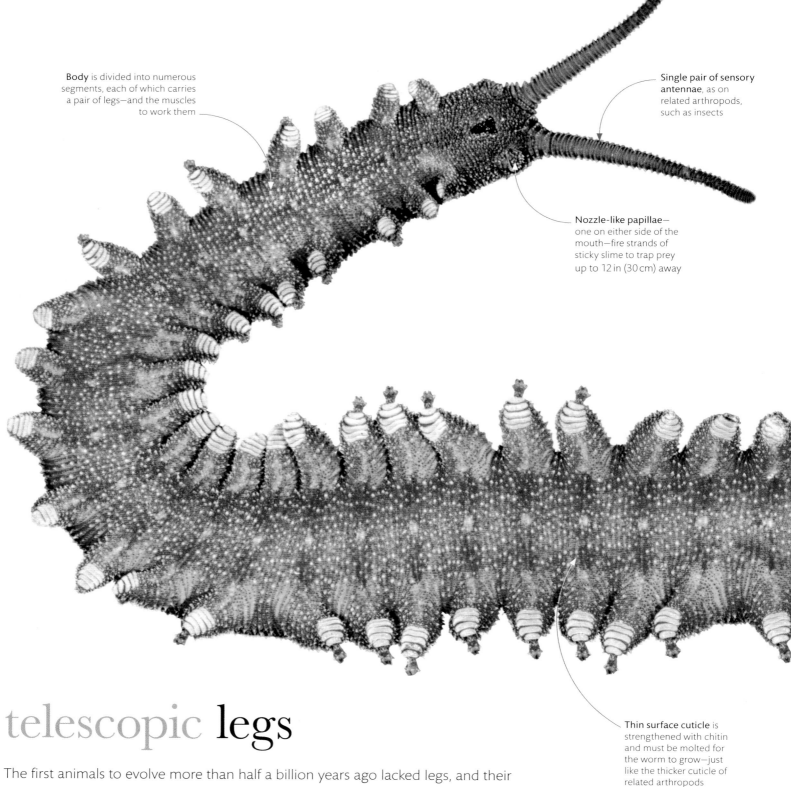

Body is divided into numerous segments, each of which carries a pair of legs—and the muscles to work them

Single pair of sensory antennae, as on related arthropods, such as insects

Nozzle-like papillae— one on either side of the mouth—fire strands of sticky slime to trap prey up to 12 in (30 cm) away

Thin surface cuticle is strengthened with chitin and must be molted for the worm to grow—just like the thicker cuticle of related arthropods

telescopic legs

The first animals to evolve more than half a billion years ago lacked legs, and their wormlike descendants still wriggle on their bellies today. But legs make moving around easier. Stubby-legged animals similar to modern velvet worms might have been the ancestors of jointed-legged arthropods, such as crustaceans, arachnids, and insects. Raised up on legs, the body experiences less friction, and muscle power can concentrate on pushing the ground with the feet rather than with the whole belly. Velvet worms lack the stiff armor of true arthropods. Their soft legs are supported by the pressure of internal body fluids, and they walk by extending and retracting these legs like telescopes.

Twin-function feet
Each leg of a velvet worm is cushioned by up to six soft foot pads. But the foot also has two hooks that help with climbing over obstacles—a characteristic that is the basis for the velvet worm taxonomic group name: Onychophora, meaning claw-bearer. SEM, x 130.

Sicklelike claw is made from extra-tough chitin

MUSCLE CONTROL

Longitudinal muscles running the length of a velvet worm's body contract to make the body shorten. This squeezes blood into the legs and causes them to telescope outward. Other muscles run from the body cavity to inside the legs, and they pull on the legs to make them swing forward or backward.

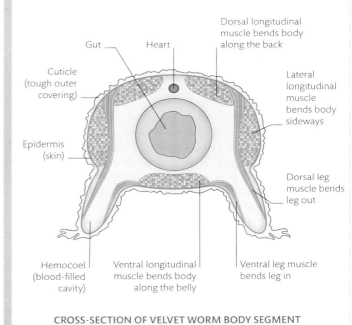

CROSS-SECTION OF VELVET WORM BODY SEGMENT

Leg retracted by the pull of the fluid pressure in the hydrostatic skeleton

Leg protracted (extended) by a combination of muscles and fluid pressure in the hydrostatic skeleton

Soft, fleshy legs are worked back and forth by muscles, but lack the hinged joints seen on arthropods

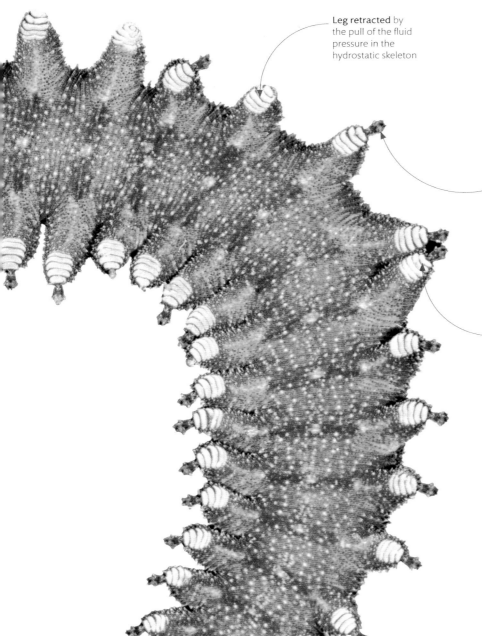

Hydrostatic legs
The soft body of a velvet worm (*Peripatus* sp.), which is on average 2 in (5 cm) long, contains a blood-filled cavity, or hemocoel, that works as a hydrostatic skeleton supporting both body and legs. The number of legs varies according to species, but in all velvet worms the movement of the legs is precisely coordinated to ripple slowly down the body and push the animal forward.

jointed legs

Insects, arachnids, and crustaceans are arthropods, meaning jointed leg—a characteristic that has been a driving force in making these animals so successful on land and in water. Stiffened with exoskeleton, but with flexible hinges, the legs are used for running, swimming, digging, and all the other limb movements needed to get around. Each joint is worked by its own pair of antagonistic muscles that pull the exoskeleton from the inside—flexing and straightening the hinges. More complex sets of muscles connect the legs to the body, so the joint can swivel like a ball-and-socket.

Digging forelegs
The rainbow scarab beetle (*Phaneus vindex*) and mole cricket (*Gryllotalpa gryllotalpa*) excavate burrows, so in both animals, part of their foreleg is flattened like a spade. The scarab digs to bury dung for its larvae, but the mole cricket, opposite, creates more permanent burrows, as it spends most of its life underground. The legs are shown at 20 times life size.

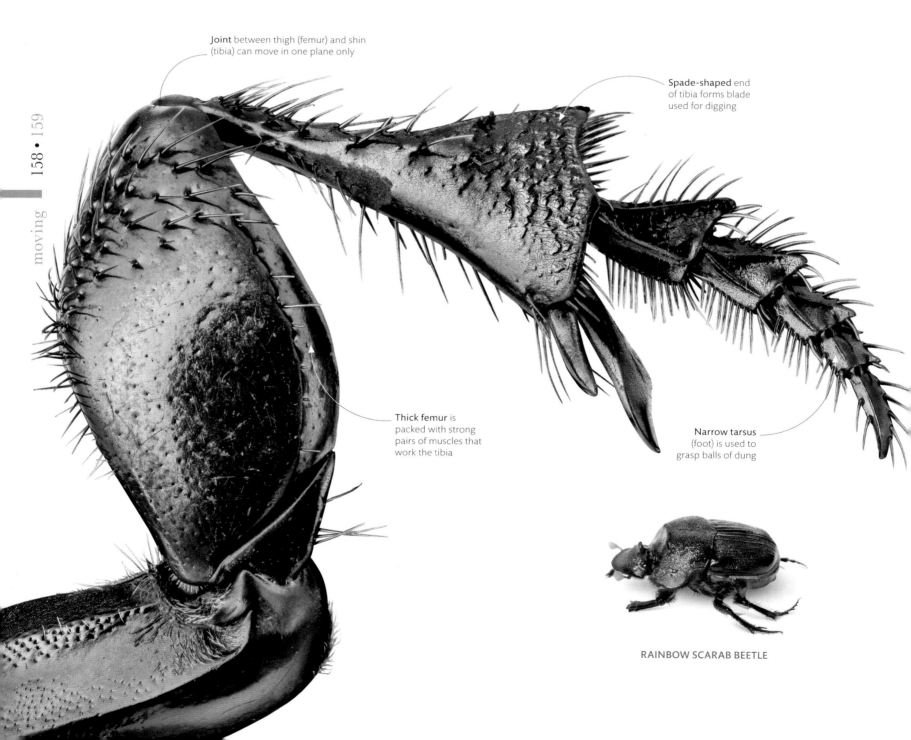

Joint between thigh (femur) and shin (tibia) can move in one plane only

Spade-shaped end of tibia forms blade used for digging

Thick femur is packed with strong pairs of muscles that work the tibia

Narrow tarsus (foot) is used to grasp balls of dung

RAINBOW SCARAB BEETLE

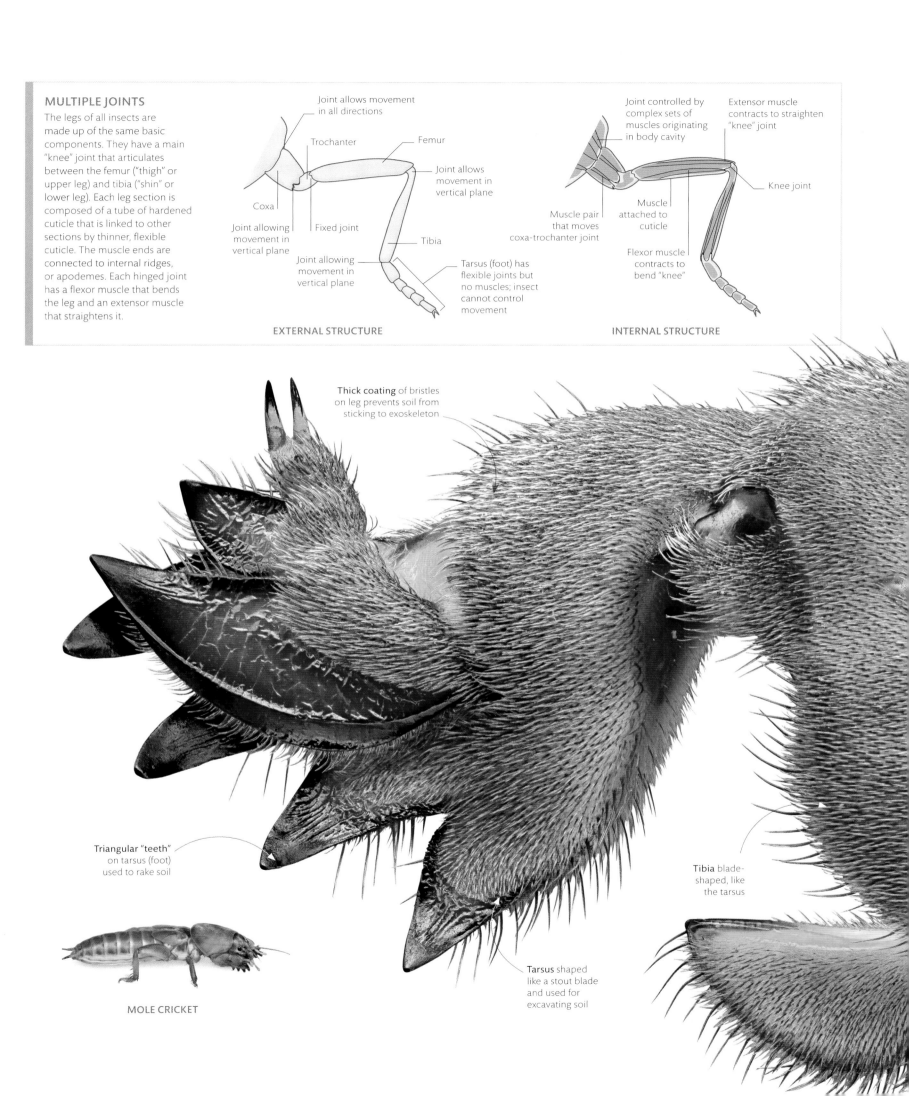

MULTIPLE JOINTS

The legs of all insects are made up of the same basic components. They have a main "knee" joint that articulates between the femur ("thigh" or upper leg) and tibia ("shin" or lower leg). Each leg section is composed of a tube of hardened cuticle that is linked to other sections by thinner, flexible cuticle. The muscle ends are connected to internal ridges, or apodemes. Each hinged joint has a flexor muscle that bends the leg and an extensor muscle that straightens it.

EXTERNAL STRUCTURE
- Joint allows movement in all directions
- Trochanter
- Femur
- Coxa
- Joint allows movement in vertical plane
- Joint allowing movement in vertical plane
- Fixed joint
- Tibia
- Joint allowing movement in vertical plane
- Tarsus (foot) has flexible joints but no muscles; insect cannot control movement

INTERNAL STRUCTURE
- Joint controlled by complex sets of muscles originating in body cavity
- Extensor muscle contracts to straighten "knee" joint
- Knee joint
- Muscle pair that moves coxa-trochanter joint
- Muscle attached to cuticle
- Flexor muscle contracts to bend "knee"

Thick coating of bristles on leg prevents soil from sticking to exoskeleton

Triangular "teeth" on tarsus (foot) used to rake soil

Tarsus shaped like a stout blade and used for excavating soil

Tibia blade-shaped, like the tarsus

MOLE CRICKET

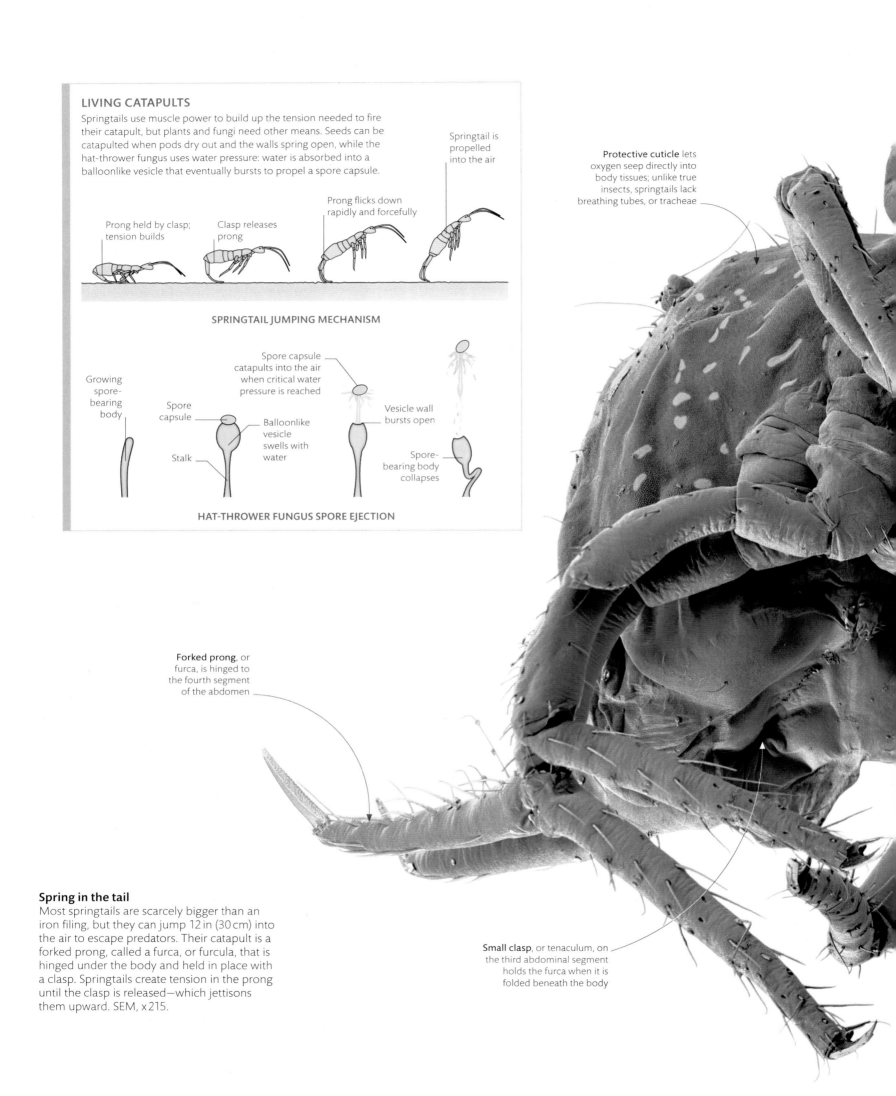

LIVING CATAPULTS

Springtails use muscle power to build up the tension needed to fire their catapult, but plants and fungi need other means. Seeds can be catapulted when pods dry out and the walls spring open, while the hat-thrower fungus uses water pressure: water is absorbed into a balloonlike vesicle that eventually bursts to propel a spore capsule.

SPRINGTAIL JUMPING MECHANISM

- Prong held by clasp; tension builds
- Clasp releases prong
- Prong flicks down rapidly and forcefully
- Springtail is propelled into the air

HAT-THROWER FUNGUS SPORE EJECTION

- Growing spore-bearing body
- Spore capsule
- Stalk
- Balloonlike vesicle swells with water
- Spore capsule catapults into the air when critical water pressure is reached
- Vesicle wall bursts open
- Spore-bearing body collapses

Protective cuticle lets oxygen seep directly into body tissues; unlike true insects, springtails lack breathing tubes, or tracheae

Forked prong, or furca, is hinged to the fourth segment of the abdomen

Small clasp, or tenaculum, on the third abdominal segment holds the furca when it is folded beneath the body

Spring in the tail

Most springtails are scarcely bigger than an iron filing, but they can jump 12 in (30 cm) into the air to escape predators. Their catapult is a forked prong, called a furca, or furcula, that is hinged under the body and held in place with a clasp. Springtails create tension in the prong until the clasp is released—which jettisons them upward. SEM, x215.

Hatlike spore capsule sits on top of the spore-bearing body

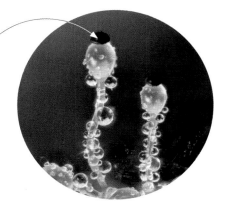

Spore thrower
The hat-thrower fungus (*Pilobolus crystallinus*) disperses its spores by catapulting them into the air. This ballistic action achieves one of the greatest known accelerations in the living world.

Terminal segments of antennae are elongated and used for touching and tasting

Walking legs are the usual means of locomotion unless danger threatens; traditionally viewed as insects because they have six jointed legs, springtails lack wings and have other unique features, so they are sometimes classified as a separate group

catapulting

Muscle power enables impressive animal feats of flying, running, jumping, and swimming. But muscles have their limitations, and some of the fastest movements in the natural world are not directly due to muscles contracting. Micro-life sometimes uses a catapult instead, gradually building up energy under tension—like an archer pulling back a bowstring—and then releasing the energy all at once, with ballistic effects. This highly effective use of muscle power can propel organisms much farther. Tiny jumping insects, such as springtails and fleas, rely on this method, and some plants and fungi use catapults too.

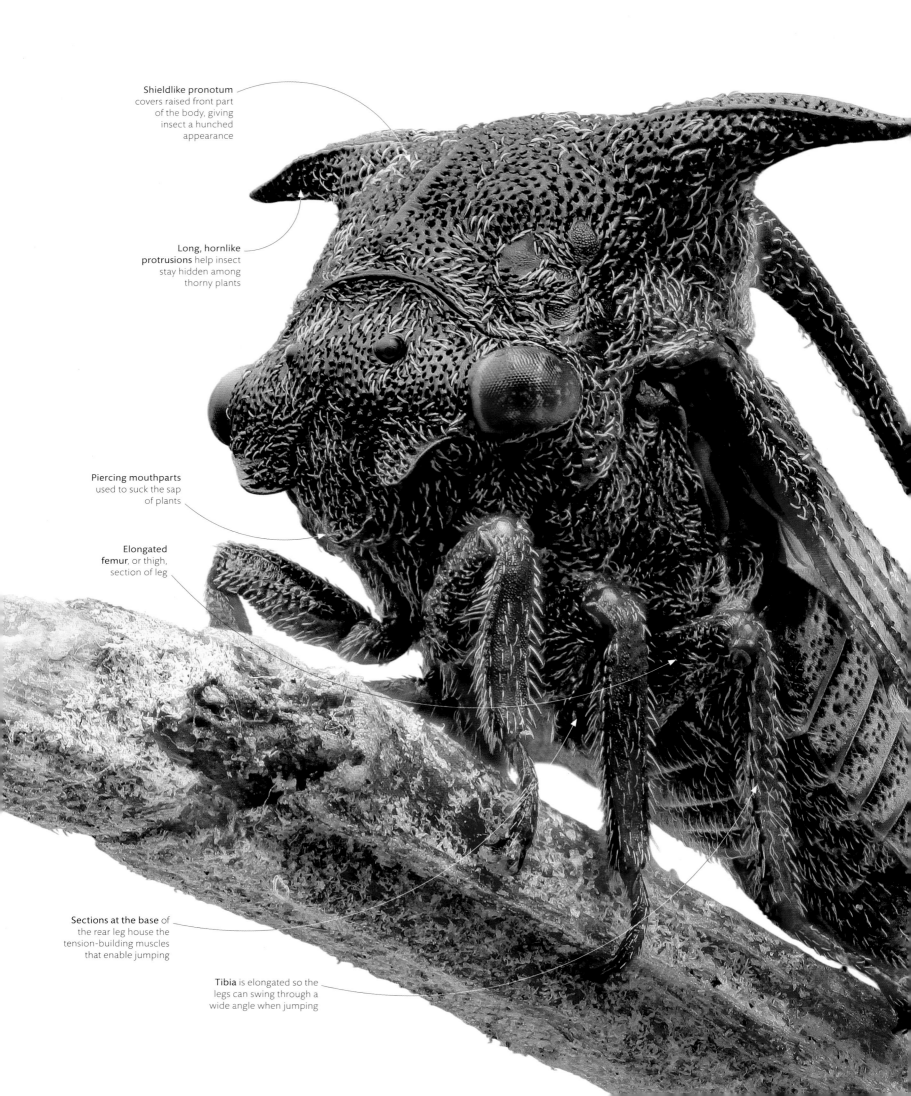

Shieldlike pronotum covers raised front part of the body, giving insect a hunched appearance

Long, hornlike protrusions help insect stay hidden among thorny plants

Piercing mouthparts used to suck the sap of plants

Elongated femur, or thigh, section of leg

Sections at the base of the rear leg house the tension-building muscles that enable jumping

Tibia is elongated so the legs can swing through a wide angle when jumping

Cryptic jumper
Treehoppers are plant-sucking bugs, related to froghoppers and singing cicadas, which rely on camouflage and speed to evade predators, such as birds. The horned treehopper (*Leptocentrus taurus*), seen here at 30 times life size, has legs that work like catapults, sending it jumping between branches.

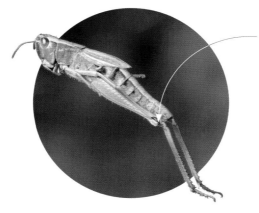

As knee joint releases, shins spring outward, flinging insect into its jump

Catapulting knees
The thigh muscles of a meadow grasshopper (*Pseudochorthippus parallelus*) contract, generating tension until a cliplike structure in the knees is released, straightening its legs.

jumping with legs

Some arthropod legs are specialized in generating such force that they launch the animal into the air. These limbs have thick muscles that pack many fibers, generating a large pulling force. The leg sections are long, so that they swing far and fast around the joints. Some crickets are champion long-jumpers, because they have legs with both these features. Other jumping insects with shorter legs leap by launching themselves like catapults. Their strong muscles work by building up tension around a fastened joint, enabling them to jump rapidly when the tension is released.

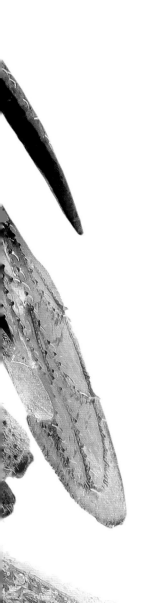

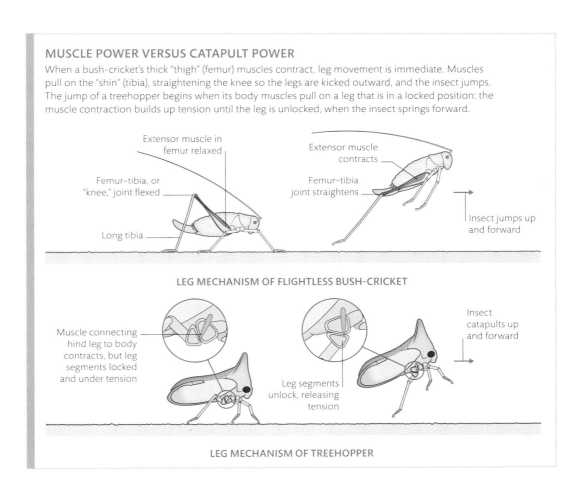

MUSCLE POWER VERSUS CATAPULT POWER
When a bush-cricket's thick "thigh" (femur) muscles contract, leg movement is immediate. Muscles pull on the "shin" (tibia), straightening the knee so the legs are kicked outward, and the insect jumps. The jump of a treehopper begins when its body muscles pull on a leg that is in a locked position: the muscle contraction builds up tension until the leg is unlocked, when the insect springs forward.

LEG MECHANISM OF FLIGHTLESS BUSH-CRICKET

LEG MECHANISM OF TREEHOPPER

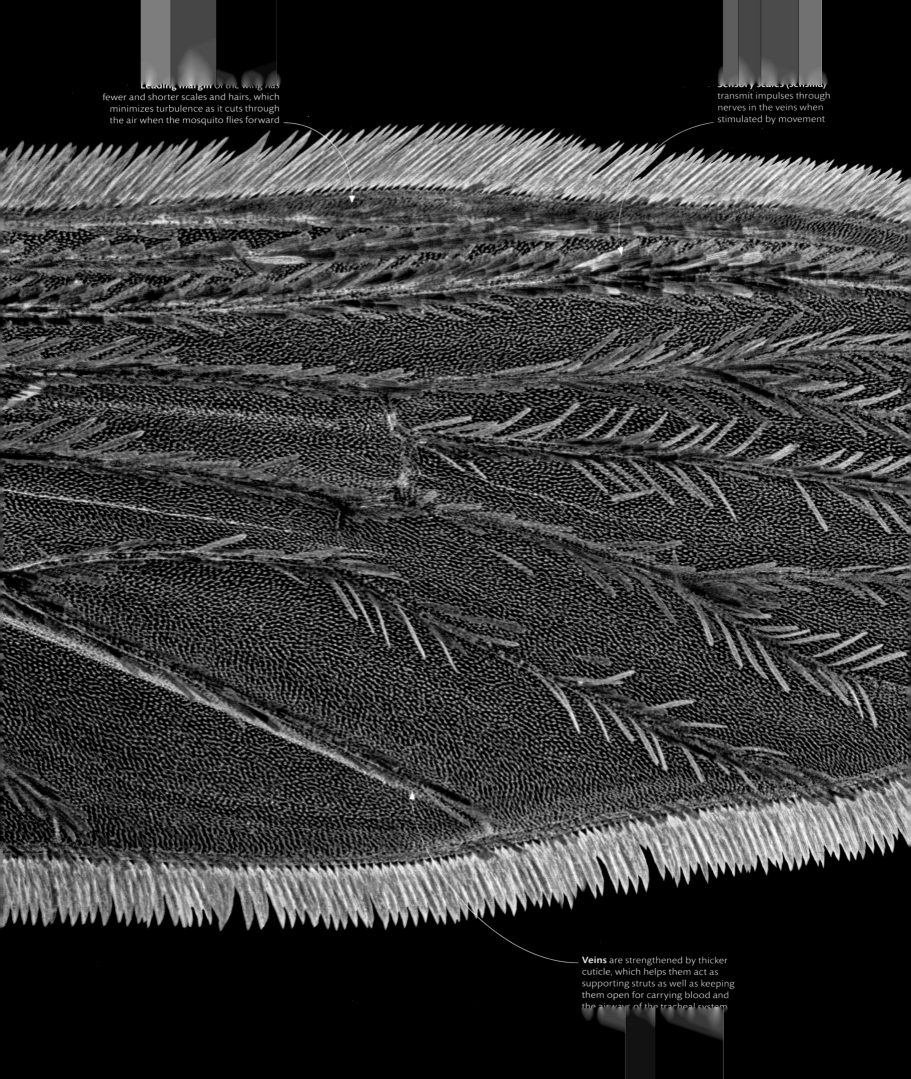

insect wings

Insects are the only invertebrates capable of flight, and their ability to take to the air undoubtedly helped them succeed in dominating life on land: up to 90 percent of all animal species are estimated to be insects. Whereas flying vertebrates—birds and bats—refashioned existing forelimbs as wings, insects evolved custom-made wings as outward flaps of their exoskeleton. Each wing maximizes the area for lift, while minimizing weight, by being little more than a thin extension of the surface cuticle, with just supporting veins sandwiched in between.

Tiny stiff hairs called microtrichia may help to repel water, keeping the wing dry

Cuticle covering the upper and lower wing membranes is composed of chitin—the same tough material from which the rest of the exoskeleton is made

Flight membrane
The wing of a mosquito (*Culex* sp.) is so thin that it appears translucent, except for a sparse coating of tiny hairs and scales along the veins and wing edge. The scales are sensilla: sensory structures that detect the movement and stress of the wing's cuticle, which gives information to help with flight control. LM, x200.

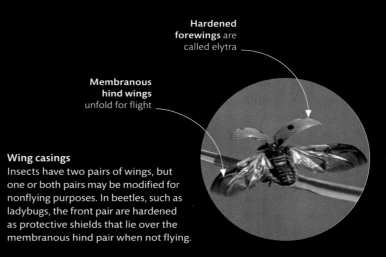

Hardened forewings are called elytra

Membranous hind wings unfold for flight

Wing casings
Insects have two pairs of wings, but one or both pairs may be modified for nonflying purposes. In beetles, such as ladybugs, the front pair are hardened as protective shields that lie over the membranous hind pair when not flying.

WING VEINS
The thin membrane of an insect's wing is supported by stiff, branching veins that cross-connect. Each main vein carries a nerve and air-filled trachea to oxygenate wing tissue. The veins divide the wing into a series of cells: the overall pattern is important in identifying taxonomic groups.

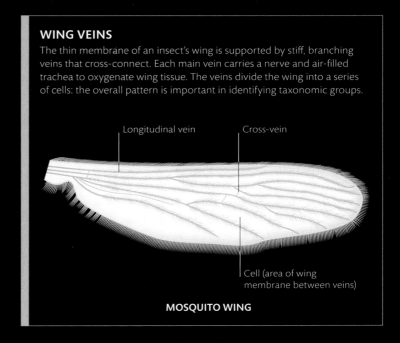

Longitudinal vein

Cross-vein

Cell (area of wing membrane between veins)

MOSQUITO WING

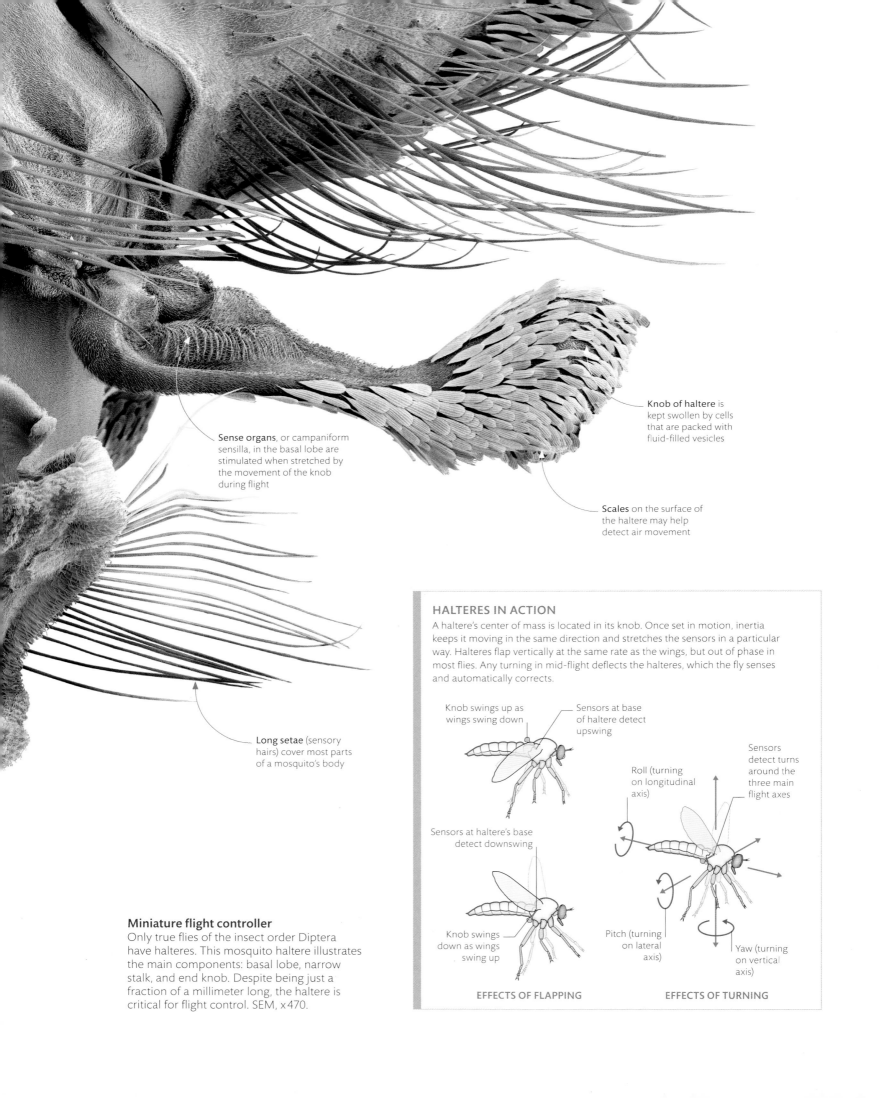

Knob of haltere is kept swollen by cells that are packed with fluid-filled vesicles

Sense organs, or campaniform sensilla, in the basal lobe are stimulated when stretched by the movement of the knob during flight

Scales on the surface of the haltere may help detect air movement

Long setae (sensory hairs) cover most parts of a mosquito's body

Miniature flight controller
Only true flies of the insect order Diptera have halteres. This mosquito haltere illustrates the main components: basal lobe, narrow stalk, and end knob. Despite being just a fraction of a millimeter long, the haltere is critical for flight control. SEM, ×470.

HALTERES IN ACTION
A haltere's center of mass is located in its knob. Once set in motion, inertia keeps it moving in the same direction and stretches the sensors in a particular way. Halteres flap vertically at the same rate as the wings, but out of phase in most flies. Any turning in mid-flight deflects the halteres, which the fly senses and automatically corrects.

Knob swings up as wings swing down — Sensors at base of haltere detect upswing

Sensors at haltere's base detect downswing

Knob swings down as wings swing up

EFFECTS OF FLAPPING

Roll (turning on longitudinal axis)

Sensors detect turns around the three main flight axes

Pitch (turning on lateral axis)

Yaw (turning on vertical axis)

EFFECTS OF TURNING

stabilizing flight

Flight involves more than strong muscles and flapping wings: it also needs fine control over the body's position while in midair. Flies have unrivaled flight control that takes them on perfectly straight paths, both forward and backward, and permits lightning-fast maneuvers to evade predators. Their skills derive from unique flight equipment—a single pair of conventional wings and a second pair reshaped into small clubs called halteres. Carried on stalks, the clubs sway in all directions. They flap up and down, like the true wings, but also swing around if the insect twists or turns. All these movements stretch sensory cells at the base of the halteres, triggering nerve impulses that enable the fly to make automatic adjustments and stabilize its flightpath.

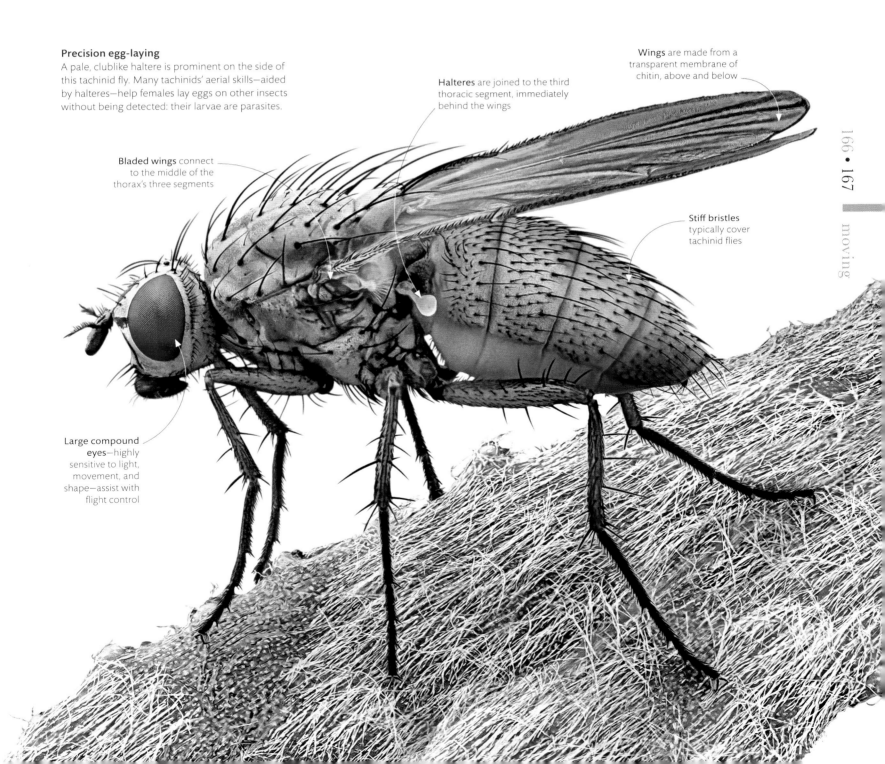

Precision egg-laying
A pale, clublike haltere is prominent on the side of this tachinid fly. Many tachinids' aerial skills—aided by halteres—help females lay eggs on other insects without being detected: their larvae are parasites.

Halteres are joined to the third thoracic segment, immediately behind the wings

Wings are made from a transparent membrane of chitin, above and below

Bladed wings connect to the middle of the thorax's three segments

Stiff bristles typically cover tachinid flies

Large compound eyes—highly sensitive to light, movement, and shape—assist with flight control

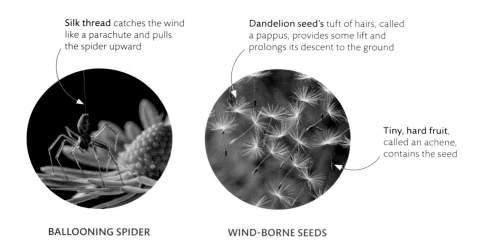

- **Silk thread** catches the wind like a parachute and pulls the spider upward
- **Dandelion seed's** tuft of hairs, called a pappus, provides some lift and prolongs its descent to the ground
- **Tiny, hard fruit,** called an achene, contains the seed

BALLOONING SPIDER   WIND-BORNE SEEDS

Aeroplankton
Some of the smallest airborne organisms drift on the wind, rather than using powered flight. This aeroplankton uses currents of air for dispersal and includes microbes, spores, pollen, seeds, and tiny spiders, which eject silk threads into the air to catch the wind.

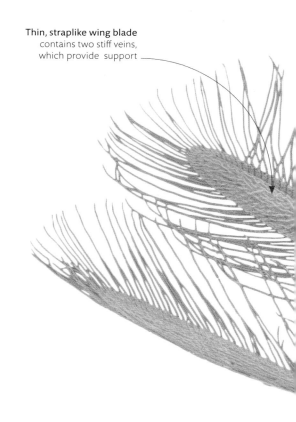

Thin, straplike wing blade contains two stiff veins, which provide support

tiniest fliers

Like anything that flies, insects stay airborne because of two forces: thrust, which pushes them forward, and lift, which keeps them up. Few insects glide, and those that do only pause their flapping for a few seconds. This is because they rely on beating wings to achieve both thrust and lift. Powerful thorax muscles work the wings with astonishing speed, reaching hundreds of beats every second. Each wing stroke pushes back on the air to propel the insect. At the same time, flapping generates tiny whirls of air, or vortices, above the wings, creating pockets of low pressure that produce lift.

Wings for a micro-world
A thrips scarcely larger than a hyphen on this page feels air resistance much more than a bigger flier—and the slightest breeze can send it off course. Thin wings with hairy fringes help to cut through this resistance. On the upthrust, the wings clap together, and when they peel apart, air rushes between them and improves lift. SEM, x130.

WING COUPLING

Most insects have two pairs of wings, which are coupled by hooks or other means so that they flap together. Wings that flap in unison provide a greater surface area for generating better thrust or lift. In some groups of flying insects, either the forewings or hindwings are modified and not involved in flying.

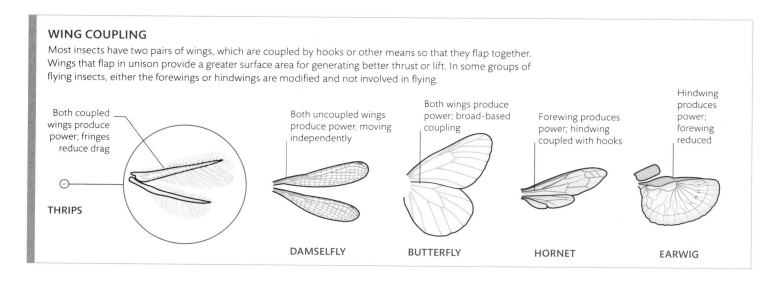

- Both coupled wings produce power; fringes reduce drag — **THRIPS**
- Both uncoupled wings produce power, moving independently — **DAMSELFLY**
- Both wings produce power; broad-based coupling — **BUTTERFLY**
- Forewing produces power; hindwing coupled with hooks — **HORNET**
- Hindwing produces power; forewing reduced — **EARWIG**

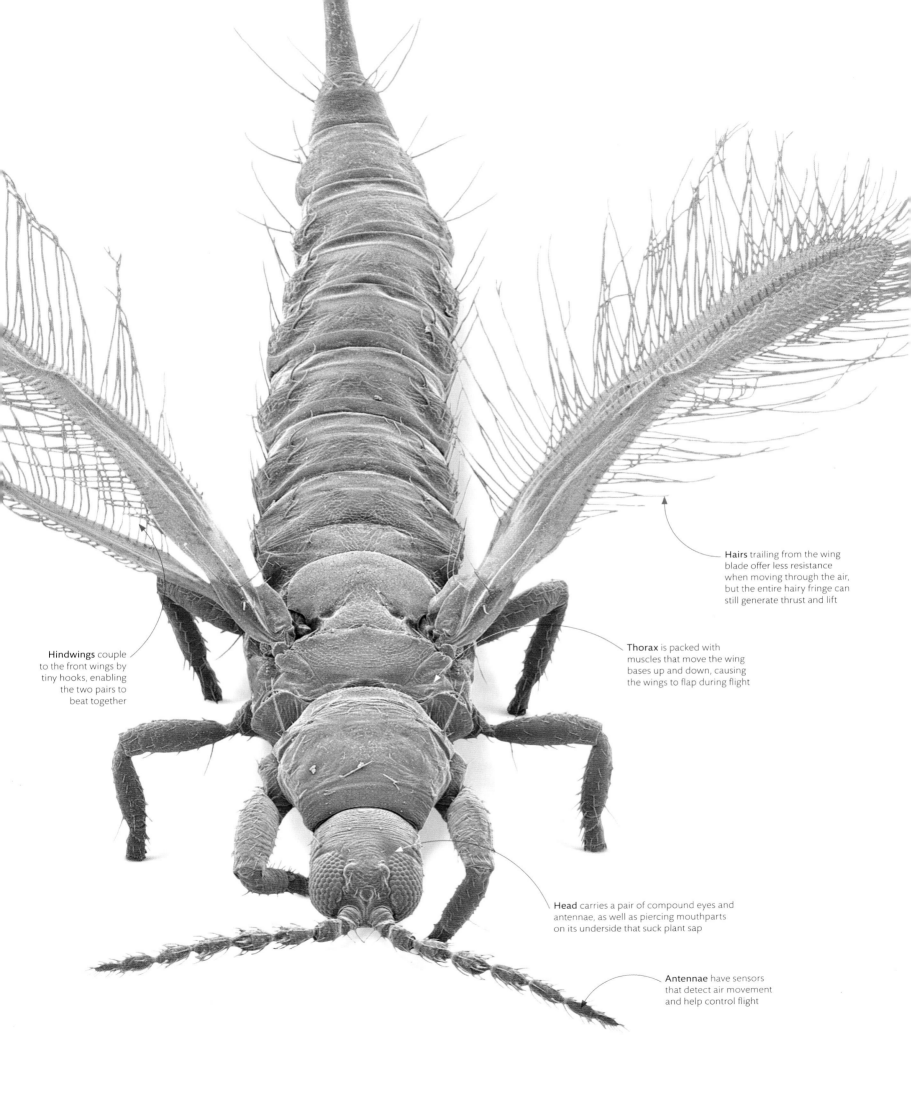

Hairs trailing from the wing blade offer less resistance when moving through the air, but the entire hairy fringe can still generate thrust and lift

Thorax is packed with muscles that move the wing bases up and down, causing the wings to flap during flight

Hindwings couple to the front wings by tiny hooks, enabling the two pairs to beat together

Head carries a pair of compound eyes and antennae, as well as piercing mouthparts on its underside that suck plant sap

Antennae have sensors that detect air movement and help control flight

MACROLIFE

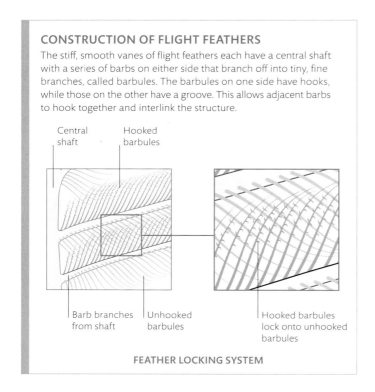

CONSTRUCTION OF FLIGHT FEATHERS
The stiff, smooth vanes of flight feathers each have a central shaft with a series of barbs on either side that branch off into tiny, fine branches, called barbules. The barbules on one side have hooks, while those on the other have a groove. This allows adjacent barbs to hook together and interlink the structure.

- Central shaft
- Hooked barbules
- Barb branches from shaft
- Unhooked barbules
- Hooked barbules lock onto unhooked barbules

FEATHER LOCKING SYSTEM

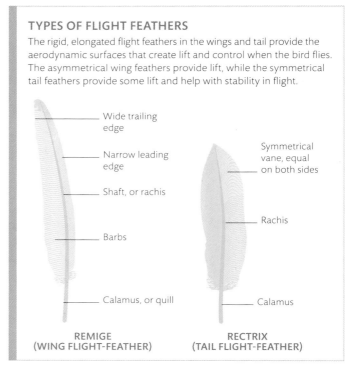

TYPES OF FLIGHT FEATHERS
The rigid, elongated flight feathers in the wings and tail provide the aerodynamic surfaces that create lift and control when the bird flies. The asymmetrical wing feathers provide lift, while the symmetrical tail feathers provide some lift and help with stability in flight.

- Wide trailing edge
- Narrow leading edge
- Shaft, or rachis
- Barbs
- Calamus, or quill

REMIGE (WING FLIGHT-FEATHER)

- Symmetrical vane, equal on both sides
- Rachis
- Calamus

RECTRIX (TAIL FLIGHT-FEATHER)

flight feathers

Feathers are made from keratin, the same protein found in mammal hair and reptile scales. The first feathered animals were dinosaurs, and feathers today are used for insulation, camouflage, display, and flight. Like hair, a feather grows from a follicle in the skin, but instead of forming a single strand, it has a stiff shaft (rachis) that branches into barbs, then into smaller barbules. Downy feathers provide insulation, but the more complex, rigid structures in the wings and the tail, some of which are attached directly to the bird's skeleton, enable flight.

Feathers need maintenance
Birds maintain their feathers by running their beak over each one to clean it of dirt and parasites, relink the barbules, and coat the surface with preen oil produced by a gland above the tail.

Barbs become disconnected when dirty, wet, or neglected

Barbules are composed of sheets of woven keratin molecules

Flight feathers close up
This SEM of a feather from a swallow (*Hirundo rustica*) shows the smallest branches in the feather vane—the barbules. Some barbules have hooks, or barbicels, on their tips that grab neighboring unhooked barbules. These microscopic linkages—along with the thick, stiff rachis—are what make the aerodynamically shaped flight feathers strong, yet flexible and light. SEM, x5,000

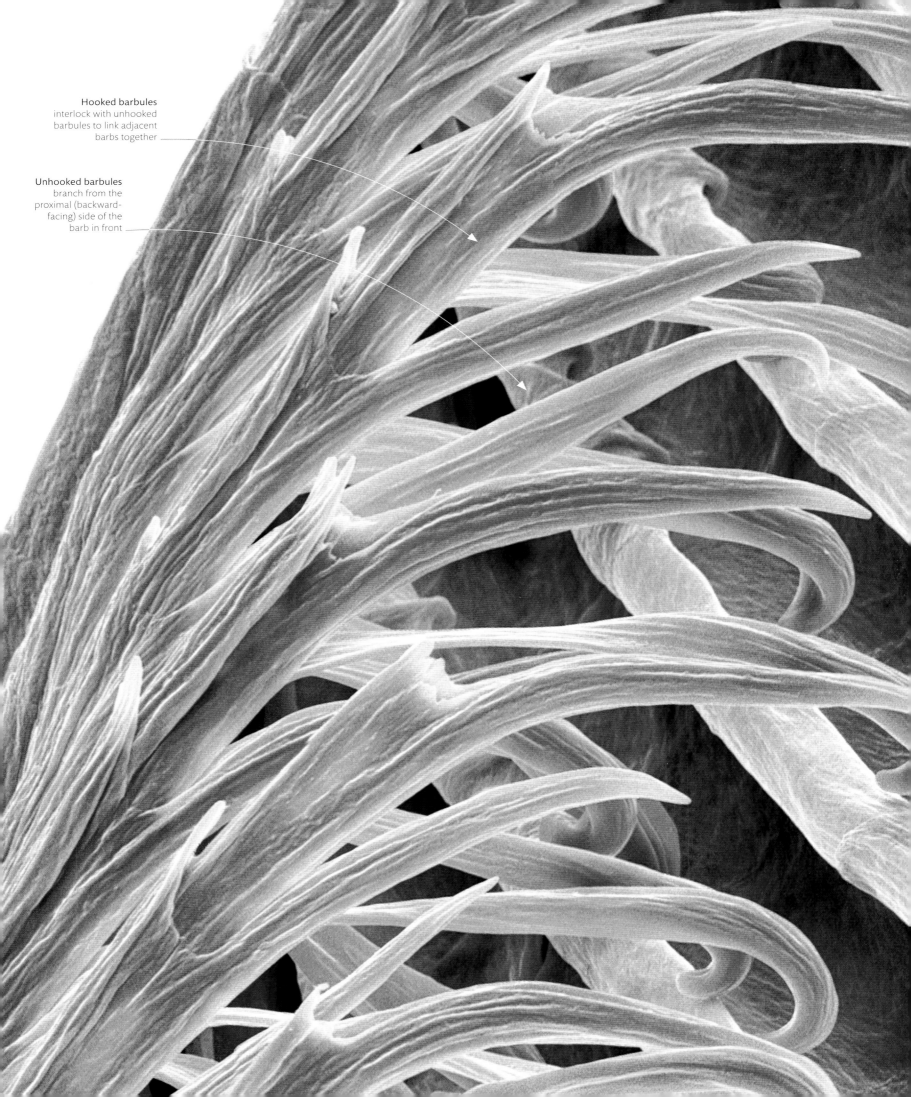

Hooked barbules interlock with unhooked barbules to link adjacent barbs together

Unhooked barbules branch from the proximal (backward-facing) side of the barb in front

Encapsulated passenger

A turtle mite (*Uropoda* sp.) is well suited to catch a ride on the underside of a dung beetle (*Aphodius prodromus*). It travels as a young nymph, attached and encased in a protective capsule, with its flat, folded limbs tucked in tightly. When the dung beetle reaches a dung pile—food for both insect and passenger—the mite disembarks as a breeding adult.

Mite hangs on with its legs and feeds using piercing mouthparts

Parasitic cargo

Not all mite passengers are harmless. The *Varroa* mite (*Varroa destructor*) sucks fat from honeybees (*Apis mellifera*). It transmits disease and is responsible for killing entire colonies.

hitchhiking mites

Animals that routinely run, fly, or swim can easily colonize new habitats, so slow movers with poor powers of dispersal sometimes take advantage by hitching a ride—a behavior called phoresy. Some tiny invertebrates, including mites and pseudoscorpions, have made hitchhiking their speciality. Clinging to insects such as beetles or ants, turtle mites get transported to new places. Most are harmless, grazing on the fungi that grow on waste matter, so their unwitting host merely has the inconvenience of carrying a tiny extra weight.

LIFE CYCLES FOR HITCHHIKING

Turtle mites are named for their shell-like bodies. Like many mites, they experience a series of molts to grow from larvae to sexually mature adults. In some species, the second nymph (deutonymph) stage can be phoretic—meaning that it hitchhikes.

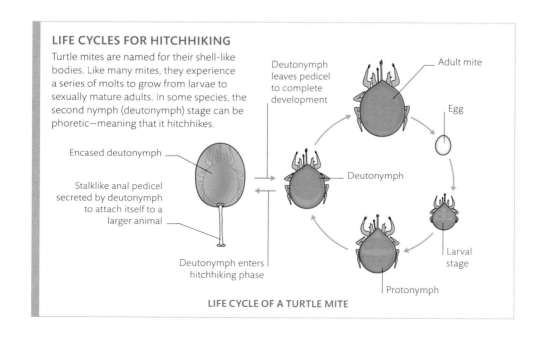

- Encased deutonymph
- Stalklike anal pedicel secreted by deutonymph to attach itself to a larger animal
- Deutonymph enters hitchhiking phase
- Deutonymph leaves pedicel to complete development
- Adult mite
- Egg
- Deutonymph
- Larval stage
- Protonymph

LIFE CYCLE OF A TURTLE MITE

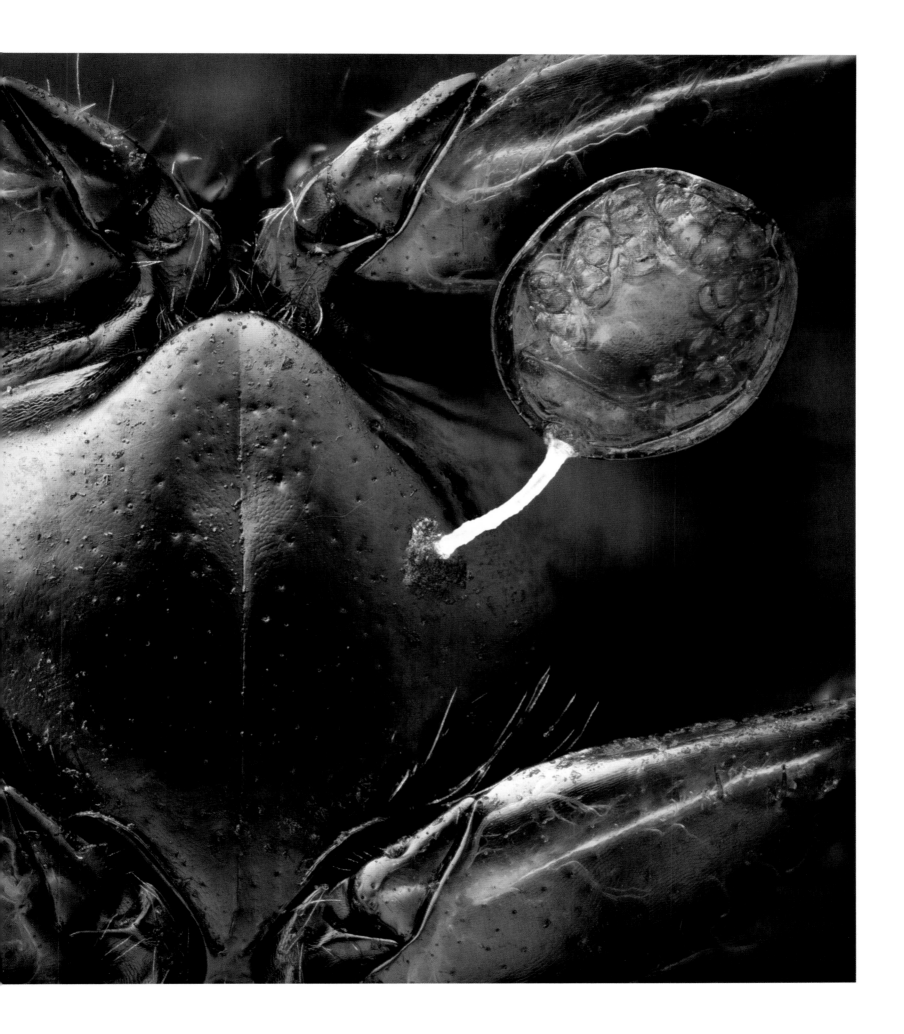

supporting and protecting

Even the smallest organisms need some kind of framework, casing, or supportive scaffolding to hold their shape or anchor their body parts while they move. Being small and vulnerable, they also need defense, whether it is physical, chemical, or simply strength in numbers.

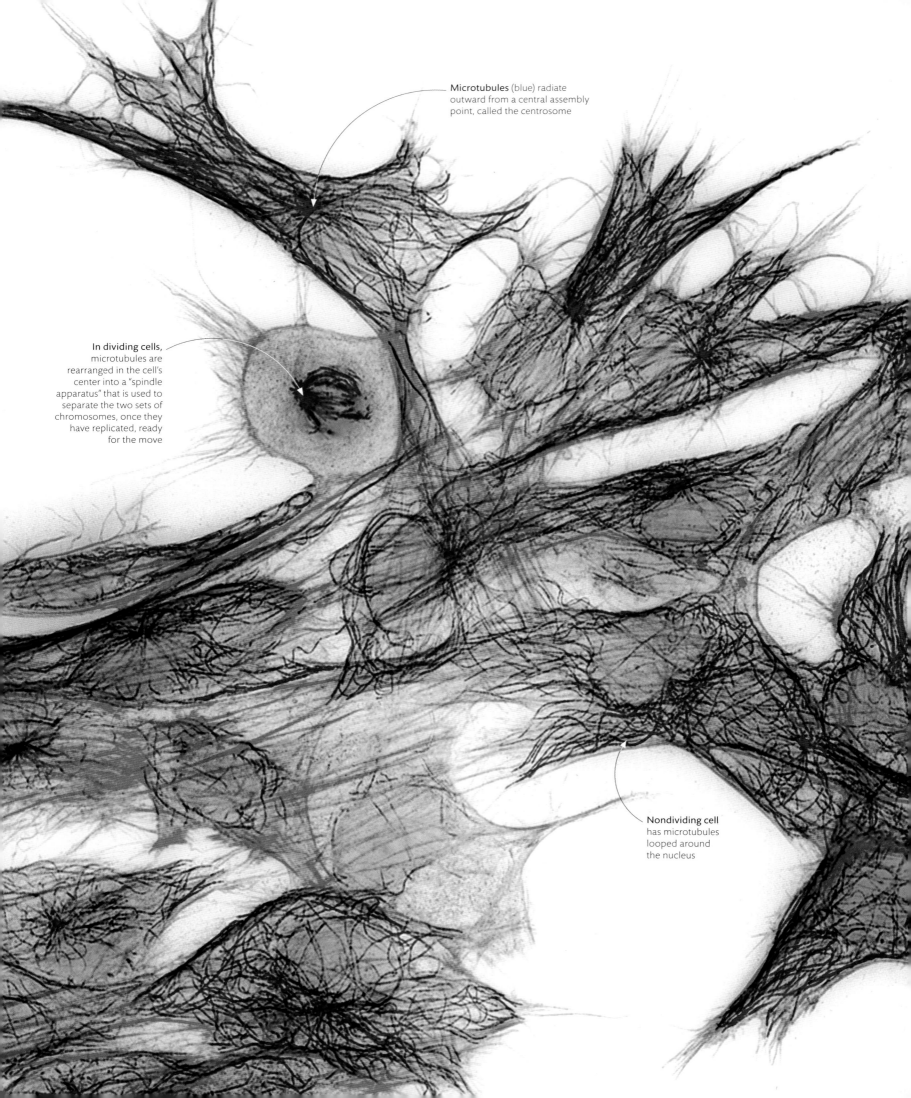

a cell's internal skeleton

While some cells, including those of plants, are enclosed by a stiff outer wall (see pp.204–05), all cells—even those without walls—are supported on the inside, too. They have a cell skeleton, or cytoskeleton, that consists of an elaborate arrangement of struts and cables made from protein. This structure maintains a cell's shape and helps to bring about movement—such as when a cell splits in two during division (see pp.254–55). The cytoskeleton also provides the fibers that make muscles contract and which help to toughen skin.

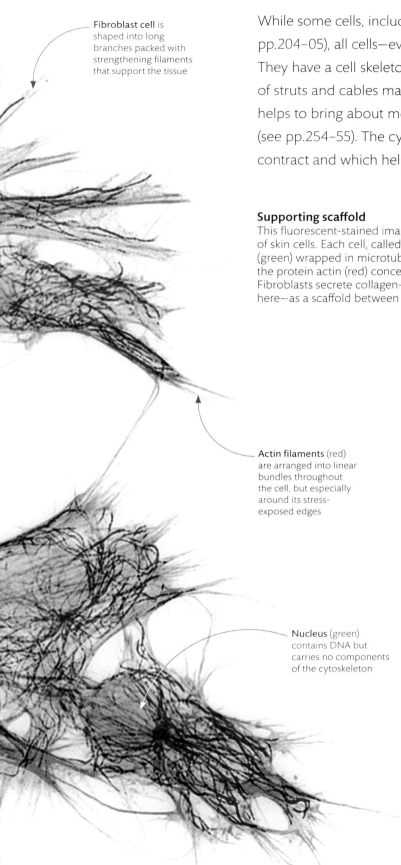

Fibroblast cell is shaped into long branches packed with strengthening filaments that support the tissue

Actin filaments (red) are arranged into linear bundles throughout the cell, but especially around its stress-exposed edges

Nucleus (green) contains DNA but carries no components of the cytoskeleton

Supporting scaffold
This fluorescent-stained image reveals the cytoskeletons of skin cells. Each cell, called a fibroblast, has a nucleus (green) wrapped in microtubules (blue), with filaments of the protein actin (red) concentrated around the edges. Fibroblasts secrete collagen—another protein, not shown here—as a scaffold between the cells. SEM, x1,500.

Assembly point
Inside the cell, microtubules are assembled at a specific point, or centrosome. In dividing animal cells, this is marked by a pair of short, tubular structures known as centrioles.

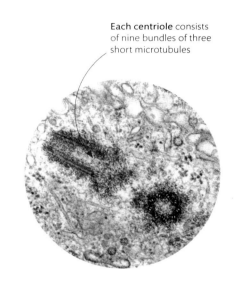

Each centriole consists of nine bundles of three short microtubules

PARTS OF A CELL'S SKELETON
A typical cytoskeleton has three components. Thin, supporting actin filaments gather in places exposed to physical stress, while thick microtubules made from long, hollow cylinders of tubulin guide the movement of chromosomes or beating cilia and flagella. Ropelike filaments of intermediate width improve mechanical strength.

- Microtubule organizing centre (centrosome)
- Tubulin microtubule (brown)
- Actin filament (red)
- Flagellum (when present) has a core of microtubules
- Intermediate filament (green) made from multiple types of protein
- Microvilli projections (when present) contain actin filaments that increase surface area for nutrient absorption

CYTOSKELETON OF AN ANIMAL CELL

Diatom valves
Every diatom's frustule is made of two halves, or valves. The upper valve is called the epitheca, and is slightly larger than the lower one, the hypotheca. The tightly fitting valves can move relative to one another, allowing the diatom to grow as it prepares to divide. During cell division, the two daughter cells each inherit one of the valves from the parent, and grow a second valve to complete the pair.

Groove, or raphe, secretes liquid that the diatom uses to attach to surfaces or glide over them

Girdle of thin silica encloses the joint between the two valves

Tubular structures project around the edge of this upper valve

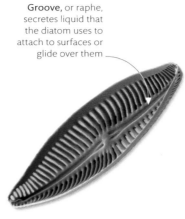

PENNATE HYPOTHECA
Navicula sp.

COMPLETE CENTRIC DIATOM
Thalassiosira sp.

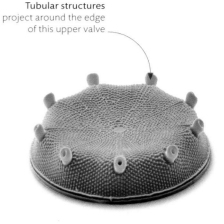

CENTRIC EPITHECA
Aulacodiscus oreganus

Pennate diatoms
These diatoms have a bilateral symmetry—the left side is a mirror image of the right. Pennate diatoms are generally elongated along one axis. They can be boat shaped or shaped like rods or needles. Pennate diatoms typically do not float well, so they tend to be found among sediments.

Boat-shaped diatoms are broader in the midsection

Rod-shaped diatoms glued together as a stack to form a colony

Rows of openings called striae allow nutrients into the cell and waste to be ejected

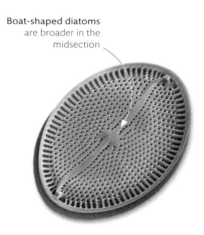

BOAT-SHAPED
Navicula sp.

COLONIAL
Achnanthidium sp.

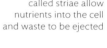
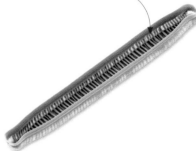

NEEDLE-SHAPED
Synedra sp.

Centric diatoms
Diatoms that survive in the water column as drifting plankton tend to have a radial symmetry, and are known as centric diatoms. This type of shape increases the cell's surface area relative to its mass, which helps photosynthesizing diatoms stay afloat in the sunlit zone near the water's surface.

Concave sides separate points of different sizes

Upper surface of centric diatoms often has a disklike appearance

Surface pores (striae) release mucuslike liquid that help it to link to nearby cells

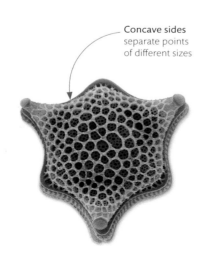

HEXAGONAL
Pseudictyota dubium

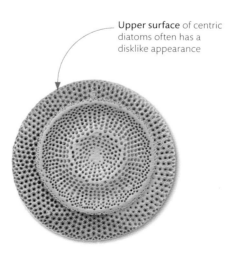

DISKLIKE
Coscinodiscus sp.

TRIANGULAR
Triceratium sp.

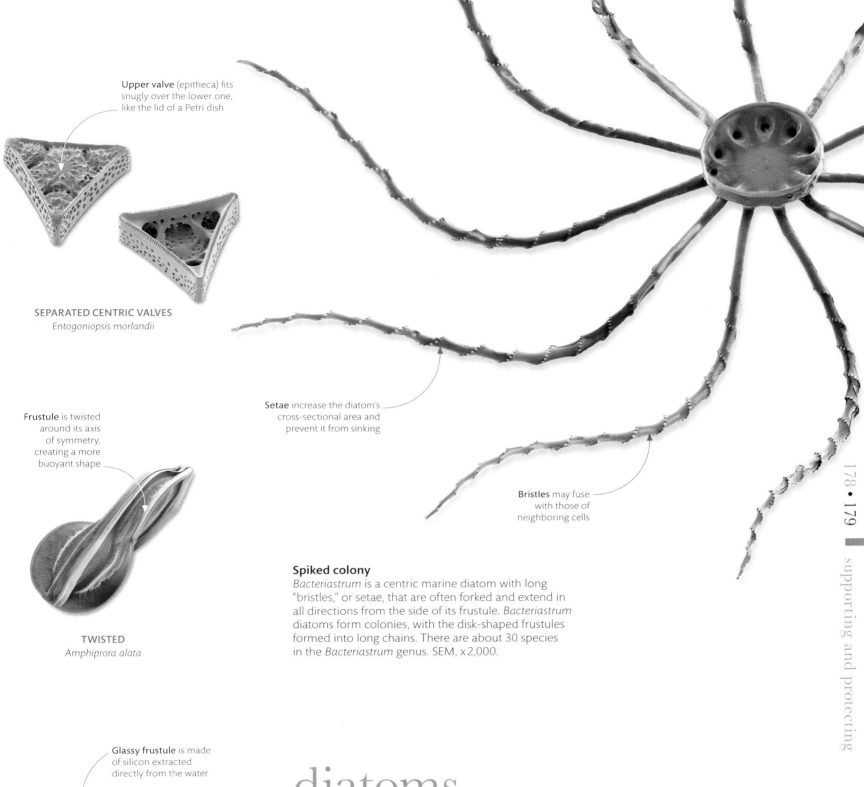

Upper valve (epitheca) fits snugly over the lower one, like the lid of a Petri dish

SEPARATED CENTRIC VALVES
Entogoniopsis morlandii

Frustule is twisted around its axis of symmetry, creating a more buoyant shape

TWISTED
Amphiprora alata

Setae increase the diatom's cross-sectional area and prevent it from sinking

Bristles may fuse with those of neighboring cells

Spiked colony
Bacteriastrum is a centric marine diatom with long "bristles," or setae, that are often forked and extend in all directions from the side of its frustule. *Bacteriastrum* diatoms form colonies, with the disk-shaped frustules formed into long chains. There are about 30 species in the *Bacteriastrum* genus. SEM, ×2,000.

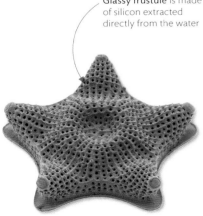

Glassy frustule is made of silicon extracted directly from the water

STAR-SHAPED
Triceratium sp.

diatoms

A major class of single-celled algae, diatoms mostly live as floating plankton in marine and freshwater habitats, although they are also found in soil. It is estimated that the photosynthesis of Earth's diatoms generates between 20 and 50 percent of the planet's free oxygen. Diatoms are classified according to the shape of their intricate cell wall, or frustule, which is made of silica and ranges from 2 to 200 micrometers in length. If conditions are ideal, a diatom may live for 6 days, reproducing by binary fission (splitting in two) every 24 hours. In some species, the algae link up to form colonies.

Diverse shells
The shells, or tests, of forams come in a variety of shapes according to species, from the simplest spheres to complex, multichambered coils. There are nearly 9,000 living species of forams, but over 40,000 more species are known only as fossils.

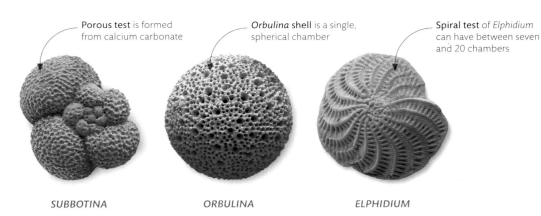

Porous test is formed from calcium carbonate

Orbulina shell is a single, spherical chamber

Spiral test of Elphidium can have between seven and 20 chambers

SUBBOTINA ORBULINA ELPHIDIUM

microscopic shells

Many single-celled organisms, such as algae and bacteria, are completely sealed within rigid protective walls. However, cell walls restrict mobility, and many microbes that have to change shape to get food inhabit tiny self-made shells instead. Marine protozoa called foraminiferans (or "forams") stretch out pseudopods—threads of cytoplasm, a thin jelly inside the cell (see pp.132–33)—to move, and to gather food and materials. Some forams use calcium from sea water to build chalky shells, while others use silica sand. In all cases, holes left in the shell allow the foram to continue extending its pseudopods.

FROM MICROBE TO ROCK

The shells of trillions of ocean forams, deposited on the seafloor over millions of years, become compacted and cemented together to form sedimentary rock such as limestone. Because different species lived at different times, foram fossils can be used to help date the rocks in which they are found.

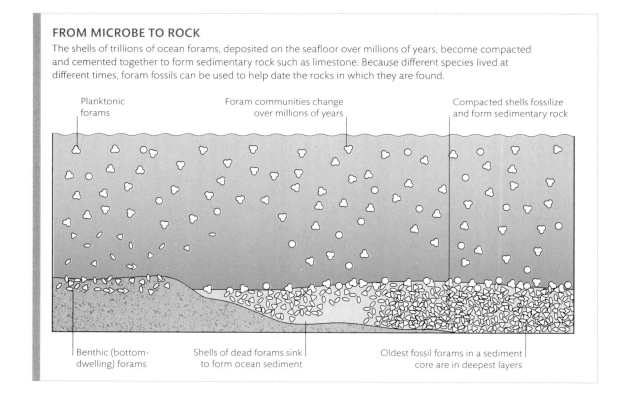

Planktonic forams

Foram communities change over millions of years

Compacted shells fossilize and form sedimentary rock

Benthic (bottom-dwelling) forams

Shells of dead forams sink to form ocean sediment

Oldest fossil forams in a sediment core are in deepest layers

Sculptured shell
The shell of a *Favulina* foram is supported by rigid cross-struts that make up a mosaic of geometric shapes. In living forams, feeding threads of thin jellylike cytoplasm extending from the organism inside the shell project through an opening in the side of the structure. Additional shell-making threads intermesh to form a granular layer, called ectoplasm, over the surface. SEM, x9,600.

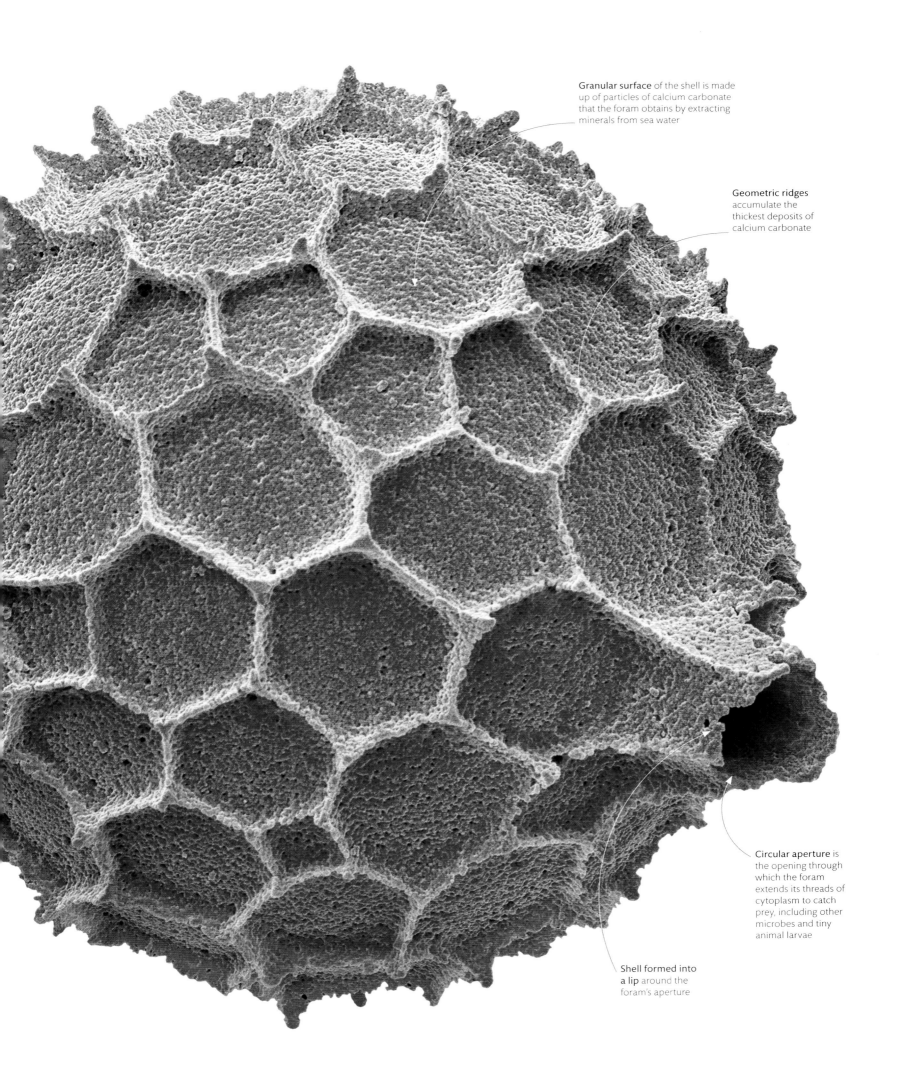

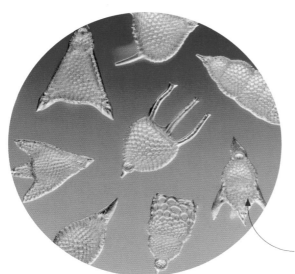

Prehistoric plankton
The shells of many types of radiolarian have been found preserved as fossils, such as these fossils seen in a light micrograph. The oldest date back half a billion years to the time when the first complex animal life was evolving in the oceans.

Different shell shapes of fossil radiolarians are used to identify their species

silica skeletons

The rich, drifting plankton of the open ocean teems with single-celled organisms. Although no more than a cell, many—including radiolarians—are far from simple. Spines radiate from the delicate internal skeleton of these tiny starlike organisms. Anchored to a cage in the center of the cell, the spines are made from glassy silica or other minerals extracted from sea water, while between them, sticky threads called axopods trap even smaller prey organisms. In warm tropical waters, the spines also help buoy the organisms upward to float near the surface.

RADIOLARIAN STRUCTURE
As well as having a hard mineral skeleton to give it shape, a radiolarian is supported by protein and oil. The protein forms an internal capsule that encloses most vital structures, such as the nucleus. Low-density oil droplets in vacuoles around the periphery of the cell form a spongy layer that aids buoyancy. In turbulent conditions, the vacuoles lose their oily contents, and the radiolarian sinks to safer depths.

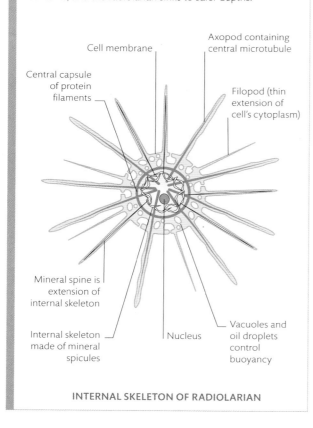

- Cell membrane
- Axopod containing central microtubule
- Central capsule of protein filaments
- Filopod (thin extension of cell's cytoplasm)
- Mineral spine is extension of internal skeleton
- Internal skeleton made of mineral spicules
- Nucleus
- Vacuoles and oil droplets control buoyancy

INTERNAL SKELETON OF RADIOLARIAN

Diversity of micro-skeletons
The chalky shells (see pp.180–81) of many organisms corrode in the sea, which becomes acidic as carbon dioxide dissolves in the seawater, but silica skeletons are more resistant to such effects. Consequently, silica skeletons accumulate in abundance—sometimes exquisitely preserved—in the ooze on the ocean floor. There are more than 1,000 species of living radiolarian—their varied forms evident in this illustration from Ernst Haeckel's *Die Radiolarien*, 1862—and many times more are known as fossils.

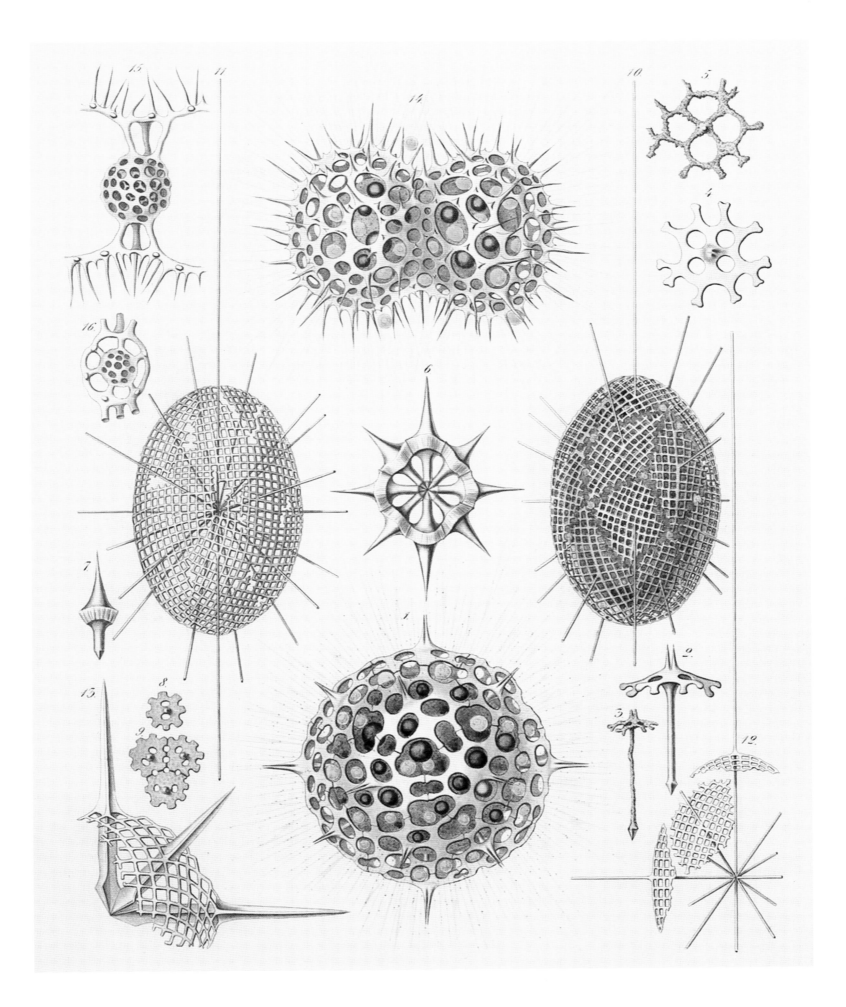

Complex shell
The theca of an armored dinoflagellate has two major plate regions—the epitheca and the hypotheca—that are divided by a furrow called the cingulum. Each of these large plate regions can be composed of up to 100 individual plates. This complex shell wall protects the dinoflagellate as it moves through the water in a corkscrew motion. This specimen (right) is shown swimming with its upper region pointing down.

Two distinct horns protrude from posterior end, which may help to stop cell sinking

Hypotheca plate region protects lower part of cell

Transverse furrow (cingulum), houses a flagellum that can maneuver the cell

Epitheca plate region protects upper part of cell

Anterior (apex) of *Protoperidinium* has a single horn

cellulose armor

Dinoflagellates are unicellular plankton that are so successful that blooms in their population can cause tidal waters to become discolored in a phenomenon called "red tides." They have two flagella—a longitudinal flagellum and a transverse flagellum—that propel the body through water in a distinctive whirling motion. The shapes of dinoflagellates are maintained by a firm layer, or pellicle, underlying their cell membrane. It is composed of tightly packed flattened sacs, filled with fluid. In many dinoflagellates these sacs contain tough cellulose—the same fibre in plant cell walls—to harden this layer into armorlike plates.

DINOFLAGELLATE STRUCTURES
A stiff outer pellicle typically divides a dinoflagellate into two parts: an upper episome and a lower hyposome, with the longitudinal flagellum carried in the latter, and the transverse flagellum in the groove, or cingulum, between these parts. In armored, or testate, dinoflagellates, these parts are supported by plates called epitheca and hypotheca.

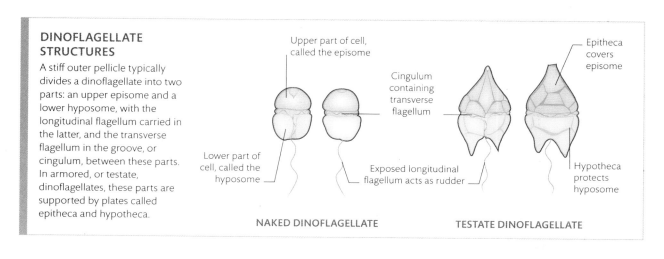

Upper part of cell, called the episome

Cingulum containing transverse flagellum

Epitheca covers episome

Lower part of cell, called the hyposome

Exposed longitudinal flagellum acts as rudder

Hypotheca protects hyposome

NAKED DINOFLAGELLATE   TESTATE DINOFLAGELLATE

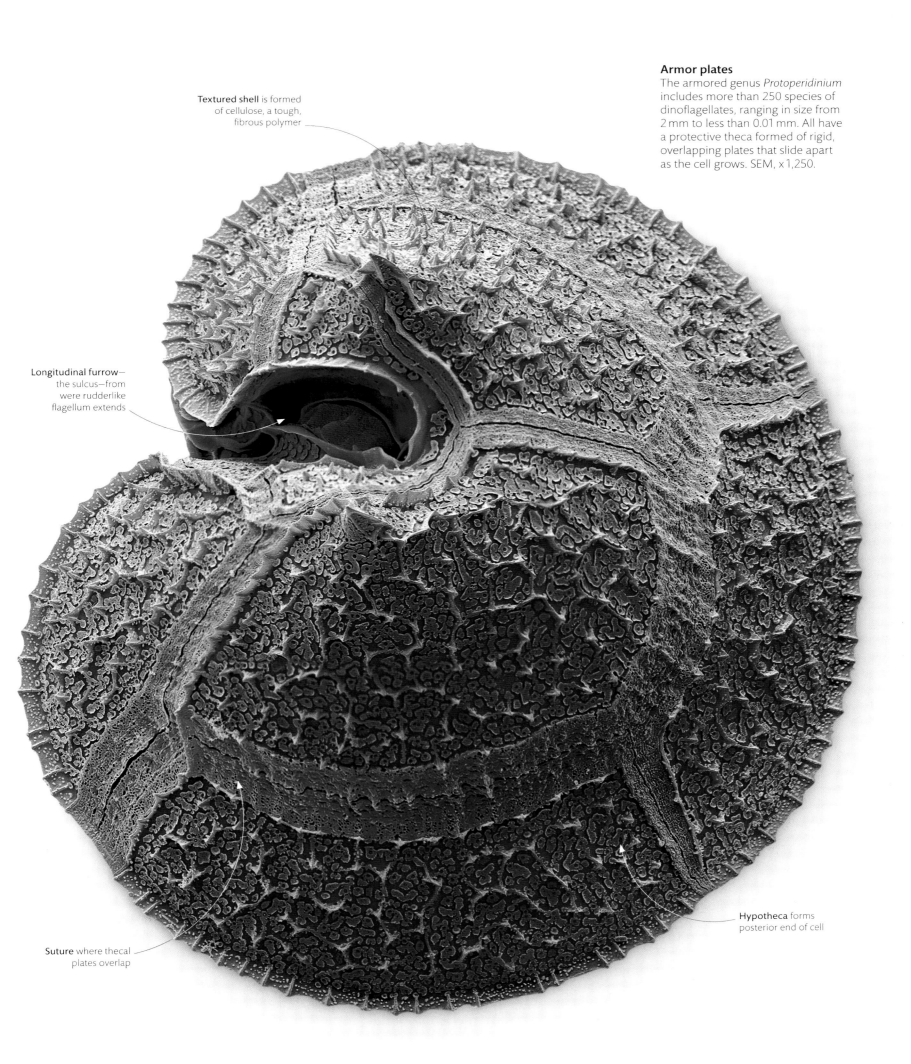

Armor plates
The armored genus *Protoperidinium* includes more than 250 species of dinoflagellates, ranging in size from 2 mm to less than 0.01 mm. All have a protective theca formed of rigid, overlapping plates that slide apart as the cell grows. SEM, x 1,250.

Textured shell is formed of cellulose, a tough, fibrous polymer

Longitudinal furrow—the sulcus—from were rudderlike flagellum extends

Suture where thecal plates overlap

Hypotheca forms posterior end of cell

Megascleres

The largest spicules, called megascleres, form the main structural units of a sponge skeleton. They are often constructed from pointed, spiky elements and range in size from 60 micrometers to 1/12 in (2 mm). Megascleres are classified by the number of axes they have branching from their center, rather than the number of tips. Monaxon spicules are simple cylinders with only one axis, while triaxons and tetraxons have three and four axes respectively. These spicules form the larger framework of the skeleton.

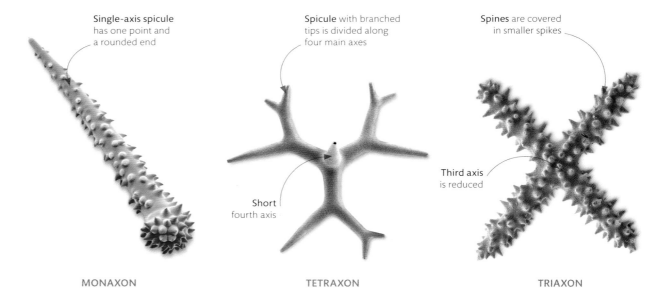

Single-axis spicule has one point and a rounded end

Spicule with branched tips is divided along four main axes

Spines are covered in smaller spikes

Short fourth axis

Third axis is reduced

MONAXON

TETRAXON

TRIAXON

Microscleres

The smallest spicules, called microscleres, are less than 60 micrometers long and are often rounded. Like megascleres, they are also classified by the number of axes they have. Athough they provide the sponge with very little structural integrity, they can act as protective armor. Microscleres are mixed in among megascleres, which are located in the mesohyl—a gel-filled layer between the sponge's inner and outer tissues. Megascleres can also be curved, especially when spicules with a single axis have become bent.

Sigma microscleres can curve into circles, "C" shapes, or "S" shapes

Small starlike spicule has short, rounded spines

Microsclere forms a curved bow shape

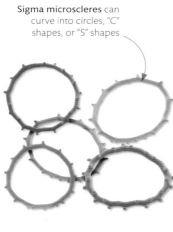

SIGMA

STRONGYLASTER

TOXA

Complex shapes

Beyond the two basic forms of spikes and rounded shapes, sponge spicules are found in a vast array of more complex shapes. These spicules display a large number of secondary details, such as shovel-shaped tips and shortened spines. Both megascleres and microscleres can exhibit stellation, where the tips are themselves divided, while in some spicules, the tips may widen into blades or form rounded knobs.

Tips are all shovel-shaped

Short spines emerge from tips and along axis

Tips of spines have become highly divided

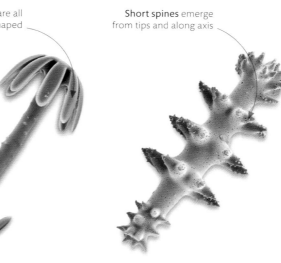

CHELAE

SANAIDASTER

STELLATE

sponge spicules

Sponges are the simplest multicelled animals—formed of a small set of cell types that build bodies in regular and irregular shapes, from chimneys to flat encrustations. These hollow bodies draw a current of water through their cores in order to filter food. The sponge body is usually supported by a rigid internal skeleton formed of hard deposits of calcium carbonate or silica, which form tiny spiked structures called spicules. These sponge spicules take many different shapes and can be used to aid the identification of sponge species.

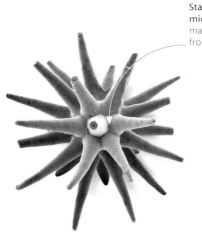

Star-shaped microsclere with many points emerging from a small center

EUASTER

Only one end has a blade or shovel shape

ANISOCHELA

Spicules are woven into a thin mesh

Gaps between spicules allow water to move in and out of sponge body

Venus's flower basket
The glass sponge *Euplectella aspergillum*, known as Venus's flower basket, has an intricate skeleton constructed from a framework of six-pointed spicules. They are constructed from silica that the sponge extracts from surrounding sea water.

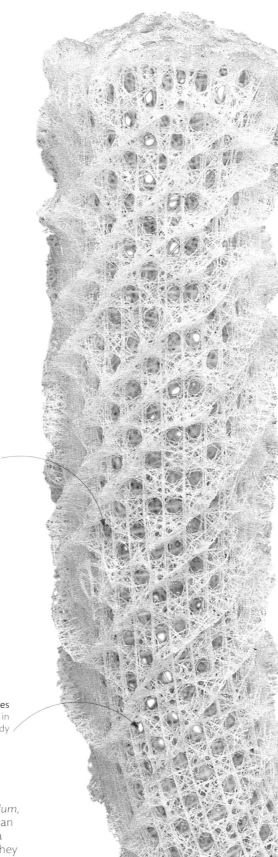

shedding water

A microhabitat moistened by rainfall can be a perilous place for a tiny animal to live. On a microscopic scale, water clings like molasses and can be a deadly trap. Springtails—tiny, flightless relatives of insects—need moisture to prevent their bodies from drying up, but breathe air directly across their skin, so can drown in just a single raindrop. Most insects channel air into their tissues through microscopic pipes called tracheae (see pp.84–85), but springtails are so small that oxygen can reach respiring cells simply by seeping through their body surface. A specially adapted surface cuticle helps to keep this surface aerated and dry.

Bumpy skin
Tiny bumps, called papillae, increase the surface area of the cuticle and are coated with microscopic nodules, which improve its water-repelling properties.

Each papilla is coated with dozens of nodules

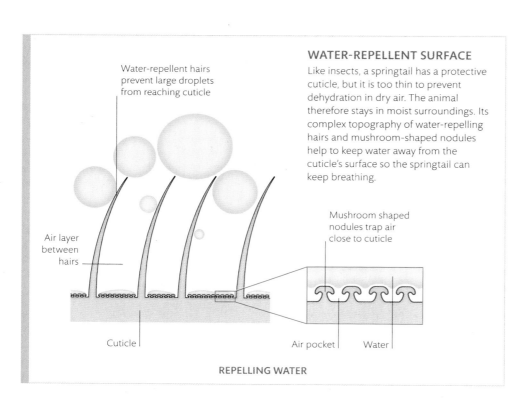

WATER-REPELLENT SURFACE
Like insects, a springtail has a protective cuticle, but it is too thin to prevent dehydration in dry air. The animal therefore stays in moist surroundings. Its complex topography of water-repelling hairs and mushroom-shaped nodules help to keep water away from the cuticle's surface so the springtail can keep breathing.

Water-repellent hairs prevent large droplets from reaching cuticle

Air layer between hairs

Cuticle

Mushroom shaped nodules trap air close to cuticle

Air pocket | Water

REPELLING WATER

A sparse covering of surface hairs called chaetae is typical of springtails that live aboveground and among leaf litter

A thin, protective cuticle, made of a hardened mixture of oil and protein, is secreted by underlying skin epidermis

Skin breather
This species of springtail (*Bilobella*) crawls rather than jumps (see pp.160–61), and must stay dry in its wet habitat of leaf litter. Its bristles and nodules prevent water from forming a continuous, suffocating film over its skin. SEM, ×560.

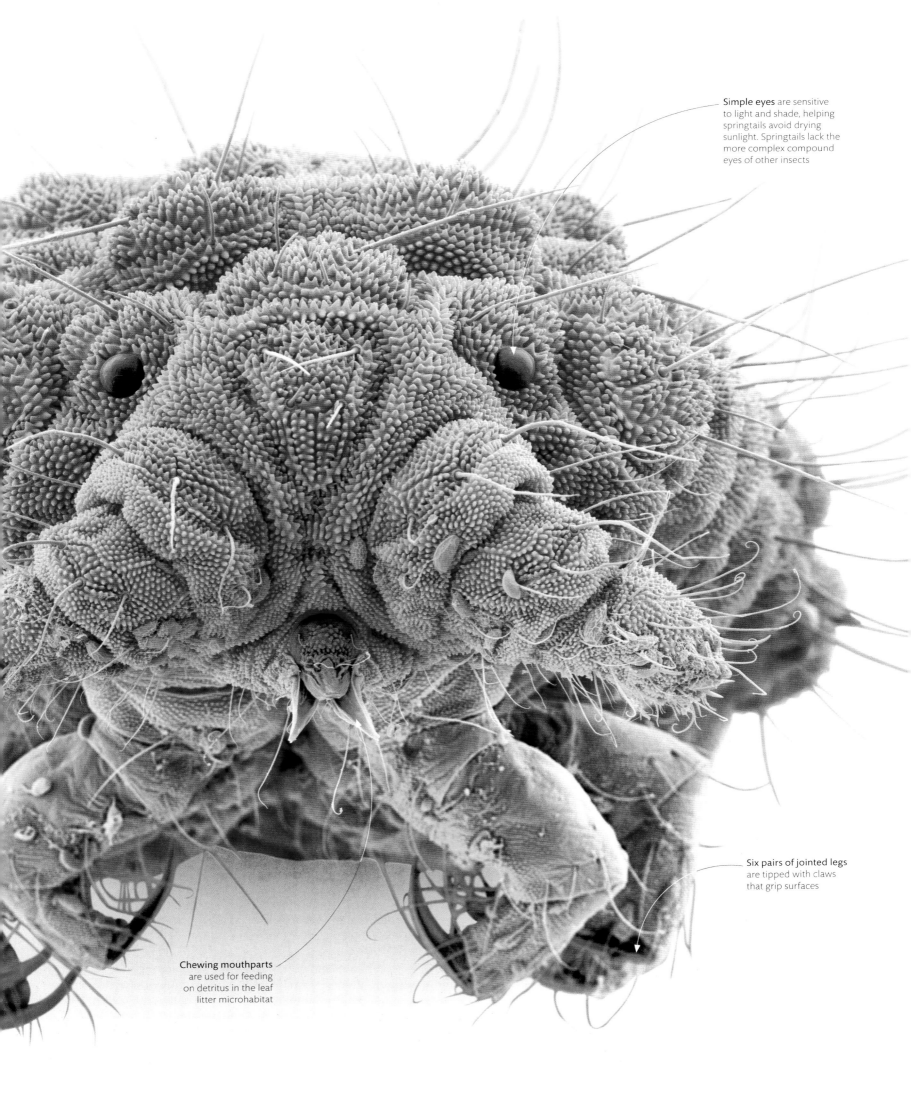

Simple eyes are sensitive to light and shade, helping springtails avoid drying sunlight. Springtails lack the more complex compound eyes of other insects

Six pairs of jointed legs are tipped with claws that grip surfaces

Chewing mouthparts are used for feeding on detritus in the leaf litter microhabitat

WATER-REPELLENT NANOSTRUCTURES

The wax-protein secretion produced by some mites and leafhoppers dries into powdery particles, or brochosomes. These intricately detailed structures form on the cuticle surface, lifting liquids clear and keeping air circulating close to the cuticle.

All-around protection

The waxy deposits on the oribatid soil mite's (*Eobrachychthonius* sp.) armorlike cuticle provide a barrier to the passage of water. This not only keeps moisture inside its desiccation-prone tiny body, but it also repels water droplets, ensuring that the surface of its cuticle is dry, so the mite can breathe. SEM, x 1,000.

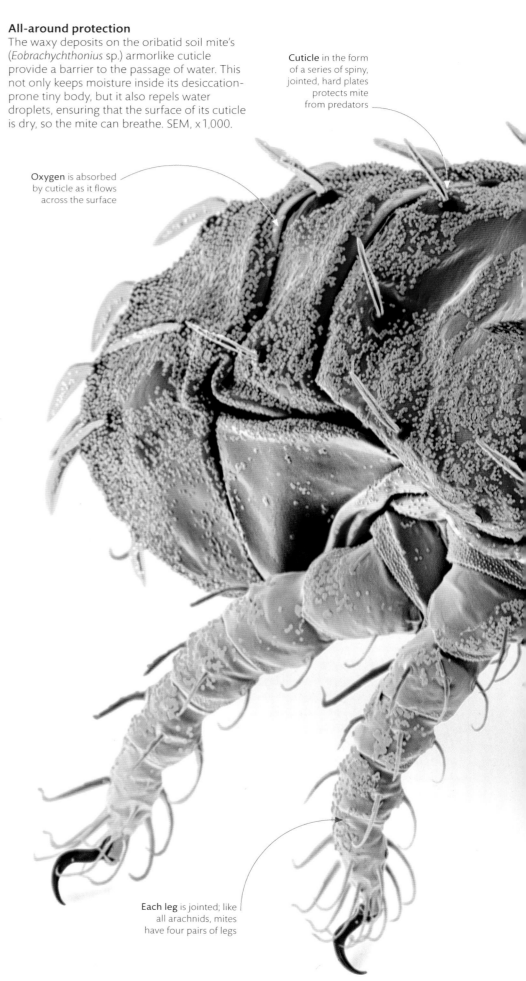

Cuticle in the form of a series of spiny, jointed, hard plates protects mite from predators

Oxygen is absorbed by cuticle as it flows across the surface

Each leg is jointed; like all arachnids, mites have four pairs of legs

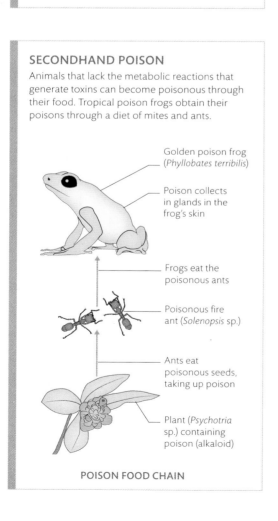

SECONDHAND POISON

Animals that lack the metabolic reactions that generate toxins can become poisonous through their food. Tropical poison frogs obtain their poisons through a diet of mites and ants.

- Golden poison frog (*Phyllobates terribilis*)
- Poison collects in glands in the frog's skin
- Frogs eat the poisonous ants
- Poisonous fire ant (*Solenopsis* sp.)
- Ants eat poisonous seeds, taking up poison
- Plant (*Psychotria* sp.) containing poison (alkaloid)

POISON FOOD CHAIN

- Water droplet
- Skin or cuticle surface

DROPLET ON UNMODIFIED SURFACE

- Brochosome particle
- Water droplet
- Layer of brochosomes repels water

DROPLET ON COATED SURFACE

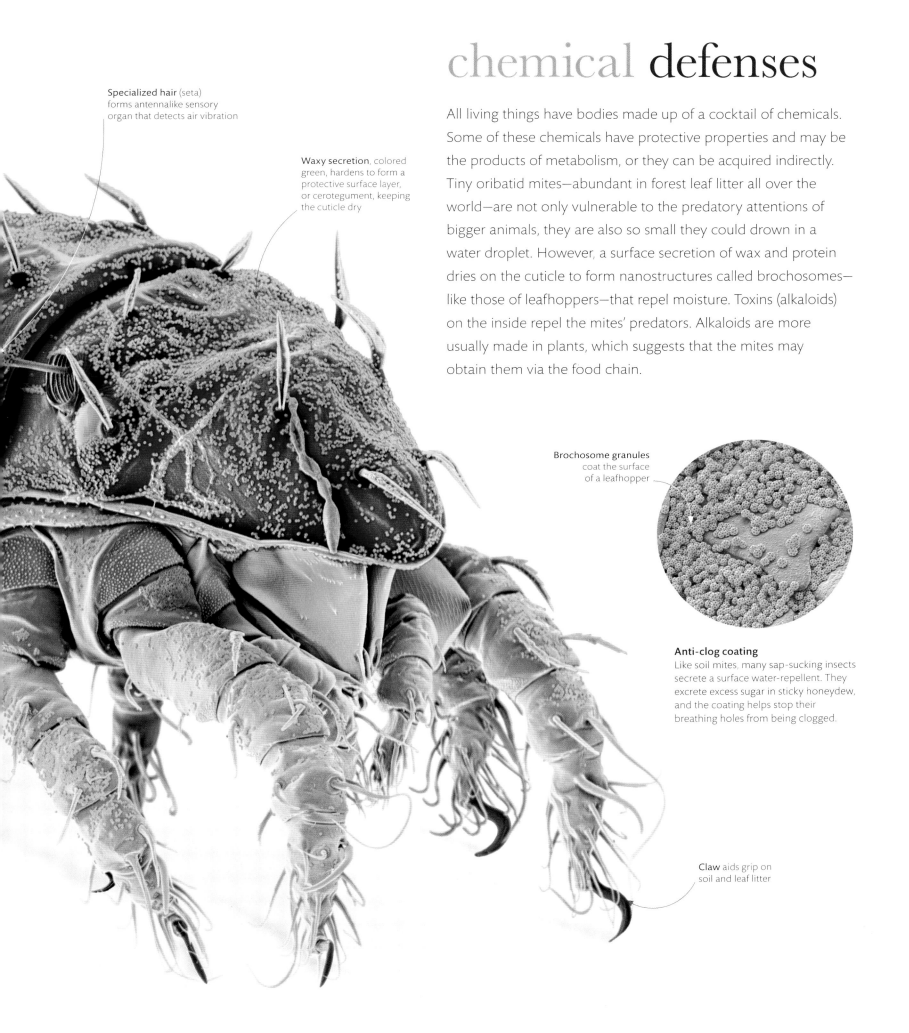

chemical defenses

All living things have bodies made up of a cocktail of chemicals. Some of these chemicals have protective properties and may be the products of metabolism, or they can be acquired indirectly. Tiny oribatid mites—abundant in forest leaf litter all over the world—are not only vulnerable to the predatory attentions of bigger animals, they are also so small they could drown in a water droplet. However, a surface secretion of wax and protein dries on the cuticle to form nanostructures called brochosomes—like those of leafhoppers—that repel moisture. Toxins (alkaloids) on the inside repel the mites' predators. Alkaloids are more usually made in plants, which suggests that the mites may obtain them via the food chain.

Specialized hair (seta) forms antennalike sensory organ that detects air vibration

Waxy secretion, colored green, hardens to form a protective surface layer, or cerotegument, keeping the cuticle dry

Brochosome granules coat the surface of a leafhopper

Anti-clog coating
Like soil mites, many sap-sucking insects secrete a surface water-repellent. They excrete excess sugar in sticky honeydew, and the coating helps stop their breathing holes from being clogged.

Claw aids grip on soil and leaf litter

Outgrowths of the cuticle develop into erect scales over the back of the body, and may help camouflage the weevil

Flexible joints of thin, unhardened cuticle allow the different parts of the exoskeleton to move relative to each other

Final armor
Invertebrates such as spiders and crabs grow bigger through successive molts, well into adulthood. But for insects like this mango seed weevil (*Sternochetus mangiferae*), most growth happens as a larva—and after their final molts to pupa and then to adult, they will grow no more.

TOP VIEW   BOTTOM VIEW

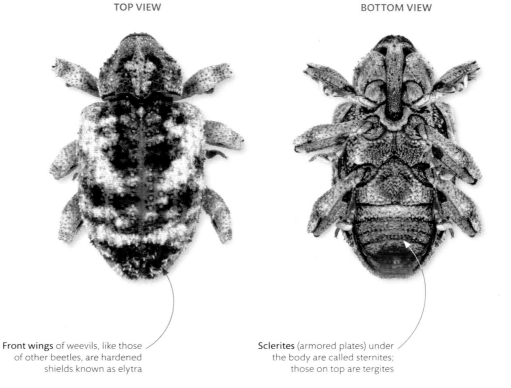

All-round protection
The exoskeleton of a weevil protects it above and below—there is no soft underbelly. Its body is covered in plates, called sclerites, while a helmetlike capsule encases the head, and tubular segments support the multi-jointed legs.

Front wings of weevils, like those of other beetles, are hardened shields known as elytra

Sclerites (armored plates) under the body are called sternites; those on top are tergites

skeleton on the outside

Many animals—including insects, spiders, and crustaceans—wear their body-supporting skeleton on the outside, like armor. Called an exoskeleton, it is made up of hard plates linked by flexible joints. Muscles beneath the exoskeleton pull on the plates to create movement. The armor is produced from substances secreted by an underlying layer of cells. Chitin, a fibrous material similar to plant cellulose, comprises the tough foundation, and is coated with wax to seal in moisture. The surface is finished with a mixture of protein and oil that hardens like biological cement. An exoskeleton is effective protection, but once formed, it cannot expand. In order to grow larger, animals supported in this way shed their armor from time to time by molting.

MULTILAYERED SKELETON
An insect's exoskeleton is a nonliving cuticle that is formed of three layers: a tough, waterproof epicuticle, a hard exocuticle, and a softer endocuticle. Below these is a layer of living cells: the epidermis and associated glands. Trichogen cells in the epidermis produce hairlike sensory setae that project from the surface of the cuticle.

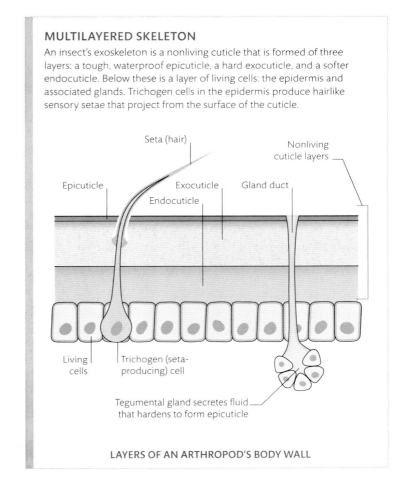

LAYERS OF AN ARTHROPOD'S BODY WALL

Spider can change between yellow and white to match the color of flower, where it has chosen to lie in wait for insects that come to feed

CRAB SPIDER
Misumena vatia

Spider sits still, disguised as a knot on a branch during the day; it will move to attend its web at nightfall

BARK SPIDER
Caerostris sp.

Cryptic predators and prey
Camouflage may simply involve the matching of color and texture to the background, as in bark spiders and crab spiders. Other animals mimic something specific in the surroundings, such as a leaf, or mask their body with objects from their environment. An orchid-mimicking flower mantis can not only hide in a flower to ambush pollinators, it may also attract them by mimicking the bloom's scent.

Carcasses of ant prey mask the bug's identity from jumping spiders, which hunt assassin bugs by sight

ANT-SNATCHING ASSASSIN BUG
Acanthaspis sp.

Praying mantis looks like an orchid and attracts pollinating insects, which it eats

ORCHID MANTIS
Hymanopus coronatus

Insect is one of many intricately camouflaged species in the katydid family

GREEN LEAF-MIMIC KATYDID
Aegimia elongata

Camouflaged forewings blend in with surroundings

When its camouflage fails, moth uses a "deimatic," or startle, display, as its wings part to reveal eyespots, or ocelli

IO MOTH
Automeris io

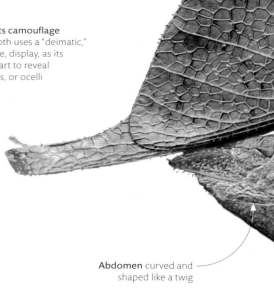

Fine ridges in insect's wing mimic veins in a leaf

Abdomen curved and shaped like a twig

staying hidden

Animals have evolved elaborate means of escaping detection, either to help them stalk prey, or help them evade predators. Crypsis, the ability to avoid detection, can involve adopting a nocturnal or undercover lifestyle, camouflage, or any combination of the three. Some animals resemble their surroundings; while others – including some assassin bugs and decorator crabs – construct their own camouflage with objects they find. In a related strategy called mimicry, some animals escape attention by looking like other life forms or objects. While defenseless prey have evolved to look hazardous; dangerous predators have evolved to look harmless.

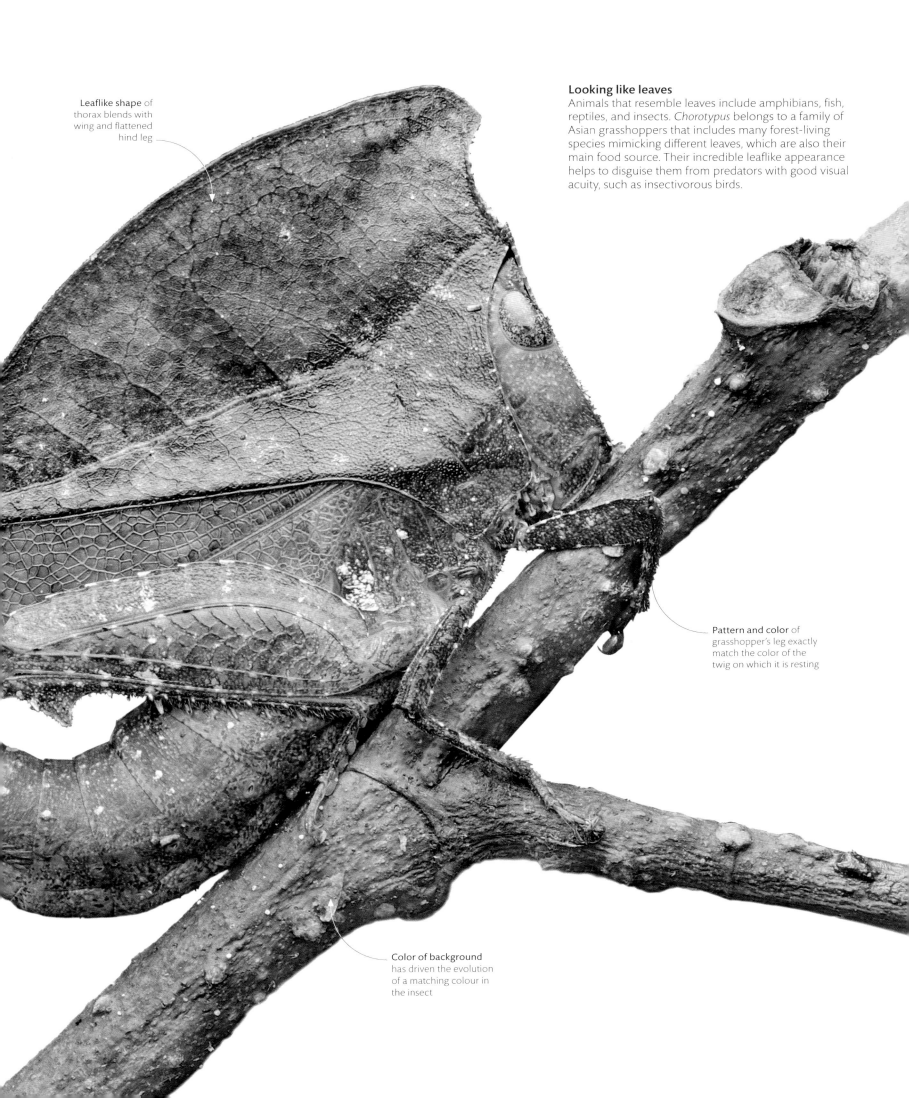

Leaflike shape of thorax blends with wing and flattened hind leg

Looking like leaves
Animals that resemble leaves include amphibians, fish, reptiles, and insects. *Chorotypus* belongs to a family of Asian grasshoppers that includes many forest-living species mimicking different leaves, which are also their main food source. Their incredible leaflike appearance helps to disguise them from predators with good visual acuity, such as insectivorous birds.

Pattern and color of grasshopper's leg exactly match the color of the twig on which it is resting

Color of background has driven the evolution of a matching colour in the insect

PARTS OF THE SKELETON

Echinoderm ossicles develop in the dermis—a thick layer just below the skin surface. As they grow, the ossicles project out of the dermis but remain covered by a continuous layer of epidermis. Some ossicles elaborate into defensive "pedicellariae" structures, which end in miniature snapping, hinged jaws.

CROSS SECTION THROUGH URCHIN BODY WALL

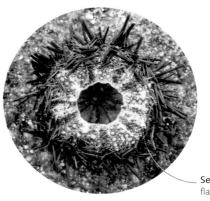

Skeletons on shore
Echinoderm skeletons generally disintegrate rapidly after death, leaving numerous microscopic calcite plates. Spines and plates can sometimes be found on the shore, occasionally still attached to an intact—but easily broken—test.

Sea urchin tests have fused, flattened ossicles, with many long spines attached

Protective spines
The bumpy appearance of a common sunstar's (*Crossaster papposus*) skin, right, is caused by underlying, unevenly sized ossicles called paxillae. Each paxillum is column-shaped and crowned with a cluster of slender spines. Round-ended papulae—thin-walled extensions of the body cavity—fill the space between and act as gills by releasing waste and absorbing oxygen. LM, x14.

echinoderm skeleton

Echinoderms—including sea urchins, sea cucumbers, and starfish—have an internal skeleton, called an endoskeleton. The endoskeleton develops in the thick dermis layer of the skin, and consists of small hard elements, or ossicles, constructed from microcrystals of calcite. Ossicles come in a variety of forms, including plates, spines, rods, and other specialized structures. In sea urchins, plates fuse to form a rigid, protective test, whereas in starfish they are movable and less tightly packed, providing flexibility.

MACROLIFE

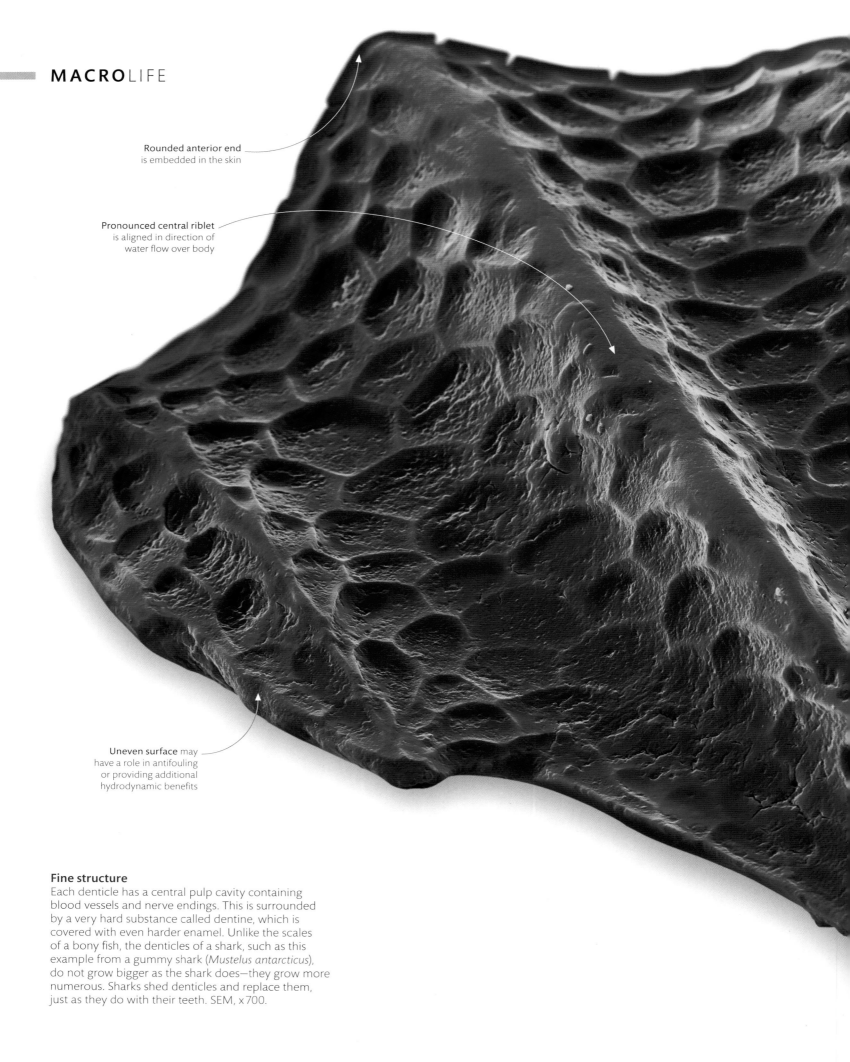

Rounded anterior end is embedded in the skin

Pronounced central riblet is aligned in direction of water flow over body

Uneven surface may have a role in antifouling or providing additional hydrodynamic benefits

Fine structure
Each denticle has a central pulp cavity containing blood vessels and nerve endings. This is surrounded by a very hard substance called dentine, which is covered with even harder enamel. Unlike the scales of a bony fish, the denticles of a shark, such as this example from a gummy shark (*Mustelus antarcticus*), do not grow bigger as the shark does—they grow more numerous. Sharks shed denticles and replace them, just as they do with their teeth. SEM, x700.

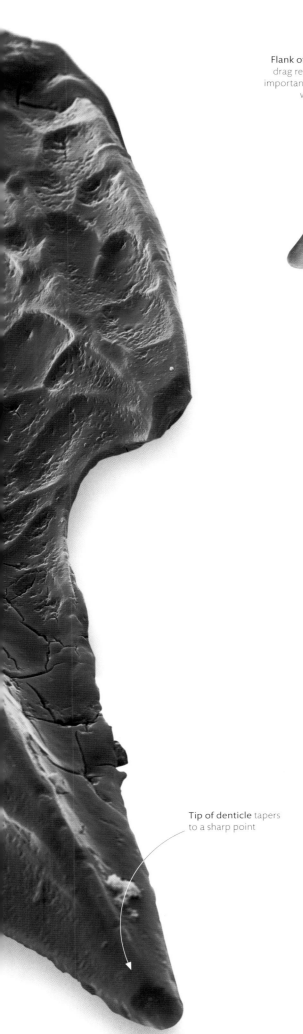

Tip of denticle tapers to a sharp point

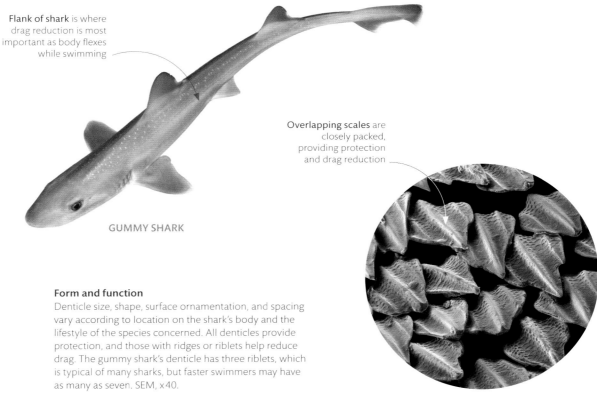

Flank of shark is where drag reduction is most important as body flexes while swimming

Overlapping scales are closely packed, providing protection and drag reduction

GUMMY SHARK

Form and function
Denticle size, shape, surface ornamentation, and spacing vary according to location on the shark's body and the lifestyle of the species concerned. All denticles provide protection, and those with ridges or riblets help reduce drag. The gummy shark's denticle has three riblets, which is typical of many sharks, but faster swimmers may have as many as seven. SEM, x40.

shark skin

Tiny, toothlike scales called denticles, embedded in shark skin, play varied and vital roles. Defensive denticles with upward-pointing cusps deter predators and help prevent settlement of fouling organisms, which can impede swimming efficiency. Thick denticles found on sharks living close to the sea bed protect the body against abrasion from rocks. Ridged denticles found on all species help control water flow, reducing drag and increasing hydrodynamic efficiency. Some deep-dwelling sharks even have concave-shaped denticles that focus light generated by bioluminescent organs.

DENTICLE BRISTLING

Denticles on some fast-swimming sharks bristle upward to angles of 50 degrees or more, resulting in increased turbulence. This helps the shark by reducing the drag associated with laminar flow (flow of liquid in smooth, undisrupted layers). The denticles lift, maintaining a thin layer of turbulent water next to the skin. This swirling layer of water reduces drag, allowing the shark to slip easily through the water.

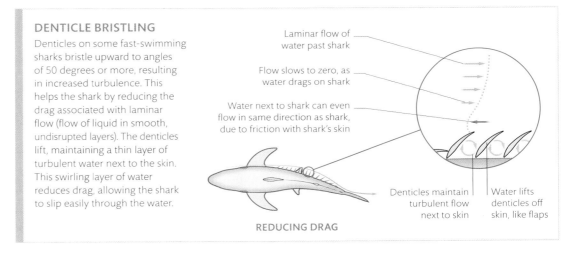

Laminar flow of water past shark

Flow slows to zero, as water drags on shark

Water next to shark can even flow in same direction as shark, due to friction with shark's skin

Denticles maintain turbulent flow next to skin

Water lifts denticles off skin, like flaps

REDUCING DRAG

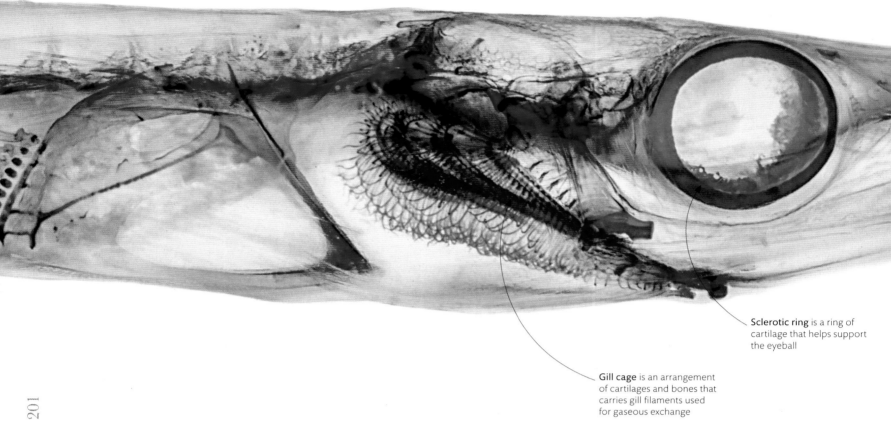

Sclerotic ring is a ring of cartilage that helps support the eyeball

Gill cage is an arrangement of cartilages and bones that carries gill filaments used for gaseous exchange

vertebrate skeletons

The skeleton of a vertebrate (backboned) animal develops inside the body and grows with the rest of the tissues, unlike the exoskeleton of insects and other arthropods, which must be molted. It is composed of stiff, rubbery cartilages and hard bones that articulate around joints worked by muscles. The earliest fish had skeletons made entirely of cartilage, and some—such as sharks—have a cartilaginous skeleton today. Most living vertebrates, however, are mainly bony. Being harder than cartilage, bone keeps its shape better and gives stronger support for living on land.

Components of the skull are drawn into a narrow snout with a small terminal mouth

Two-part skeleton

Most bones develop from cartilages formed in the embryo. However, even an adult bony vertebrate, such as this tube-snout (*Aulorhynchus flavidus*) fish, retains cartilaginous parts. Bone, here stained purple and red, reinforces the main framework of the fish's body, especially the spine. Cartilage, stained blue, has elastic properties and greater flexibility than bone.

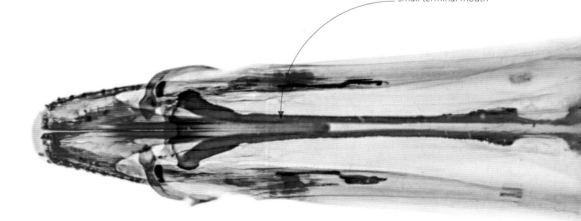

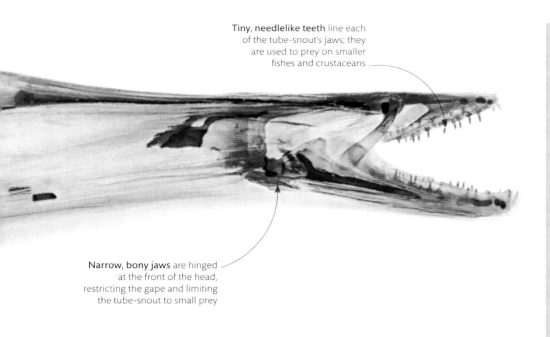

Tiny, needlelike teeth line each of the tube-snout's jaws; they are used to prey on smaller fishes and crustaceans

Narrow, bony jaws are hinged at the front of the head, restricting the gape and limiting the tube-snout to small prey

CARTILAGE AND BONE COMPARED

Cartilage and bone are both made from cells separated in a background matrix. Cartilage has a flexible protein matrix that supplies the cells with food and oxygen. In bone, the protein matrix is hardened with calcium minerals, so its cells need to build "lifelines" of cytoplasm to interact with other cells and blood vessels.

Chondrocyte (cartilage-forming cell) — Flexible protein matrix — Lacuna (fluid-filled space around chondrocyte)

CARTILAGE TISSUE

Osteocyte (bone-forming cell) — Hard mineral-protein matrix — Cytoplasm threads penetrate hard matrix

BONE TISSUE

Preserved in rock
Bone resists decomposition for longer than soft tissues, giving time for skeletons to become fossilized over many years.

Fossilized skeleton formed over time after fish was buried in sediment

Bones of the cranium (part of the skull) form a protective helmet around the brain

Spine is composed of a chain of interlocking bones called vertebrae that enclose the spinal cord and provide the main bony support for the long body

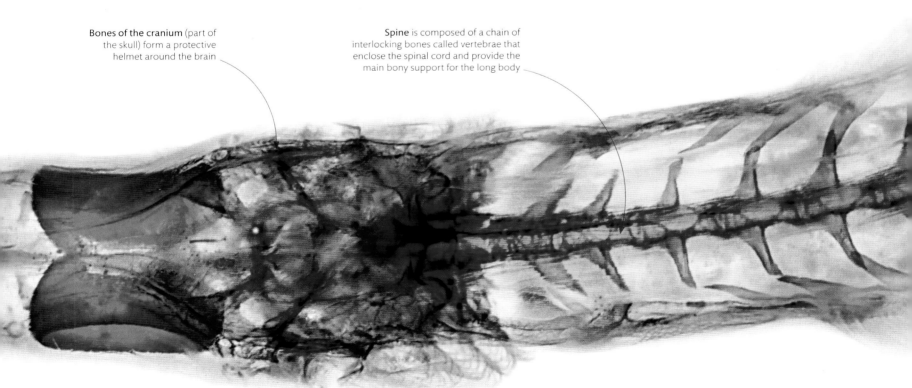

MACROLIFE

mammal hair

Hair—one of the defining mammalian features—is made of keratin, the same protein that forms fingernails and claws, stiffens feathers, and waterproofs reptilian scales. Mammal hair, in its various forms, has many uses: whiskers are touch sensors, spines and quills are defensive weapons, and nasal hairs work as air filters. Hair can also boldly declare a mammal's presence or subtly camouflage it. But the most basic function of hair is as insulation: a layer of short hairs can trap air near the skin and reduce heat loss. Wet hair cannot do this as effectively, so a degree of waterproofing is achieved by longer, oiled guard hairs that cover the insulating hairs beneath.

HAIR ANATOMY

Beneath a hair's transparent outer layer, or cuticle, is the cortex—a lattice of dead cells composed primarily of the protein keratin. Microscopic keratin filaments coil together to form bigger microfibril strands, which in turn bundle together to form macrofibrils—the main structure of cortex cells.

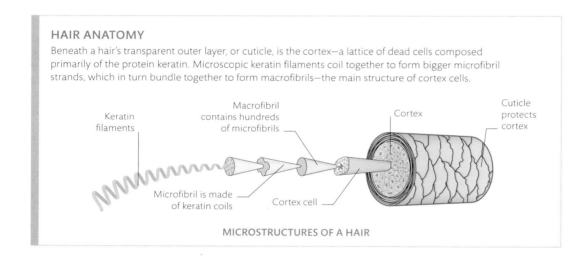

MICROSTRUCTURES OF A HAIR

THERMOREGULATION IN HUMANS

Even in relatively hairless mammals, such as humans, body hair is used to maintain body temperature. When skin is warm, tiny muscles attached to the hairs relax and the hairs flop over, allowing the skin to be cooled by sweat. In cold conditions, the muscles pull hairs erect to trap a thicker layer of air against the body, though it is not effective in humans.

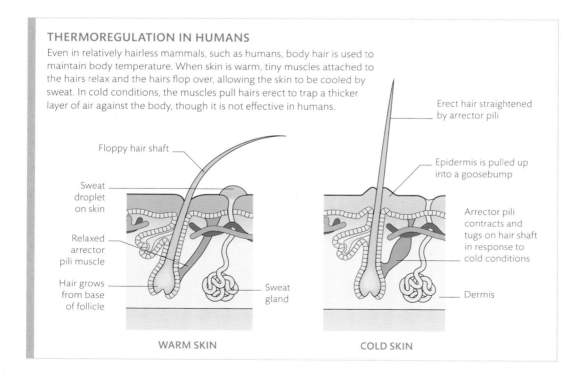

WARM SKIN | COLD SKIN

Cuticle cells overlap like roof tiles, protecting the cortex layer beneath

Human hair shafts
Hair cells develop at the bottom of pits, called follicles, in the skin's epidermis. The follicles extend down into the dermis layer beneath. As hair cells divide and multiply they fill with the protein keratin (keratinization) and die. This process gradually builds a hair shaft. Only the hair cells in the follicles are alive. SEM, x470.

Top layers of epidermis (the skin's surface layer) are made of dead cells, like the hair shaft

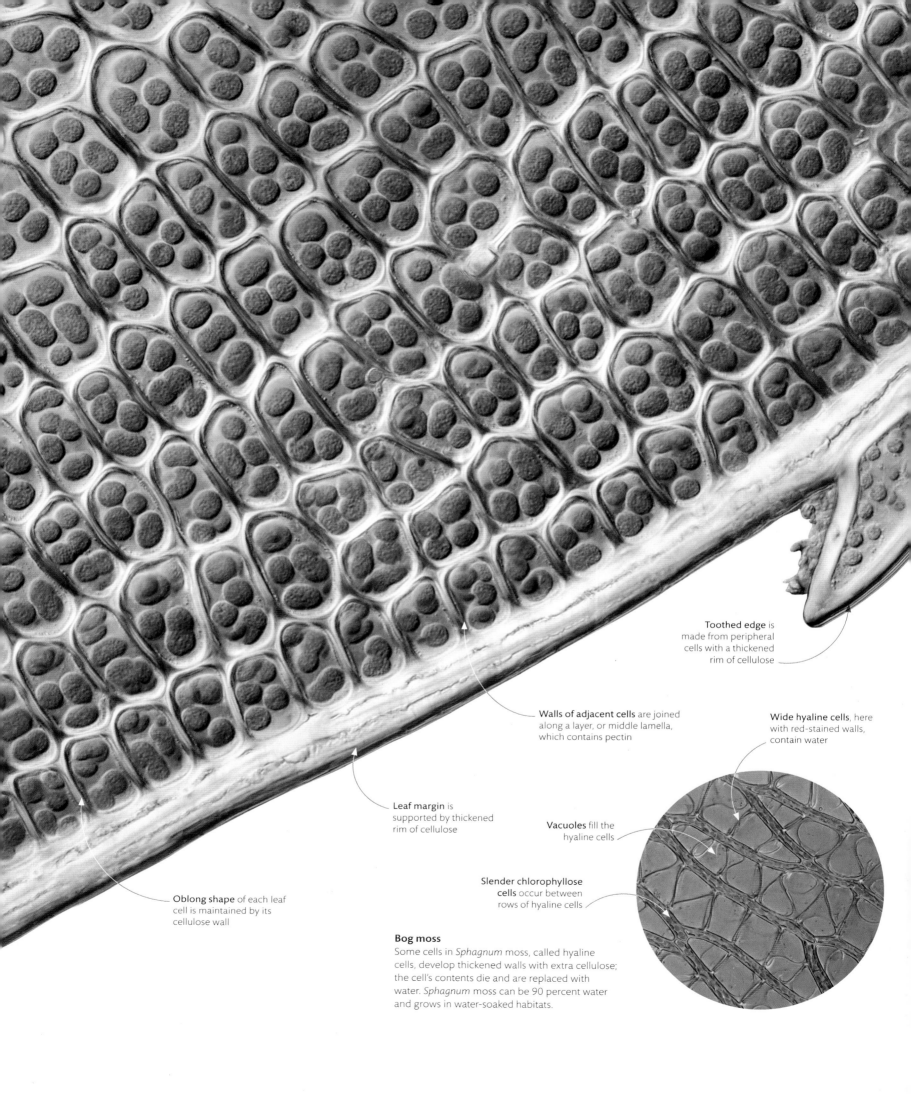

Toothed edge is made from peripheral cells with a thickened rim of cellulose

Walls of adjacent cells are joined along a layer, or middle lamella, which contains pectin

Wide hyaline cells, here with red-stained walls, contain water

Leaf margin is supported by thickened rim of cellulose

Vacuoles fill the hyaline cells

Slender chlorophyllose cells occur between rows of hyaline cells

Oblong shape of each leaf cell is maintained by its cellulose wall

Bog moss

Some cells in *Sphagnum* moss, called hyaline cells, develop thickened walls with extra cellulose; the cell's contents die and are replaced with water. *Sphagnum* moss can be 90 percent water and grows in water-soaked habitats.

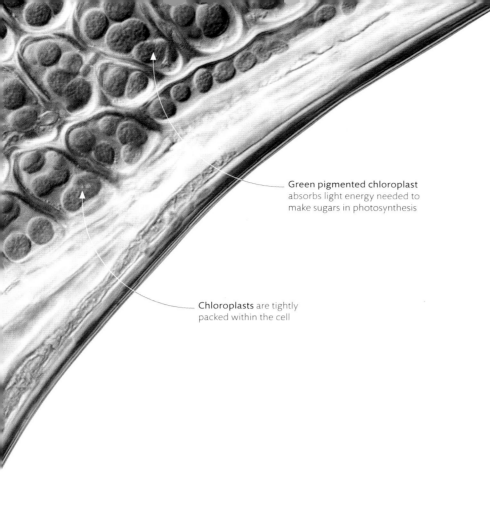

Green pigmented chloroplast absorbs light energy needed to make sugars in photosynthesis

Chloroplasts are tightly packed within the cell

Cell walls in moss
A moss leaf is very thin—most of its blade is just one cell thick—and it lacks the waxy cuticles of other plants. Each cell, packed with green chloroplasts that carry out photosynthesis, is glued tightly to its neighbors by pectin—the material that makes jelly set. Cells can exchange vital substances with their surroundings through their permeable cellulose walls. LM, x 1,000.

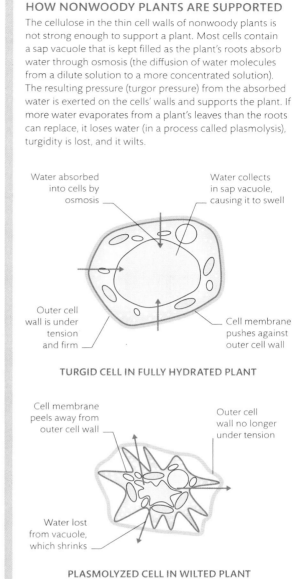

HOW NONWOODY PLANTS ARE SUPPORTED
The cellulose in the thin cell walls of nonwoody plants is not strong enough to support a plant. Most cells contain a sap vacuole that is kept filled as the plant's roots absorb water through osmosis (the diffusion of water molecules from a dilute solution to a more concentrated solution). The resulting pressure (turgor pressure) from the absorbed water is exerted on the cells' walls and supports the plant. If more water evaporates from a plant's leaves than the roots can replace, it loses water (in a process called plasmolysis), turgidity is lost, and it wilts.

Water absorbed into cells by osmosis

Water collects in sap vacuole, causing it to swell

Outer cell wall is under tension and firm

Cell membrane pushes against outer cell wall

TURGID CELL IN FULLY HYDRATED PLANT

Cell membrane peels away from outer cell wall

Outer cell wall no longer under tension

Water lost from vacuole, which shrinks

PLASMOLYZED CELL IN WILTED PLANT

cell walls

Plant cells have well-defined walls, which makes them much easier to distinguish than the cells of animals. All plant cell walls are made from cellulose, which is the material responsible for most of the fiber in plants. The walls of some plant cells, such as those in woody tissue, are reinforced with a tough substance called lignin, which seals them so well that their contents die. The dead framework that remains provides a microscopic structure that can support the biggest and heaviest living plants—trees. Nonwoody plants have much thinner walls compared to those of trees, but contain high-pressure, water-filled vacuoles that keep the cells firm and swollen, or turgid, helping to keep the plant upright.

MACROLIFE

supporting stems

The stem and branches of a plant can creep along the ground or tower high into the air, but in all plants, they provide the supporting framework for exposing the leaves they carry to sunlight for photosynthesis. Taller stems can overshadow competitors—the main reasons trees invest so much energy in reaching upward. Strength comes from the pipes—called xylem vessels—used in transporting water up from the ground: their walls are reinforced with cellulose and woody material called lignin. Along with phloem vessels that carry food-rich sap, they are typically arranged in vascular bundles that stretch from roots to leaves.

Epidermal cells form a single outer layer and help reduce water loss

Inner cortex (brown) forms a structural layer

Stems will scramble up surfaces for support while growing

Reaching heights
Even though a clematis can grow a sturdy woody stem, it is a climber: it uses a tree to help with its support. In this way, it can reach sunlight in a woodland habitat without investing in a trunk of its own.

Hollow stems, or culms, allow bamboo to grow taller than other grasses

VASCULAR TISSUE
Xylem and phloem—the two types of vascular tissue that transport vital liquids in a plant—both consist of vessels formed by interlinked cells. Xylem, which carries water and minerals from roots to leaves, has larger vessels of varying diameter formed by dead cells that lack end walls. In phloem, which conducts food throughout the plant, the vessels are narrow and composed of living cells with sievelike, perforated end walls.

Giant grass
Grasses cannot thicken their stem by making more xylem vessels, as clematis can—but bamboo, a grass, achieves a similar effect by building woody fibers between its vascular bundles.

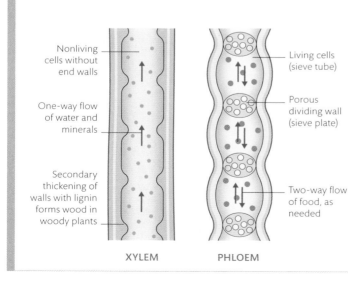

Nonliving cells without end walls

One-way flow of water and minerals

Secondary thickening of walls with lignin forms wood in woody plants

Living cells (sieve tube)

Porous dividing wall (sieve plate)

Two-way flow of food, as needed

XYLEM PHLOEM

Vascular bundles
Bundles of xylem (colored blue) and phloem (yellow) vessels are arranged in a ring in this cut section of a clematis (*Clematis* sp.) stem. The growing plant will add more xylem vessels to make the stem get thicker and stronger, so that it can support more leaves.

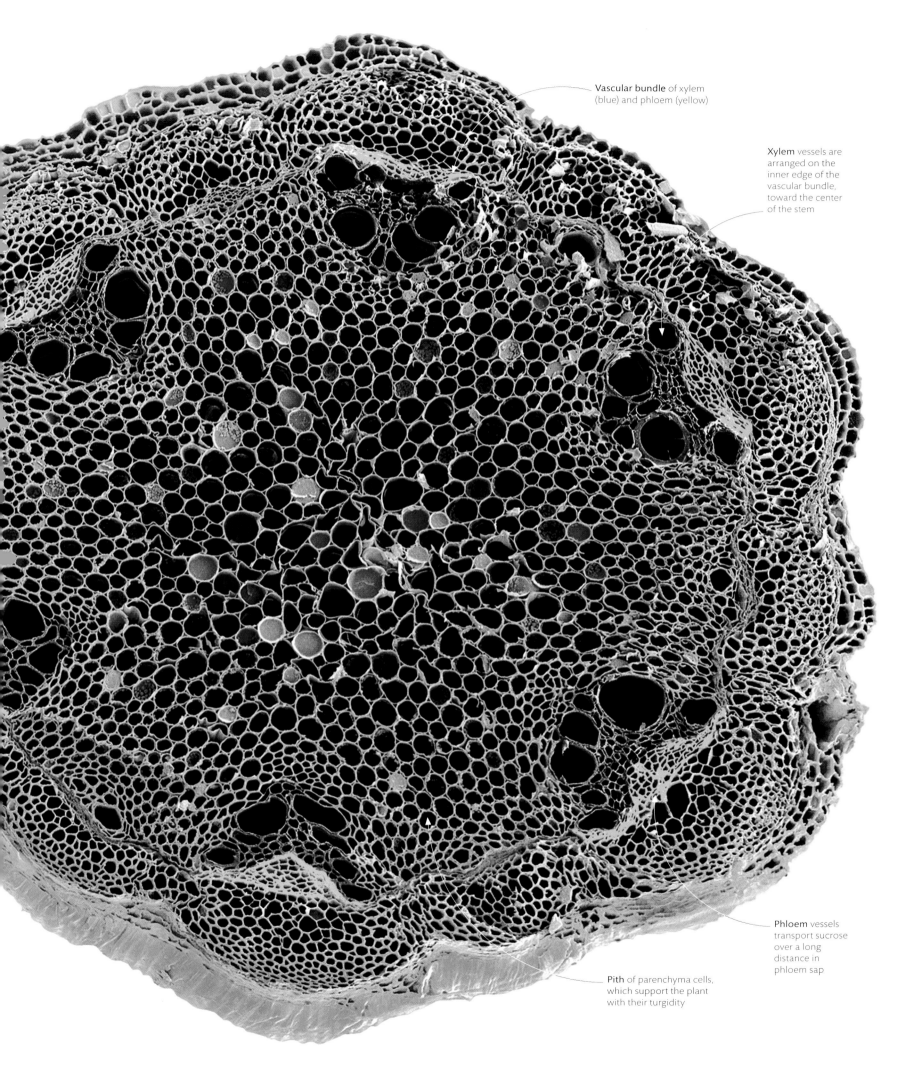

Vascular bundle of xylem (blue) and phloem (yellow)

Xylem vessels are arranged on the inner edge of the vascular bundle, toward the center of the stem

Phloem vessels transport sucrose over a long distance in phloem sap

Pith of parenchyma cells, which support the plant with their turgidity

Indumentum

The outer surface of a leaf—its indumentum—is the combination of fine "hairs" called trichomes, that cover it; leaves may have several types of trichomes. The surface from which they grow is known as the upper epidermis (such terms can also apply to other parts of the plant). There are about 10 loosely defined types of leaf indumentum with a variety of functions, from helping a plant retain moisture to detering pests.

Peltate trichomes give leaf a scaly appearance

SCALY
Air plant
Tillandsia sp.

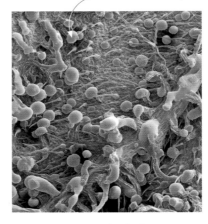

Glands exude mucilage that makes leaf sticky

MUCILAGINOUS
Cannabis
Cannabis sativa

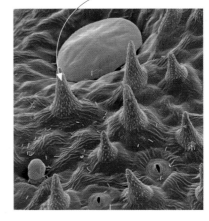

Short papilla make leaf feel textured

SCABROUS
Marjoram
Origanum majorana

Leaf hairs

The trichomes of a leaf develop into a wide range of shapes with different functions—many leaves have more than one type. Formed of a single cell in some cases, or multiple cells in others, trichomes can be defensive prickles that deter predators. Other trichomes are secretory glands, which release bitter oils or sticky gels. Some function just like animal hairs, blanketing the leaf surface against the cold and preventing frost damage. Yet more are there to gather tiny droplets of dew as a valuable source of water.

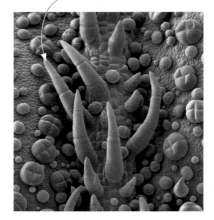

Long, unbranched hairs formed from stacks of epidermal hair cells

MULTICELLULAR
Basil
Ocimum basilicum

Bladelike structure formed from single epidermal hair cell

UNICELLULAR
Geranium
Pelargonium Crispum

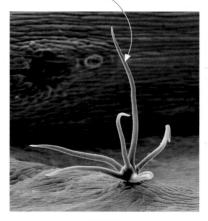

Stalkless hairs start from a single point

SESSILE
Witch hazel
Hamamelis virginiana

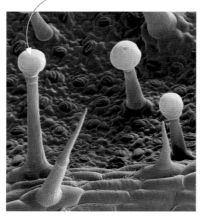

Secretory bulb forms at tip of tall hair

GLANDULAR PILATE
Lemon balm
Melissa officinalis

Mushroom-shaped structures prevent water loss

PELTATE
Olive tree
Olea europaea

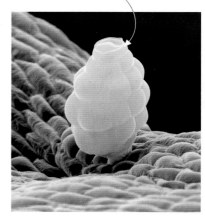

Oil gland dominates trichome structure

GLANDULAR CAPITATE
Hibiscus
Hibiscus sp.

leaf surfaces

The leaf is a plant's solar panel, capturing light and harnessing its energy by photosynthesis, and many appear flat with a wide, seemingly smooth surface. But when viewed under a microscope, the surface, or epidermis, of a leaf can be seen to have a diversity of structures—often very different on its upper and lower surfaces. Epidermal cells fall into two types. Flat cells create the leaf structure and are generally coated in a waxy cuticle that helps reduce water loss, while epidermal hairs, or trichomes, create a surface called an indumentum.

STRIGOSE
Soybean
Glycine max

Long, straight trichomes point in similar direction

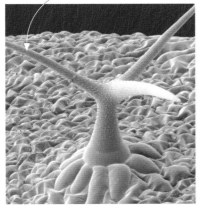

STALKED
Thale cress
Arabidopsis thaliana

Starlike hairs develop from a single tall stalk

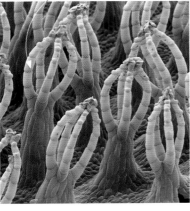

DENDRITIC
Floating fern
Salvinia natans

Treelike branches emerge from a common base

Protective spikes
This scanning electron micrograph of the first "true" leaves (see p.265) of a cannabis (*Cannabis sativa*) seedling reveals the claw-shaped trichomes that cover the surface. The leaves also develop glandular trichomes that secrete foul-tasting oils.

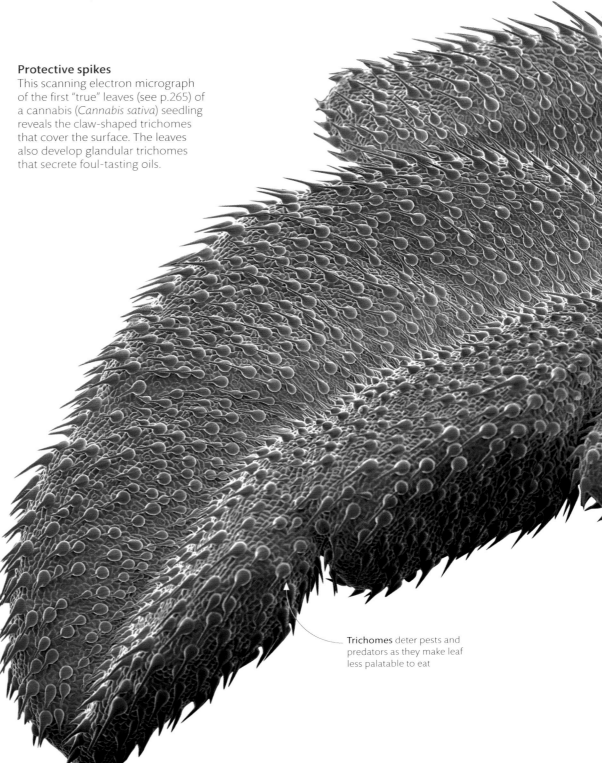

Trichomes deter pests and predators as they make leaf less palatable to eat

STINGERS COMPARED

The venom systems of related wasps, bees, and ants carry similar components: a gland that produces the venom, a sac that stores it, a lancet that delivers the toxic dose, and a Dufour's gland that makes accessory chemicals, such as pheromones. Bees famously die when their sting is used on mammalian skin, as the stinger is ripped out. But when used against a bee's usual target, arthropods, the stinger slides safely in and out without harming the bee.

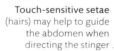

SOCIAL WASP (VESPIDAE) — Venom gland, Sting lancet (stinger)

BEE (APIDAE) — Valves help pump venom out in bees and ants

ANT (FORMICIDAE) — Venom sac, Pheromones in Dufour's gland guide wingless ants to and from their nest

Ant stinger
The visible part of an ant's stinging apparatus is a curved lancet, which is hard enough to penetrate the thickest cuticle—or even the skin of a human. The venom itself is made in a gland at the rear of the abdomen, just behind the base of the lancet.

insect stingers

A painful—even deadly—sting is an effective means of repelling enemies or disabling prey. Foremost among stinging insects are the social hymenopterans, a group that includes ants, bees, and wasps. Their stinging involves injecting venom through a sharp lancet. Since the lancet is a modified egg-laying tube, or ovipositor, all the stinging individuals are female. These insects live in complex societies, typically made up of stinging female workers that defend the nest, stingless reproductive males, and a breeding queen who lays her eggs via an opening at the base of her stinger.

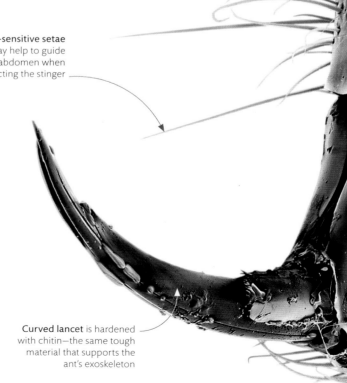

Touch-sensitive setae (hairs) may help to guide the abdomen when directing the stinger

Curved lancet is hardened with chitin—the same tough material that supports the ant's exoskeleton

Dual-action weaponry
A specialist in hunting springtails, the trap-jawed ant (*Acanthognathus teledectus*) uses its long jaws to grab a victim, then its sting to make the kill.

Abdomen coils under the thorax so that the ant can jab forward with its stinger

irritating hairs

Vulnerable animals can employ novel defense mechanisms to ward off predators. Some types of tarantulas and caterpillars have detachable hairs—modified sensory structures called setae (see pp.154–55)—with needlelike tips or microscopic barbs. Called urticating hairs, they embed themselves in the skin, eyes, or airways. The irritation they cause is enough to repel large predators and may even kill smaller ones. Among tarantulas, only American species—about 90 percent of them—possess urticating hairs. Many rub their bodies with their legs to flick a cloud of weaponized hairs in an enemy's face.

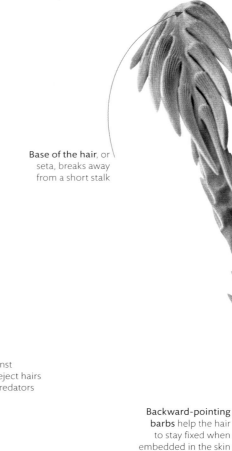

Base of the hair, or seta, breaks away from a short stalk

Backward-pointing barbs help the hair to stay fixed when embedded in the skin

Legs rub against abdomen to eject hairs at potential predators

Colorful source
The Mexican red-knee tarantula (*Brachypelma smithi*) is one of many tropical tarantulas that use urticating hairs for defense.

Hairs up close
A tarantula's urticating hairs are modified to break off in a narrow zone of weakness near their base. They have forward- or backward-facing spines that get caught in the flesh of a predator, making them difficult to dislodge. These hairs are usually produced in patches on the tarantula's body—particularly on the sides and back of the abdomen. SEM, x2,300.

HAIR TYPES IN TARANTULAS

Non-urticating hairs stay rooted in the body, but tarantulas produce at least six kinds of detachable urticating hairs. It is thought that each type has evolved to achieve a different kind of defense, although the details are not clear. Type I hairs, for instance, are built into egg sacs and may help repel parasitic flies or marauding ants, while type III—with longer barbs—may be more effective against larger vertebrate predators.

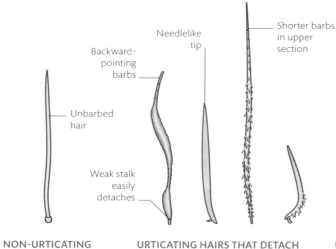

Backward-pointing barbs

Needlelike tip

Shorter barbs in upper section

Unbarbed hair

Weak stalk easily detaches

NON-URTICATING HAIR

URTICATING HAIRS THAT DETACH FROM BASAL STALK (TYPE I–IV)

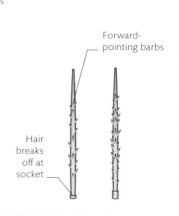

Forward-pointing barbs

Hair breaks off at socket

URTICATING HAIRS THAT DETACH FROM BASAL SOCKET (TYPE V–VI)

Hard, ridged, bristlelike structure of hair is strengthened with chitin—the same hard material that makes up the skeleton of arthropods

Sharp, pointed tip of hair can penetrate deep into skin

stinging hairs

Like the larvae of many insects, the slow-moving caterpillars of moths and butterflies can be easy targets for predators, but many have evolved ways to avoid danger using chemical warfare. The hairs and spines that help provide physical defense in other insects are modified for delivering venom. Unlike the stinger of a wasp or scorpion, which has muscles that pump venom (see p.210), caterpillar venom is delivered passively: the brittle tips of their hairs break off when touched, releasing their contents into the predator's skin. Depending upon species, the effect varies from irritating discomfort to severe pain. Venomous caterpillars are typically brightly colored or patterned so that predators, once stung, soon learn to leave them alone.

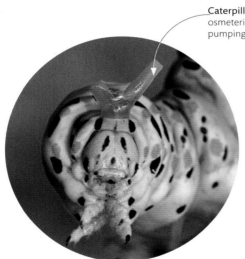

Caterpillar inflates osmeterium by pumping blood into it

Defense by stench
The caterpillar of the Old World swallowtail butterfly (*Papilio machaon*) relies on smell to fend off danger. It projects a fleshy, forked organ, known as an osmeterium, which releases a pineapplelike scent that repels enemies such as spiders and ants.

Bristles are branched, which maximizes the number of breakable venom-releasing hair tips

COMPARING DEFENSIVE HAIRS AND SPINES IN CATERPILLARS
The venomous equipment of a caterpillar develops as modified "hairs" (called setae) or spines. In setae, hair cells (trichogens) in the epidermis secrete venom into the shaft, but venomous spines develop as outgrowths of the cuticle. In both cases, venom is released if the tip is broken.

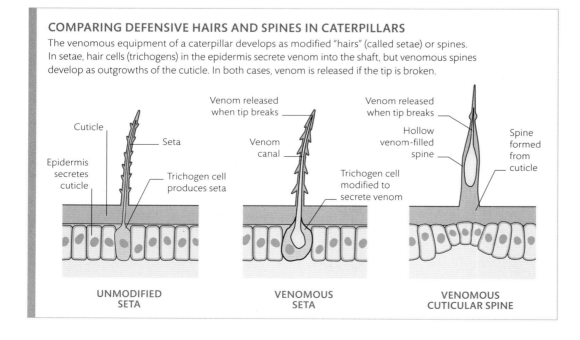

UNMODIFIED SETA — Cuticle, Seta, Epidermis secretes cuticle, Trichogen cell produces seta

VENOMOUS SETA — Venom released when tip breaks, Venom canal, Trichogen cell modified to secrete venom

VENOMOUS CUTICULAR SPINE — Venom released when tip breaks, Hollow venom-filled spine, Spine formed from cuticle

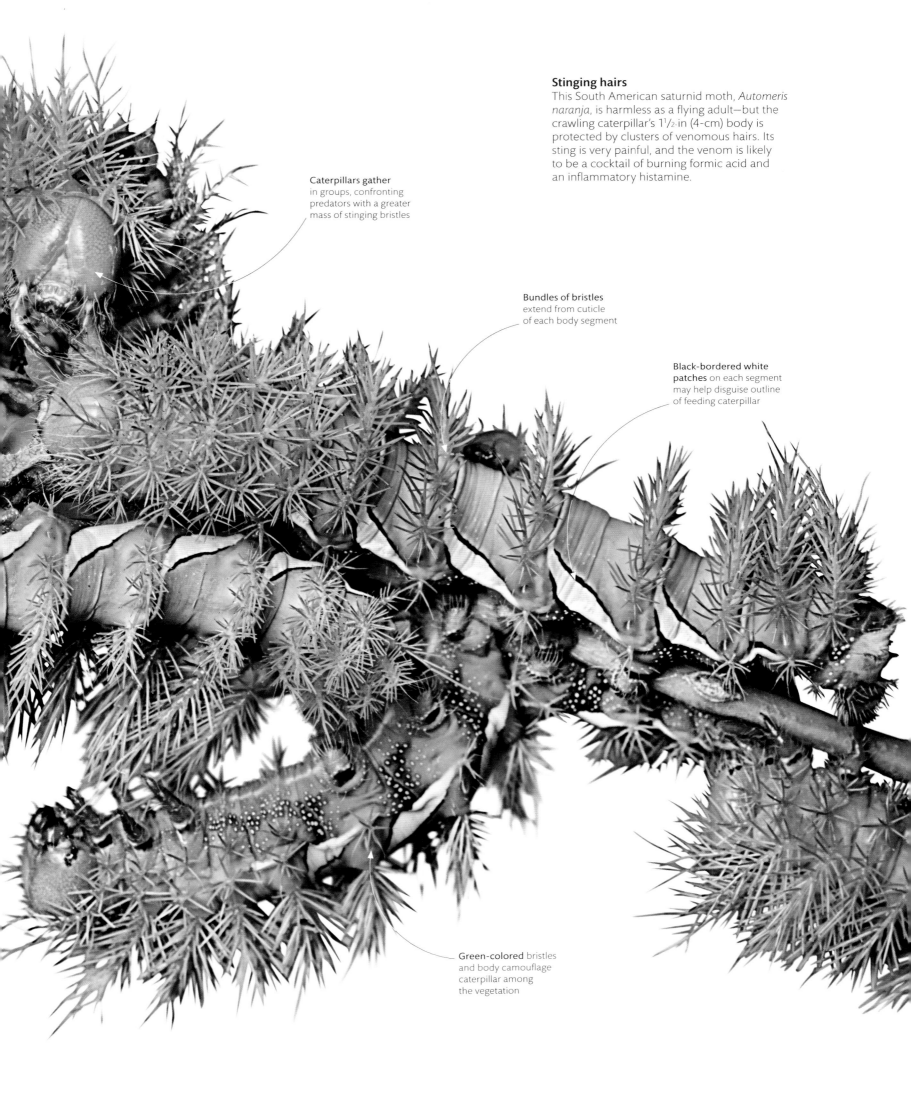

Stinging hairs
This South American saturnid moth, *Automeris naranja*, is harmless as a flying adult—but the crawling caterpillar's 1½-in (4-cm) body is protected by clusters of venomous hairs. Its sting is very painful, and the venom is likely to be a cocktail of burning formic acid and an inflammatory histamine.

Caterpillars gather in groups, confronting predators with a greater mass of stinging bristles

Bundles of bristles extend from cuticle of each body segment

Black-bordered white patches on each segment may help disguise outline of feeding caterpillar

Green-colored bristles and body camouflage caterpillar among the vegetation

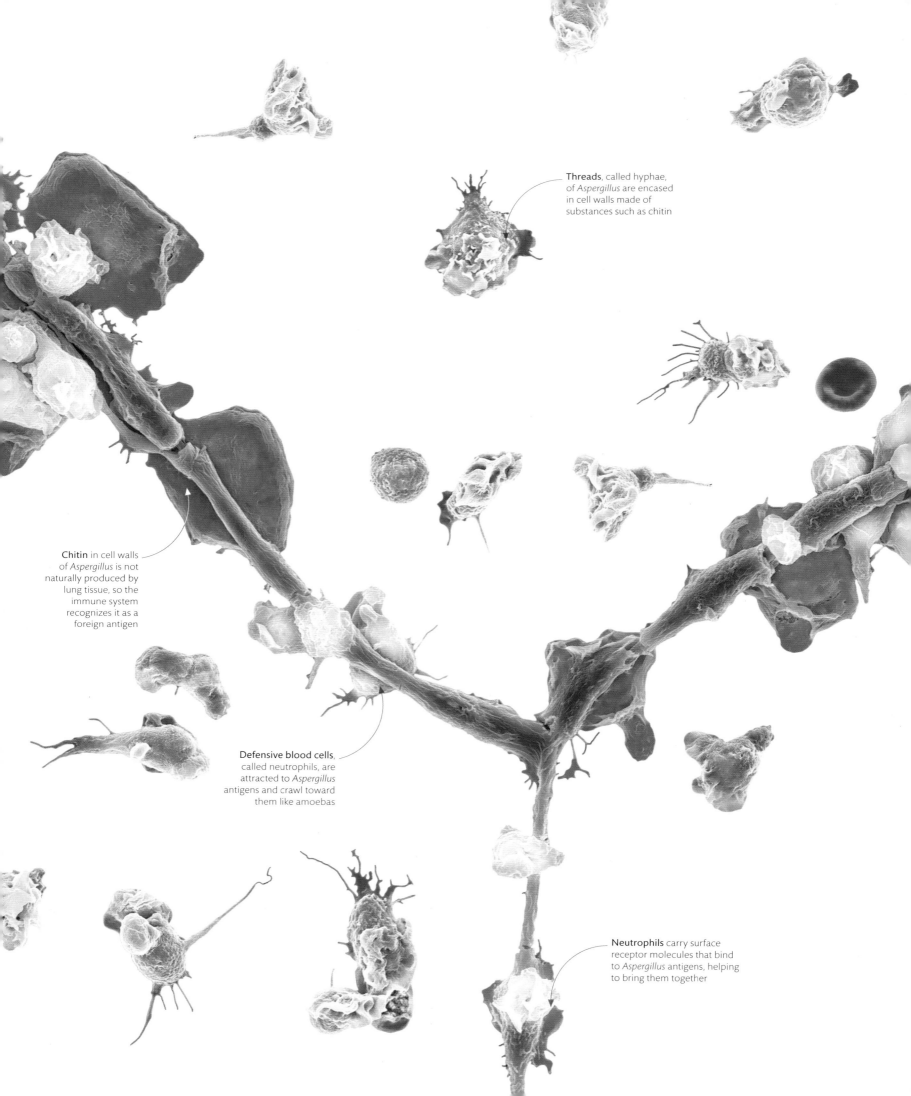

Macrophages and neutrophils, once bound, start to destroy the fungus by phagocytosis

Predatory defenders
Neutrophils can digest entire animal parasites, such as this microfilarial worm, by engulfing them with cellular threads called pseudopods and releasing digestive enzymes from sacs called lysosomes—a process known as phagocytosis (see pp.38–39).

Parasitic nematode worm
Wuchereria bancrofti can be spread between humans through mosquito bites

Larger defensive cells, called macrophages, which reside permanently in the lungs, also target *Aspergillus*

BUILDING UP IMMUNITY
Digestive defence cells, such as neutrophils, are largely indiscriminate, but white blood cells called lymphocytes are more focused. First exposure to a particular antigen stimulates lymphocytes to release an antigen-specific antibody and build a memory of the infection. A second exposure to the same antigen triggers a bigger release of the antibody, which stops the infection.

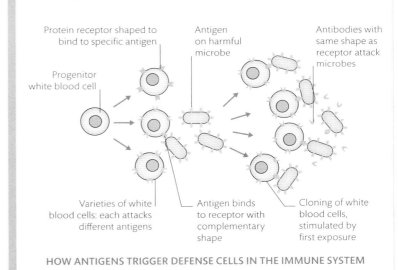

Protein receptor shaped to bind to specific antigen
Antigen on harmful microbe
Antibodies with same shape as receptor attack microbes
Progenitor white blood cell
Varieties of white blood cells: each attacks different antigens
Antigen binds to receptor with complementary shape
Cloning of white blood cells, stimulated by first exposure

HOW ANTIGENS TRIGGER DEFENSE CELLS IN THE IMMUNE SYSTEM

Attacking fungus
The threads of an infectious *Aspergillus* mold can grow inside the lungs after germinating from inhaled spores. But defense cells of the immune system, called neutrophils and macrophages, bind to the foreign walls of the fungus and help to destroy it with their digestive enzymes. SEM, ×2,000.

internal defense

All living things are a potential source of nourishment for parasites, and are therefore targets for infectious organisms that might cause harm. If an organism's outer defenses are breached, it can actively fight intruders to prevent an infection from taking hold by using its "immune" system, which can recognize anything that is not a natural component of the body. The immune system picks out foreign, or "non-self," molecules (called antigens), carried by an intruder, and can target a counterattack. Most animals have microbe-eating cells that smother the enemy and digest it, but backboned animals (vertebrates) can also flood their bloodstream with disease-preventing chemicals called antibodies.

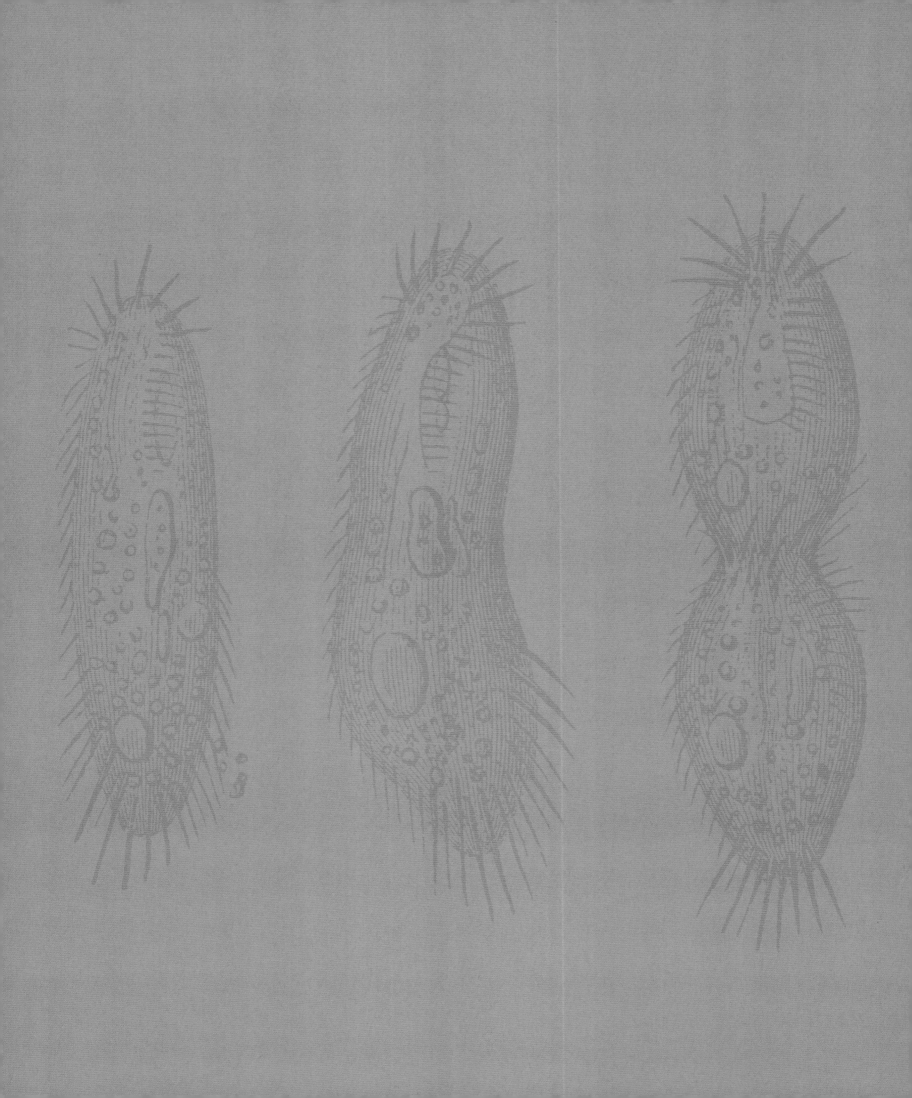

reproducing

The life-forms living today are those that have prevailed, after 4 billion years of evolution, by being good at leaving offspring. They have done this either by simply replicating themselves, or by mixing their DNA with others of their species, producing new combinations of genes—some of which will win out in future battles for survival.

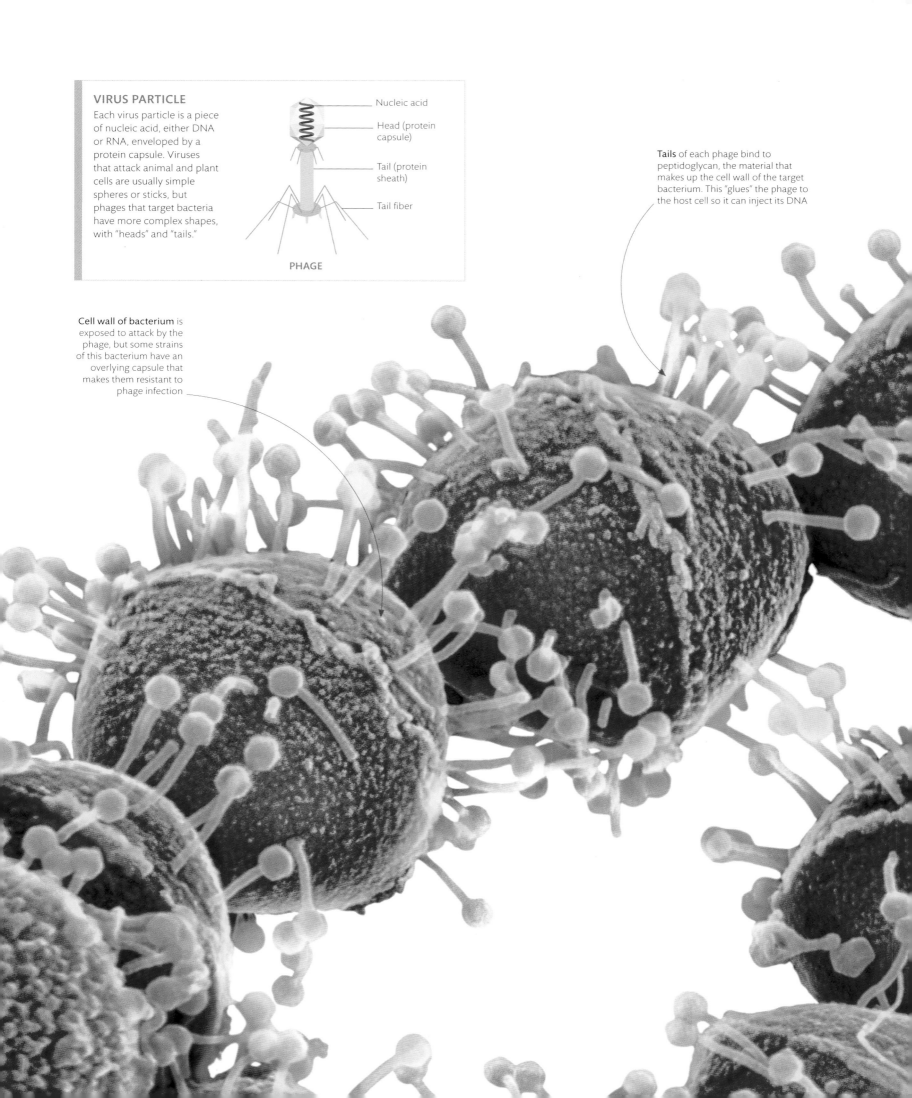

VIRUS PARTICLE
Each virus particle is a piece of nucleic acid, either DNA or RNA, enveloped by a protein capsule. Viruses that attack animal and plant cells are usually simple spheres or sticks, but phages that target bacteria have more complex shapes, with "heads" and "tails."

- Nucleic acid
- Head (protein capsule)
- Tail (protein sheath)
- Tail fiber

PHAGE

Tails of each phage bind to peptidoglycan, the material that makes up the cell wall of the target bacterium. This "glues" the phage to the host cell so it can inject its DNA

Cell wall of bacterium is exposed to attack by the phage, but some strains of this bacterium have an overlying capsule that makes them resistant to phage infection

sabotaging cells

Viruses are the ultimate infectious agents. Stripped down to the minimal equipment needed to invade and replicate, viruses are little more than self-copying chemical particles—dozens of times smaller than the tiniest bacterium. Unlike all true organisms, they are not made up of complex cells and lack the enzymes and other critical ingredients of life needed to trigger their own replication. Consequently, viruses are entirely dependent on infecting living cells to replicate. As they replicate and produce millions more infectious viruses, they can end up destroying the host cell.

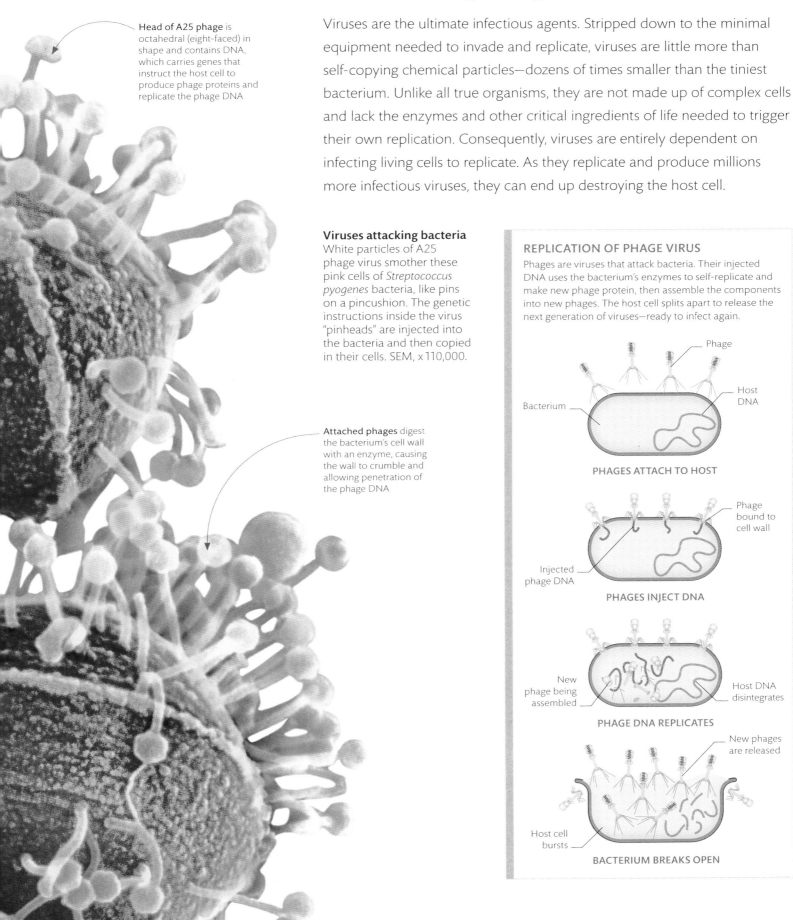

Head of A25 phage is octahedral (eight-faced) in shape and contains DNA, which carries genes that instruct the host cell to produce phage proteins and replicate the phage DNA

Attached phages digest the bacterium's cell wall with an enzyme, causing the wall to crumble and allowing penetration of the phage DNA

Viruses attacking bacteria
White particles of A25 phage virus smother these pink cells of *Streptococcus pyogenes* bacteria, like pins on a pincushion. The genetic instructions inside the virus "pinheads" are injected into the bacteria and then copied in their cells. SEM, x 110,000.

REPLICATION OF PHAGE VIRUS

Phages are viruses that attack bacteria. Their injected DNA uses the bacterium's enzymes to self-replicate and make new phage protein, then assemble the components into new phages. The host cell splits apart to release the next generation of viruses—ready to infect again.

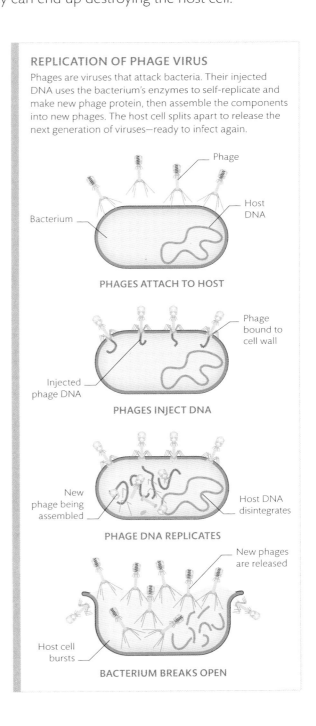

PHAGES ATTACH TO HOST

PHAGES INJECT DNA

PHAGE DNA REPLICATES

BACTERIUM BREAKS OPEN

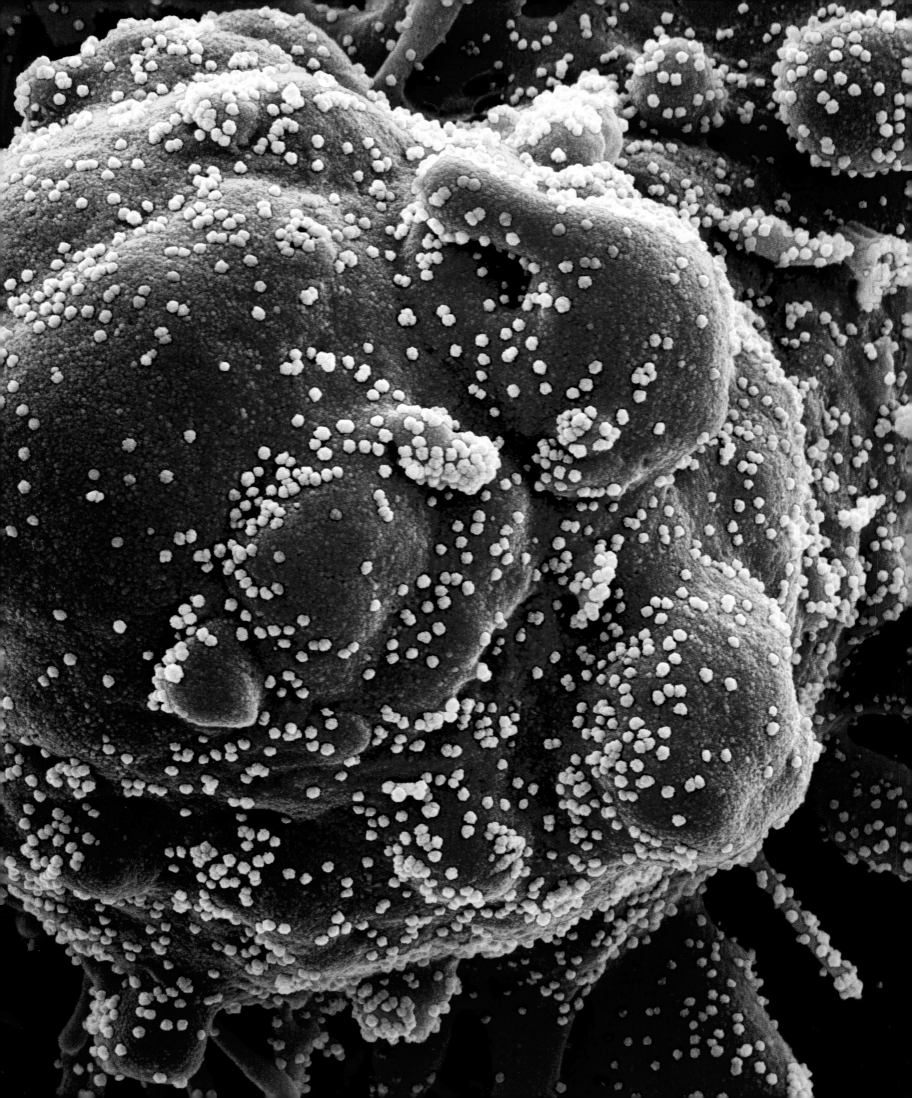

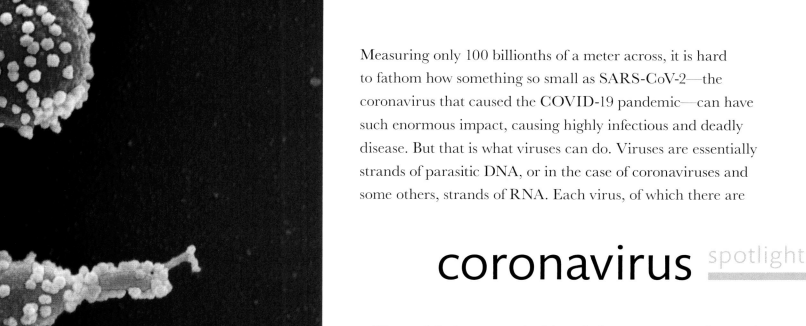

Measuring only 100 billionths of a meter across, it is hard to fathom how something so small as SARS-CoV-2—the coronavirus that caused the COVID-19 pandemic—can have such enormous impact, causing highly infectious and deadly disease. But that is what viruses can do. Viruses are essentially strands of parasitic DNA, or in the case of coronaviruses and some others, strands of RNA. Each virus, of which there are

coronavirus spotlight

millions of distinct types, is able to infect one or more forms of life, from a tree or an ant colony to a human being. A viral genome—the unique genetic sequence of a virus—codes for proteins that surround and protect the parasitic DNA or RNA. In the case of coronaviruses, parts of this protein coating jut out as spikes. When coronaviruses were first described in the 1960s, virologists studying them through microscopes initially saw the spike proteins as a halo around the virus, so this new virus type was named coronavirus, meaning "crown virus."

The spike proteins of a coronavirus are shaped so they can bond to structures on the surface of a host cell. This connection opens a route for the viral RNA to enter the cell. Once inside, it hijacks the cell's genetic apparatus, forcing it to churn out thousands of copies of itself and produce fresh protein jackets for each new virus—which eventually kills the cell. Most often, coronaviruses infect the respiratory systems of mammals and birds. While they usually cause trivial infections, such as the common cold, SARS-CoV-2, which emerged in 2019, is able to go on to attack other body systems and has proven far more dangerous.

Spike proteins in the protein coat of the virus particle

coronavirus particle (model)

Heavy load
The cell in this scanning electron micrograph is dying. The attack of thousands of SARS-CoV-2 virus particles—seen here in yellow, attached to its surface—has destroyed the cell's internal machinery. The cell's outer membrane is gradually collapsing, leaving dozens of protrusions that will eventually fragment. SEM, ×30,000.

swarming bacteria

Despite being equipped with chemical tricks that are impossible in more complex plants and animals—such as using nitrogen gas or generating methane—bacteria are almost universally single-celled. They lack the signals and receptors needed to keep cells together to make tissues, organs, and multicelled bodies. But one extraordinary group of bacteria come close: myxobacteria live in soil, where they glide toward dead and waste matter to feed. When supplies run low, the starved cells release chemical distress signals that draw cells together so they can pool their digestive enzymes. If food runs out, they bunch more tightly to form mushroomlike reproductive structures, which burst and disperse spores.

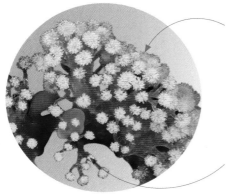

Mushroomlike fruiting body contains spores for dispersal

Stalked colonies are elaborate by bacterial standards

Bacterial pack
The gliding cells of myxobacteria stimulate one another when they make contact. This makes them join into a "pack" so they can cooperate in making surprisingly complex fruiting bodies.

Dispersing cluster
This cluster of hard-walled spores is ready to be blown by the wind to new sources of food, where each will grow into a new bacterium. Cells of the myxobacterium *Myxococcus xanthus* cooperate in this way when stressed by overcrowding and diminishing resources. About 100,000 of them have gathered to make this spore-releasing fruiting body. SEM, x 2,300.

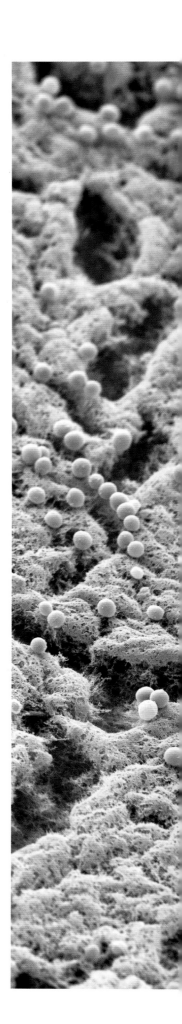

DEALING WITH ADVERSITY

Many organisms, from slime molds (see p.244) and fungi (see p.230) to aphids (see p.243), respond to diminishing food resources by quickly producing many offspring that then disperse to new habitats. In myxobacteria, a fruiting body produces cells that become encapsulated in a hard wall that can withstand dry conditions, enabling the organism to disperse as spores on the wind.

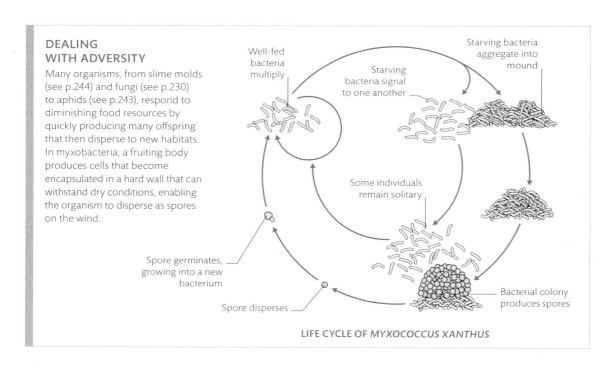

LIFE CYCLE OF *MYXOCOCCUS XANTHUS*

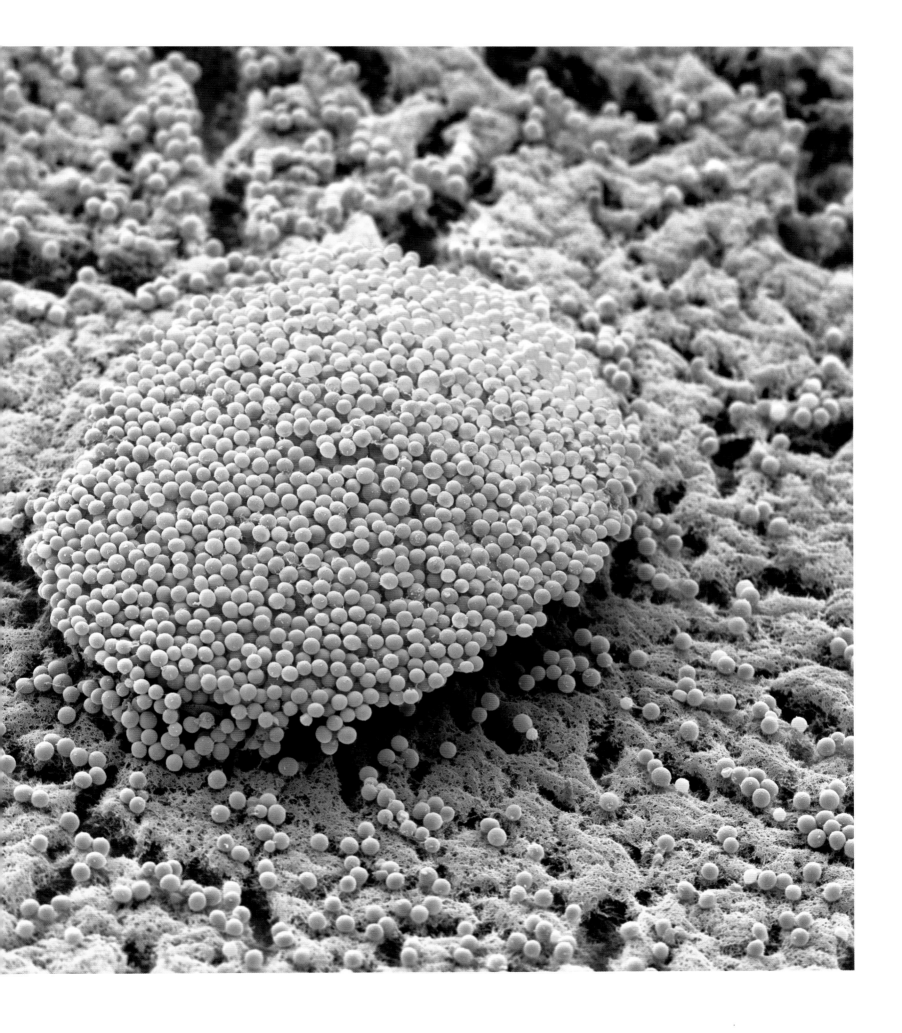

Doubling cells
The cell of the desmid *Micrasterias thomasiana*—a freshwater alga—consists of two mirror-equal halves, with a nucleus in the middle. During asexual reproduction, the nucleus is copied into two genetically identical nuclei. Each half then produces a new bud that grows to restore the cell's symmetry. LM, x450.

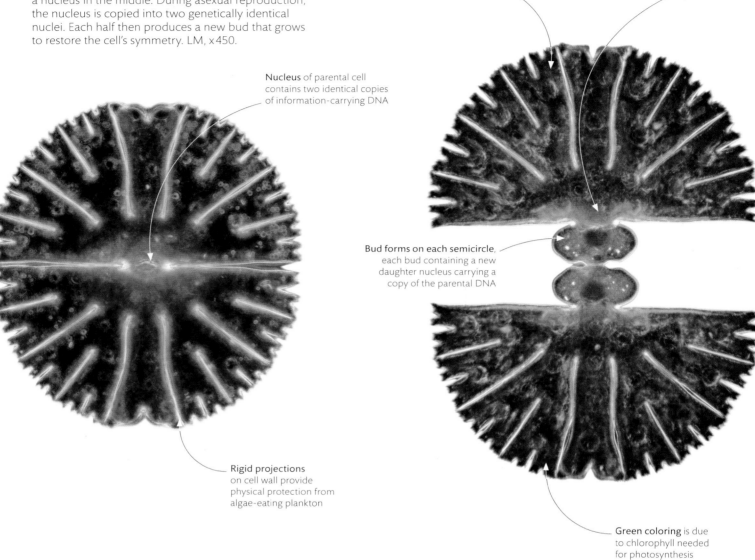

Semicircular halves of parent cell split from each other

Bridge, or isthmus, connects new bud to parent cell

Nucleus of parental cell contains two identical copies of information-carrying DNA

Bud forms on each semicircle, each bud containing a new daughter nucleus carrying a copy of the parental DNA

Rigid projections on cell wall provide physical protection from algae-eating plankton

Green coloring is due to chlorophyll needed for photosynthesis

Bud will detach when fully formed

asexual reproduction

Single-celled organisms, such as algae, can rapidly produce many offspring simply by dividing, in an asexual cycle that produces genetically identical clones. While algae can have sexual stages, in which cells mix their genes before reproducing, asexual reproduction is a more efficient way of exploiting bountiful resources, such as a surge in nutrients—however brief. Many multicelled algae and plants can also split asexually if each fragment develops a shoot. But the bodies of most animals—except for simple invertebrates, such as hydras—are too complex to reproduce this way.

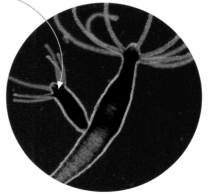

Budding animal
A hydra, a freshwater relative of anemones and jellyfish, has such a simple body structure that it can grow a bud that is a genetic clone.

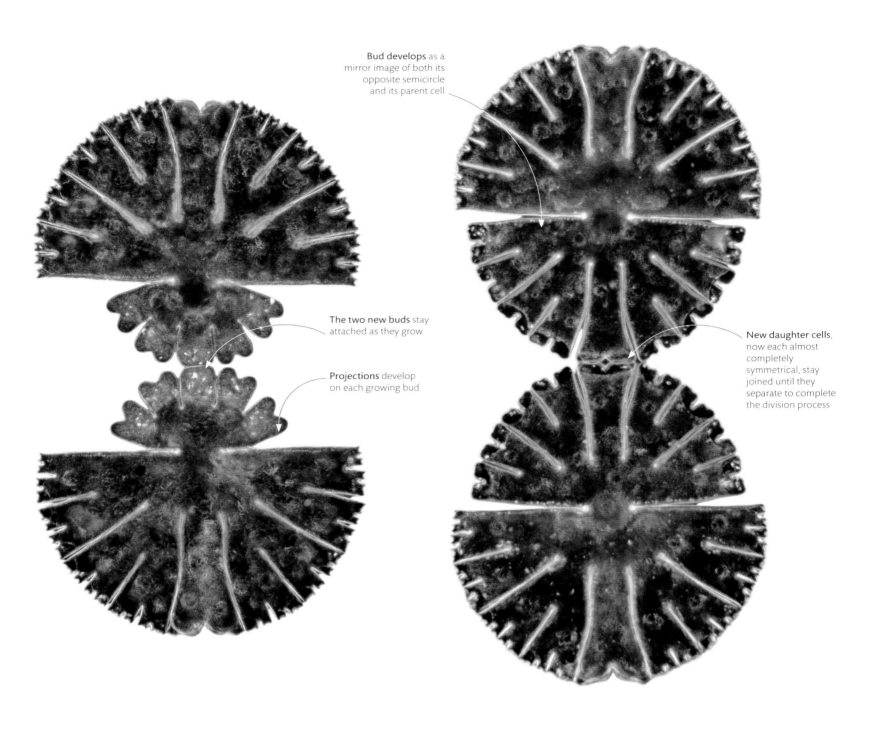

Bud develops as a mirror image of both its opposite semicircle and its parent cell

The two new buds stay attached as they grow

Projections develop on each growing bud

New daughter cells, now each almost completely symmetrical, stay joined until they separate to complete the division process

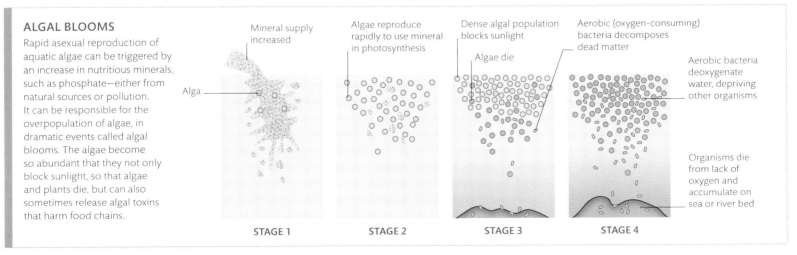

ALGAL BLOOMS

Rapid asexual reproduction of aquatic algae can be triggered by an increase in nutritious minerals, such as phosphate—either from natural sources or pollution. It can be responsible for the overpopulation of algae, in dramatic events called algal blooms. The algae become so abundant that they not only block sunlight, so that algae and plants die, but can also sometimes release algal toxins that harm food chains.

Mineral supply increased

Alga

Algae reproduce rapidly to use mineral in photosynthesis

Dense algal population blocks sunlight

Algae die

Aerobic (oxygen-consuming) bacteria decomposes dead matter

Aerobic bacteria deoxygenate water, depriving other organisms

Organisms die from lack of oxygen and accumulate on sea or river bed

STAGE 1   STAGE 2   STAGE 3   STAGE 4

SPERM AND EGG COMPARED

An egg is packed with almost all the cellular material needed for the growth of an embryo: it only lacks the set of the male parent's genes needed to start the process. The sperm, in contrast, is adapted for swimming and carries minimal baggage. Even its nucleus is condensed in size—despite it carrying about the same number of genes as the egg.

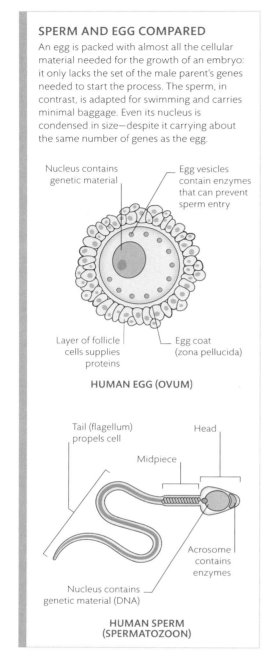

STAGES OF FERTILIZATION

When a sperm reaches its target, it burrows between the follicle cells that helped nourish the egg in the ovary, until it reaches the slimy egg coating beneath—then uses digestive enzymes to tunnel further. When it injects its nucleus, containing its DNA, egg vesicles release enzymes to harden the coat, stopping other sperm from gaining entry.

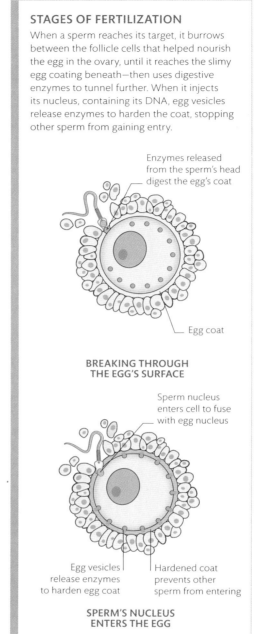

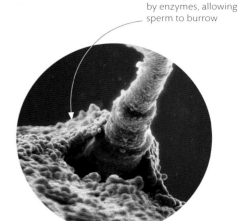

Breaking through
A sac of enzymes, called the acrosome, in the nose of a sperm breaks open to digest its way through the egg's outer coat.

fertilizing an egg

Fertilization is not only the beginning of new life, it is a union of different bloodlines. In any animal, when a sperm joins an egg, it mixes DNA from a male and a female parent, creating a new combination of genes that has never before existed—the essential advantage of sexual reproduction over simple replication. It means that the new life that grows from every fertilized egg is genetically unique. But it takes opportunity and chance for events to succeed: swimming sperm have to travel to reach a receptive egg and then must penetrate the egg's coat to deliver their nucleus's cargo of genes. Once this fuses with the egg's nucleus (containing the female parent's DNA), cellular division begins to create the growing embryo (see pp.256–57).

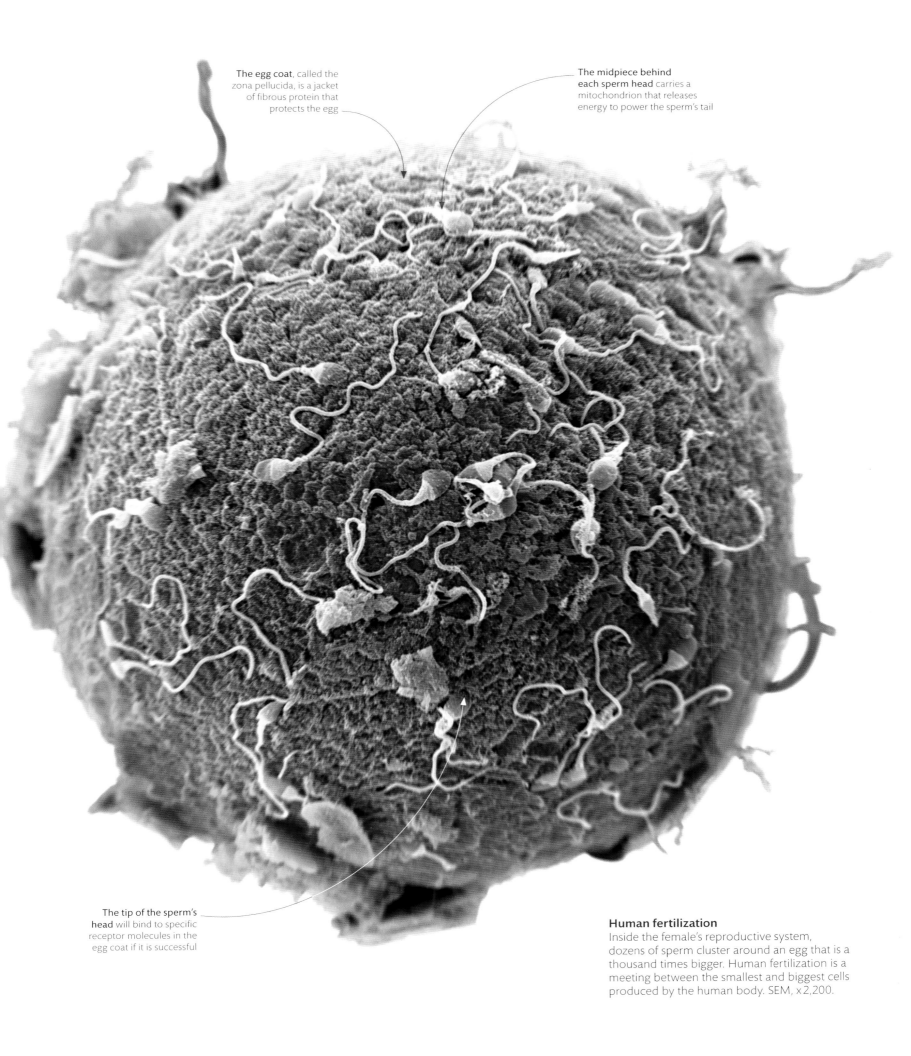

The egg coat, called the zona pellucida, is a jacket of fibrous protein that protects the egg

The midpiece behind each sperm head carries a mitochondrion that releases energy to power the sperm's tail

The tip of the sperm's head will bind to specific receptor molecules in the egg coat if it is successful

Human fertilization
Inside the female's reproductive system, dozens of sperm cluster around an egg that is a thousand times bigger. Human fertilization is a meeting between the smallest and biggest cells produced by the human body. SEM, x2,200.

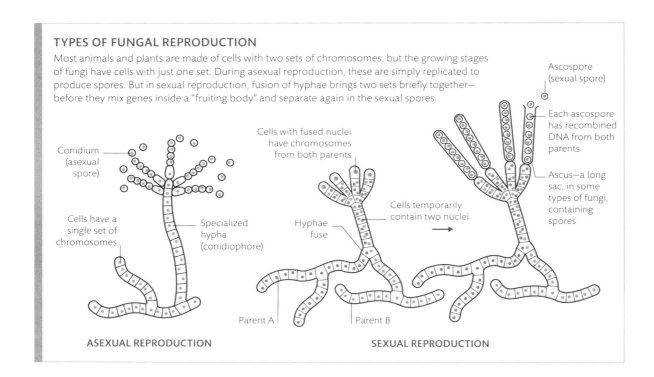

TYPES OF FUNGAL REPRODUCTION

Most animals and plants are made of cells with two sets of chromosomes, but the growing stages of fungi have cells with just one set. During asexual reproduction, these are simply replicated to produce spores. But in sexual reproduction, fusion of hyphae brings two sets briefly together—before they mix genes inside a "fruiting body" and separate again in the sexual spores.

Spores from mushrooms
Mushrooms, such as the velvet shank mushroom (*Flammulina velutipes*) are large fruiting bodies that produce sexual spores resulting from fusion of different fungal "mating types."

fungus reproduction

Fungi have a unique way of reproducing sexually: instead of fertilizing eggs with swimming sperm (see pp.228–29), their body filaments—called hyphae—fuse in a process called conjugation. They lack distinct male and female sexes, but the hyphae coming together can belong to genetically different "mating types." The fused hyphae then grow into spore-producing bodies—such as a mushroom or toadstool. Fungal spores can also be produced asexually, but all disperse in the air to germinate into new hyphae elsewhere.

Producing spores
Aspergillus niger is a black soil mold that causes spoilage of fruit and vegetables. Like other fungi, its network of microscopic hyphae occasionally sprouts reproductive structures that make spores. Those in this image have been produced asexually from single hyphae, so the spores they release will germinate into genetic clones. SEM, x 1,400.

Swollen tip of conidiophore carries phialides and their developing conidia

Chains of conidia are produced by the phialides; the conidia are shed and disperse on air currents

Conidiophore is a specialized hypha on which conidia are produced

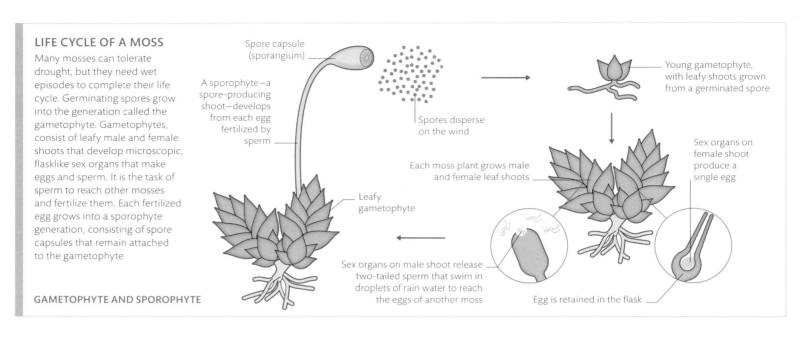

LIFE CYCLE OF A MOSS

Many mosses can tolerate drought, but they need wet episodes to complete their life cycle. Germinating spores grow into the generation called the gametophyte. Gametophytes, consist of leafy male and female shoots that develop microscopic, flasklike sex organs that make eggs and sperm. It is the task of sperm to reach other mosses and fertilize them. Each fertilized egg grows into a sporophyte generation, consisting of spore capsules that remain attached to the gametophyte.

GAMETOPHYTE AND SPOROPHYTE

Two-tiered plant
Moss spore capsules tower above the plant's moisture-loving leafy shoots, which hug the ground. Their height gives the spores the best chance of being caught by the wind.

Capsule stalks of common cord moss grow to a height of 3 in (8 cm)

alternating generations

Single-celled spores and gametes (sperm and eggs) are both ways of reproducing, and plants have life cycles that alternate between both. A moss is confined to damp places like its algal ancestors, partly because its propagules are gametes: swimming sperm that are splashed towards the eggs of other individuals by raindrops. The resulting generation of mosses, however, sprouts capsules that scatter drought-resistant spores into the wind, to germinate on wet ground. This alternation between spore- and gamete-producing generations is common to all plants, but in most plants, fertilization occurs in a cone or flower, to make seeds.

Moss sporophytes
The spore-producing stage of a moss, the sporophyte, consists of a stalk and a capsule. The sporophyte turns green with chlorophyll and starts photosynthesizing as it matures.

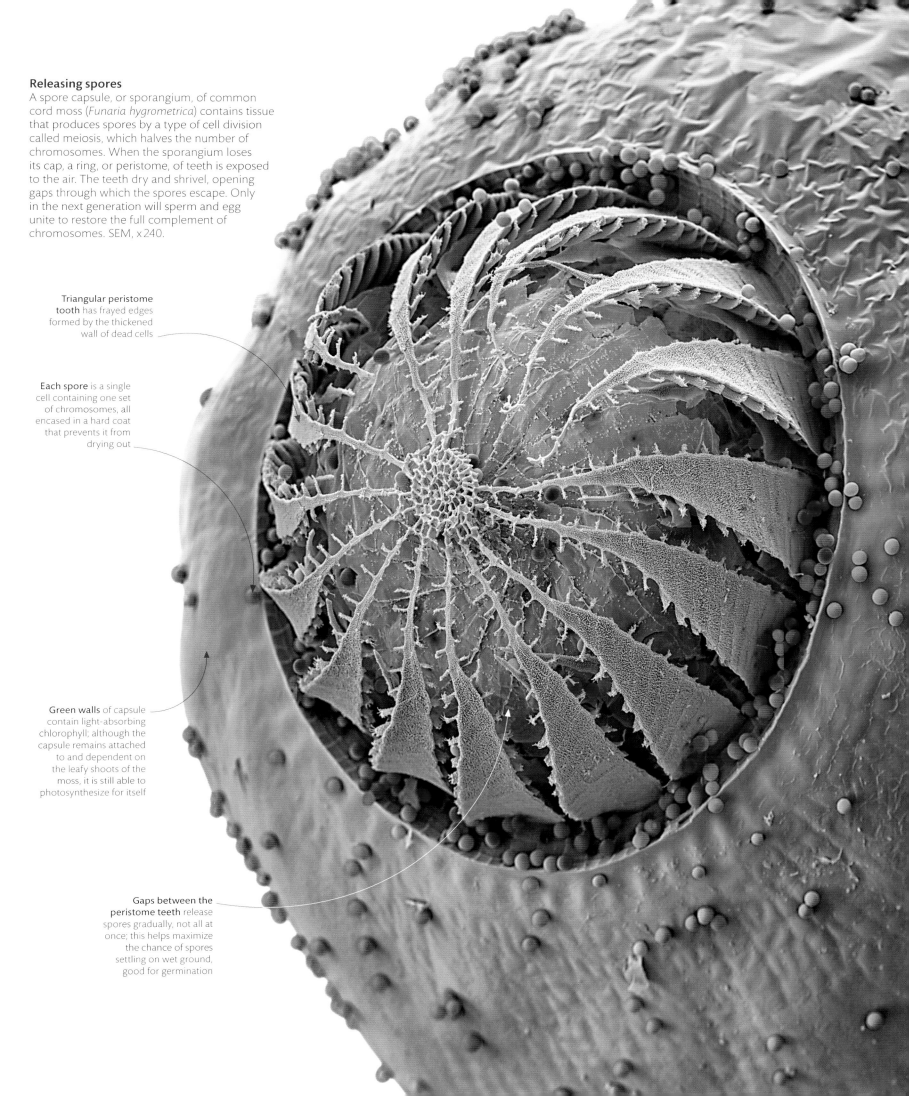

Releasing spores

A spore capsule, or sporangium, of common cord moss (*Funaria hygrometrica*) contains tissue that produces spores by a type of cell division called meiosis, which halves the number of chromosomes. When the sporangium loses its cap, a ring, or peristome, of teeth is exposed to the air. The teeth dry and shrivel, opening gaps through which the spores escape. Only in the next generation will sperm and egg unite to restore the full complement of chromosomes. SEM, x240.

Triangular peristome tooth has frayed edges formed by the thickened wall of dead cells

Each spore is a single cell containing one set of chromosomes, all encased in a hard coat that prevents it from drying out

Green walls of capsule contain light-absorbing chlorophyll; although the capsule remains attached to and dependent on the leafy shoots of the moss, it is still able to photosynthesize for itself

Gaps between the peristome teeth release spores gradually, not all at once; this helps maximize the chance of spores settling on wet ground, good for germination

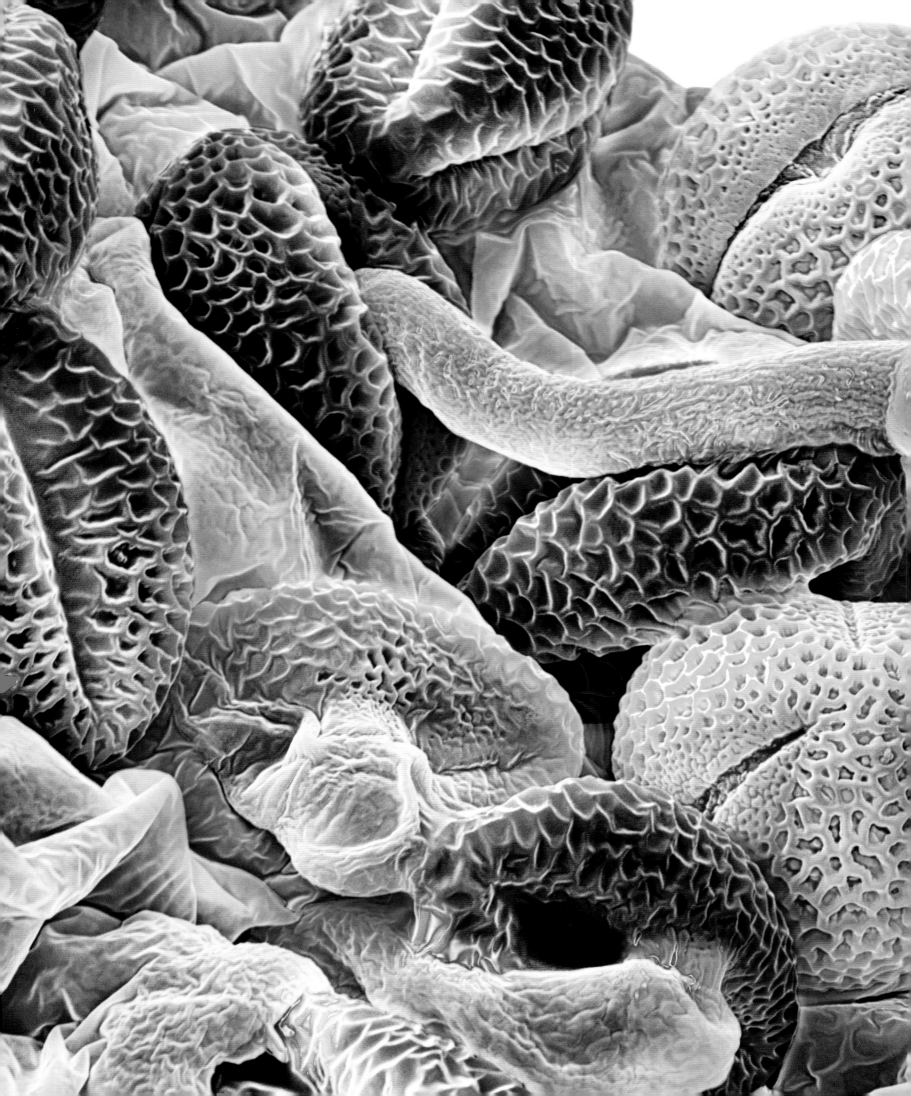

MACROLIFE

Exine, the tough, sculpted surface of a pollen grain, helps it stick to both flower stigmas and insect pollinators

Pollen tube contains a nucleus, which controls growth toward the ovule; the tube acts as a conduit for the male gametes toward the egg

Pollen grain from a plant of a different species (yellow) is incompatible, and the pollen tube fails to grow, due to chemical inhibitors from the stigma

sex in flowering plants

In flowering plants, male gametes (sex cells) are not sperm, but simply sperm nuclei containing DNA. They are carried in pollen grains from the anthers of one plant to the female flower parts on a second plant. There, fertilization mixes the genes from both parents, ensuring the variability that may help some offspring survive; while some plants can self-fertilize, this results in offspring with lower genetic diversity. Pollen can be spread in vast quantities by wind, as occurs in grasses and many trees, or it can be carried in a more targeted way by pollinating animals that visit only a limited range of flowers.

Attracting insects
Bees, butterflies, and other insects are attracted to the open flowerheads of this shrub, Laurestinus (*Viburnum tinus*), and transfer pollen between its flowers.

Blue-black berries contain seeds to be dispersed

Response to pollen
In this image, insects have carried pollen to a receptive stigma in the center of a flower of *Viburnum* sp. When pollen from another flower of the same species (gray) is deposited on a flower, it begins to swell. One grain has begun to absorb moisture from the plant, driving the pollen tube to emerge through an opening, called a germ pore, in the pollen's exine—its tough outer coat. Pollen grains germinating on a compatible stigma are guided by chemical signals to grow down to the ovule (see right). SEM, x6,000.

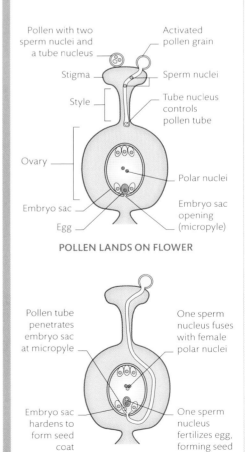

DOUBLE FERTILIZATION

On landing on a stigma (the top of the female reproductive parts of a flower) a pollen tube penetrates the style below, and grows down into the ovary. Sperm nuclei (male gametes) pass down the tube: one fuses with the egg to form a seed; the other fuses with two polar nuclei, forming the endosperm (food store).

POLLEN LANDS ON FLOWER

FERTILIZATION

Surface
The outer surface of a pollen grain is called the exine. One of the chief characteristics used to classify pollen is the fine structure of this surface when viewed under a microscope. The exine is made of a tough material called sporopollenin, which is more or less impervious to chemical attack, so the grain's internal structures are protected. Grains that have been preserved in deposits are useful in forensic identification of species.

Tiny but distinct **pits** pepper the grain's surface

FOVEOLATE
Mosquito grass
Bouteloua gracilis

Large **holes** give the exine a netlike appearance

RETICULATE
Passionflower
Passiflora caerulea

Ridges running in the same general direction swathe the exine

STRIATE
Angel's trumpet
Brugmansias sp.

Pores
It is common for a pollen grain to have openings on its surface, or at least regions where the exine thins out. These openings allow the pollen grain to swell and shrink without cracking the protective coating as it gains and loses moisture, and also provide openings for pollen tubes to extend out of the grain during germination. Porate pollens have round openings called pores; such pollens are classified by the number of pores their grains possess.

Raised sections surround the pores in the exine

FENESTRATE
Chicory
Cichorium intybus

Multiple pores spread evenly over the surface

PANTOPRATE
Dianthus "pink" flower
Dianthus sp.

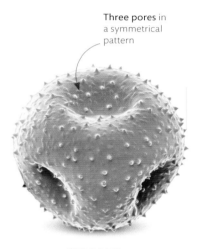

Three pores in a symmetrical pattern

TRIPORATE
Christmas cactus
Schlumbergera sp.

Colpi
Instead of round pores, some pollens have a groovelike opening that runs most of the length of the grain. This feature, known as a colpus (plural: colpi), is generally found on nonspherical pollen grains. Colpate pollens are classified by the number and location of their colpi. Some pollen grains are colporate, having a mix of both colpi and pores.

Single furrowlike **opening** runs along the oval grain

MONOCOLPATE
Lily
Liliaceae

Regularly spaced **grooves** around grain's central axis

ZONOCOLPATE
California poppy
Eschscholzia californica

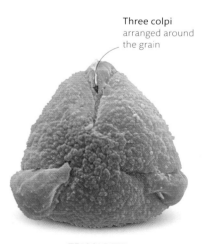

Three colpi arranged around the grain

TRICOLPATE
Pedunculate oak
Quercus robur

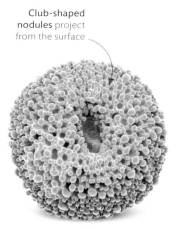

CLAVATE
Geranium
Geranium sp.

Club-shaped nodules project from the surface

Single pore on the surface of this simple grain

MONOPORATE
Grass
Poaceae

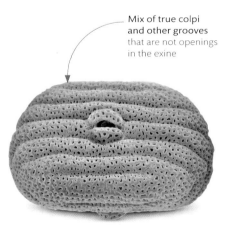

Mix of true colpi and other grooves that are not openings in the exine

HETEROCOLPATE
Flowering herb
Mimulopsis sp.

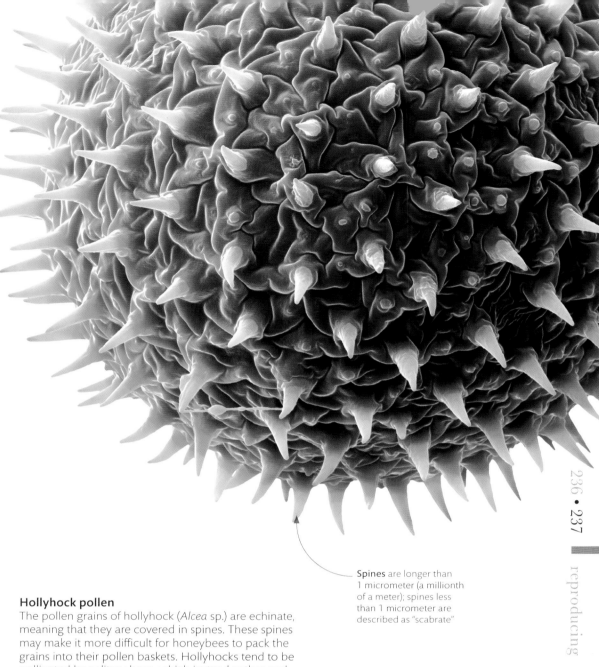

Hollyhock pollen
The pollen grains of hollyhock (*Alcea* sp.) are echinate, meaning that they are covered in spines. These spines may make it more difficult for honeybees to pack the grains into their pollen baskets. Hollyhocks tend to be pollinated by solitary bees, which instead gather and transport the pollen using bushy hairs on their back legs. SEM, x2,400.

Spines are longer than 1 micrometer (a millionth of a meter); spines less than 1 micrometer are described as "scabrate"

pollen grains

Flowering plants and conifers produce pollen, a fine powder of microscopic grains used in sexual reproduction. The pollen grains are transferred by wind, water, or an animal vector—such as a bee—to another member of the plant species. Grains spread by wind are smoother, lightweight, and small—generally less than 0.01 mm across. They are produced by small flowers, and dustlike clouds of pollen can sometimes be seen billowing in the wind during spring and summer. By contrast, animal-borne pollen is larger—sometimes more than 0.1 mm across. The grains often have spiked, hooked, or sticky surfaces to cling to animals visiting colorful or scented flowers.

The bee might seem a very familiar insect—honeybees have been reared for millennia. Honeybee society consists of an egg-laying queen and a hive of female workers, and the males—called drones—die soon after impregnating the queen. Worker bees look after the eggs and larvae, maintain a honeycomb nest, and collect pollen and nectar from flowers that they turn into a honey store to sustain a colony of hundreds.

bees spotlight

However, honeybees represent just a fraction of the world's bee species. There are about 20,000 species of bee, which comprise the group, or clade, Anthophila, meaning "flower lovers"—a member of the order Hymenoptera, which includes wasps and ants. Like their relatives, all bees have a characteristic narrow waist between the thorax and abdomen, although bees are also often so heavily bristled that this feature is obscured.

Most bee species live solitary lives and few manufacture honey, but all members of the Anthophila eat a mixture of pollen and nectar, and hoard food in nests to provide for their newly hatched larvae. Flowers attract bees using strong colors and scents and, in return for providing free food, the bees transfer pollen caught on their body hairs to other flowers—an essential requirement of reproduction and seed production (see pp.234–35). One of the most effective pollinators, it is estimated that bees pollinate two-thirds of the world's crop plants. But this crucial link is seriously threatened by an alarming drop in bee numbers caused by loss of habitat, pesticide poisoning, and climate change.

Sticky pollen trapped by dense leg hairs

laden with pollen

Pollen collector
A foraging honeybee slurps up nectar and carries it to the hive in a throat pouch, or "honey stomach." Bees also transport pollen in basketlike structures (corbiculae) on their hind legs, but a great deal of pollen is trapped on a bee's bristly hairs, and much will be transferred to flowers.

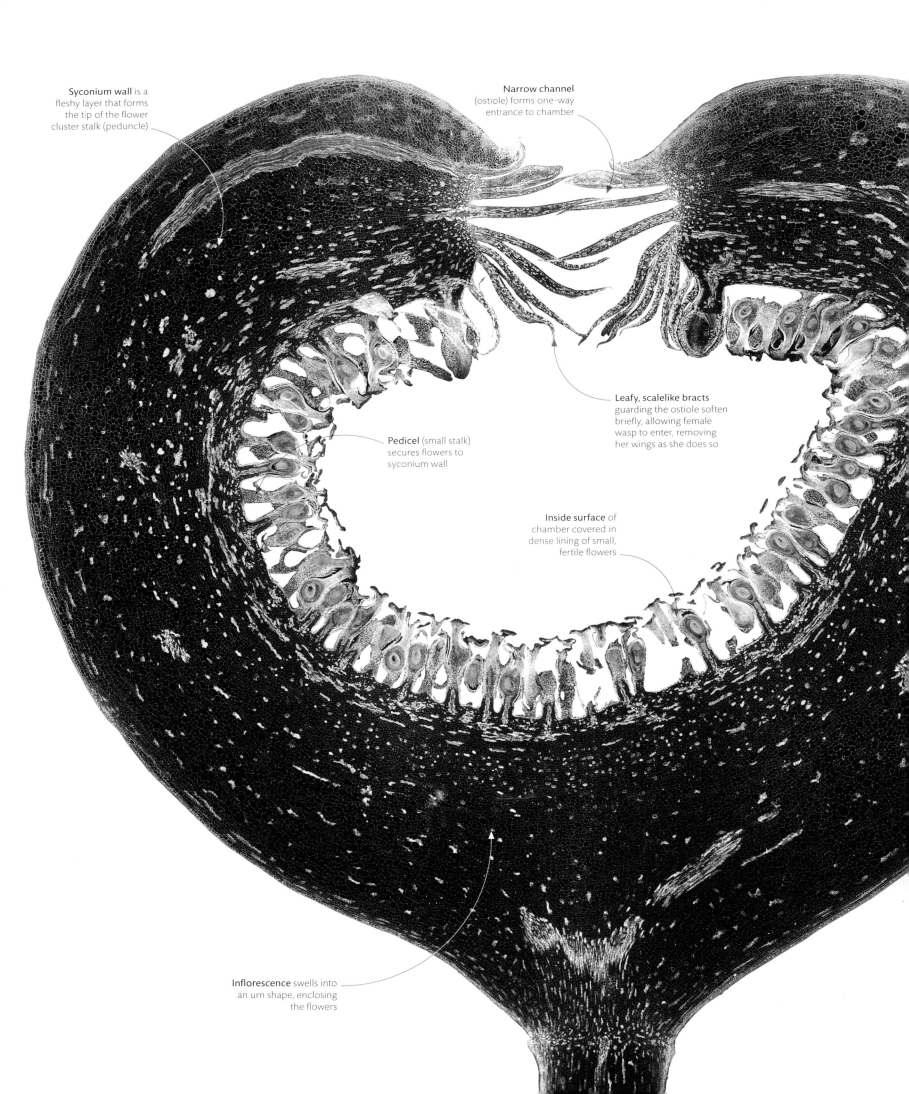

Hidden flowers
The flowers of fig trees (Ficus sp.), as seen in this vertical section through a fig, are hidden inside a hollow chamber formed from the swollen base of the inflorescence. Before the fig can ripen, the unripe fruit must be invaded by a female fig wasp, which lays her eggs in some flowers and pollinates others.

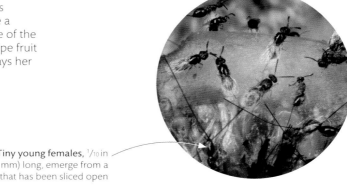

Tiny young females, 1/10 in (2–3 mm) long, emerge from a fig that has been sliced open

Prospecting females
The fig wasp (Blastophaga psenes) only lays eggs in the common fig (Ficus carica). The females may carry pollen up to 6 miles (10 km) in search of a fig to invade.

hidden pollination

Many plants depend on insects to pollinate their flowers. A few rely on a single kind of insect. However, the association between fig trees and fig wasps is especially close, as neither could survive without the other. This association gives the wasps a place to raise their young and, while doing so, pollinate the flowers inside the unripe fig. Most fig trees rely on a particular species of fig wasp: there are about 850 types of figs growing in the tropics, and 900 species of fig wasps. Although the wasps die after pollinating the fig, figs are not full of dead wasps—they secrete an enzyme that digests the wasps, which helps nourish the ripening fruit.

LIFE CYCLE OF A FIG WASP
A female fig wasp enters an unripe fig through the ostiole, a squeeze so tight that her wings are torn off. Inside the fig, she lays eggs into hundreds of flowers, which then develop into galls (see pp.300–01) that protect the larvae. Stored pollen from sacs under her legs is dusted onto unharmed flowers so they develop normally, then she dies. Wingless males hatch from the eggs first and mate with females while they are inside their galls. The young females then emerge, collect pollen, climb out of the fig, and fly off, ready to begin the cycle anew in another fig.

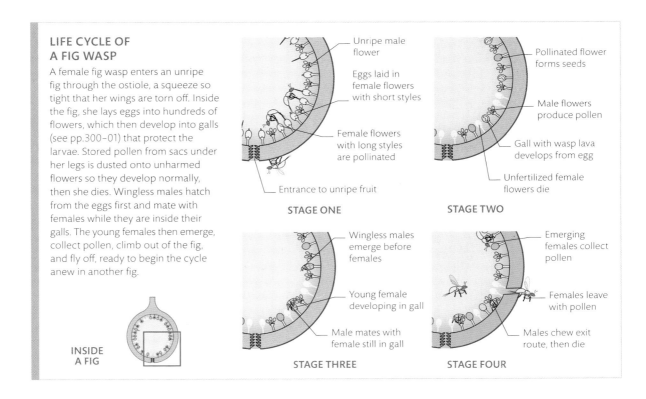

INSIDE A FIG

STAGE ONE
- Unripe male flower
- Eggs laid in female flowers with short styles
- Female flowers with long styles are pollinated
- Entrance to unripe fruit

STAGE TWO
- Pollinated flower forms seeds
- Male flowers produce pollen
- Gall with wasp lava develops from egg
- Unfertilized female flowers die

STAGE THREE
- Wingless males emerge before females
- Young female developing in gall
- Male mates with female still in gall

STAGE FOUR
- Emerging females collect pollen
- Females leave with pollen
- Males chew exit route, then die

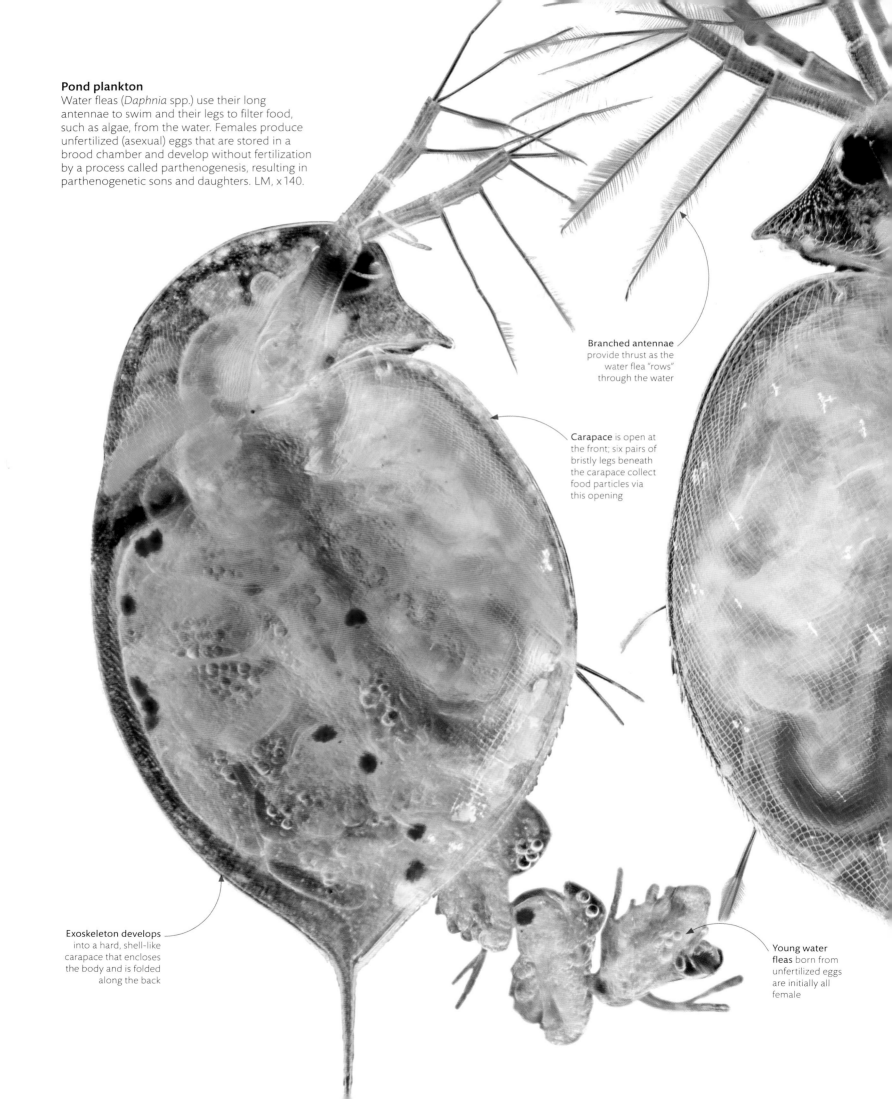

Pond plankton
Water fleas (*Daphnia* spp.) use their long antennae to swim and their legs to filter food, such as algae, from the water. Females produce unfertilized (asexual) eggs that are stored in a brood chamber and develop without fertilization by a process called parthenogenesis, resulting in parthenogenetic sons and daughters. LM, x 140.

Branched antennae provide thrust as the water flea "rows" through the water

Carapace is open at the front; six pairs of bristly legs beneath the carapace collect food particles via this opening

Exoskeleton develops into a hard, shell-like carapace that encloses the body and is folded along the back

Young water fleas born from unfertilized eggs are initially all female

boom and bust

When habitats or resources are short-lived, many tiny animals can mature and breed very fast. Aphids plague new plant growth in spring and summer, just as water fleas swarm ponds that will dry up at the end of summer. In both cases, forgoing sex is key: aphids and water fleas can give birth without fertilization, which speeds up reproduction. But after a boom comes a bust, as food supplies dwindle. Aphids develop wings and fly to new places. Water fleas stay put and reproduce sexually, making eggs that overwinter and hatch when conditions improve the next year.

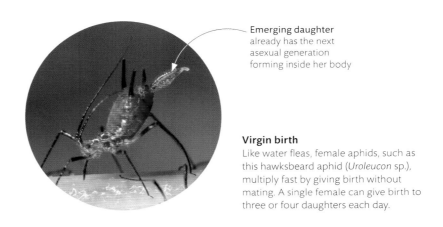

Emerging daughter already has the next asexual generation forming inside her body

Virgin birth
Like water fleas, female aphids, such as this hawksbeard aphid (*Uroleucon* sp.), multiply fast by giving birth without mating. A single female can give birth to three or four daughters each day.

LIFE CYCLE OF A WATER FLEA
The change from asexual to sexual reproduction in water fleas is triggered by deteriorating, overcrowded conditions after a summer boom. Males are produced and after mating, sexual (fertilized) eggs are released by the females. A tough casing protects the eggs during winter, and new females hatch the following spring.

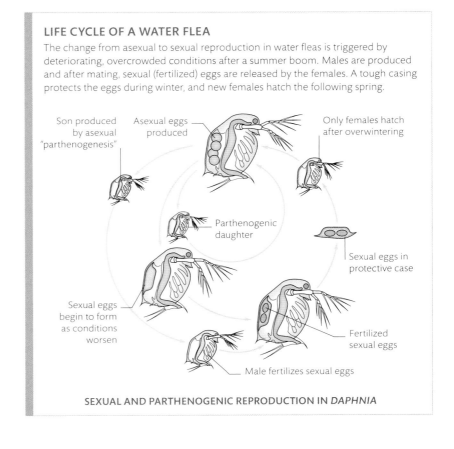

Son produced by asexual "parthenogenesis"

Asexual eggs produced

Only females hatch after overwintering

Parthenogenic daughter

Sexual eggs in protective case

Sexual eggs begin to form as conditions worsen

Fertilized sexual eggs

Male fertilizes sexual eggs

SEXUAL AND PARTHENOGENIC REPRODUCTION IN *DAPHNIA*

Brood chamber beneath transparent carapace holds asexual eggs that can develop without fertilization

Unfertilized eggs hatch within a day at summer temperatures; offspring take 10 days to reach egg-producing maturity

escaping starvation

Slime molds are unusual organisms that can live as single cells, but can also join to form aggregations and reproductive structures, as do some bacteria (see p.224). They are neither animal nor fungus, but have characteristics of both, and some are relatives of amoebas (see pp.132–33). Slime molds thrive on bacteria and fungi found, for example, in dead wood and in leaf litter. When food supplies run low, slime molds can trigger a new stage in the life cycle that boosts their chances of survival.

Fruiting bodies, or sporangia, burst, releasing mass of tangled "hairs" called a capillitium

Spreading the spores
These fruiting bodies are sprouting from a living mass, or plasmodium, of the slime mold *Trichia decipiens*. Some of the fruiting bodies have burst, so the tiny spores can be carried by the wind to new food sources where they can mature into new amoebalike cells. SEM, x 140.

Spores from burst sporangia cover surface of an intact sporangium

SURVIVAL OF A SLIME MOLD
Slime molds spend most of their lives as single-celled organisms called myxamoebas. When food is scarce, they release a chemical signal that attracts others, and they merge to form a mass called a plasmodium. Stalked spore bodies (sporangia) develop, then split, sending spores into the wind to seek new food supplies.

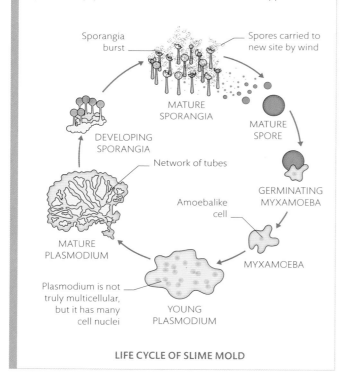

LIFE CYCLE OF SLIME MOLD

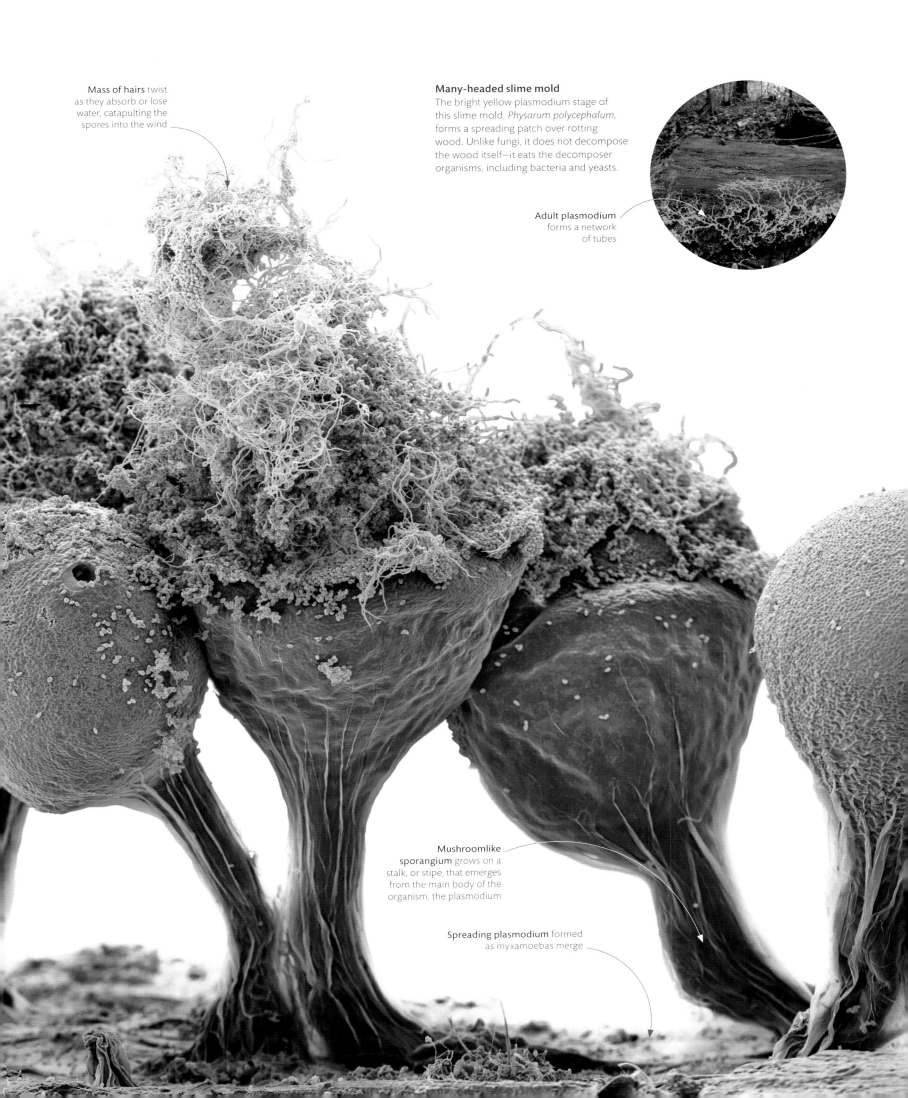

Mass of hairs twist as they absorb or lose water, catapulting the spores into the wind

Many-headed slime mold
The bright yellow plasmodium stage of this slime mold, *Physarum polycephalum*, forms a spreading patch over rotting wood. Unlike fungi, it does not decompose the wood itself—it eats the decomposer organisms, including bacteria and yeasts.

Adult plasmodium forms a network of tubes

Mushroomlike sporangium grows on a stalk, or stipe, that emerges from the main body of the organism, the plasmodium

Spreading plasmodium formed as myxamoebas merge

Eyespan of male stalk-eyed fly can be up to ¼ in (8 mm) wide

Antenna is located near eye, not center of head as is common among flies

Extreme eyespan
The eyestalks of a stalk-eyed fly—*Diasemopsis meigenii*, and others like it—may have evolved because they improved peripheral vision, giving a wider field of view, allowing it to spot predators or competitors. But the fact that males have longer stalks than females suggests that sexual selection is at work: that females are more likely to mate with males with the longest stalks. In some species, the stalks of males are so long that they seem to be a physical burden—but this is outweighed by the benefits they get in reproductive output.

Compound eye has a large field of vision

competing for mates

Many animals have adaptations of behavior or anatomy that are driven by mate choice (sexual selection), rather than the struggle for existence (natural selection). Where reproductive success is based on competition between either males or females, a particular trait shown by individuals of one sex, such as bright colors or elaborate antlers, may cause the opposite sex to choose mating with them over a competitor. These mating choices may be linked to producing fit, healthy offspring more likely to survive to adulthood, but sometimes, the value is simply in producing attractive offspring more likely to be chosen by the next generation of animals looking for mates. One unusual example of a reproductive trait is the excessively widely spaced eyes of male stalk-eyed flies.

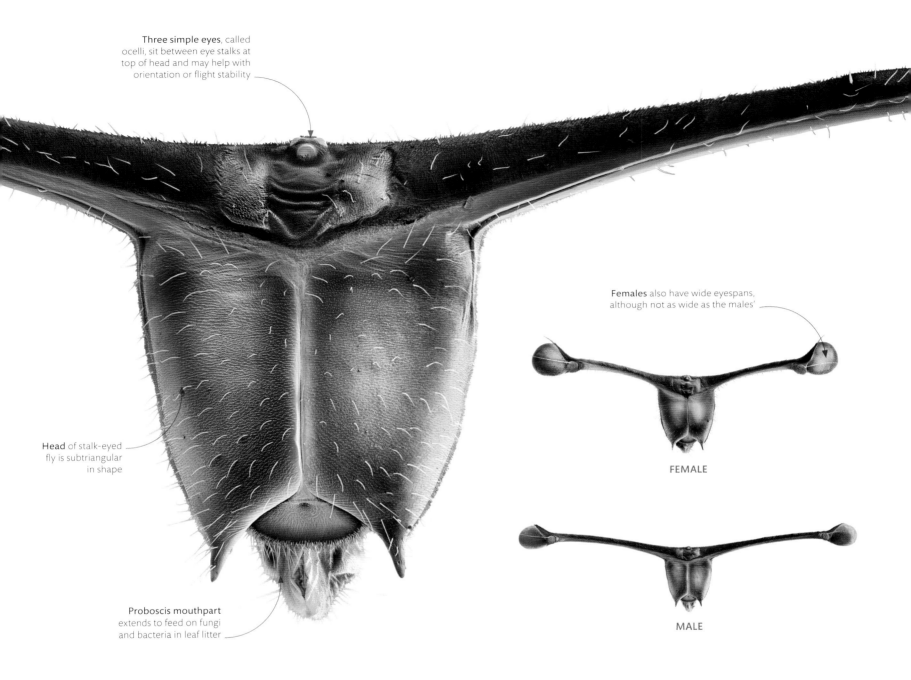

Three simple eyes, called ocelli, sit between eye stalks at top of head and may help with orientation or flight stability

Head of stalk-eyed fly is subtriangular in shape

Proboscis mouthpart extends to feed on fungi and bacteria in leaf litter

Females also have wide eyespans, although not as wide as the males'

FEMALE

MALE

Pairs of male stalk-eyed flies compete face-to-face for mating rights

Eyeing up each other
Male stalk-eyed flies compete in a courtship ritual, called lekking, where they face each other and compare eyespan. Females appear to prefer to mate with the males that have the widest eyespan. An extreme eyespan may indicate to the female that the male has greater fertility or genetic quality.

SEXUAL SELECTION
Reproductive traits, such as the stalk-eyed fly's eyespan, help animals find mates that will improve their ability to leave offspring—this is called sexual selection. It has lead to sexual dimorphism, where males and females look or behave differently. Male birds of paradise (*Cicinnurus magnificus*), for example, have bright plumage to attract a mate, but females do not.

MALE BIRDS OF PARADISE COMPETE FOR A MATE

Wasp builds nest from wet mud that hardens as it dries

BUILDING NEST

Single egg is laid directly into the nest

LAYING EGG

Wasp paralyzes prey with her sting, then places it in the nest

DELIVERING PREY

Nursery and pantry provided
An orange potter wasp (*Delta latreillei*) provides shelter as well as food for her larva. She collects mud to build a potlike nest, then lays an egg, before supplying paralyzed caterpillars and other insects to feed the carnivorous larva.

Female spiders of this species are $1/3–2/3$ in (8.5–15 mm) long

parental care

Every kind of living thing does whatever it can to pass on its genes in the most effective way. For many, this involves generating huge numbers of spores, seeds, eggs, or babies, so that chance alone ensures that a few will escape predators and other dangers and survive. But some animals raise the odds by being good parents. The micro-world has some remarkable examples of this kind of family life: spiders that carry their eggs, earwigs that tend and keep their young clean, wasps that feed their larvae. In nature's economy, parents that invest time and energy like this compensate by producing fewer offspring, but each one is more likely to survive.

Spider grasps egg sac with fangs, or chelicerae

Egg sac is made from spider's silk and contains more than 100 eggs, although fewer than 50 percent are fertilized

Feelerlike pedipalps are angled downward to help spider hold the egg sac against the underside of the body

From predator to parent
For 3 weeks, a female nursery-web spider (*Pisaura mirabilis*) will not feed while she carries her egg sac in her fangs. Days before she produced her eggs, the male offered her a silk-wrapped fly as a nuptial gift—which distracted her long enough for him to mate. When the eggs are ready to hatch, the female spider builds a tentlike web to protect her brood until they disperse.

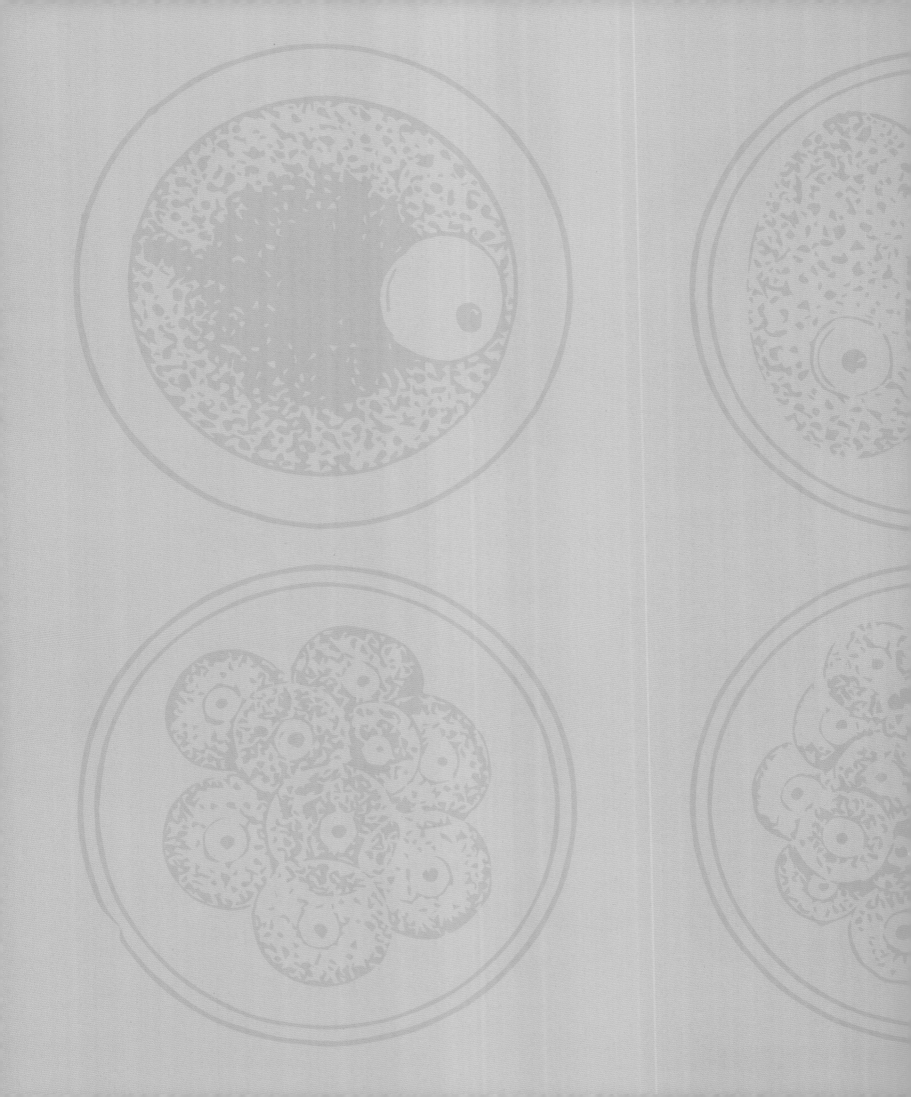

growing
and changing

Living creatures grow—growth is a key attribute that defines life—but multicelled life grows in a complex, organized way involving cell division, divergence, and specialization. The start of an individual's life can also be the beginning of a multistage life cycle with phases devoted to growth, reproduction, or dispersal.

Single-celled alga
Cells of the alga *Chlamydomonas* sp. are similar to those of *Volvox* in having two flagella, a chloroplast, and an eyespot. However, this species never forms colonies and stays single-celled.

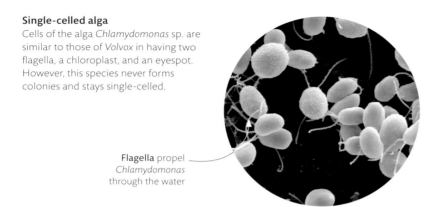

Flagella propel *Chlamydomonas* through the water

Cells working together
Each cell in a colony of *Volvox* sp. carries a green chloroplast that absorbs light energy for photosynthesis and two invisible beating flagella (threads) for propulsion; some develop bigger light-sensitive eyespots than others. The colony moves closer to light detected by the eyespots, while new colonies produced inside bud from reproductive cells. LM, x500.

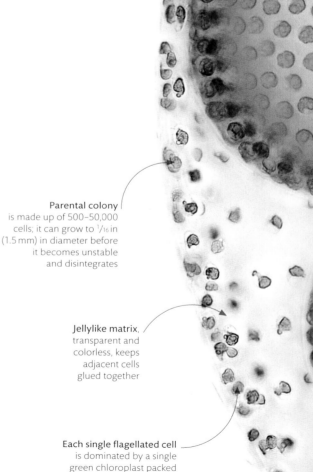

Cells of smaller embryonic colonies that form inside the parental colony are more closely packed than those of the parent; all cells communicate with their neighbors via fine threads of cytoplasm

colonies of cells

While many microbes are single-celled, animals and plants have bodies that are made up of countless cells that work together and are specialized to perform particular tasks. However, some living things seem to bridge the gap between these levels of organization. *Volvox* is a tiny spherical pond alga just visible to the naked eye. It is a hollow ball of cells, which are glued together in a thin layer of jelly around a watery interior. Although it is a colony, it has differentiated parts: some cells are better at sensing light, while others help with reproduction.

Parental colony is made up of 500–50,000 cells; it can grow to $1/16$ in (1.5 mm) in diameter before it becomes unstable and disintegrates

Jellylike matrix, transparent and colorless, keeps adjacent cells glued together

Each single flagellated cell is dominated by a single green chloroplast packed with chlorophyll

ASEXUAL REPRODUCTION OF NEW VOLVOX COLONIES
Development of new *Volvox* colonies involves intricate movements of cells. Surface cells lose their flagella and divide to produce a cup that must then turn inside out to ensure that new colonies can keep growing their flagella on the outside.

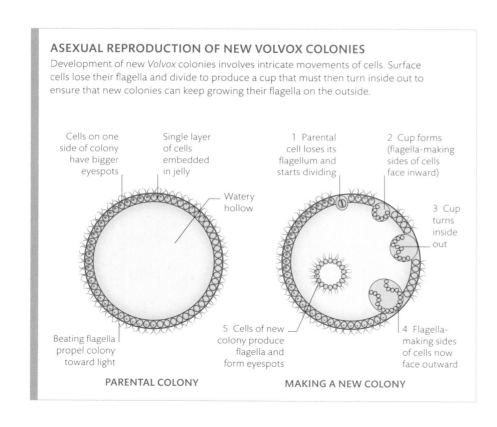

Cells on one side of colony have bigger eyespots

Single layer of cells embedded in jelly

1 Parental cell loses its flagellum and starts dividing

2 Cup forms (flagella-making sides of cells face inward)

Watery hollow

3 Cup turns inside out

Beating flagella propel colony toward light

5 Cells of new colony produce flagella and form eyespots

4 Flagella-making sides of cells now face outward

PARENTAL COLONY

MAKING A NEW COLONY

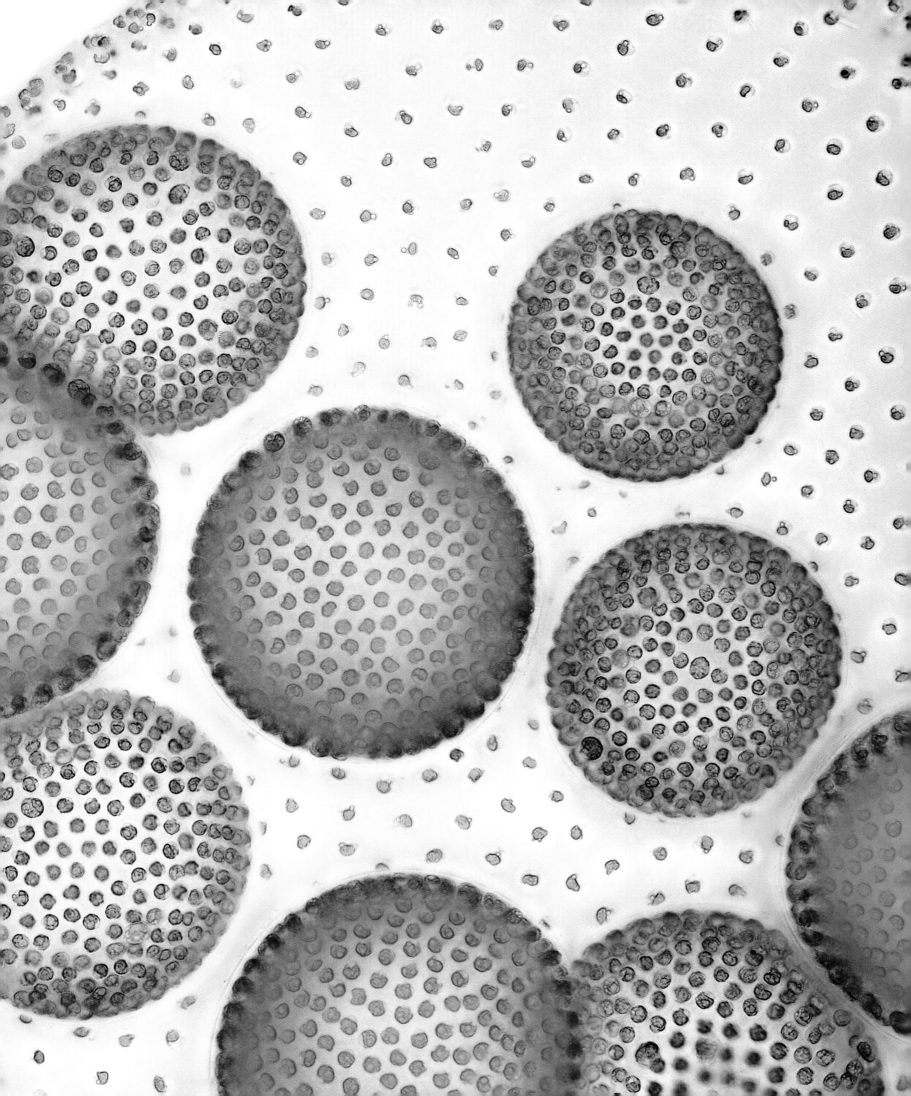

cell division

Life is made up of cells, and new cells originate when older ones divide. The growing body of an animal or plant has trillions of them and increases in size by building protein and other new material, and through dividing its cells. This repeated cycle of building and division is responsible for growth, a body's ability to repair damage, and for development. Every cell carries a copy of the genes from the single fertilized egg, and the genes can be switched on or off in different parts of the body so they can develop in different ways.

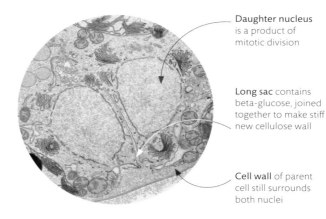

Daughter nucleus is a product of mitotic division

Long sac contains beta-glucose, joined together to make stiff new cellulose wall

Cell wall of parent cell still surrounds both nuclei

Dividing plant cells
Plant and algae cells are surrounded by a rigid cell wall, so lack the flexibility to split by simple constriction. Instead, adjacent daughter cells assemble a wall between them.

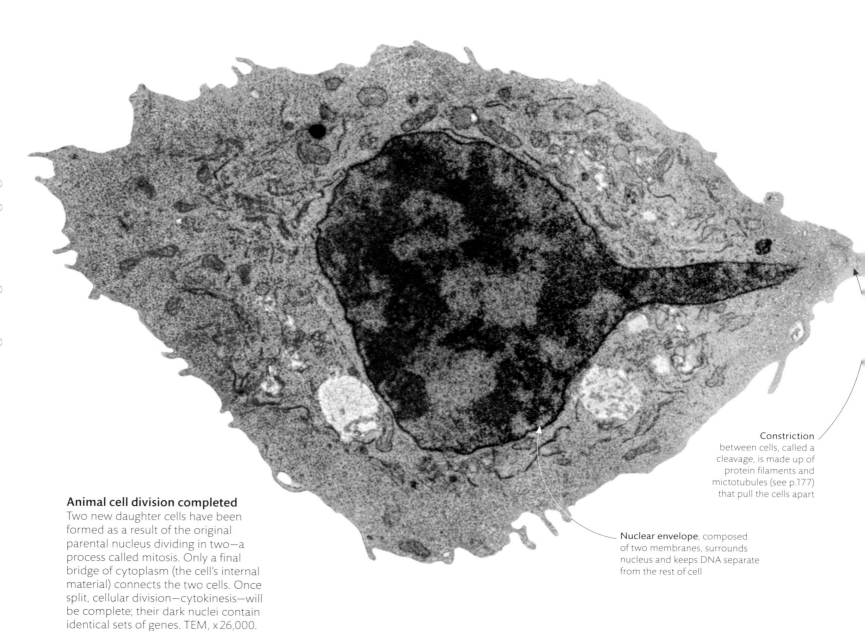

Animal cell division completed
Two new daughter cells have been formed as a result of the original parental nucleus dividing in two—a process called mitosis. Only a final bridge of cytoplasm (the cell's internal material) connects the two cells. Once split, cellular division—cytokinesis—will be complete; their dark nuclei contain identical sets of genes. TEM, x26,000.

Constriction between cells, called a cleavage, is made up of protein filaments and mictotubules (see p.177) that pull the cells apart

Nuclear envelope, composed of two membranes, surrounds nucleus and keeps DNA separate from the rest of cell

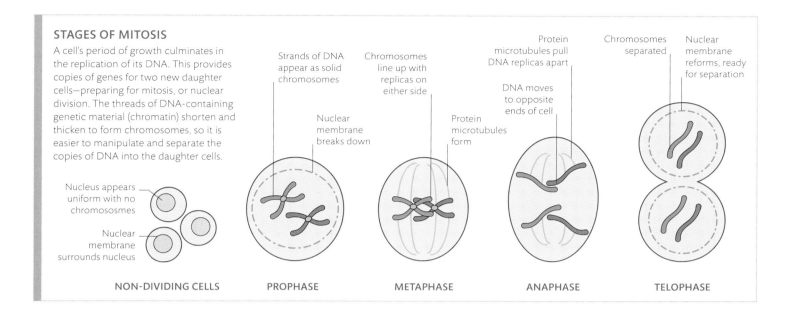

STAGES OF MITOSIS

A cell's period of growth culminates in the replication of its DNA. This provides copies of genes for two new daughter cells—preparing for mitosis, or nuclear division. The threads of DNA-containing genetic material (chromatin) shorten and thicken to form chromosomes, so it is easier to manipulate and separate the copies of DNA into the daughter cells.

- Nucleus appears uniform with no chromososmes
- Nuclear membrane surrounds nucleus

NON-DIVIDING CELLS

- Strands of DNA appear as solid chromosomes
- Nuclear membrane breaks down

PROPHASE

- Chromosomes line up with replicas on either side
- Protein microtubules form

METAPHASE

- Protein microtubules pull DNA replicas apart
- DNA moves to opposite ends of cell

ANAPHASE

- Chromosomes separated
- Nuclear membrane reforms, ready for separation

TELOPHASE

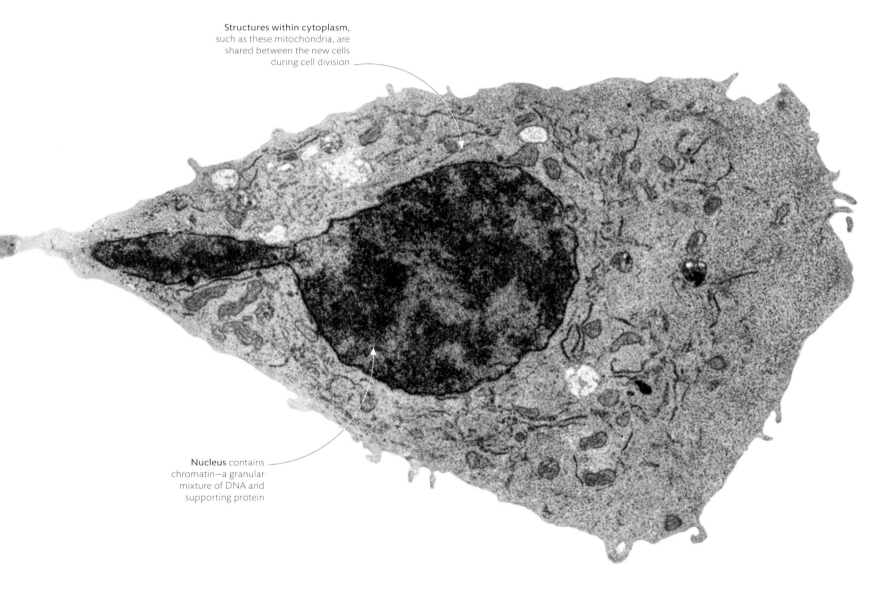

Structures within cytoplasm, such as these mitochondria, are shared between the new cells during cell division

Nucleus contains chromatin—a granular mixture of DNA and supporting protein

developing embryo

Unless it develops asexually (see pp.226–27), plant and animal life begins as a fertilized egg that contains all the genetic material needed for an adult body—made up of trillions of cells—to form. Immediately after fertilization, cells start to divide and take up the positions that will seal their fate as they develop into the tissues and organs of the different parts of the adult. Although the cells in the fertilized egg start as genetic clones, they are all destined to do different jobs and the genes they carry are switched on, or off, to program the formation of different parts of the body.

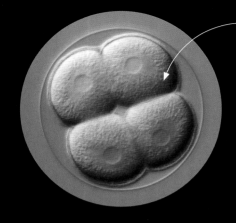

Blastula made of four equally sized cells, created by two cell divisions

Ball of cells
The fertilized egg cell of a sea urchin (*Paracentrotus lividus*) begins cell division. All animal embryos start life like this, as a ball of cells called a blastula.

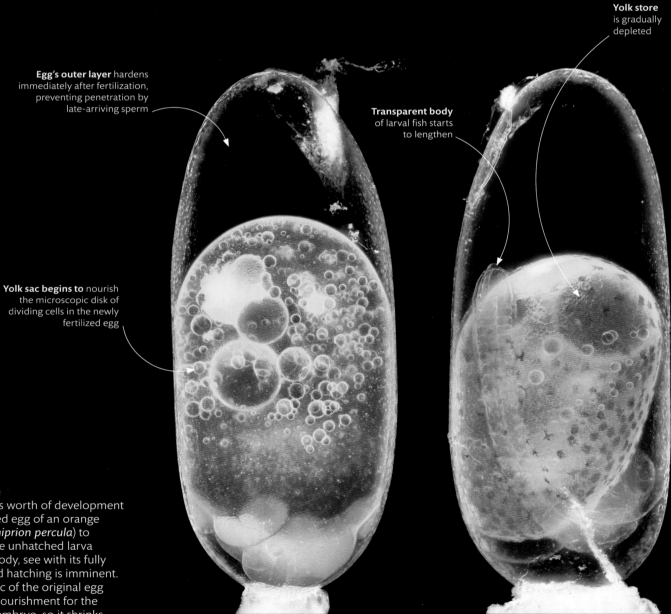

Yolk store is gradually depleted

Egg's outer layer hardens immediately after fertilization, preventing penetration by late-arriving sperm

Transparent body of larval fish starts to lengthen

Yolk sac begins to nourish the microscopic disk of dividing cells in the newly fertilized egg

Becoming a fish
Just over a week's worth of development takes the fertilized egg of an orange clownfish (*Amphiprion percula*) to a stage where the unhatched larva can wriggle its body, see with its fully formed eyes, and hatching is imminent. The large yolk sac of the original egg supplies all the nourishment for the developing fish embryo, so it shrinks as the fish grows. LM, x150.

DAY ONE: FERTILIZATION

DAY THREE, MORNING

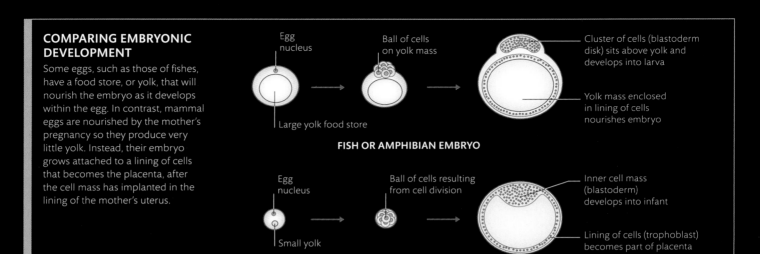

COMPARING EMBRYONIC DEVELOPMENT

Some eggs, such as those of fishes, have a food store, or yolk, that will nourish the embryo as it develops within the egg. In contrast, mammal eggs are nourished by the mother's pregnancy so they produce very little yolk. Instead, their embryo grows attached to a lining of cells that becomes the placenta, after the cell mass has implanted in the lining of the mother's uterus.

FISH OR AMPHIBIAN EMBRYO — Egg nucleus; Large yolk food store; Ball of cells on yolk mass; Cluster of cells (blastoderm disk) sits above yolk and develops into larva; Yolk mass enclosed in lining of cells nourishes embryo

PLACENTAL MAMMAL EMBRYO — Egg nucleus; Small yolk; Ball of cells resulting from cell division; Inner cell mass (blastoderm) develops into infant; Lining of cells (trophoblast) becomes part of placenta

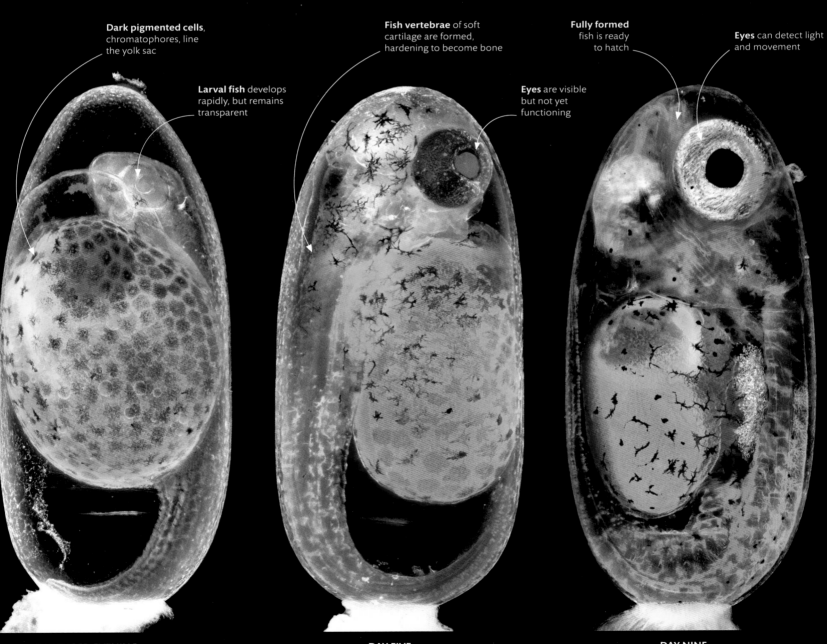

DAY THREE, EVENING
- **Dark pigmented cells**, chromatophores, line the yolk sac
- **Larval fish** develops rapidly, but remains transparent

DAY FIVE
- **Fish vertebrae** of soft cartilage are formed, hardening to become bone
- **Eyes** are visible but not yet functioning

DAY NINE
- **Fully formed** fish is ready to hatch
- **Eyes** can detect light and movement

insect eggs

EGG PODS

Some kinds of insects lay clusters of eggs that are united within a single protective pod, called an ootheca. The oothecae of cockroaches are characteristically purse-shaped and deposited on the ground, while many grasshoppers produce tubular oothecae that are buried below the surface. Praying mantises attach their oothecae to stems. The number of eggs per ootheca varies from just a few to several hundred, depending upon the insect species.

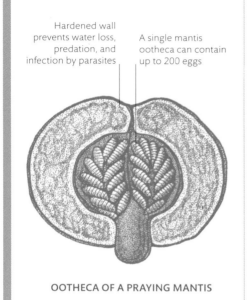

Hardened wall prevents water loss, predation, and infection by parasites

A single mantis ootheca can contain up to 200 eggs

OOTHECA OF A PRAYING MANTIS

To reproduce on land, animals mate so their sperm can swim to the egg inside the female's body—the fertilized eggs may then be laid in a shell that stops them from drying out. Insects have been doing this for more than 300 million years, being among the first animals to live on land. A protein envelope, or chorion, around the eggs hardens to form a protective coat, often with an intricate pattern of sculpturing. The chorion traps water inside, but is also sufficiently porous to breathe—while the egg's store of yolk supplies nourishment. This provides perfect conditions for the embryo to develop, until it hatches—as a nymph (a young version of the adult) or larva.

Sculptured egg

The egg of a zebra longwing butterfly (*Heliconius charithonia*) is patterned with interconnecting ridges. The geometric shapes—common on many insect eggs—result from a similar honeycomb pattern of secretory cells in the butterfly's ovaries that leave an impression as the egg shell forms. Tiny pores are concentrated on the ridges, which helps oxygen reach the embryo inside. SEM, x300.

A place to hatch

Many insects produce gluelike secretions from their abdomen that can stick their eggs to places that benefit their young. Shield bugs lay eggs on plants that can nourish the hatchlings, while eggs that hang from threads may be protected from leaf-crawling predators.

Eggs are laid on leaf, which will become food for hatchlings

SHIELD BUG

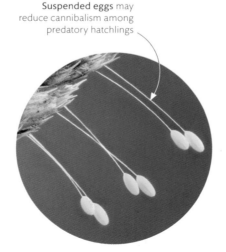

Suspended eggs may reduce cannibalism among predatory hatchlings

LACEWING

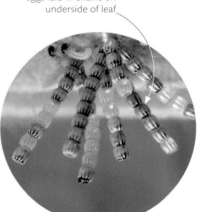

Larvae hatch from eggs laid in chains on underside of leaf

MAP BUTTERFLY

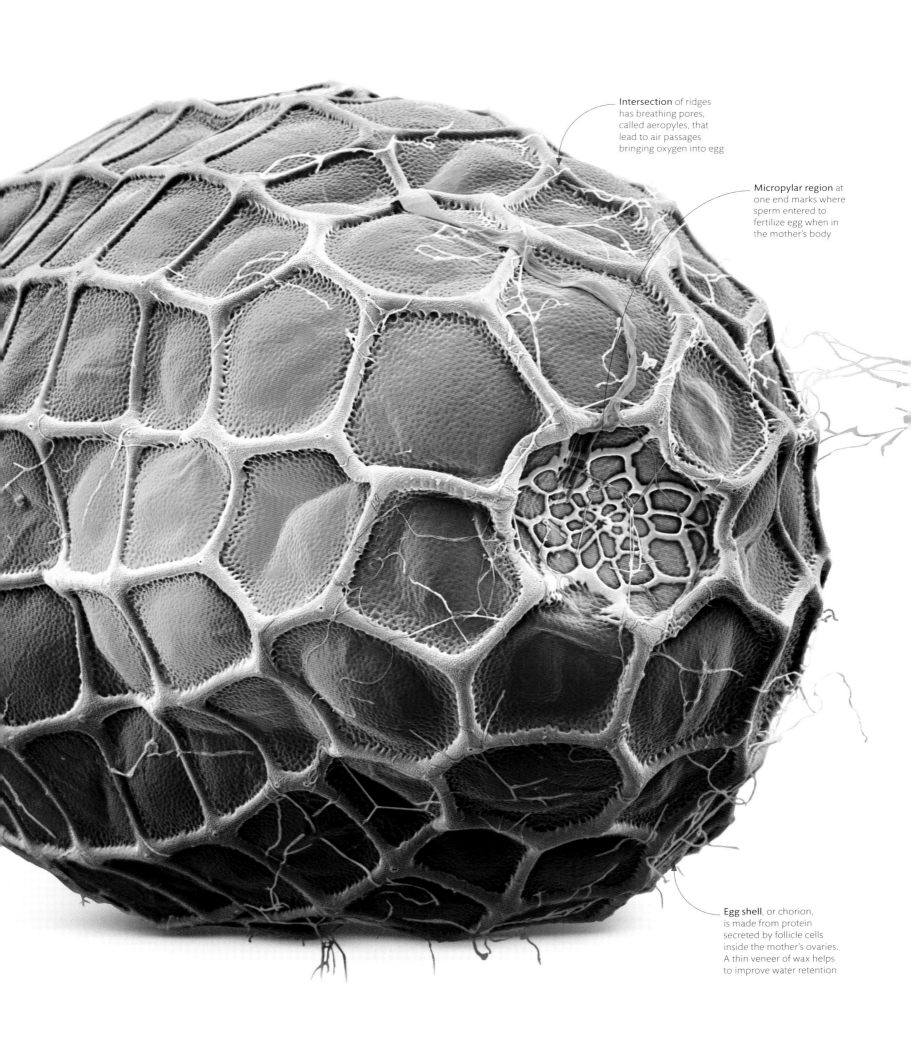

MACROLIFE

how ferns grow

Plants and animals grow from single cells—spores or fertilized eggs—that contain the genetic blueprint needed to develop an adult body. Fertilized eggs are products of two individuals, a male and female parent (see pp.228–29), but spores come from one parent (see pp.232–33). Ferns produce spores in tiny capsules on the underside of their fronds. The capsules dry, split, and disperse the spores, which will germinate if they land where moisture and light can stimulate them. They pass through a tiny leafless stage that produces eggs and sperm before a leafy fern grows from a fertilized egg.

The fronds of most ferns are divided into leaflets called pinnae

FROND

Fertile frond
Fern spores are usually produced beneath the fronds. In most species, such as this *Dryopteris filix-mas*—commonly known as the "male fern"—each cluster of spore capsules, called a sorus, lies under a scale. The scale, or indusium, peels back when the spores are ready to be released.

Sori are arranged in rows along each leaflet; here, the indusium has flaked away to expose the spores beneath

Each capsule encloses developing spores and is attached to the fern leaf by a short stalk beneath

TWO-STAGE GROWTH

As in mosses (see pp.232–33), two alternating stages, or generations, of a fern's life cycle begin with a single cell. A spore is a cell with a single, or haploid, set of chromosomes. It develops into a simple "prothallus," and it is this that produces sperm and eggs. Eggs, once fertilized, develop into larger, leafy ferns, and—because they result from fusion of sperm and egg—their cells carry a double, or diploid, set of chromosomes. Diploid chromosomes occur in many animals and plants, and may help to mask malfunctioning genes by having a "backup" set.

Female sex organ with egg

Male sex organs produce sperm and cross-fertilize other prothalli

UNDERSIDE OF PROTHALLUS

Spore divides to form a haploid prothallus

Single-celled spore contains a haploid set of chromosomes

Prothallus produces haploid egg cells

Fertilized diploid egg

Diploid leafy fern

Adult plant produces haploid spores

Sorus on underside of mature frond

×10

LIFE SIZE

THE TWO GENERATIONS OF GROWTH IN A FERN

Spore capsules
These spore capsules, or sporangia, of a sword fern (*Nephrolepis* sp.) are ripe for releasing their contents. Each has a ridged rim, or annulus, composed of thin-walled cells that easily lose water. As each annulus dries, it shrivels and breaks away from the rest of the sporangium, snapping open and letting the spores disperse. LM, ×700.

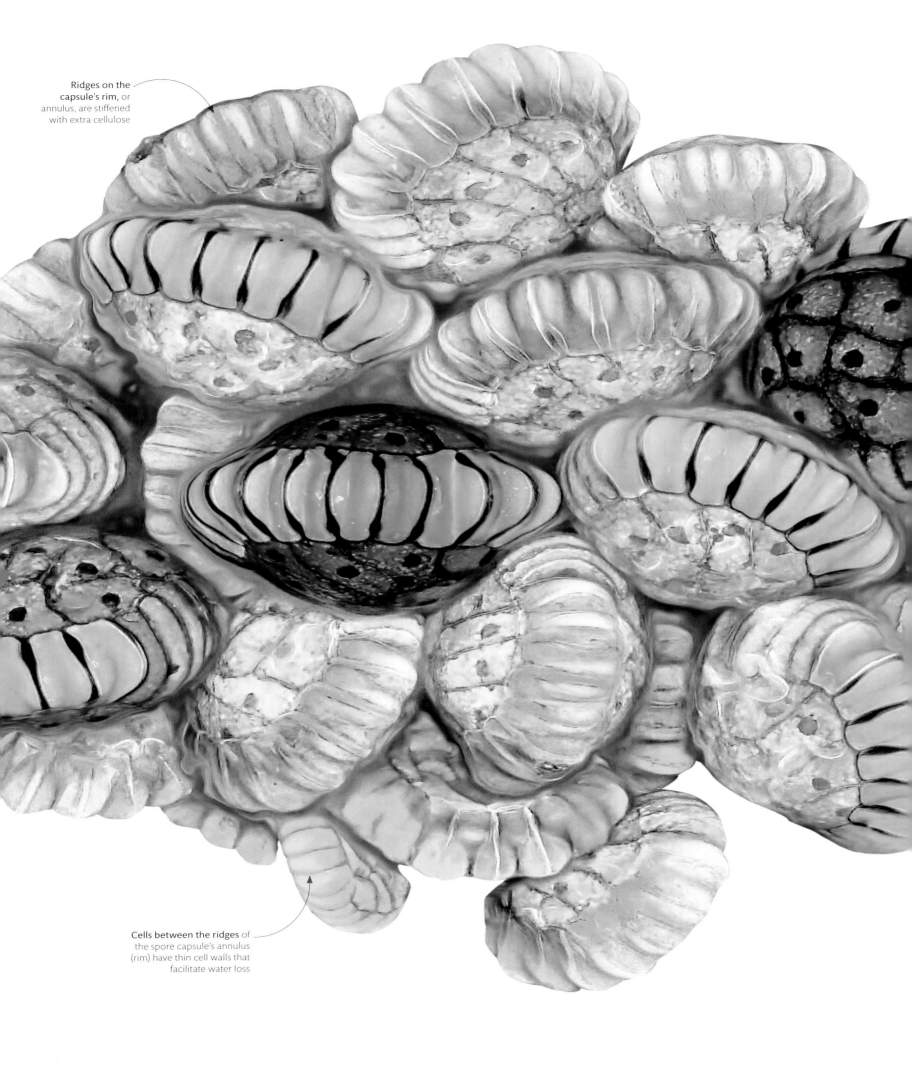

Ridges on the capsule's rim, or annulus, are stiffened with extra cellulose

Cells between the ridges of the spore capsule's annulus (rim) have thin cell walls that facilitate water loss

Gymnosperm
Conifers and their relatives are nonflowering plants called gymnosperms, which means "naked seeds." Their seeds lack the fruit casing possessed by the seeds of flowering plants, or angiosperms ("coated seeds"). Gymnosperm seeds grow inside female cones (male cones produce pollen), and are released into the air. In the case of pine cones, the seeds have wings to help them disperse when they fall to the ground as the cone dries out.

Comparatively large seeds, up to 1½ in (4 cm) in size, are a favorite food of possums

BUNYA PINE
Araucaria bidwillii

Nutlike seed about ¾ in (2 cm) in size develops inside a fleshy, outer covering (not shown) with a rancid smell

GINKGO
Ginkgo biloba

Moist orange coating needs to dry out before the seed can sprout

SAGO PALM
Cycas revoluta

Monocots
About one-fifth of flowering plants are monocots (short for monocotyledons). Their seeds contain a single cotyledon, or embryonic leaf. Monocots include grasses, sedges, palm trees, and the amaryllis family. The seeds of monocot plants generally contain a starchy food store known as an endosperm. In grain crops, such as wheat and rice, the endosperm makes up most of the seed.

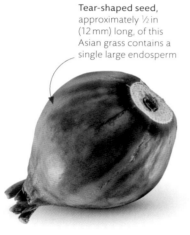

Tear-shaped seed, approximately ½ in (12 mm) long, of this Asian grass contains a single large endosperm

JOB'S TEARS
Coix lacryma

Glossy black seed is roughly triangular in cross section

ONION
Allium cepa

Sac (perigynium) enveloping the seed-containing fruit is formed from a specialized leaf called a bract

SEDGE
Carex buekii

Eudicots
Plants with two embryonic leaves in their seeds are known as eudicots. Unlike a monocot seed, a eudicot seed has a reduced endosperm. Instead, the cotyledons themselves may store the food needed to fuel the seed's germination, or they may open shortly after germination and produce it by photosynthesis. The majority of flowering plants—about 200,000 species—are eudicots.

Hard, grooved fruit casing called a schizocarp surrounds the seed

FENNEL
Foeniculum vulgare

Several flattened, oval seeds are dispersed inside a capsule

SESAME
Sesamum indicum

Lightweight seed with pitted surface is easily carried on the wind

COMMON POPPY
Papaver rhoeas

seeds

A seed is the product of plant reproduction, when a male gamete delivered by a pollen grain fuses with its female counterpart inside a flower or cone. The single, fertilized cell develops into a plant embryo that is supplied with nutrients and sealed in a protective casing. Seeds are adapted to a particular mode of dispersal. Airborne seeds are small and lightweight, while those spread by water have a low density and thus are buoyant. Seeds hidden inside sweet fruits are eaten by animals and then excreted elsewhere.

Hard shell contains a rich store of fats, proteins, and sugars in this ¾ in (2 cm) seed

STONE PINE
Pinus pinea

Tiny seed, up to 0.9 mm in length, has no endosperm and relies on nutrients from soil fungi to fuel growth

ORCHID
Neottia ovata

Ridges on the seed's surface cling to animal fur to aid dispersal

CARROT
Daucus carota

Sedge seed head
The seed head of a sedge (*Carex* sp.) develops from a cluster of flowers on a single stem—a structure called an inflorescence. The seed head gradually dries out, and when ready it will "shatter"—the slightest knock or puff of wind will release the loosened seeds, scattering them to the ground.

Seed head forms from dozens of tiny individual flowers, each of which produces a single seed

Each tubular sac, or perigynium, houses a hard fruit called an achene, which itself contains a seed

Large food stores

Bread wheat (*Triticum aestivum*) seeds have an extra-large food store, as can be seen in these scanning electron micrographs of a germinating grain. Germination is triggered by warmth, moisture, and light, and the starch in the body of the seed provides the seedling with energy. Wheat is classed as a monocot, as it only produces one cotyledon (seed leaf); plants that produce two are called eudicots. SEM, x30.

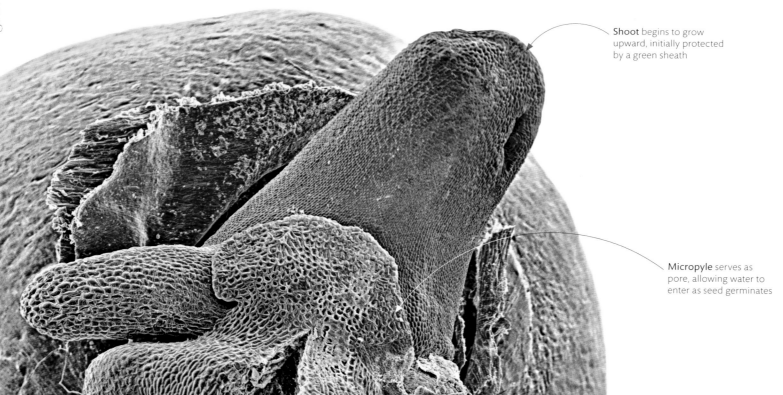

Micropyle is the weakened part of seed coat, from which the radicle emerges

Primary root, or radicle, grows downward, anchoring the seedling in the soil

Seed coat forms protective outer layer around seed's large starch store

Hairs on the radicle increase the root's ability to absorb moisture and minerals from the soil

Shoot begins to grow upward, initially protected by a green sheath

Micropyle serves as pore, allowing water to enter as seed germinates

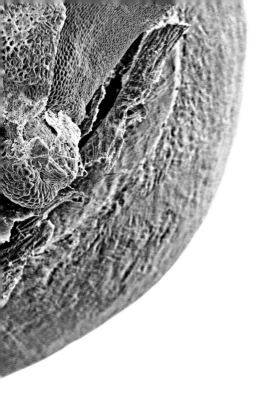

GERMINATING ABOVE OR BELOW GROUND

Flowering plant embryos contain cotyledons that act as early food storage organs. When plants such as French beans germinate, the stem below the cotyledons, called the hypocotyl, elongates and lifts the cotyledons above ground (epigeal germination). They wither once their food store is depleted. In wheat, in contrast, the upper stem (epicotyl) expands, leaving the cotyledon underground (hypogeal germination).

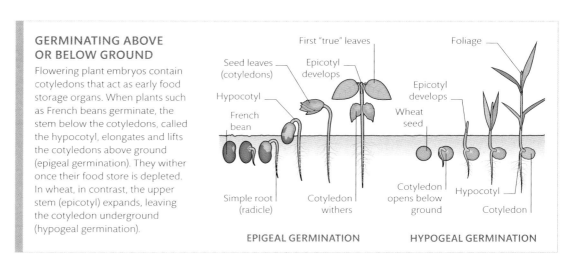

EPIGEAL GERMINATION | HYPOGEAL GERMINATION

seed germination

Most plant seeds are encased in a fleshy or papery fruit that encourages dispersal by animals or the wind. Every seed contains an embryo from which the plant develops, plus starch and proteins that provide an initial food supply for the seedling. Wind-dispersed seeds must be small and light, but the presence of lots of food for the embryo makes seeds big and heavy. Wheat (*Triticum* sp.) seeds, although light when dispersed by wind in the wild, now contain a huge volume of starch due to domestication by humans. Such large, heavy seeds cannot be dispersed by the wind, and wheat must rely on human help to scatter its seed.

Breeding bread

The small, light seeds of wild einkorn are dispersed easily by the wind. This plant naturally crossed with a wild goat grass, resulting in a plumper hybrid—emmer wheat, triggering some of the first cultivation by humans. Emmer crossed with another goat grass to produce the even larger hybrid, bread wheat, which has such fat seeds that it is wholly reliant on humans for seed dispersal.

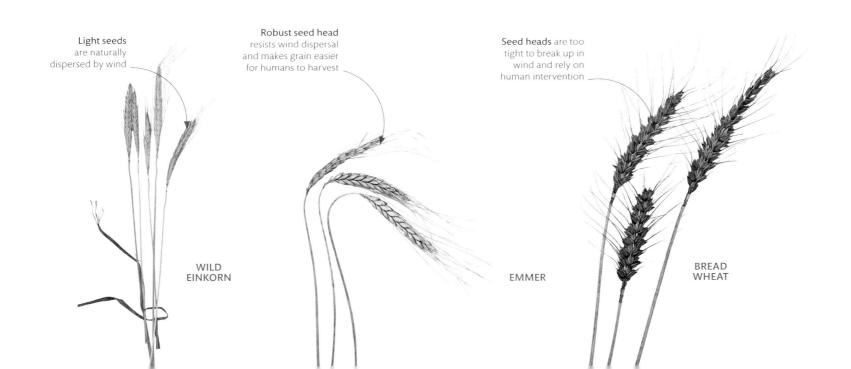

Light seeds are naturally dispersed by wind

WILD EINKORN

Robust seed head resists wind dispersal and makes grain easier for humans to harvest

EMMER

Seed heads are too tight to break up in wind and rely on human intervention

BREAD WHEAT

simple plants

Large plants develop tough stems that hold their leaves high into the air and roots that penetrate deep into the ground—adaptations that help them absorb light and carbon dioxide from above, and water and minerals from below. But the simplest plants—among the first to colonize land nearly half a billion years ago—hug the ground. Liverworts and mosses lack the complex layers of tissues needed to produce big leaves and roots. Their simple leaves and roots, called respectively phyllids and rhizoids, are just one cell thick. And without the stiff reinforcement present in more complex plants, they cannot grow very tall.

Underside of a leafy liverwort
Many liverworts, such as this creeping fingerwort (*Lepidozia reptans*) and most mosses, develop tiny leaflike phyllids that are very thin and lack veins. Their simple tissues easily dry out but are also highly absorbent, so most liverworts and mosses thrive best in moist habitats. LM, x530.

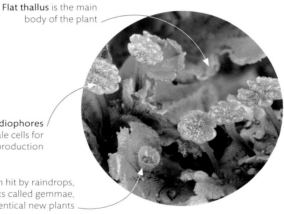

Flat thallus is the main body of the plant

Spreading its offspring
A common liverwort (*Marchantia* sp.) lacks even phyllids—it is a flat, leafless sheet, or thallus. The only upright structures are the reproductive bodies—this one shows both asexual and sexual structures. The plant is anchored to the ground by rhizoids—fine, rootlike threads.

Parasol-shaped antheridiophores produce male cells for sexual reproduction

Gemma cup, when hit by raindrops, scatters asexual disks called gemmae, which form identical new plants

Colors generated by a stain that fluoresces in ultraviolet light, highlighting the liverwort's cellular structure

SIMPLE AND COMPLEX LEAVES

Mosses and liverworts have phyllids made of a single cell layer used for both photosynthesis and exchange of gases. Some mosses have a midrib that carries narrow transport vessels. Ferns and other "vascular" plants (see pp.206–07) have thicker leaves with reinforced vessels, tissue layers specialized either in photosynthesis or gaseous exchange, and a waxy cuticle that reduces dehydration.

Phyllid consisting of single layer of photosynthetic cells that requires constant dampness

Supportive, pliable, midrib

Simple water transport vessels that lack secondary thickening by lignin

SECTION THROUGH PHYLLID (NONVASCULAR "LEAF") OF A MOSS

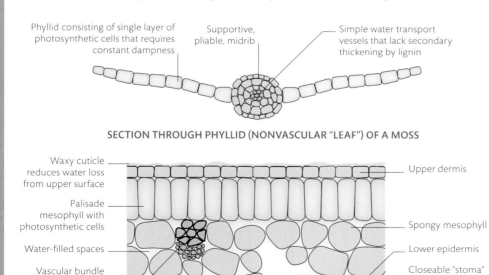

Waxy cuticle reduces water loss from upper surface

Upper dermis

Palisade mesophyll with photosynthetic cells

Spongy mesophyll

Water-filled spaces

Lower epidermis

Vascular bundle of transport vessels, thickened with lignin

Closeable "stoma" opening controls water loss by transpiration

SECTION THROUGH LEAF OF A VASCULAR PLANT

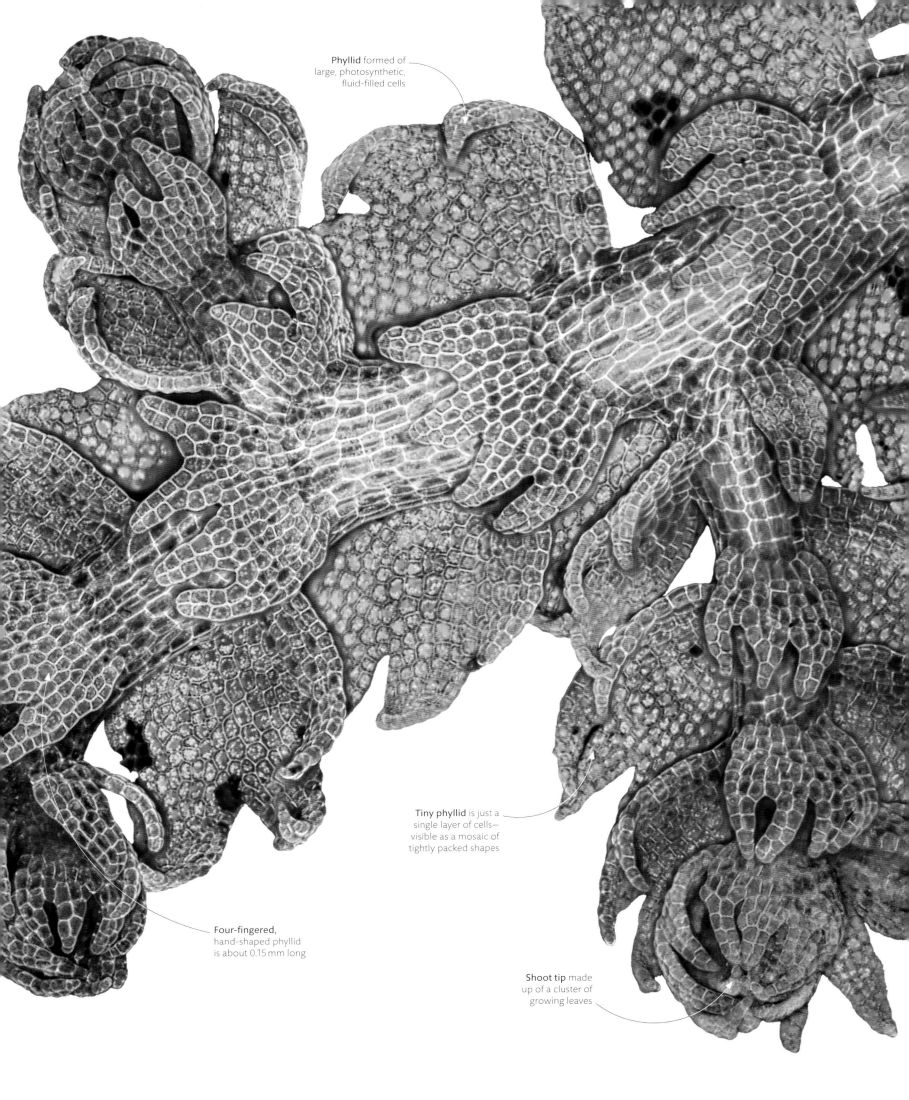

Phyllid formed of large, photosynthetic, fluid-filled cells

Tiny phyllid is just a single layer of cells—visible as a mosaic of tightly packed shapes

Four-fingered, hand-shaped phyllid is about 0.15 mm long

Shoot tip made up of a cluster of growing leaves

growing up as plankton

Many marine animals begin life as larvae that drift in the plankton (see pp.286–87), but ultimately descend to the sea bed and develop into urchins, crabs, snails, and other bottom-living invertebrates. At the mercy of predators and currents, larvae often have only spines or shells for protection and a fringe of tiny beating hairs, or cilia, that provide propulsion for swimming and catching food. Over time, sometimes with multiple changes of shape, they metamorphose into sexually mature adults capable of producing eggs and larvae.

Drifting larva
A five-day-old larva of a sea potato (*Echinocardium cordatum*)—a type of sea urchin—is called an echinopluteus. Eight long arms carry stiff, calcareous rods that make up its skeleton. The arms, and parts of the rest of the body, bear the beating cilia. LM, x400.

LARVAE IN THE PLANKTON

The planktonic larvae of jointed-legged crustaceans such as shrimps, lobsters, and crabs occur in a bewildering variety of forms. Painstaking studies of development, where each larva is tracked over time, have enabled zoologists to link larvae to their adult forms.

PLANKTONIC CRUSTACEAN LARVAE; ENGRAVING BY ERNST HAECKEL (1904)

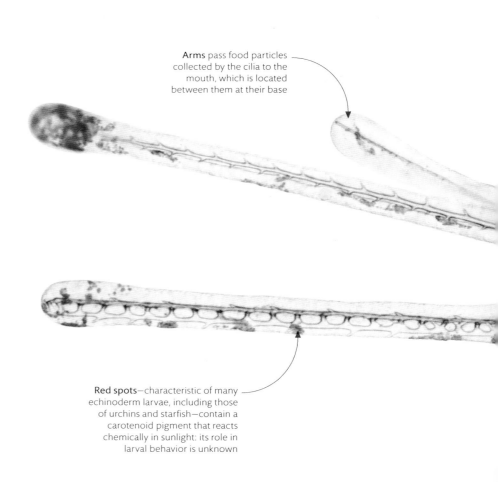

Spinelike arms protect against small planktonic predators and are sites where muscles attach; they are lost during metamorphosis, but the adult grows new arms

Arms pass food particles collected by the cilia to the mouth, which is located between them at their base

Red spots—characteristic of many echinoderm larvae, including those of urchins and starfish—contain a carotenoid pigment that reacts chemically in sunlight: its role in larval behavior is unknown

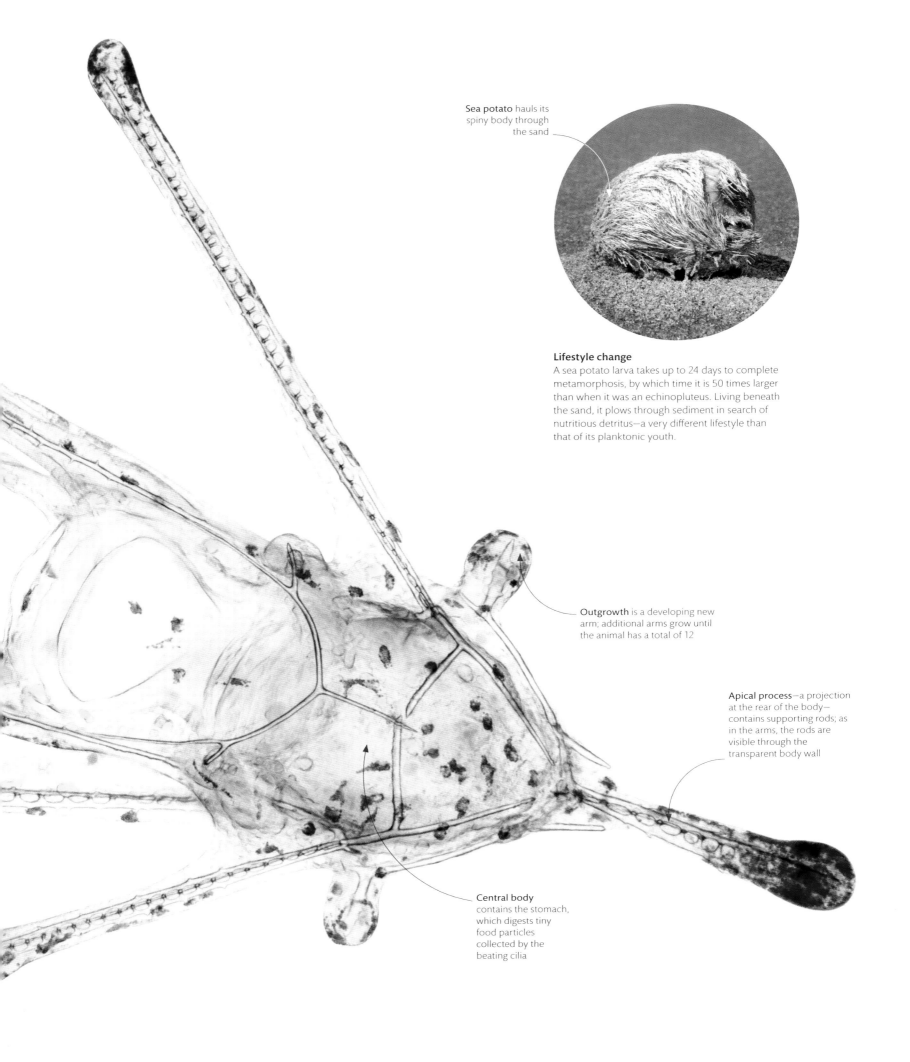

Sea potato hauls its spiny body through the sand

Lifestyle change
A sea potato larva takes up to 24 days to complete metamorphosis, by which time it is 50 times larger than when it was an echinopluteus. Living beneath the sand, it plows through sediment in search of nutritious detritus—a very different lifestyle than that of its planktonic youth.

Outgrowth is a developing new arm; additional arms grow until the animal has a total of 12

Apical process—a projection at the rear of the body—contains supporting rods; as in the arms, the rods are visible through the transparent body wall

Central body contains the stomach, which digests tiny food particles collected by the beating cilia

growing in steps

The armorlike cuticle, or exoskeleton, of an insect or other arthropod is a tough outer layer, but since it is not made from living cells, it cannot grow. Consequently, an insect must repeatedly shed its cuticle and make a new one to get bigger. As tissues build on the inside—including a new, as yet softer, cuticle beneath the old one—they trigger a pair of glands at the front of the body to release a hormone that starts a molting process. The old cuticle is loosened and splits, and the insect wriggles free. The new cuticle then expands before it hardens around its enlarged owner.

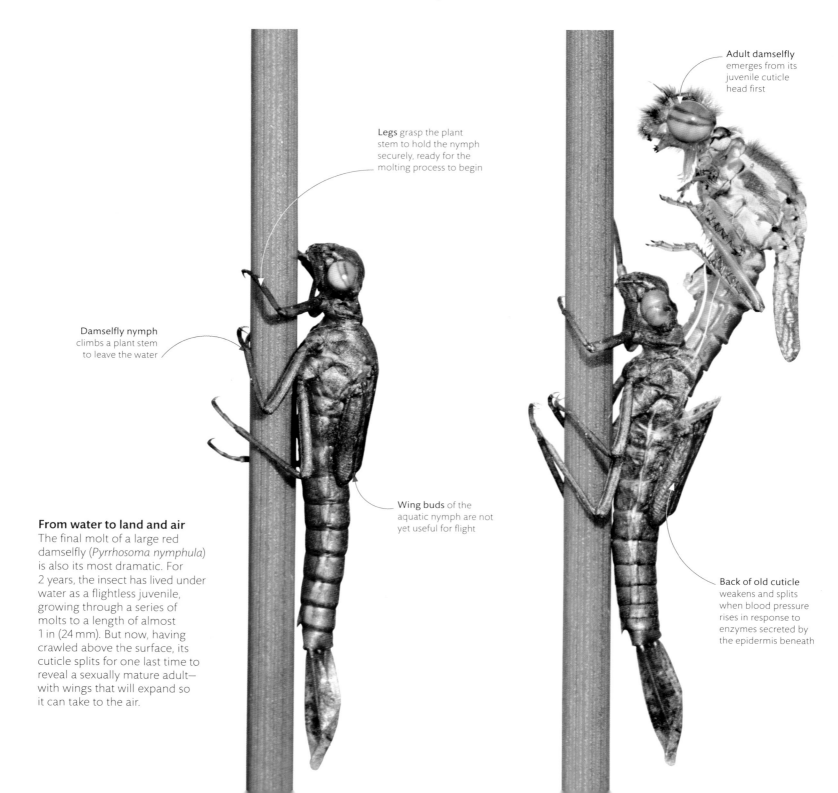

Adult damselfly emerges from its juvenile cuticle head first

Legs grasp the plant stem to hold the nymph securely, ready for the molting process to begin

Damselfly nymph climbs a plant stem to leave the water

Wing buds of the aquatic nymph are not yet useful for flight

Back of old cuticle weakens and splits when blood pressure rises in response to enzymes secreted by the epidermis beneath

From water to land and air
The final molt of a large red damselfly (Pyrrhosoma nymphula) is also its most dramatic. For 2 years, the insect has lived under water as a flightless juvenile, growing through a series of molts to a length of almost 1 in (24 mm). But now, having crawled above the surface, its cuticle splits for one last time to reveal a sexually mature adult—with wings that will expand so it can take to the air.

GROWING WITH METAMORPHOSIS

Many insects, such as grasshoppers, undergo the kind of incomplete metamorphosis seen in a damselfly: they grow as miniature versions of adults but lack fully formed wings. Other insects, such as beetles, go through a more complete metamorphosis: the juvenile larva is very different from the adult and passes through a pupa stage, during which its body structure is drastically reorganized.

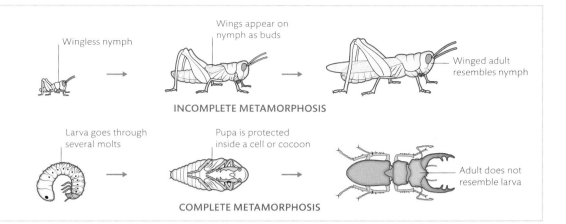

INCOMPLETE METAMORPHOSIS
- Wingless nymph
- Wings appear on nymph as buds
- Winged adult resembles nymph

COMPLETE METAMORPHOSIS
- Larva goes through several molts
- Pupa is protected inside a cell or cocoon
- Adult does not resemble larva

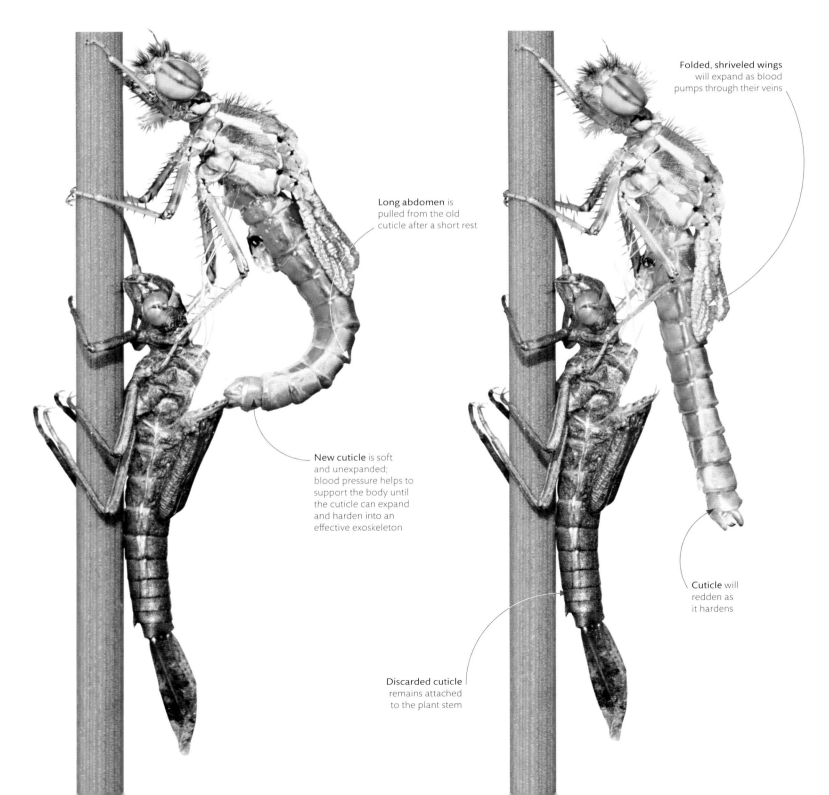

Long abdomen is pulled from the old cuticle after a short rest

New cuticle is soft and unexpanded; blood pressure helps to support the body until the cuticle can expand and harden into an effective exoskeleton

Folded, shriveled wings will expand as blood pumps through their veins

Cuticle will redden as it hardens

Discarded cuticle remains attached to the plant stem

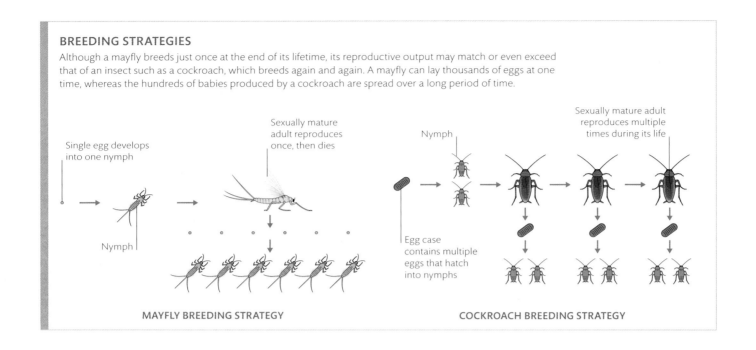

BREEDING STRATEGIES

Although a mayfly breeds just once at the end of its lifetime, its reproductive output may match or even exceed that of an insect such as a cockroach, which breeds again and again. A mayfly can lay thousands of eggs at one time, whereas the hundreds of babies produced by a cockroach are spread over a long period of time.

MAYFLY BREEDING STRATEGY

COCKROACH BREEDING STRATEGY

long and short lifespans

A life spent in the micro-world can be very short or surprisingly long. Some tiny animals, such as rotifers, may only live a few days; others, like tardigrades, can survive for years in a dormant state. While a long life can produce many offspring if it is mostly spent as a sexually mature adult, some organisms mature late and breed only once—but prolifically—at the end of their life. Mayflies can live for years as aquatic juveniles, or nymphs. Their metamorphosis into flying adults is the beginning of the end: within hours of swarming to mate and scatter eggs, all have perished.

Short-lived adult

Having lived as a wingless aquatic juvenile until now, a newly emerged adult mayfly has one goal: to reproduce. Juveniles have chewing mouthparts, so they can feed and grow, but the mouth of an adult can only imbibe water. Unable to refuel its body, it has just enough sufficient energy for its mating flights.

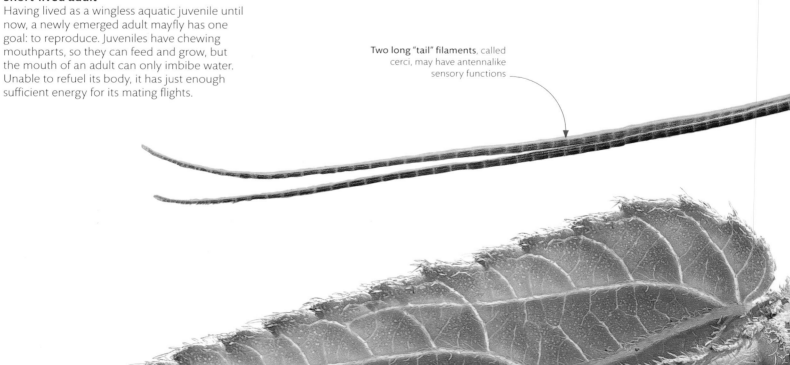

Two long "tail" filaments, called cerci, may have antennalike sensory functions

Aquatic youth
For the majority of their existence, mayflies live under water as juveniles called nymphs. To build up energy stores for their few hours as flying adults, most nymphs feed on algae, diatoms, and detritus; a few species are predatory.

Leaflike gills on a nymph's abdomen beat to help absorb plenty of oxygen from surrounding water

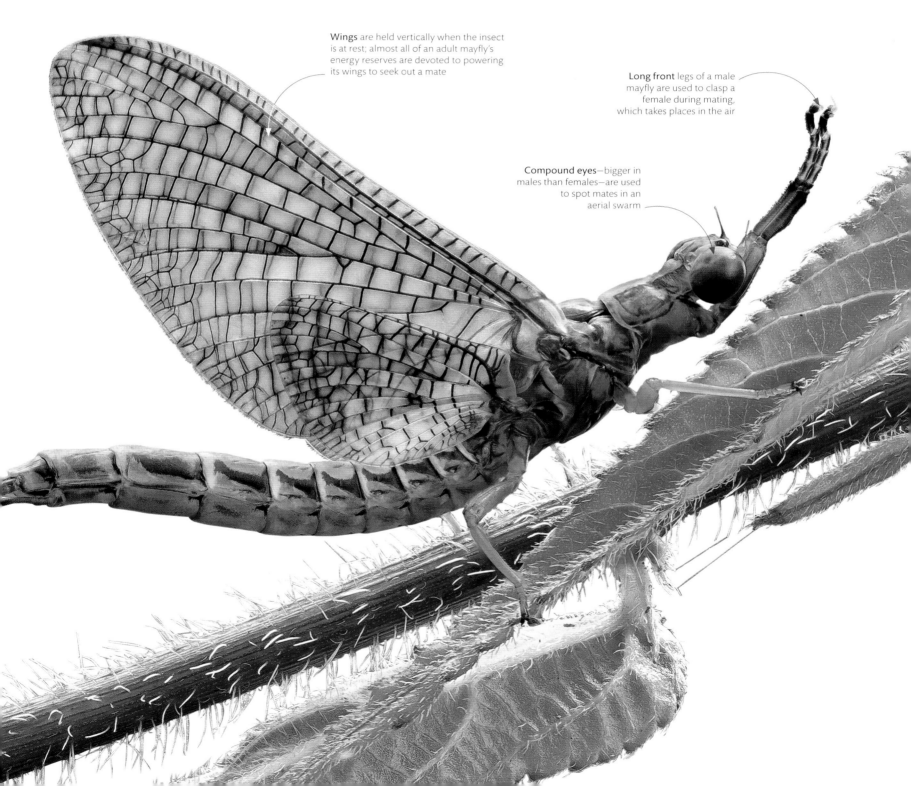

Wings are held vertically when the insect is at rest; almost all of an adult mayfly's energy reserves are devoted to powering its wings to seek out a mate

Long front legs of a male mayfly are used to clasp a female during mating, which takes places in the air

Compound eyes—bigger in males than females—are used to spot mates in an aerial swarm

habitats and lifestyles

Micro-life occupies every imaginable habitat, including some that seem fatally inhospitable. Ocean surface waters are full of clouds of plankton, and microorganisms fill the ooze of deep-sea sediments. Tiny animals live between particles of soil, and minute parasites live on the skin—and in the guts, blood, and even the brains—of larger organisms.

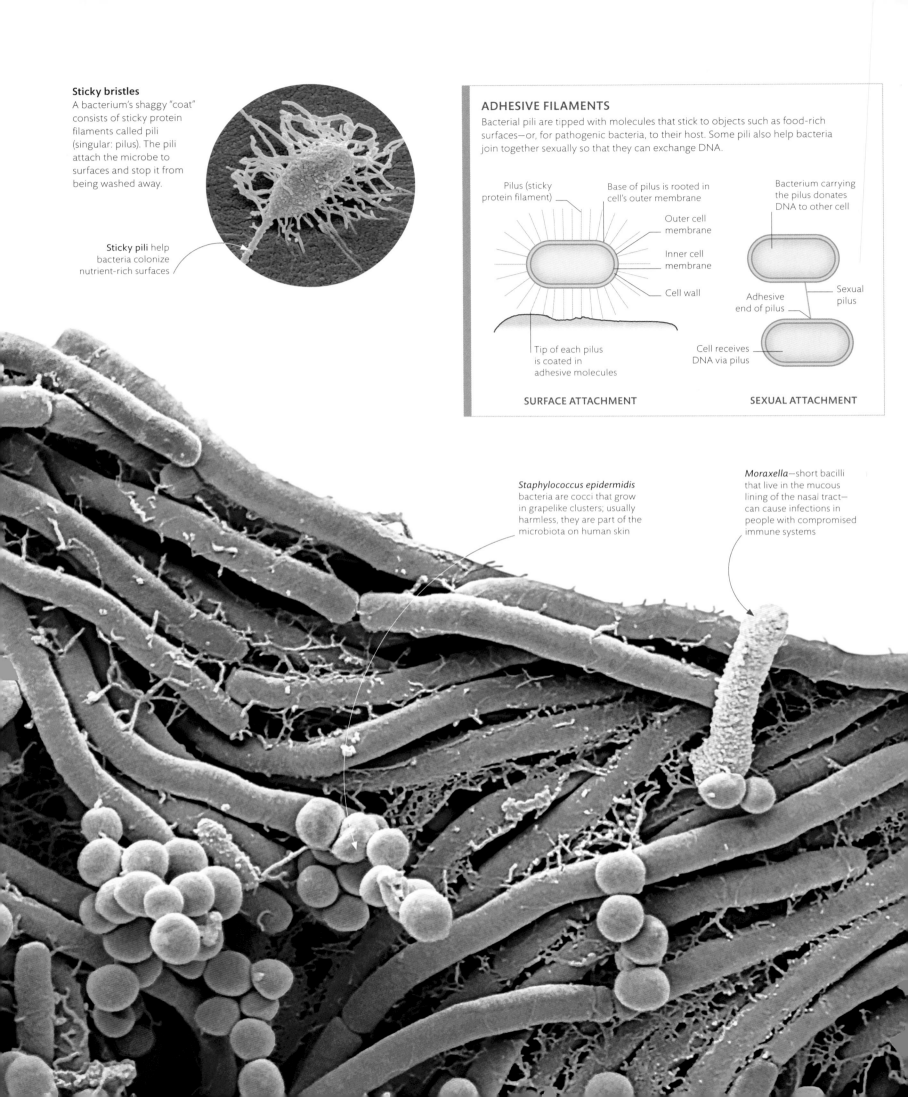

Sticky bristles
A bacterium's shaggy "coat" consists of sticky protein filaments called pili (singular: pilus). The pili attach the microbe to surfaces and stop it from being washed away.

Sticky pili help bacteria colonize nutrient-rich surfaces

ADHESIVE FILAMENTS

Bacterial pili are tipped with molecules that stick to objects such as food-rich surfaces—or, for pathogenic bacteria, to their host. Some pili also help bacteria join together sexually so that they can exchange DNA.

Pilus (sticky protein filament)
Base of pilus is rooted in cell's outer membrane
Bacterium carrying the pilus donates DNA to other cell
Outer cell membrane
Inner cell membrane
Cell wall
Adhesive end of pilus
Sexual pilus
Tip of each pilus is coated in adhesive molecules
Cell receives DNA via pilus

SURFACE ATTACHMENT   **SEXUAL ATTACHMENT**

Staphylococcus epidermidis bacteria are cocci that grow in grapelike clusters; usually harmless, they are part of the microbiota on human skin

Moraxella—short bacilli that live in the mucous lining of the nasal tract—can cause infections in people with compromised immune systems

ubiquitous bacteria

Bacteria are equalled in their widespread distribution only by viruses. They live in the ground, on surfaces, and as airborne spores. A teaspoon of soil can contain many times more bacteria than the number of people who have ever lived. Bacteria even occupy our bodies, inside and out, where they equal, or even exceed, the number of human cells. Being the smallest of organisms, bacteria are the perfect colonists: a bacterium less than one-tenth the size of a human skin cell can settle secured—with room for others—in a microscopic crack or crater, and most bacteria have clinging protein fibers to help them "stick." For bacteria that thrive on organic nutrients, invisible contaminants of oil, sweat, and other human secretions provide the food they need to multiply.

Touch-screen microbes
This tangled mass of bacteria is a culture from a single swab of a mobile phone touch-screen. It contains a mixture of species, including rod-shaped bacilli and spherical cocci. Tests show that an average handset carries 18 times more potentially harmful bacteria than the flush handle of a toilet. SEM, ×21,000.

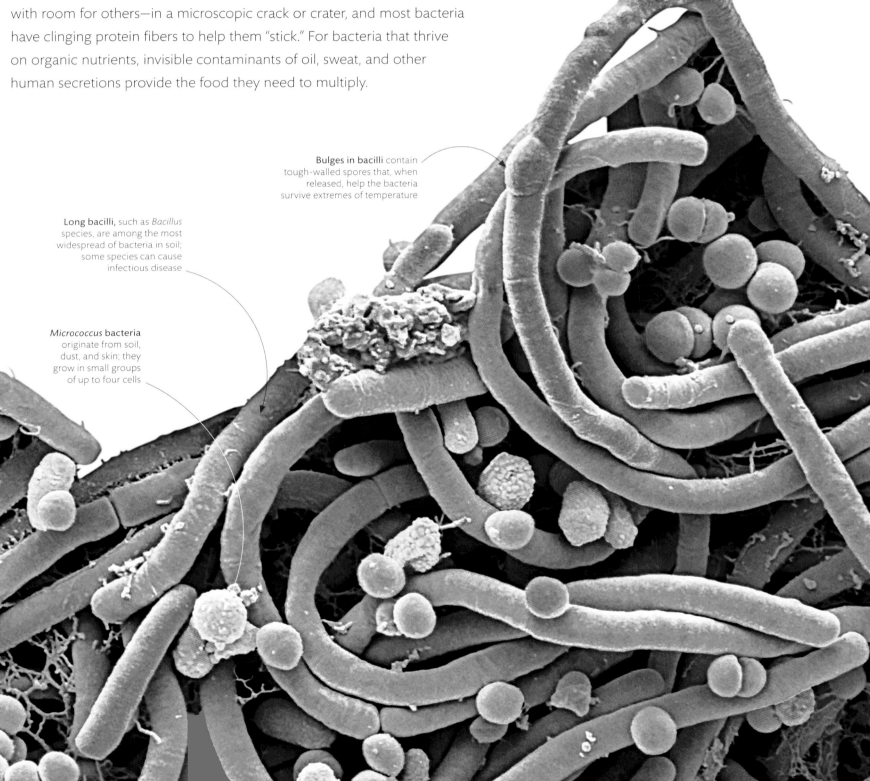

Bulges in bacilli contain tough-walled spores that, when released, help the bacteria survive extremes of temperature

Long bacilli, such as *Bacillus* species, are among the most widespread of bacteria in soil; some species can cause infectious disease

Micrococcus **bacteria** originate from soil, dust, and skin; they grow in small groups of up to four cells

surviving extremes

Many places on Earth are considered "extreme" from a human perspective, but they are completely normal for life forms called extremophiles. Most are unicellular microorganisms belonging to a group called Archaea, but some bacteria and even a few multicellular organisms are extremophiles. They fall into categories: thermophiles are heat-loving while psychrophiles, or cryophiles, thrive in extreme cold. Acidophiles live in extremely acidic conditions and halophiles in high salinity. An organism that is adapted to multiple extreme conditions is known as a polyextremophile. Many archaeans obtain energy from inorganic compounds, such as sulfur or ammonia, which are abundant in their particular extreme environment.

Archaellar filaments, like bacterial flagella, enable high-speed swimming, but also function in surface attachment

HYPERTHERMOPHILE

The "rushing fireball" *Pyrococcus furiosus* is a hyperthermophile, living at temperatures even higher than those of thermophiles. As an archaean, it has similarities with bacteria as well as its own unique characteristics.

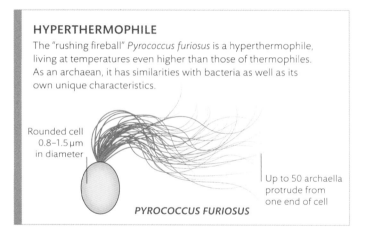

Rounded cell 0.8–1.5 μm in diameter

Up to 50 archaella protrude from one end of cell

PYROCOCCUS FURIOSUS

Life in a boiling-hot vent

Pyrococcus furiosus lives between hot fluid streaming from volcanic "hydrothermal" vents on the sea floor and the surrounding cold sea water—thriving at 212°F (100°C) but dying below 158°F (70°C). Such heat unravels and destroys the proteins of most organisms, but hyperthermophilic archaeans such as *Pyrococcus* have heat-resistant enzymes, and their DNA is protected by heatproof proteins. SEM, ×24,000.

WHERE DO EXTREMOPHILES LIVE?

Extremophiles thrive in seemingly inhospitable places both on land and in the sea, in places such as hot springs and volcanoes, hot arid deserts, freezing ice caps, and permafrost. Others live in hydrothermal vents, deep sea trenches, and anoxic lakes (lakes without dissolved oxygen). They even grow at sites contaminated with organic solvents, heavy metals, acid mine residues, and nuclear waste.

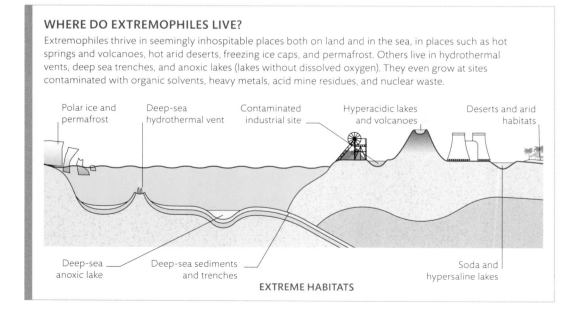

Polar ice and permafrost | Deep-sea hydrothermal vent | Contaminated industrial site | Hyperacidic lakes and volcanoes | Deserts and arid habitats

Deep-sea anoxic lake | Deep-sea sediments and trenches | Soda and hypersaline lakes

EXTREME HABITATS

Cells form a living layer, or biofilm, and communicate with the help of a network of archaellar filaments

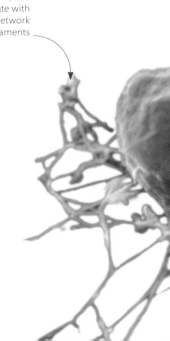

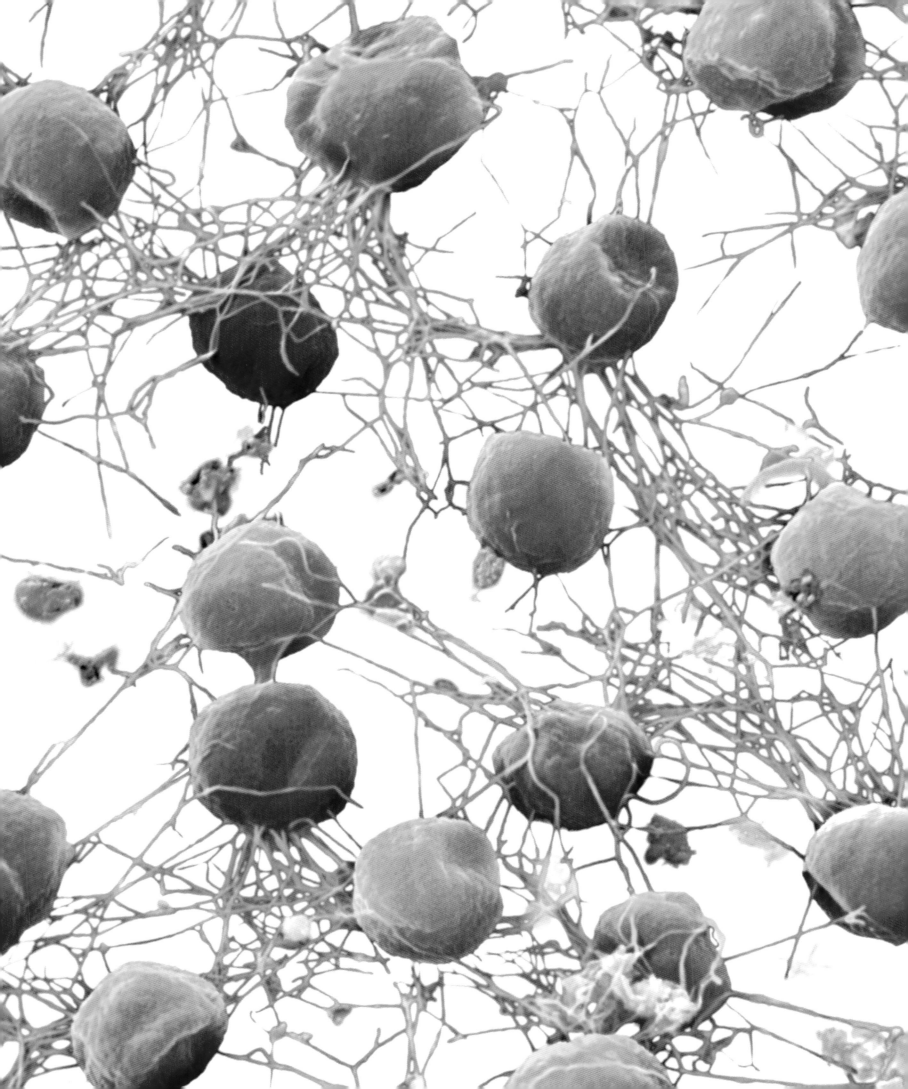

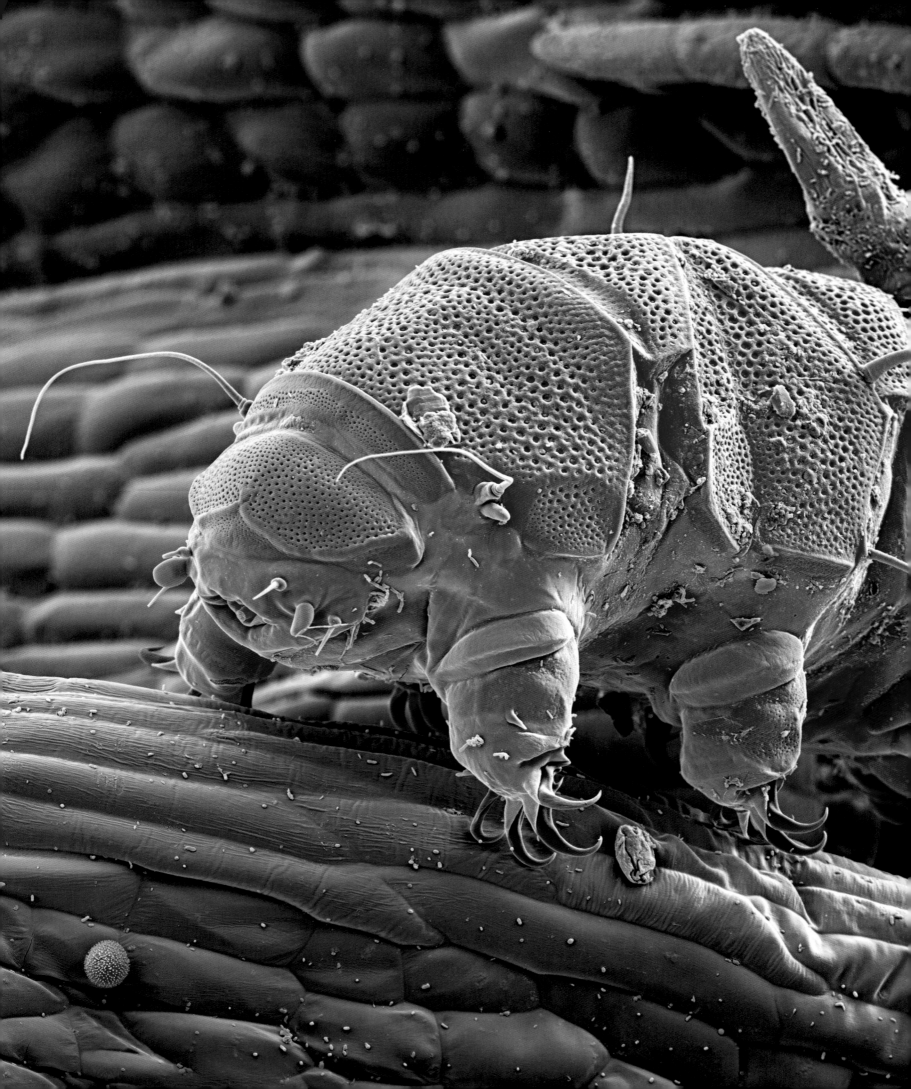

Nicknamed "water bears" for their habit of walking on stubby, claw-tipped legs with a lumbering gait, tardigrades are microscopic animals with astonishing resilience. They must be surrounded by water to grow and stay active, but whether living between tangles of detritus or sand grains, tardigrades thrive in vast numbers. Wet moss, a favored habitat, may home a million tardigrades in a patch the size of a tv tray.

tardigrades spotlight

In extreme conditions tardigrades can shut down almost all vital bodily functions in order to survive. A tardigrade has a remarkable response to such unsuitable conditions: it withdraws its legs, loses water, and secretes a tough casing to turn itself into a dormant husk called a tun (seen in *Paramacrobiotus kenianus* below). Its metabolism all but stops and outward signs of life vanish. Scientists studying the dormant state, known as cryptobiosis, have plunged tardigrades into liquid helium, blasted them with ionizing radiation, and even sent them into outer space. Each time, the tardigrades have survived.

Why these animals can be so durable is not fully understood. Certainly, as tuns, they can disperse like dust on the wind, giving many species a wide global distribution. This durability allows some tardigrades to tolerate habitats too extreme for most other life—such as the Greenland ice sheet or the ocean floor. A bizarre cocktail of genes may explain these adaptations: nearly one-fifth of a tardigrade's DNA seems foreign, obtained from microbes—some of which thrive in extreme conditions.

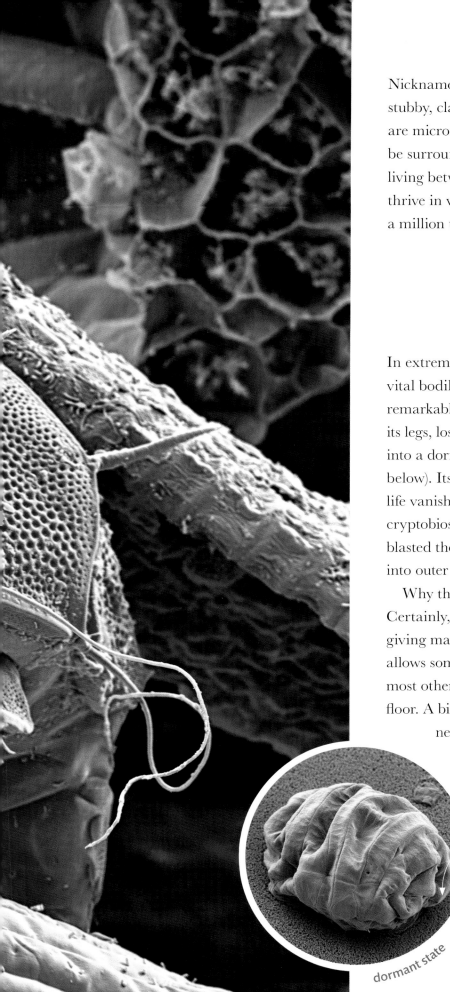

dormant state

Tardigrade's head and legs retract as it dessicates to survive extremely dry conditions

Miniature armor
Most tardigrades are practically colorless, but the cuticle of *Echiniscus granulatus* is arranged into hard plates that are chestnut brown. Spines and filaments may provide some further protection from tiny predators, while four claws on each foot help grip vegetation. SEM, x 1,500.

surviving cold

Lichens are a living partnership between a fungus and a green alga or cyanobacterium. The fungus provides the structure and helps store water, while the alga lives within its body, producing food for both through photosynthesis (see pp.316–17). Together they survive much tougher conditions than either partner could alone. Lichens can photosynthesize at –4°F (–20°C), as they have a natural antifreeze in their cells. It is estimated that lichens occur on 80 percent of the Earth's land surface, but they are a dominant species in polar tundra; there are eight recorded species within 416 miles (670 km) of the South Pole.

Lichen close up
Reindeer lichen (*Cladonia rangiferina*) grows commonly in open Arctic tundra. Naturally pale gray in color, this is classed as a fruticose lichen from its branched structure. SEM, x250.

Reindeer eat up to 4½ lbs (2 kg) of lichen per day

Algal partner is hidden within the lichen structure

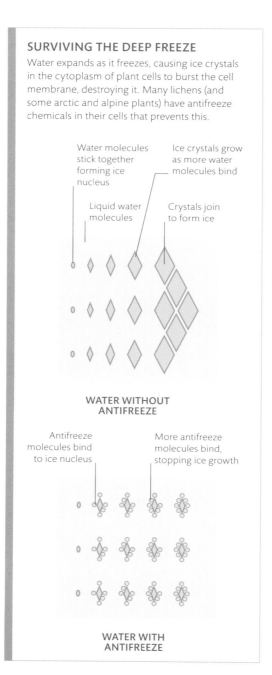

SURVIVING THE DEEP FREEZE
Water expands as it freezes, causing ice crystals in the cytoplasm of plant cells to burst the cell membrane, destroying it. Many lichens (and some arctic and alpine plants) have antifreeze chemicals in their cells that prevents this.

- Water molecules stick together forming ice nucleus
- Ice crystals grow as more water molecules bind
- Liquid water molecules
- Crystals join to form ice

WATER WITHOUT ANTIFREEZE

- Antifreeze molecules bind to ice nucleus
- More antifreeze molecules bind, stopping ice growth

WATER WITH ANTIFREEZE

Reindeer food
Although lichens have a low nutritive value, they are readily available in the Arctic. In winter they provide 60–70 percent of the diet of reindeer, which will paw beneath the snow to find it.

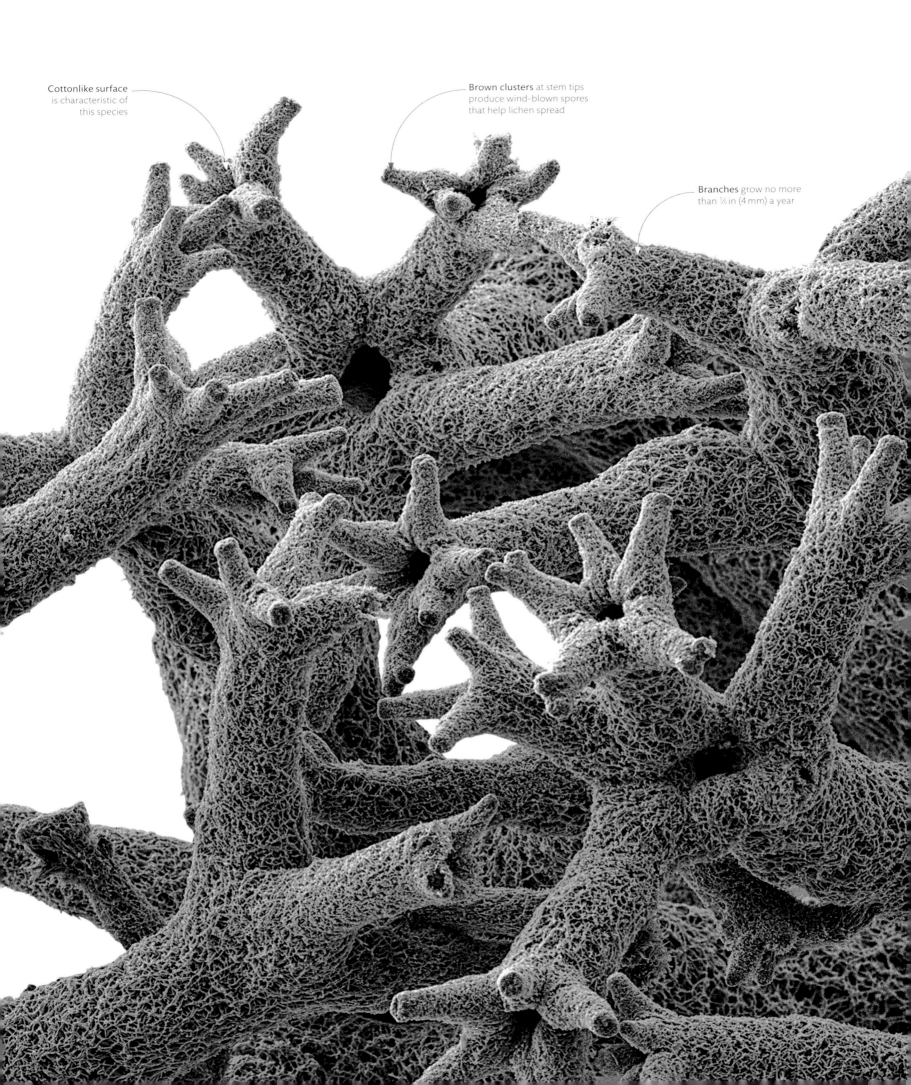

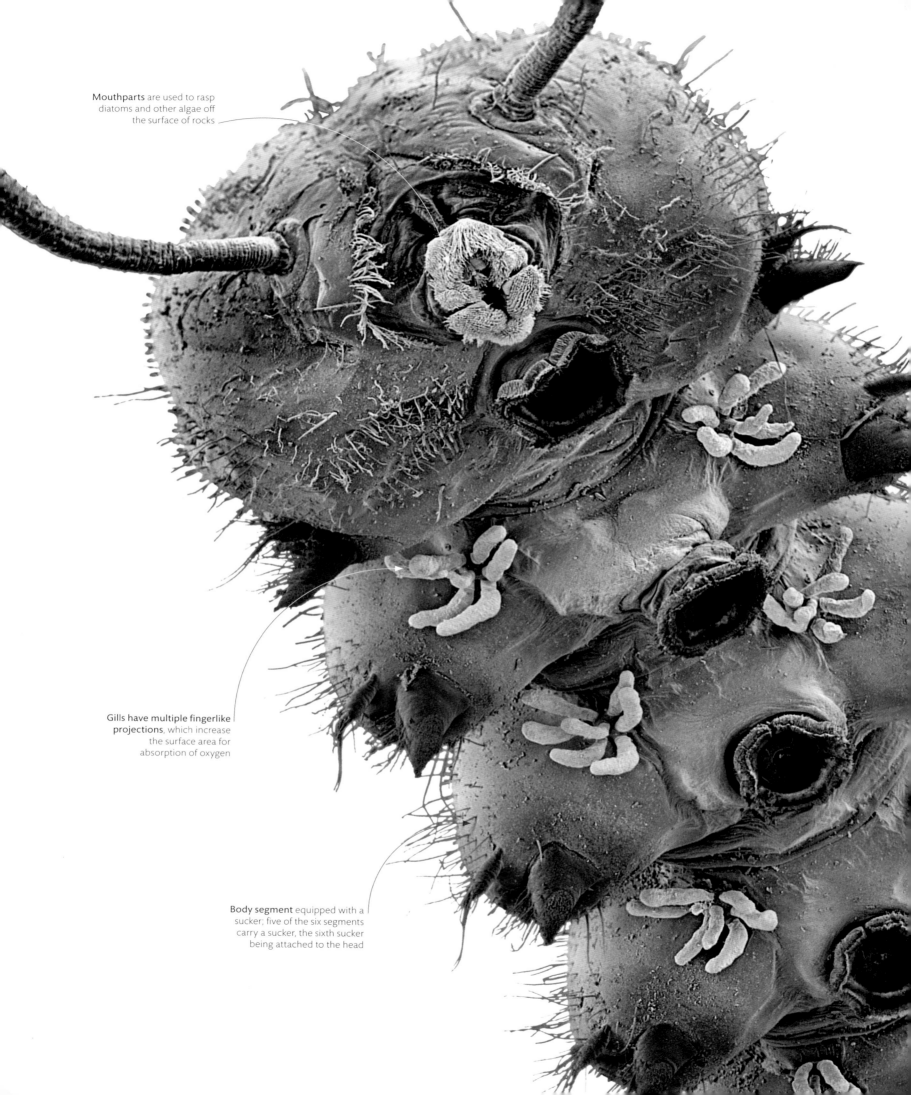

Mouthparts are used to rasp diatoms and other algae off the surface of rocks

Gills have multiple fingerlike projections, which increase the surface area for absorption of oxygen

Body segment equipped with a sucker; five of the six segments carry a sucker, the sixth sucker being attached to the head

Micro-suckers
The front end of a torrent larva—the larval stage of a net-winged midge (*Liponeura cinerascens*)—shows that it is a superbly adapted rheophile: an animal that thrives in fast-flowing water. Claws and suckers grip, while white gills collect plenty of oxygen from the aerated water. The oxygen powers muscles that work hard to keep the body hugging the riverbed. SEM, x40.

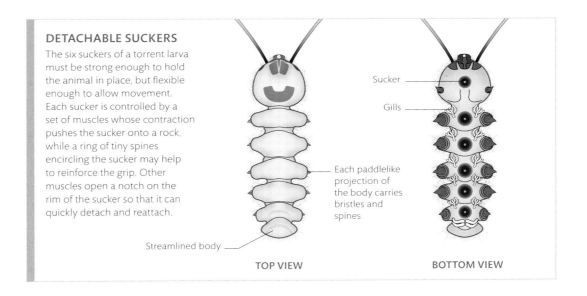

DETACHABLE SUCKERS
The six suckers of a torrent larva must be strong enough to hold the animal in place, but flexible enough to allow movement. Each sucker is controlled by a set of muscles whose contraction pushes the sucker onto a rock, while a ring of tiny spines encircling the sucker may help to reinforce the grip. Other muscles open a notch on the rim of the sucker so that it can quickly detach and reattach.

- Sucker
- Gills
- Each paddlelike projection of the body carries bristles and spines
- Streamlined body

TOP VIEW — BOTTOM VIEW

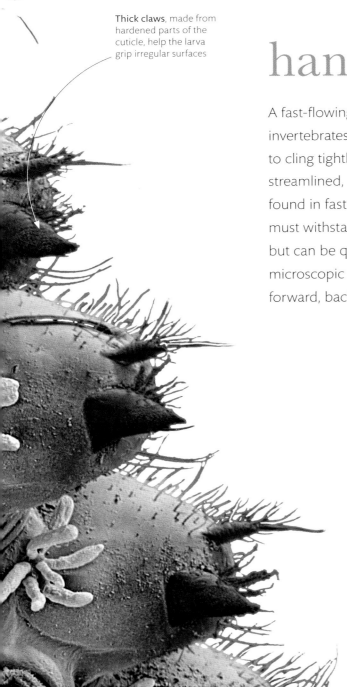

Thick claws, made from hardened parts of the cuticle, help the larva grip irregular surfaces

hanging on

A fast-flowing stream provides aquatic animals with plenty of oxygen, but tiny invertebrates risk being swept away. To live here permanently, they need to be able to cling tightly to rocks and weeds. Many have clawed, grasping feet, and a flattened, streamlined, shape that offers minimal resistance to the flow. Net-winged midges—found in fast alpine rivers—are fragile-looking flying insects, but their torrent larvae must withstand rushing waters. The larvae have six suckers that grip riverbed stones but can be quickly detached and reattached as they move around when grazing on microscopic algae, called diatoms. In this violent habitat, these tiny larvae can crawl forward, backward, and even sideways against the flow.

Alpine midge
Adult net-winged midges lay their eggs in upland streams during summer. The torrent larvae that hatch from the eggs metamorphose first into aquatic pupae and then into flying midges.

Net-winged midges are so-called because of the pattern of fine creases on their wings

habitats and lifestyles

COMPONENTS OF OCEANIC PLANKTON

The tens of thousands of planktonic species can be grouped according to size, from the large, free-floating sargassum seaweeds and jellyfishlike hydrozoa to the minute cyanobacteria. The largest organisms—the mega-, macro-, and mesoplankton—are mostly zooplankton with some floating, photosynthetic macroalgae. The smallest plankton—the picoplankton—include the smallest known and most abundant photosynthetic organisms on Earth.

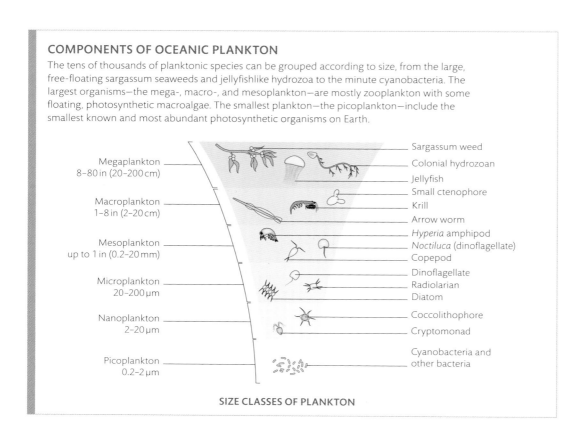

SIZE CLASSES OF PLANKTON

- Megaplankton 8–80 in (20–200 cm)
- Macroplankton 1–8 in (2–20 cm)
- Mesoplankton up to 1 in (0.2–20 mm)
- Microplankton 20–200 μm
- Nanoplankton 2–20 μm
- Picoplankton 0.2–2 μm

Sargassum weed · Colonial hydrozoan · Jellyfish · Small ctenophore · Krill · Arrow worm · *Hyperia* amphipod · *Noctiluca* (dinoflagellate) · Copepod · Dinoflagellate · Radiolarian · Diatom · Coccolithophore · Cryptomonad · Cyanobacteria and other bacteria

Myriad of shapes

Features of some larval plankton (meroplankton) hint at their adult forms, as this early 20th-century print shows. At bottom-left is a sea urchin, while at top-right is a brittle star larva, with a crab "zoea" larva below it, and an oyster larva at bottom-right. Other features are plankton-specific, such as spiny or flattened arms that protect them from predators or act like sails or stabilizers in the currents.

marine plankton

The lifeblood of the oceans, marine plankton includes minute organisms that form a key part of the marine food web and the global carbon cycle. Plankton drifts on the ocean currents and relies on them for dispersal—those plankton that can swim lack the power to overcome the currents. Some organisms, known as phytoplankton, make food by photosynthesis and live in the sunlit surface waters. Zooplankton is composed of small animals and microbes such as radiolarians and foraminiferans; some graze on phytoplankton, while others prey on smaller zooplankton.

Planktonic lifestyles

Phytoplankton, or plantlike plankton, belongs to numerous groups including dinoflagellates, diatoms, and cyanobacteria. The flattened oval shape of many of these organisms maximizes their exposure to sunlight. Zooplankton, or animallike plankton, can be divided into holoplankton, whose members spend their whole life in plankton, and meroplankton, which are planktonic only as larvae or juveniles. They may take up a seafloor existence as adults, or grow into powerful swimmers.

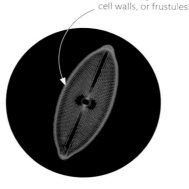

Single-celled alga has patterned, glasslike silica cell walls, or frustules

PHYTOPLANKTON
Diatom
Navicula febigerii

Planktonic snail flaps winglike parapodia to swim

HOLOPLANKTON
Sea butterfly
Limacina helicina

Juvenile fish drifts on ocean currents, moving to deep waters in its adult form

MEROPLANKTON
Mediterranean dealfish
Trachipterus trachypterus

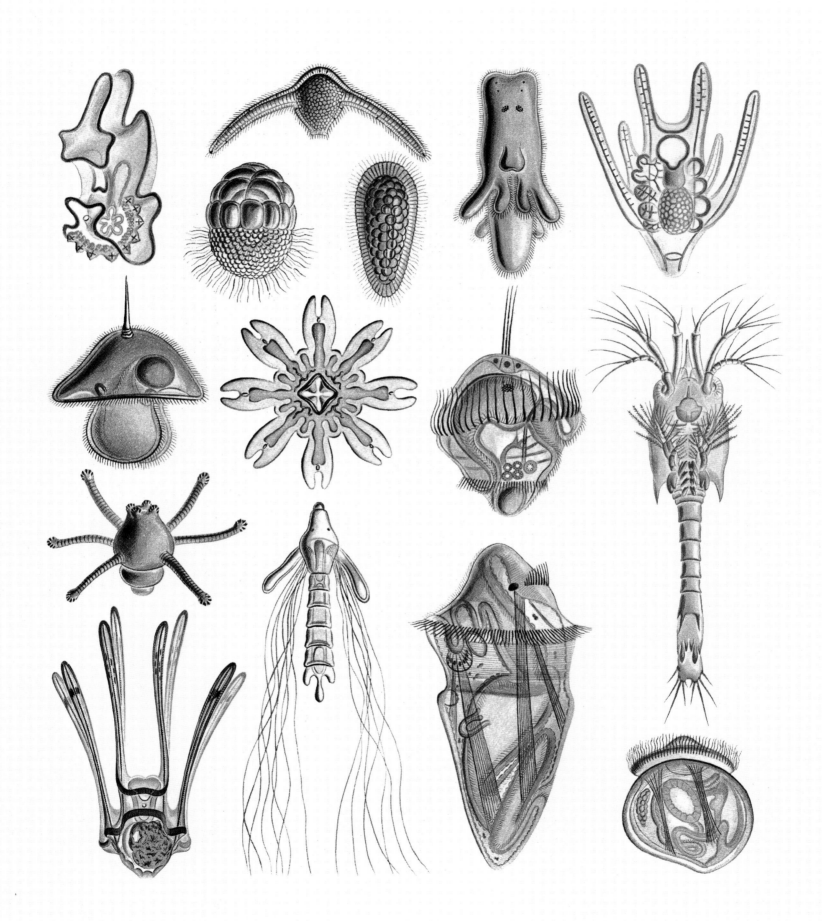

Aquatic habitats from ponds to the deep ocean are all home to microscopic crustaceans called copepods, meaning "oar feet." The name refers to the way many copepods are found in the water column paddling around as plankton, using their feathery appendages. Of the estimated 12,000 species of copepod, about 9,000 live in salt water. As well as being planktonic, copepods are widely represented in the sediments at the bottom of oceans,

spotlight copepods

found in damp leaf litter, and even in the tiny ponds trapped by the leaves of succulent plants growing in rainforest trees.

Mostly measuring less than 0.08 in (2 mm) across, copepods generally have a teardrop-shaped body covered in a thin, transparent exoskeleton, through which oxygen in the water is absorbed directly into the body. Most species have a single red compound eye on top of the head—which lends a common freshwater genus the scientific name *Cyclops*—and a pair of large antennae (plus a smaller, less obvious pair). There are between four and five pairs of limblike appendages on the underside of the body, plus another pair used as mouthparts to sift food from the water. Copepods feed on microscopic food items, chiefly phytoplankton or organic detritus. Some copepods are parasitic and attach themselves to fish, whales, and other marine animals.

During mating, the male grips the female by her antennae as he transfers sperm to her using his limbs. The eggs hatch into a larval form called a nauplius, which has a body composed entirely of a head with a series of appendages and a simple tail. Through successive molts, the larval form steadily develops the thorax and abdomen of its adult form.

Female releases eggs from sacs attached to abdomen

pair of *Cyclops* copepods

Copepod diversity
Many copepods drift in plankton, such as this calanoid copepod (top left; LM, x150) and copepod nauplius larva (top right, LM; x600). Others, such as this harpacticoid copepod (bottom; LM, x300), live in the bottom sediment.

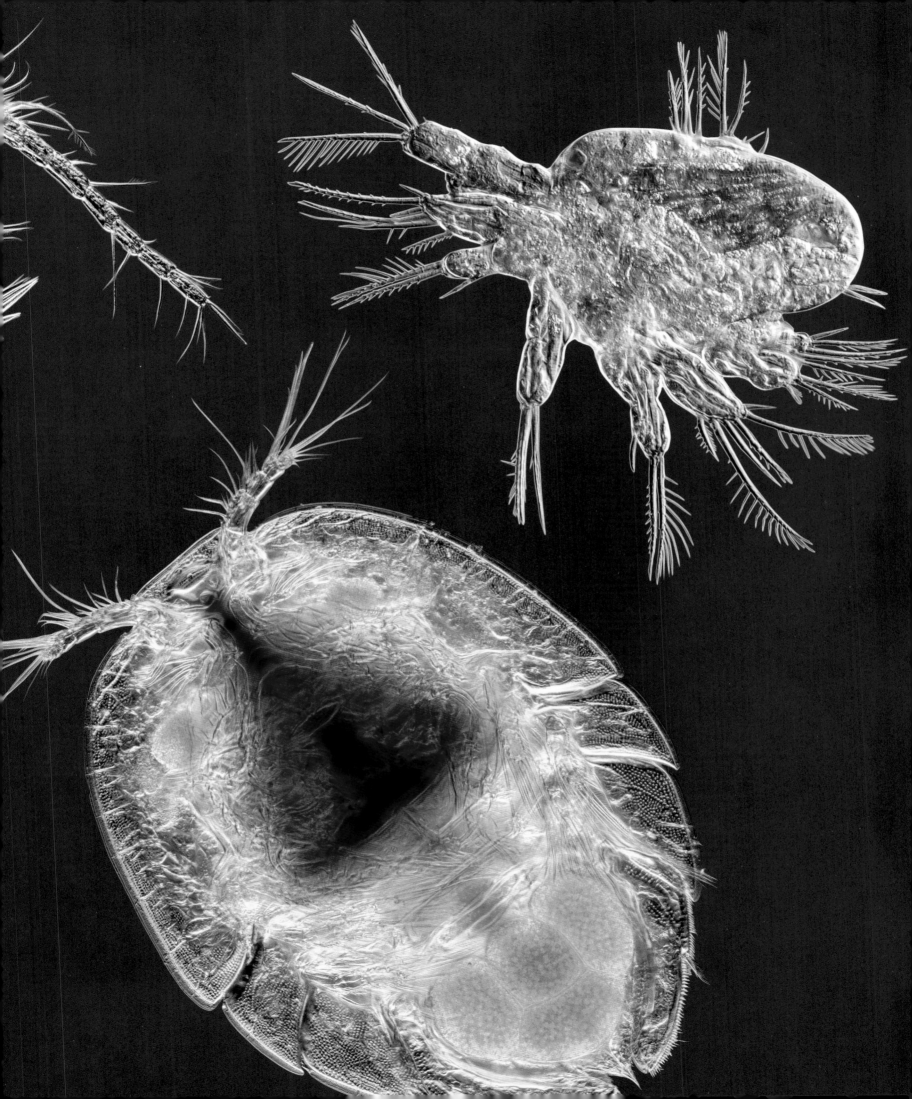

Algae

Like true plants, algae contain green pigments—called chlorophylls—that absorb light energy in photosynthesis to make sugars and other organic food. But algae lack the complex bodies, with leaves and roots, of true plants. Many are single-celled, or have the cells that clump together into colonies—such as filaments or spheres. The smallest and simplest algae are cyanobacteria: like other bacteria, their cells lack a nucleus (they are prokaryotic). Other algae have complex (eukaryotic) cells, with their DNA packaged into a nucleus and their chlorophyll contained in chloroplasts.

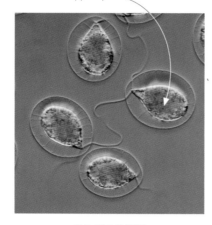

Rounded red cells can make shallow water appear pink

CHLOROPHYTE
Haematococcus pluvialis

Wedge-shaped golden diatoms share a stalk attached to bed of pond

DIATOM
Licmophora flabellata

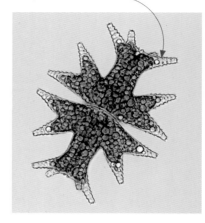

Star-shaped cell is formed of two symmetrical semicells

DESMID
Micrasterias sp.

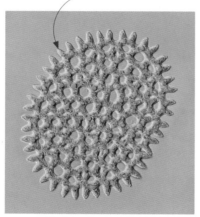

Green alga forms a disk-shaped colony as its cells divide

CHLOROPHYTE
Pediastrum duplex

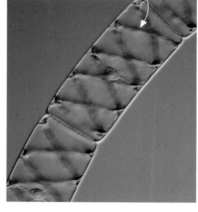

Green chloroplasts form a spiral inside chain of transparent cells

CHAROPHYTE
Spirogyra sp.

Coenobia, an algal colony where cells interconnect in a fixed arrangement

CHLOROPHYTE
Scenedesmus sp.

Protozoans

Single-celled or colonial, protozoans have complex (eukaryotic) cells containing a nucleus and other internal structures. They consume organic food—sometimes as predators or parasites. Since protozoans need to seek out sources of food, they are typically more mobile than most algae. Some use hairlike cilia or flagella for propulsion or to generate feeding currents; others creep on projections of cytoplasm called pseudopods.

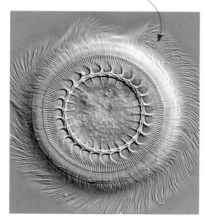

Disk-shaped parasite clings to host and uses fringe of cilia to waft food into mouth

PARASITE
Trichodina pediculus

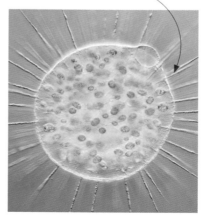

Tubular projections are used for buoyancy and catching food

PREDATOR
Actinosphaerium eichhorni

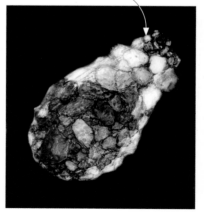

Cell is encased in specks of mineral dust glued into shell

PREDATOR
Difflugia sp.

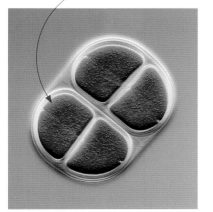

Blue chemical (phycobilin) and green chlorophyll creates teal coloring

CYANOBACTERIUM
Chroococcus turgidus

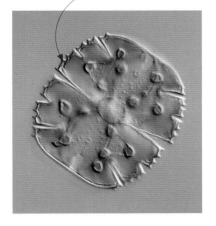

Rounded cell is divided into two semicells with a shared nucleus at center

DESMID
Micrasterias truncata

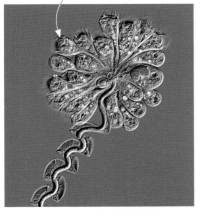

Flower-shaped colony of ciliated cells attached to a single stalk

COLONIAL FILTER-FEEDER
Apocarchesium sp.

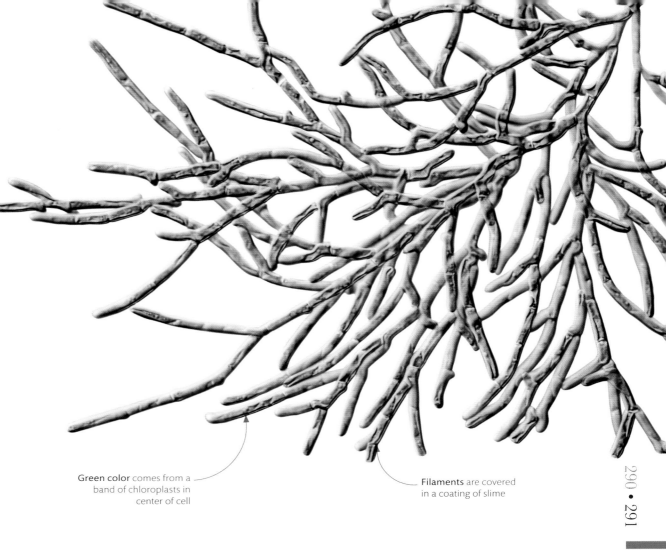

Green color comes from a band of chloroplasts in center of cell

Filaments are covered in a coating of slime

Branching algae
This branched green alga, *Microthamnion* sp., is found in cold and clear fresh water. Its complex structure has a deceptively simple plan. Each of the branching fronds is a filament chain of cells.

pond microorganisms

Freshwater ponds are rich with microscopic life. Single-celled and multi-celled algae photosynthesize where water is illuminated by sunlight. Microbes called protozoans graze on greenery and dead matter, or hunt prey. Many members of this complex underwater community float in mid-water as plankton, while others live on the bottom, where a rich source of organic food accumulates as dead matter. Microorganisms are critical components of ecosystems. Photosynthesis makes algae producers of food for the pond's food chain, while many of the grazing protozoans help decompose waste.

Beating antennae provide propulsion for swimming

Snail uses tongue, or radula, to rasp algae from surfaces

Larva uses long, pointed jaws to grab fish, or other large prey

PLANKTONIC GRAZERS
Seed shrimp
Cypris sp.

BOTTOM GRAZER
Ramshorn snail
Planorbis planorbis

MESO-PREDATOR
Diving beetle
Dytiscus sp.

Community members
Invertebrates dominate pond ecosystems. Swimming crustaceans, such as seed shrimp, collect plankton in open water. Grazing herbivores, such as snails, forage on algae-coated surfaces, while predatory diving-beetle larvae hunt along the bottom.

freshwater communities

No living thing can exist in isolation. Every one depends on other organisms, from plants using minerals recycled by decomposers, to predators that hunt prey, and parasites living in their hosts. In a freshwater pond, the tiniest creatures play out these interactions in microhabitats, where a community living in a drop of water can begin a complex food chain. Ponds are alive with large animals, such as frogs and fish, but they rely on the rich diversity of micro-life—worms, crustaceans, and mites—that pass on energy and nutrients from photosynthesizing algae to herbivores and carnivores.

Tiny predator
Despite being scarcely visible to the naked eye, most freshwater mites are predators and can subdue prey, such as insect larvae, much bigger than themselves. This image reveals the extraordinary anatomy of the underside of *Frontipoda*. Its legs are positioned in front of its eyes—seen here as red spots—and fused together at their bases. SEM, x350.

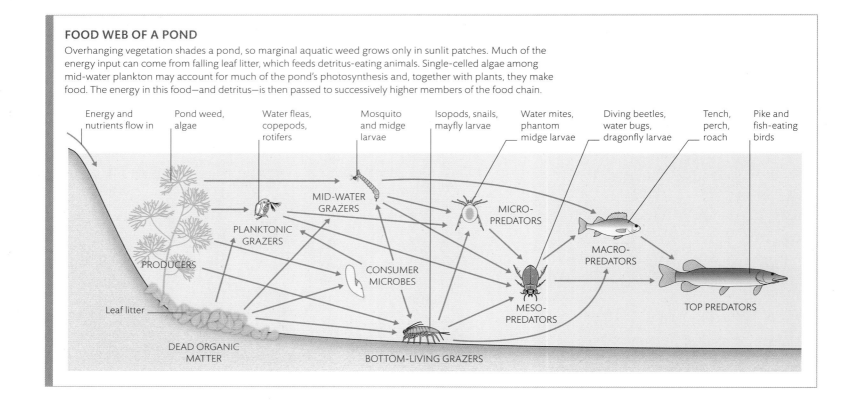

FOOD WEB OF A POND
Overhanging vegetation shades a pond, so marginal aquatic weed grows only in sunlit patches. Much of the energy input can come from falling leaf litter, which feeds detritus-eating animals. Single-celled algae among mid-water plankton may account for much of the pond's photosynthesis and, together with plants, they make food. The energy in this food—and detritus—is then passed to successively higher members of the food chain.

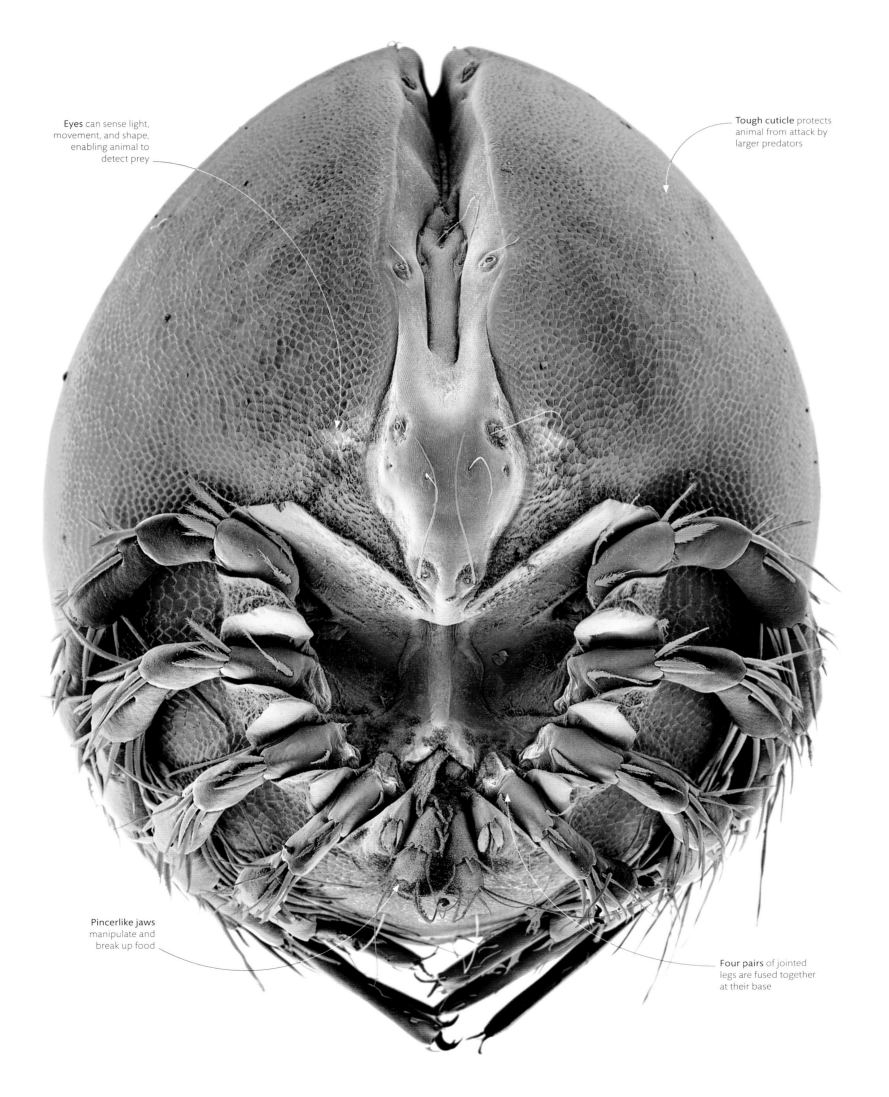

Nematodes, meaning "threadlike", are everywhere. A greater number of these worms exists on Earth than all the other animals put together. It is estimated that there could be about a million different species of nematode, and about 60 billion individual worms for every human alive today. If our planet's minerals and rocks somehow vanished, it is conjectured, the Earth's surface would still be visible as a seething mass of

spotlight nematodes

nematodes. They thrive in a huge variety of conditions in water, on land, and in other animals: they have been found living in mines about 2 miles (3 km) below the surface and can survive in the heat of deserts and hot springs, the cold of polar deserts, and the pressure of deep sea sediments.

Nematodes, also known as roundworms and threadworms, are thought to represent something close to the primitive form of the animal group that later also gave rise to insects, spiders, and crustaceans. Most nematodes are only a few micrometers (millionths of a meter) wide and grow no longer than $^1/_{10}$ in (2.5 mm). However, some parasitic nematodes can be much bigger, growing to 14 in (35 cm) in humans and up to 28 ft (8.4 m) in sperm whales. Free-living nematodes are important detritivores—animals that feed on dead organic matter. Many have adopted parasitic ways of life and are carried by, and harbored in, other animals. While many infections are largely harmless and easy to treat, others can cause long-term health problems—such as river blindness and elephantiasis—or can even be fatal.

Sense organs help nematodes smell, taste, touch, and sense temperature

ring of sense organs around mouth

Soil worm
Unlike some other worms, a nematode's body is unsegmented. This male *Caenorhabditis elegans* in its natural soil habitat has an elastic cuticle covering with ridges used for grip. Muscles run along the length of the worm's body causing it to move in rhythmic contractions. SEM, x650.

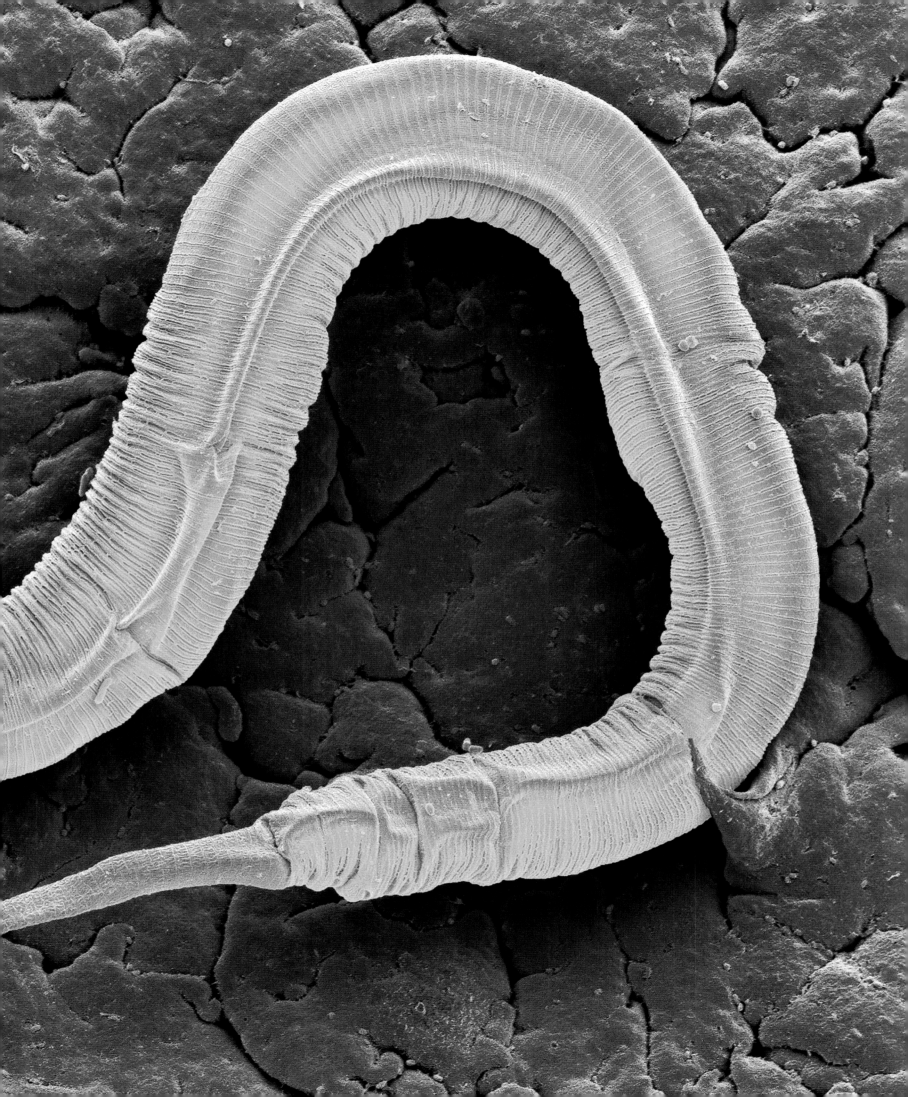

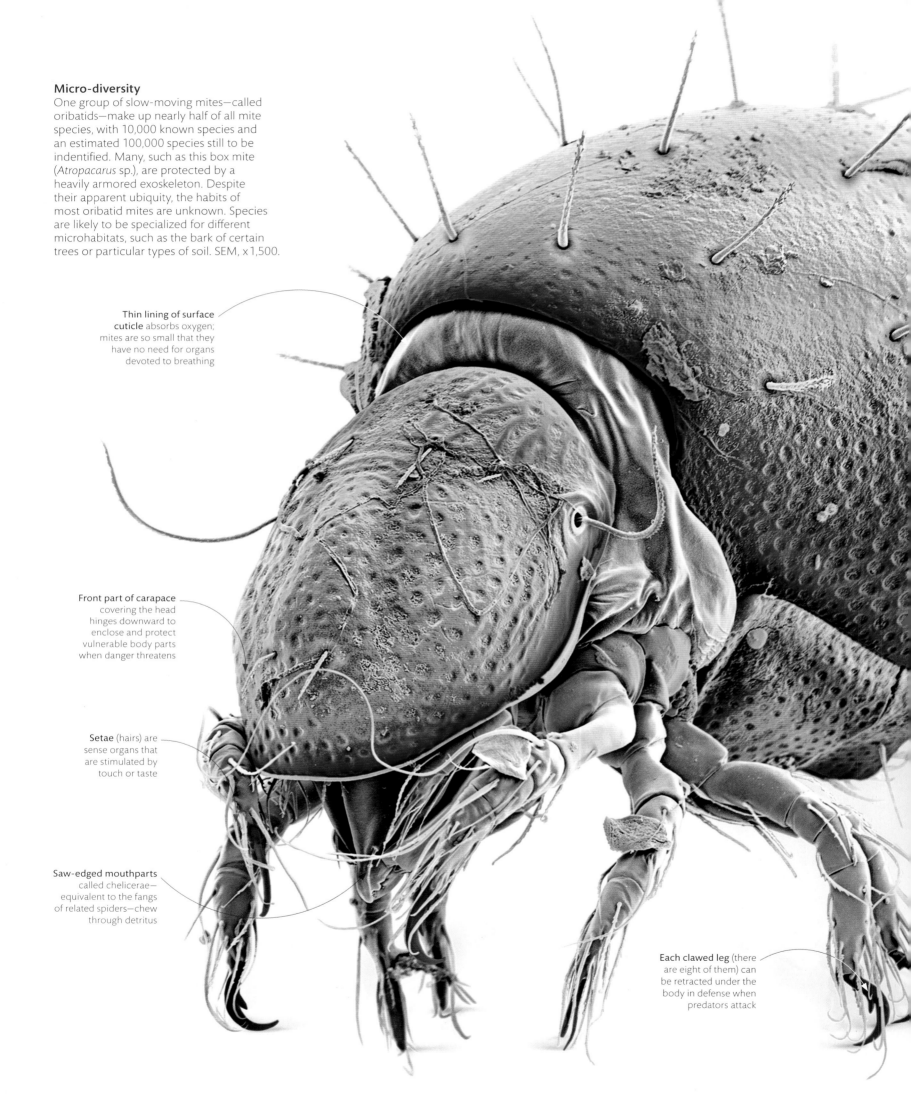

Micro-diversity

One group of slow-moving mites—called oribatids—make up nearly half of all mite species, with 10,000 known species and an estimated 100,000 species still to be indentified. Many, such as this box mite (*Atropacarus* sp.), are protected by a heavily armored exoskeleton. Despite their apparent ubiquity, the habits of most oribatid mites are unknown. Species are likely to be specialized for different microhabitats, such as the bark of certain trees or particular types of soil. SEM, x 1,500.

Thin lining of surface cuticle absorbs oxygen; mites are so small that they have no need for organs devoted to breathing

Front part of carapace covering the head hinges downward to enclose and protect vulnerable body parts when danger threatens

Setae (hairs) are sense organs that are stimulated by touch or taste

Saw-edged mouthparts called chelicerae—equivalent to the fangs of related spiders—chew through detritus

Each clawed leg (there are eight of them) can be retracted under the body in defense when predators attack

Arthropod community
Mites share their micro-level habitat of soil and leaf-litter with a wide variety of other arthropods, from waste-recycling springtails to plant-eating symphylans, as well as predatory spiders, centipedes, and insects.

Symphylans have many jointed legs—just like related millipedes and centipedes

Two antennae are springtail's main sense organs for navigating leaf litter

SYMPHYLAN

SPRINGTAIL (*SINELLA CURVISETA*)

Particles of soil or detritus sticking to the exoskeleton may help camouflage the mite, protecting it against larger predators that locate prey by vision

recycling matter

The natural world relies on animals that consume dead matter, or detritus, but most of them are much too small to be noticed. Their actions, together with those of decomposer microbes and fungi, keep nutrients recycling through habitats. These invisible detritivores include mites—tiny relatives of spiders. There can be as many as half a million mites in a single square yard of forest litter. Some mites are predatory, parasitic, or suck plant juices, but a great many eat microscopic particles of detritus or graze on the fungi that grow on waste material.

MITES IN A FOOD CHAIN
Because of the diversity of their feeding behavior, mites and other arthropods (jointed-legged invertebrates) can play multiple roles in a complex food chain. In a forest habitat they are responsible for channeling chemical energy both from photosynthetic plants and dead matter.

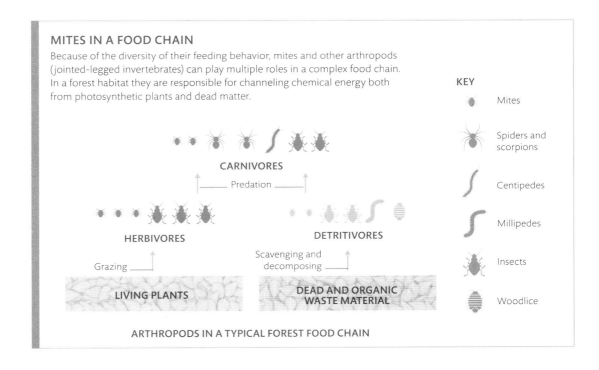

between sand grains

Many animals are so small that they are only just visible to the naked eye—and a few are even dwarfed by some single-celled organisms. Those that are small enough to live between particles of sand and soil—smaller than 1 mm but bigger than 0.05 mm—are called the meiofauna. They include tiny worms and arthropods, as well as bizarre animals belonging to minor groups, such as rotifers, kinorhynchs, and gastrotrichs. Smaller still are the microfauna—including protozoan microbes, such as amoebas—while animals larger than 1 mm, such as jumping springtails, make up the more conspicuous elements of the macrofauna.

Diversity in sediment
Despite their small size, the faunas that live in sand and soil represent a huge range of biodiversity. As shown in these SEM images, many are jointed-legged arthropods—including crustaceans (copepods and mystacocarids) and springtails (flightless relatives of insects). The remaining wormlike animals—such as gastrotrichs and rotifers—each belong to entirely different taxonomic groups of unrelated animals. Many of these animals graze on detritus—particles of dead and waste organic matter—that can accumulate in large amounts in sand and soil.

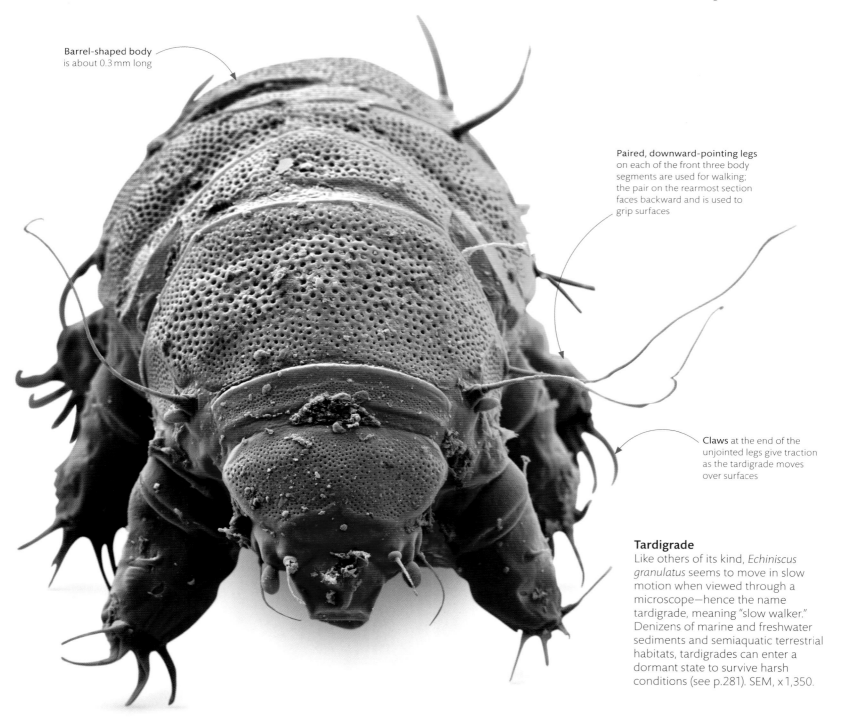

Barrel-shaped body is about 0.3 mm long

Paired, downward-pointing legs on each of the front three body segments are used for walking; the pair on the rearmost section faces backward and is used to grip surfaces

Claws at the end of the unjointed legs give traction as the tardigrade moves over surfaces

Tardigrade
Like others of its kind, *Echiniscus granulatus* seems to move in slow motion when viewed through a microscope—hence the name tardigrade, meaning "slow walker." Denizens of marine and freshwater sediments and semiaquatic terrestrial habitats, tardigrades can enter a dormant state to survive harsh conditions (see p.281). SEM, x 1,350.

Coat of hairlike cilia, used for locomotion, earns this animal the nickname "hairybelly"

GASTROTRICH
Gastrotricha
Freshwater sediments, x280

Rotifer forms a tun when water is scarce to survive dessication

ROTIFER
Bdelloidea
Freshwater sediments, x750

Long legs help with movement through sea bed

COPEPOD
Leptastacus macronyx
Marine sediments, x390

Springtail is not a member of meiofauna but its slender, segmented body lets it live between grains of soil

SPRINGTAIL
Folsomia candida
Soil, x30

Mouth swallows sand, from which food is extracted

Nut-shaped bulge forms when worm retracts its body into itself

PEANUT WORM
Phascolion sp.
Marine sediments, x80

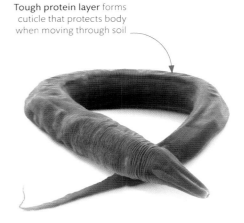

Tough protein layer forms cuticle that protects body when moving through soil

NEMATODE
Caenorhabditis elegans
Soil, x100

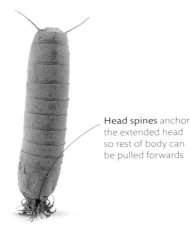

Head spines anchor the extended head so rest of body can be pulled forwards

KINORHYNCH
Kinorhyncha
Marine sediments, x125

Crown of feathery tentacles, which this segmented worm uses for filter-feeding and respiration

POLYCHAETE
Augeneriella dubia
Marine and tidal sediments, x45

Long filamentous mandibles and antennae both split into two branches

MYSTACOCARID
Derocheilocaris typica
Sandy beaches, x340

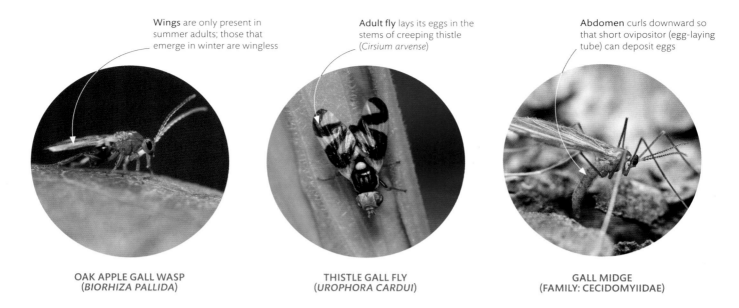

Wings are only present in summer adults; those that emerge in winter are wingless

Adult fly lays its eggs in the stems of creeping thistle (*Cirsium arvense*)

Abdomen curls downward so that short ovipositor (egg-laying tube) can deposit eggs

OAK APPLE GALL WASP
(*BIORHIZA PALLIDA*)

THISTLE GALL FLY
(*UROPHORA CARDUI*)

GALL MIDGE
(FAMILY: CECIDOMYIIDAE)

Gall makers
Most galls are caused by wasps, flies, midges, aphids, and scale insects, or by mites. Bacteria can also form galls, and fungi cause smut and rust diseases.

forming galls

There are many mutually beneficial relationships between plants and animals, but galls mainly seem to benefit the gall-causer. They typically form when an insect lays its egg into a bud, leaf tip, root, or stem—or occasionally in a flower. A larva hatches from the egg and begins to eat the material around it, stimulating the plant to produce wound tissue. This tissue forms a bump or elaborate swelling, which is the gall. The gall provides the hatched larva with food and a secure home; the plant may get some benefit, since the larva is prevented from feeding more widely on its leaves.

Oak gall diversity
This vintage engraving shows a range of galls caused by insect invasion of oak trees: (B) shows the gall produced by the oak apple gall wasp, with an adult wasp flying above; (I) depicts the root galls caused by the same species. Several of the other galls, induced by the eggs of different insects, are named after their distinctive shapes: (A) cherry, (C) blister, (E) marble, (F) artichoke, and (H) currant.

LIFE CYCLE OF THE OAK APPLE GALL WASP

The oak apple gall wasp (*Biorhiza pallida*) produces two generations—and two different types of gall—each year. In summer, a mated female flies to the ground to find the rootlet of an oak tree, in which she lays her eggs. Root galls develop around the hatched larvae. Wingless "parthenogenetic" females, which produce eggs without mating, emerge in midwinter. They climb the trunk and lay eggs in dormant leaf buds. These develop into oak apple galls, from which winged, breeding males and females emerge in midsummer.

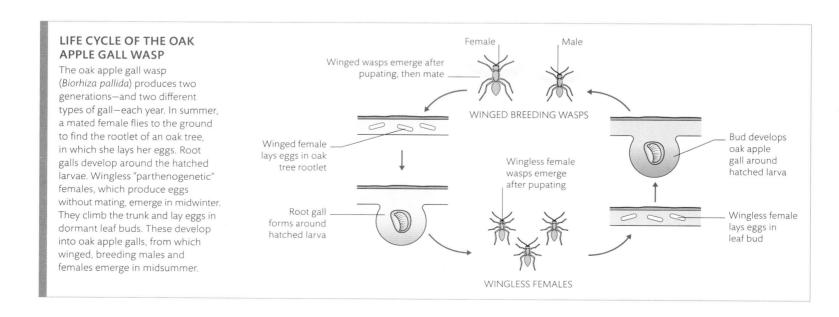

For mosquitoes—much reviled for their irritating bites and as vectors of serious disease—sucking blood is critical to producing young. Sweet plant liquids, such as nectar and sap, are the main energy sources for both female and male mosquitoes, but the female must drink blood to obtain protein to build her eggs. In most species, an initial meal of blood is necessary to trigger the release of the hormones that begin egg development.

spotlight mosquitoes

After mating in a swarm, a female mosquito uses her antennae to detect the scent and heat of a suitable host. The long palps flanking her proboscis are sensitive to exhaled carbon dioxide, helping her to zero in on the victim. She pierces the host's skin with her syringelike proboscis and pumps saliva into the wound. Anticoagulants in the saliva mix with the blood and prevent clotting so that it flows freely. As she drinks, excess liquid is excreted to allow more room in her gut for solid nutrients; even so, the meal may swell her abdomen to several times its usual size. Depending on the species, once the blood is digested and its protein used for egg production, the female lays her eggs on or close to still water, or arranged as a floating raft.

The aquatic larvae hang from the water's surface, breathing via a snorkellike tube called a siphon and feeding on bacteria, algae, and other plankton. Though they are limbless, their sides have bristles that help them swim to deeper water with wiggles of the body when disturbed. Pupation occurs at the surface, with adults emerging from between a few days to several weeks after egg-laying, according to species and temperature.

Comma-shaped pupa mostly remains at the surface but is able to swim with flicks of its abdomen

mosquito pupa

Disease carrier
This SEM shows a female *Anopheles* mosquito posed as if she is about to puncture human skin with her needlelike proboscis. Mosquito saliva can transmit parasitic and viral diseases to the host. As they feed, some *Anopheles* species pass on the *Plasmodium* protozoan—the cause of malaria. SEM, x40.

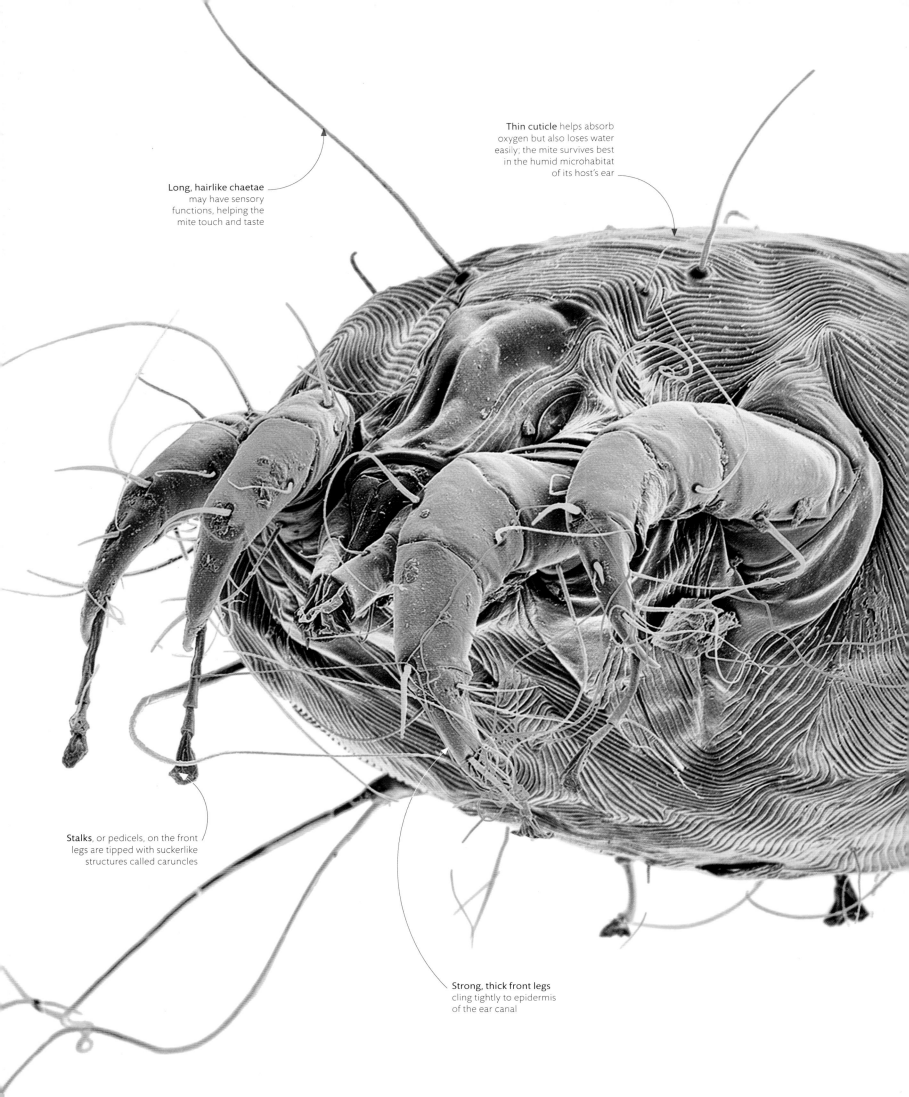

living on skin

The skin of an animal is a fruitful microhabitat for organisms small enough to live there. Not only can secreted oils—and even the skin itself—supply food, but the landscape of wrinkles and pores also offers abundant places to shelter and hide. Microbes such as bacteria can complete their entire life cycle on skin, as can tiny animals such as mites. Many mites go unnoticed by their host, grazing on oil and hidden in hair follicles. Parasitic mites, however, are more invasive and cause harm. Some use their biting mouthparts to chew the surface epidermis; others burrow deeper, causing irritation to the host and possibly infection.

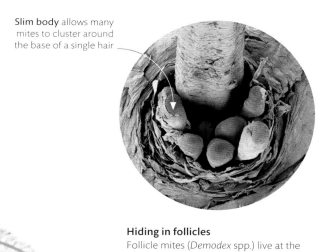

Slim body allows many mites to cluster around the base of a single hair

Hiding in follicles
Follicle mites (*Demodex* spp.) live at the base of skin hairs, feeding on oils or epidermal cells. Rarely, they trigger allergic reactions, causing hair loss or acne.

SKIN MICROHABITATS
Skin mites occupy their microhabitat in a variety of ways. Non-burrowers live either on the exposed epidermis or in follicles. Scabies mites (*Sarcoptes* sp.) soften skin with their saliva, which helps them to sink in. They then use their legs to excavate burrows, in which they lay their eggs.

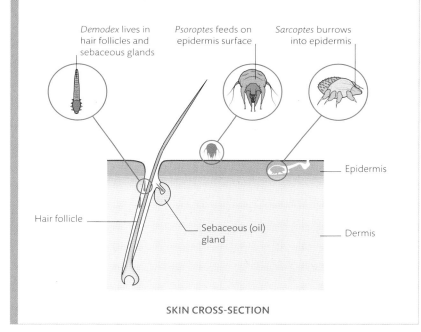

Demodex lives in hair follicles and sebaceous glands

Psoroptes feeds on epidermis surface

Sarcoptes burrows into epidermis

Hair follicle — Sebaceous (oil) gland — Epidermis — Dermis

SKIN CROSS-SECTION

Ear resident
The rabbit ear mite (*Psoroptes cuniculi*) does not burrow into the skin of its host, but instead punctures the epidermis in the ear canal and feeds on the leaking tissue fluid. As the ear canal becomes inflamed, the rabbit scratches and draws blood, which provides extra food for the parasite. The lesions the mite creates are prone to harmful infection. SEM x400.

Long, tapering rear legs extend far behind the body

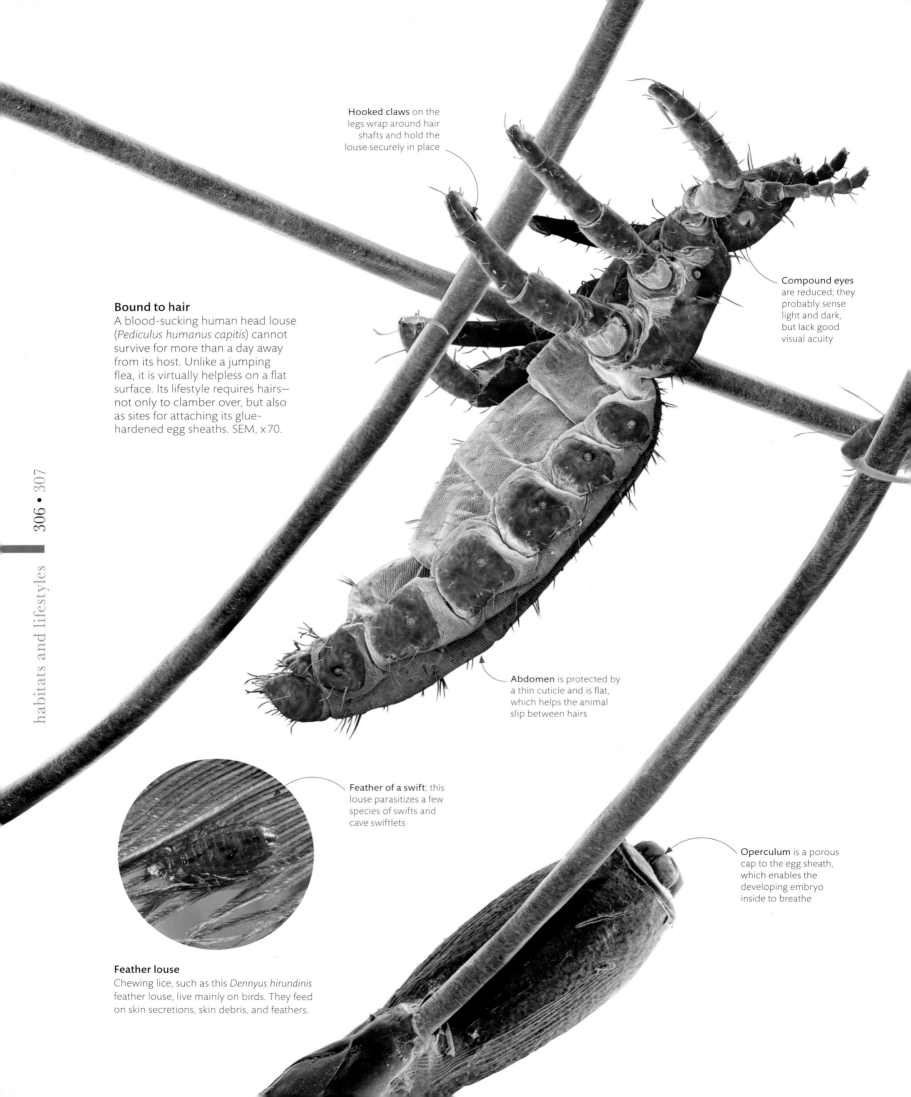

Bound to hair
A blood-sucking human head louse (*Pediculus humanus capitis*) cannot survive for more than a day away from its host. Unlike a jumping flea, it is virtually helpless on a flat surface. Its lifestyle requires hairs—not only to clamber over, but also as sites for attaching its glue-hardened egg sheaths. SEM, x70.

Hooked claws on the legs wrap around hair shafts and hold the louse securely in place

Compound eyes are reduced; they probably sense light and dark, but lack good visual acuity

Abdomen is protected by a thin cuticle and is flat, which helps the animal slip between hairs

Feather of a swift; this louse parasitizes a few species of swifts and cave swiftlets

Operculum is a porous cap to the egg sheath, which enables the developing embryo inside to breathe

Feather louse
Chewing lice, such as this *Dennyus hirundinis* feather louse, live mainly on birds. They feed on skin secretions, skin debris, and feathers.

living in hair

One group of parasites—the lice—is specialized to live in the microhabiats provided by mammal hairs and bird feathers. These flat, flightless insects can slip through these thickets and they chew skin or suck blood; many even cement their eggs to hairs or feathers. Most species live only on a particular species of host and have evolved with them. In humans, the closely related head and body lice share an origin with the lice of our closest living relative, the chimpanzee. And when humans began wearing animal skins, 100,000 years ago, our body lice evolved to lay eggs in clothing—no other louse lays eggs away from its host's body.

Splayed legs are strong enough to pull the louse through hair, but not effective for walking

Snoutlike mouthparts are armed with tiny cutting blades, or stylets, that the louse uses to puncture skin in order to feed on blood

Hard egg sheath, formed by a gluelike secretion from the female's abdomen, contains an embryo; after the nymph (young louse) hatches, the empty white sheath becomes visible as a "nit"

HOW GUT BACTERIA HELP HUMANS

"Friendly" gut bacteria interact with their hosts in several ways that help humans. They compete with harmful bacteria (pathogens) for space and nutrients. They produce chemicals, such as acids, that help reduce the growth of pathogens. They also contact immune cells in the epithelium (gut lining) called dendritic cells, and in so doing, both stimulate and regulate the host's immune response to pathogens.

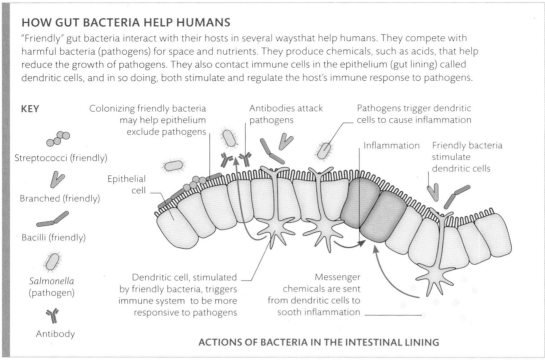

KEY
- Streptococci (friendly)
- Branched (friendly)
- Bacilli (friendly)
- *Salmonella* (pathogen)
- Antibody

Colonizing friendly bacteria may help epithelium exclude pathogens
Antibodies attack pathogens
Pathogens trigger dendritic cells to cause inflammation
Inflammation
Friendly bacteria stimulate dendritic cells
Epithelial cell
Dendritic cell, stimulated by friendly bacteria, triggers immune system to be more responsive to pathogens
Messenger chemicals are sent from dendritic cells to sooth inflammation

ACTIONS OF BACTERIA IN THE INTESTINAL LINING

Therapeutic microorganisms
Probiotics are microbes that confer human health benefits if we injest enough of them, such as in "live" yoghurt. They include bacteria and yeasts that must tolerate the stomach's acidity to reach and colonize the intestine.

Lactobacillus (pink and red) may be taken for digestive and urogenital problems

Saccharomyces boulardii (yellow) is a yeast often used to treat gastrointestinal infections

Biota in the human gut
The vast majority of bacteria in guts do not cause disease. They help their hosts to digest food and help fight off pathogenic microbes. The microbiome of humans—the bacteria living in or on our bodies—contains as many cells as there are human cells. When they leave the gut, as in this sample, the dry weight of human feces can be about 50 percent bacteria. SEM, x 15,000.

Filamentous bacteria (pink) can result from rod-shaped bacilli growing without dividing

Intestinal bacteria coat human feces after they have left the digestive tract

Bacilli, or rod-shaped bacteria, include *Lactobacillus*, which helps to suppress growth of harmful microbes and generates useful nutrients

Different shapes of bacteria reflect the variety of species living in the gut

gut communities

Hundreds of bacterial species make their home in the nutrient-rich guts of animals, from termites to humans. Densities just shy of 1 trillion (10^{12}) bacterial cells per 1/3 fl oz (milliliter) have been recorded in the human colon, making it one of the most densely populated microbial habitats on Earth. Nutrients aside, animal guts are far from hospitable, presenting challenges from extreme pH and powerful digestive enzymes, to immune systems bent on destroying bacteria. "Friendly" bacteria have coevolved with their hosts, evading destruction and helping termites to digest wood and cows to digest grass.

Parasites in human blood

Once malaria parasites have entered their host's red blood cells, they begin a self-perpetuating, and increasingly destructive, cycle of growth, reproduction, and reinfection. An invading merozoite becomes a trophozoite inside the red blood cell, then matures into a structure called a schizont. The schizont splits into many new merozoites, which rupture the cell to reinfect further cells. These images show cells from infected blood. SEM, x10,000.

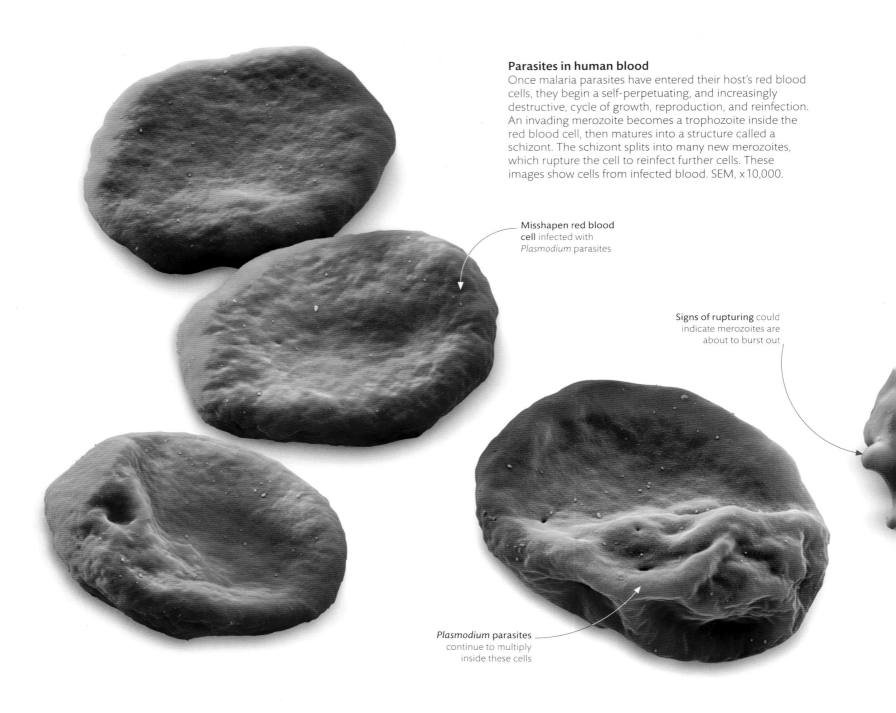

Misshapen red blood cell infected with *Plasmodium* parasites

Signs of rupturing could indicate merozoites are about to burst out

***Plasmodium* parasites** continue to multiply inside these cells

PARASITIC LIFE CYCLE

In mosquitoes, the *Plasmodium* life cycle begins when females take a blood meal containing *Plasmodium* cells called gametocytes. Fertilization takes place between male and female gametes, formed from the gametocytes, in the mosquito's stomach. The resulting zygotes then develop into oocysts, which grow and rupture, releasing sporozoites, which move to the mosquito's salivary glands. Sporozoites are then injected into humans, where they complete two further phases of their life cycle, first in liver cells, then in the blood.

LIFE CYCLE IN TWO HOSTS

infecting blood cells

Parasitic organisms do not live only in the skin (see pp.304–07) or the guts (see pp.62–63) of their hosts—some microscopic parasites are injected directly into a host's tissues, overcoming its defenses to multiply there, often causing disease. *Plasmodium* is a si ngle-celled parasite that infects two hosts: a female mosquito and a vertebrate, such as a human, where it causes malaria. Transmission occurs as the mosquito feeds—either parasites living in the mosquito's salivary gland are injected into a human, or parasites living in human blood are ingested by the mosquito when it takes a blood meal. There are many stages in the malaria parasites' life cycle, but it is when they periodically burst out of red blood cells that they cause disease symptoms.

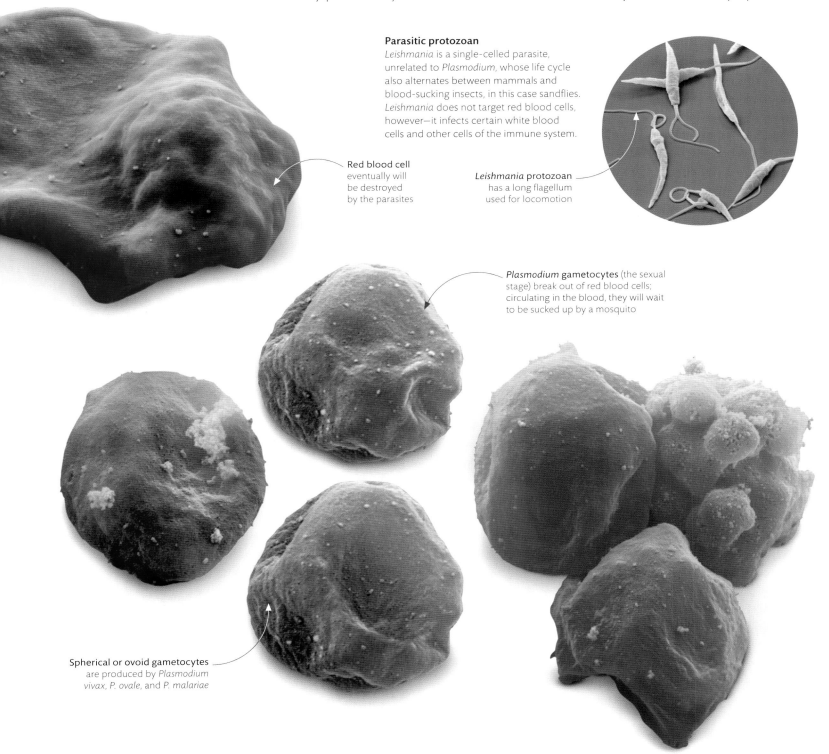

Parasitic protozoan
Leishmania is a single-celled parasite, unrelated to *Plasmodium*, whose life cycle also alternates between mammals and blood-sucking insects, in this case sandflies. *Leishmania* does not target red blood cells, however—it infects certain white blood cells and other cells of the immune system.

Red blood cell eventually will be destroyed by the parasites

Leishmania **protozoan** has a long flagellum used for locomotion

***Plasmodium* gametocytes** (the sexual stage) break out of red blood cells; circulating in the blood, they will wait to be sucked up by a mosquito

Spherical or ovoid gametocytes are produced by *Plasmodium vivax*, *P. ovale*, and *P. malariae*

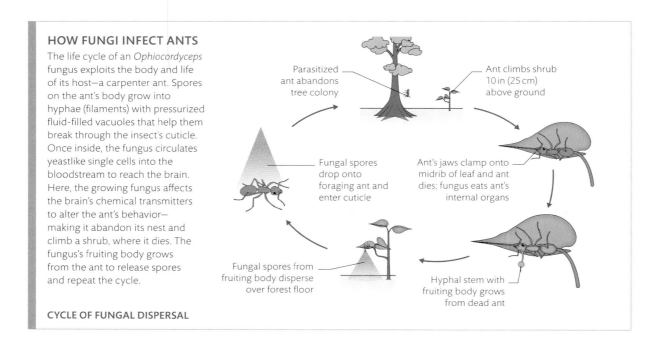

HOW FUNGI INFECT ANTS

The life cycle of an *Ophiocordyceps* fungus exploits the body and life of its host—a carpenter ant. Spores on the ant's body grow into hyphae (filaments) with pressurized fluid-filled vacuoles that help them break through the insect's cuticle. Once inside, the fungus circulates yeastlike single cells into the bloodstream to reach the brain. Here, the growing fungus affects the brain's chemical transmitters to alter the ant's behavior—making it abandon its nest and climb a shrub, where it dies. The fungus's fruiting body grows from the ant to release spores and repeat the cycle.

CYCLE OF FUNGAL DISPERSAL

brain parasites

Two challenges drive the lives of fungi, like all living organisms: finding enough food to survive and grow, and reproducing to ensure the replication of their genes. It is especially beneficial when a food supply can be the vector for dispersal too, which explains the success of the entomopathogenic fungi—literally, fungi that cause disease in insects. Spores become attached to an insect, then germinate, penetrate the cuticle, or outer casing, and spread throughout the insect's body to its brain. In so doing, the fungus releases chemicals that manipulate the insect's behavior, ensuring that it comes to rest in the optimum conditions for spore dispersal.

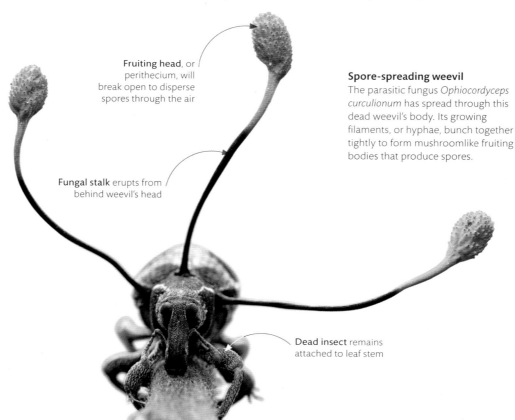

Fruiting head, or perithecium, will break open to disperse spores through the air

Fungal stalk erupts from behind weevil's head

Spore-spreading weevil
The parasitic fungus *Ophiocordyceps curculionum* has spread through this dead weevil's body. Its growing filaments, or hyphae, bunch together tightly to form mushroomlike fruiting bodies that produce spores.

Dead insect remains attached to leaf stem

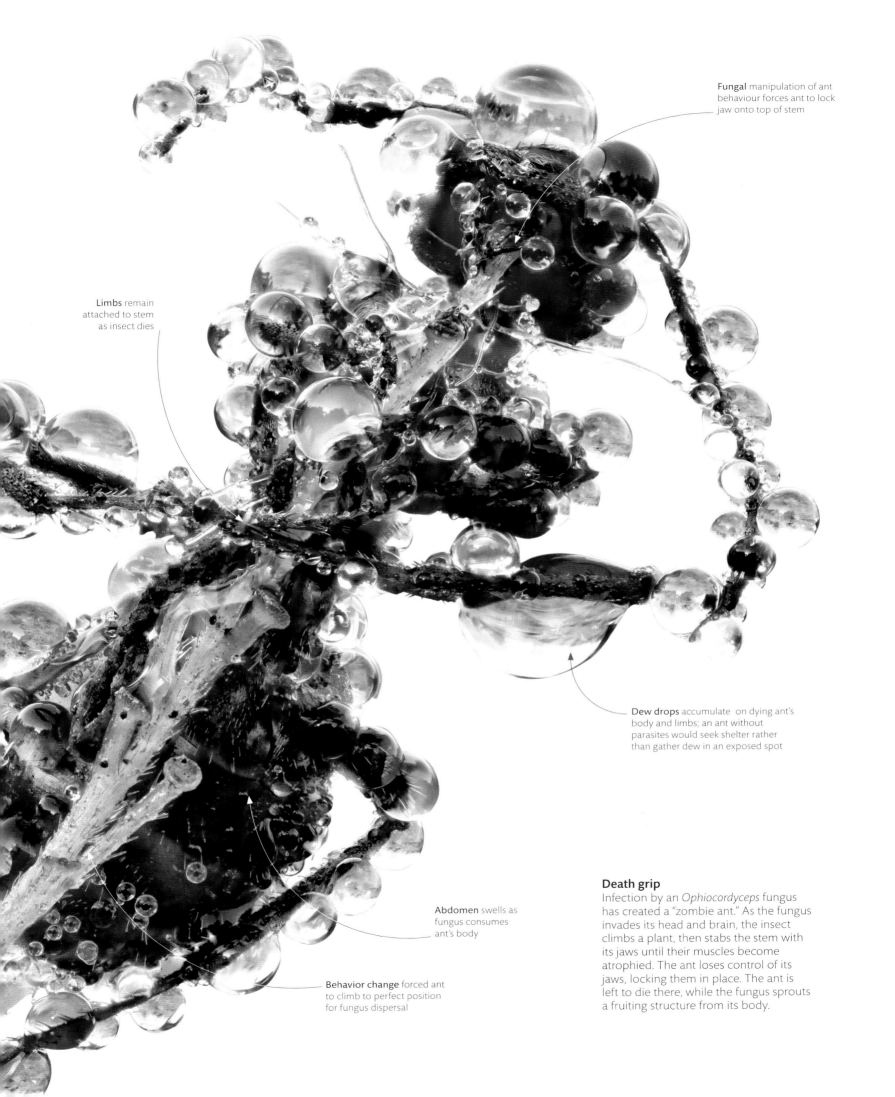

Fungal manipulation of ant behaviour forces ant to lock jaw onto top of stem

Limbs remain attached to stem as insect dies

Dew drops accumulate on dying ant's body and limbs; an ant without parasites would seek shelter rather than gather dew in an exposed spot

Abdomen swells as fungus consumes ant's body

Behavior change forced ant to climb to perfect position for fungus dispersal

Death grip
Infection by an *Ophiocordyceps* fungus has created a "zombie ant." As the fungus invades its head and brain, the insect climbs a plant, then stabs the stem with its jaws until their muscles become atrophied. The ant loses control of its jaws, locking them in place. The ant is left to die there, while the fungus sprouts a fruiting structure from its body.

plant-fungus partnership

An intimate and mutually beneficial relationship has evolved between plants and fungi. This partnership involves the absorption and exchange of nutrients. Each life-form relies on an underground network—of roots in plants and of threads called hyphae in fungi—to absorb nutrients from the soil over a large area. By spreading its hyphal network far and wide, a root-associated fungus, or mycorrhiza, captures nitrogen and phosphorus from rotting organic matter and shares these vital elements with the plant. In return, the plant supplies carbon to the fungus by sharing the sugar it makes by photosynthesis.

Surface fungal hyphae germinate from spores in the soil and penetrate the outer layers of the root

Arbuscules are branching ends of the internal fungal hyphae; they exchange nutrients with the surrounding cytoplasm of the host plant cell

Internal fungal hyphae grow through the root cells by pushing the plant's cell walls and membranes inward

Fungus-infected root
Looking like neatly packed paper boxes, many cells of this grass root contain white hyphae of mycorrhizal fungus. Different strains of fungus infect the roots of different grass species. As well as supplementing the nutrient uptake of the grass, the fungi may improve its host's resistance to herbivores and toxic metals. SEM, x 1,800.

Bright red caps and white spots of veil tissue are characteristic of this common toxic species

Mycorrhizal mushroom
Fly agaric (*Amanita muscaria*) fungus often grows as mycorrhizae on the roots of trees. Its fruiting bodies sprout beneath host species, especially birch.

TYPES OF MYCORRHIZAE
Most root-associated fungi are ectomycorrhizae, whose hyphal threads form a sheath around each root and grow between the root cells. The more invasive threads of endomycorrhizae penetrate root cells with bunches called arbuscules. Both types occur in trees and shrubs, but endomycorrhizae are more common in nonwoody plants.

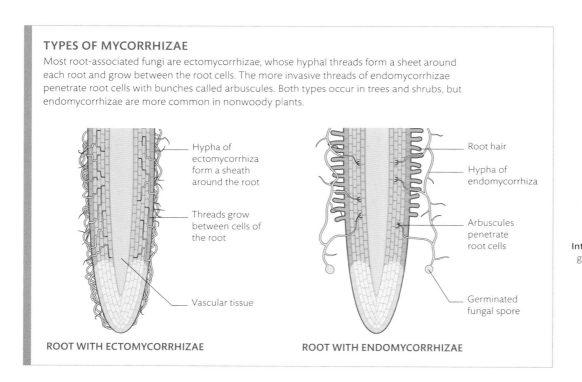

- Hypha of ectomycorrhiza form a sheath around the root
- Threads grow between cells of the root
- Vascular tissue

ROOT WITH ECTOMYCORRHIZAE

- Root hair
- Hypha of endomycorrhiza
- Arbuscules penetrate root cells
- Germinated fungal spore

ROOT WITH ENDOMYCORRHIZAE

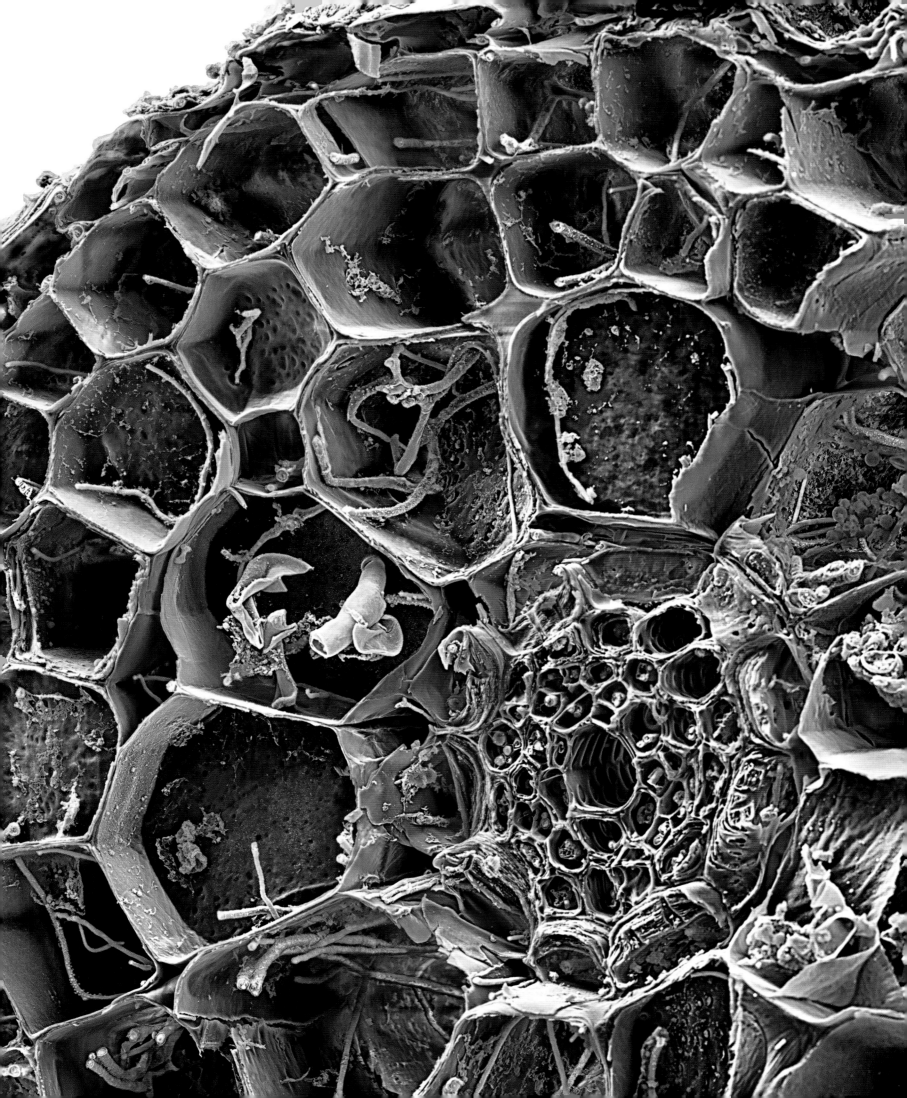

part fungus, part alga

About one-fifth of fungus species live in partnership with algae, forming an organism called a lichen. Many of these partners can survive and grow on their own, but when they come together, the resulting lichen develops very differently. Some lichens are flat and leafy, and others are tufted or resemble featureless dabs of color. In all varieties, the lichen consists of nutrient-absorbing fungal filaments that cling to rock, bark, or other anchoring places. Algal cells below the fungal surface absorb light, turning the energy into food through photosynthesis.

Intimate cohabitation
In cross-section, the two cohabiting partners that make a hammered shield lichen (*Parmelia sulcata*) are clearly visible. The brown fibers—called hyphae—are fungus. The green spheres of single-celled *Trebouxia* algae are packed with chlorophyll for photosynthesis. As well as photosynthesizing, the algae extract nitrogen gas from the atmosphere and convert it into extra nourishment that is shared with the fungus, helping the partnership prosper. SEM, x 2,590.

Upper layer is made up of compacted, non-absorbing, fungal hyphae that shield the lichen from the drying effects of sunshine

Algal cells absorb light to make sugar from carbon dioxide and water. They also help convert nitrogen from the air into compounds from which proteins are made

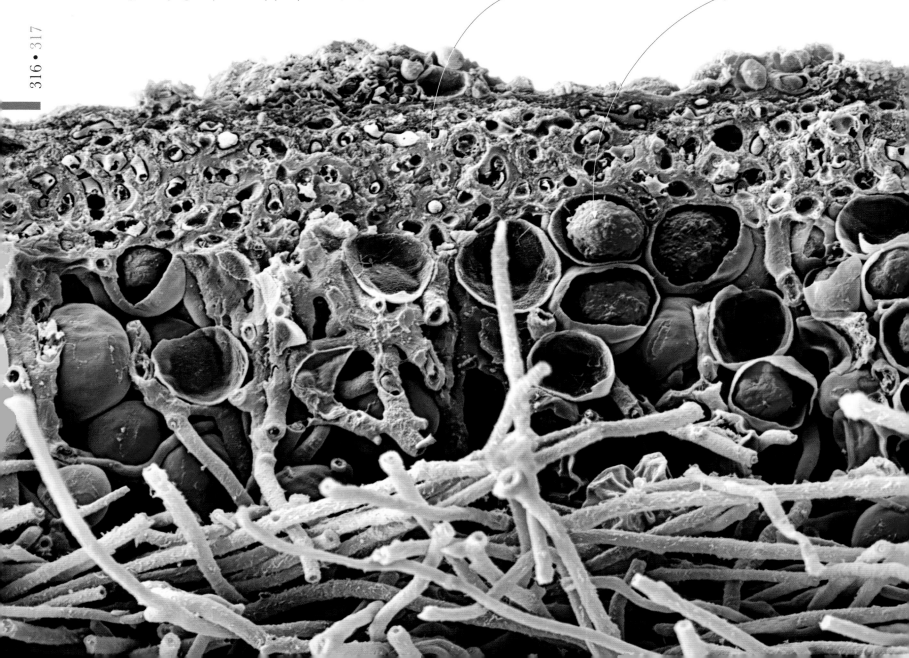

Growing on rock
Gray *Lecanora* and yellow *Caloplaca* growing on bare rock show how lichens can thrive even in the most unpromising of locations.

Lichen "crust" is firmly sealed to rock

THRIVING TOGETHER

A lichen consists of two very different organisms: the fungus absorbs food, while the alga makes its food. Both organisms usually succeed better together than alone, suggesting this relationship is mutually beneficial. The upper and lower layers are mainly composed of fungal filaments, while algal cells are mixed with filaments in the middle of the lichen. The fungal hyphae serve to attach and absorb minerals and water, whereas the algae produce sugars—a winning arrangement that mirrors the roots and leaves of a plant.

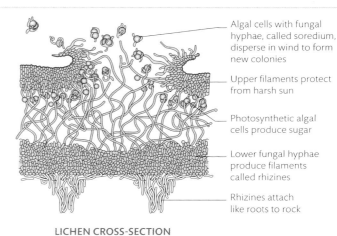

Algal cells with fungal hyphae, called soredium, disperse in wind to form new colonies

Upper filaments protect from harsh sun

Photosynthetic algal cells produce sugar

Lower fungal hyphae produce filaments called rhizines

Rhizines attach like roots to rock

LICHEN CROSS-SECTION

Connections, called haustoria, between fungal hyphae and algal cells, may help nutrients to be exchanged for mutual benefit

Loose fungal hyphae of lower layer provide a large surface area for absorbing water, organic nutrients, and minerals from rain that washes over the lichen

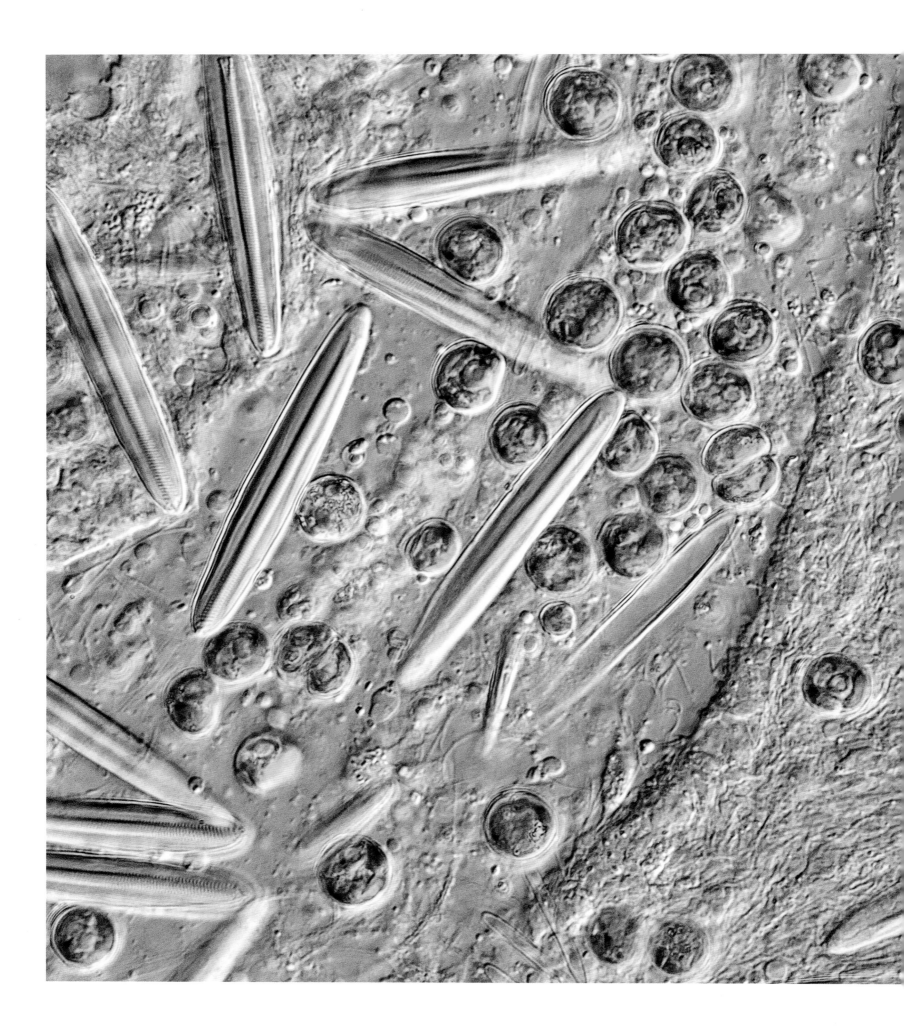

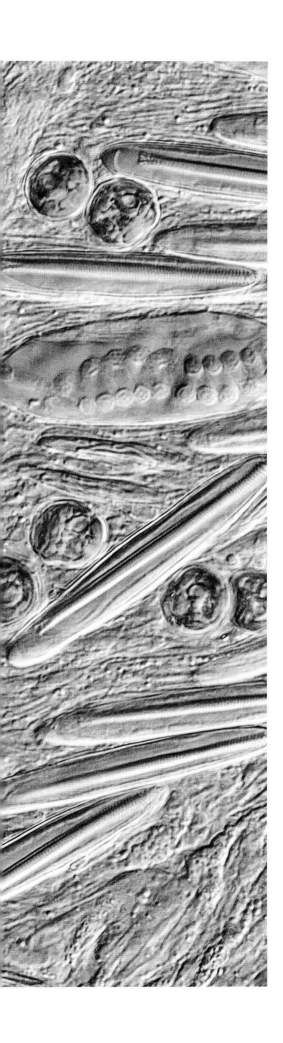

photosynthetic helpers

Some single-celled algae live in partnership with tropical reef corals, as well as various soft corals, sea anemones, and clams. Living inside these larger organisms, the algae photosynthesize and pass sugars, oxygen, and other nutrients to their host. This is crucial for the success of coral reefs, because it boosts calcification of coral skeletons, ensuring that reef growth outpaces natural erosion. In return, the algae—which are dinoflagellates known as zooxanthellae—have a safe refuge from organisms that would readily devour them. They also receive guaranteed supplies of carbon dioxide and various inorganic nutrients that are essential for photosynthesis.

Inside a glass anemone
Many yellow zooxanthellae live inside the tentacle of an *Aiptasia* glass anemone (left), alongside elongated stinging cells used for feeding and defense. The cells have a coccoid (spherical) form, without flagella. Each is about 10 micrometers wide and is colored by photosynthetic pigments contained in the dinoflagellate's chloroplasts. LM, x2,000.

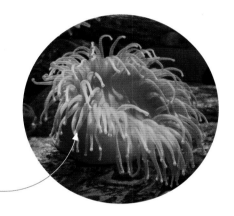

Tentacles surround mouth, almost hiding column of polyp

Additional support
Sea anemones are predatory, but they also obtain energy and oxygen from zooxanthellae. This helps the anemone, which usually lives in shallow or tidal waters, to survive when prey is scarce.

LIFE STAGES OF CORAL-DWELLING ZOOXANTHELLAE

The life cycle of *Symbiodinium* zooxanthellae includes a free-living form, where each zooxanthella has flagella to propel its movements, and a coccoid phase inside the host. When the flagellated cells enter a coral polyp via its mouth and pass into the body wall they lose their flagella, becoming coccoid, but continue to photosynthesize.

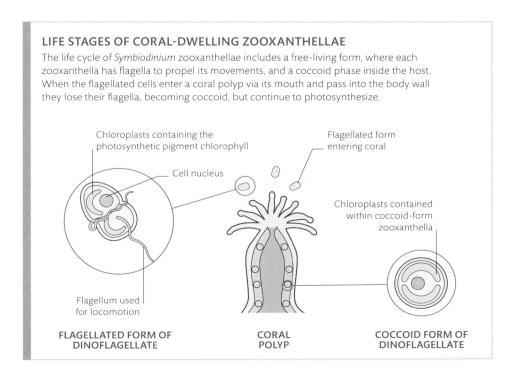

Chloroplasts containing the photosynthetic pigment chlorophyll

Cell nucleus

Flagellated form entering coral

Chloroplasts contained within coccoid-form zooxanthella

Flagellum used for locomotion

FLAGELLATED FORM OF DINOFLAGELLATE   **CORAL POLYP**   **COCCOID FORM OF DINOFLAGELLATE**

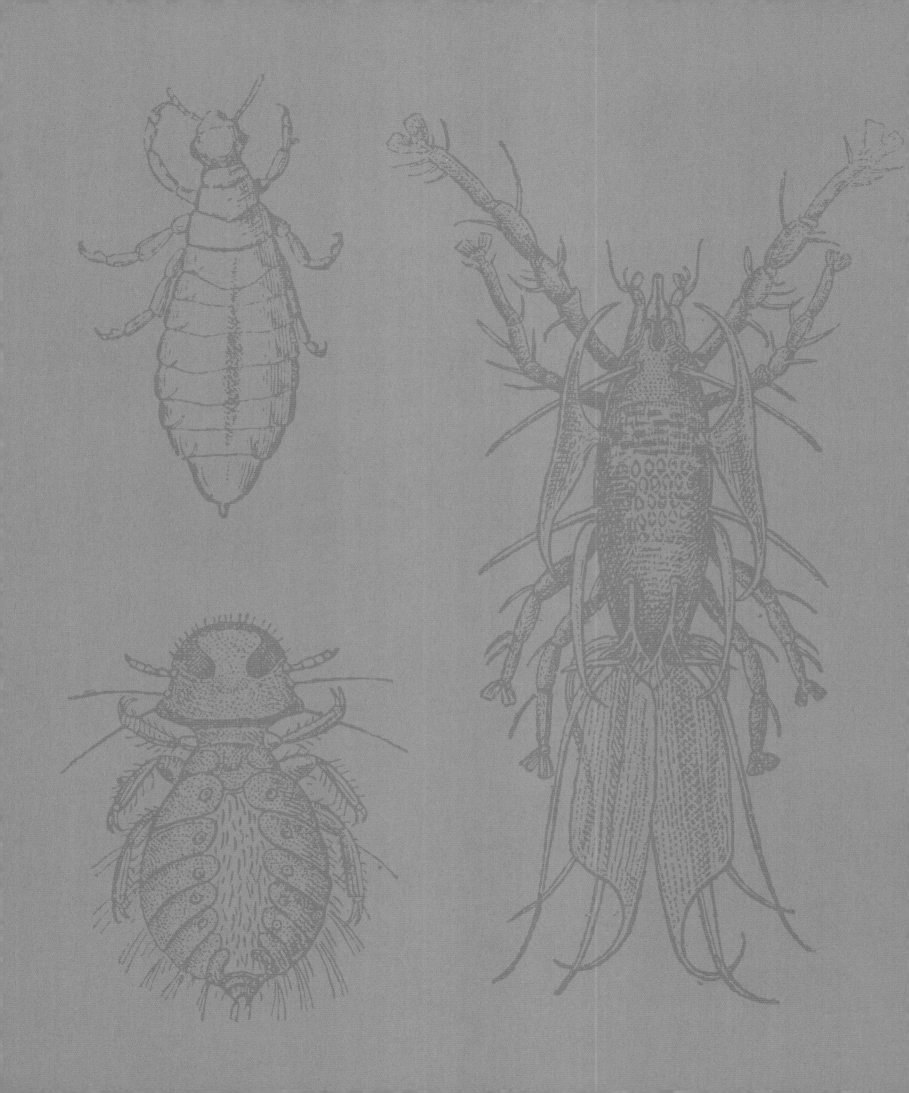

classification

Humans have imposed order on the natural world for the practical reason of easy reference—information storage and recall. However, organizing species into groups has taken on a deeper meaning—that of uncovering their evolutionary relationships.

classifying life

The hierarchical system of biological classification used today, which divides larger groups of organisms into smaller ones, was formalized by the 18th-century Swedish naturalist Carl Linnaeus. He also devised the two-part method of Latinizing species names, such as *Homo sapiens*. However, most of the rich diversity of micro-life was discovered with ever-improving microscopes after Linnaeus, and it was not until biologists knew more about cells, genetics, and biochemistry that organisms' true affinities began to be appreciated. Since Charles Darwin established that all life is a product of evolution, the relatedness of life forms has guided their classification. Today, although Linnaeus's naming system is still used, life is classified into evolutionary groups called clades, each of which includes all the descendants of a common ancestor. These groups, and the numbers of species they contain, are continually revised.

Naming a beetle
The French entomologist Henri Jekel described this beetle in 1860 from specimens earlier collected in Madagascar. Its scientific name, *Trachelophorus giraffa*, means "giraffelike one bearing a neck."

CLASSIFYING A GIRAFFE-NECKED WEEVIL
Animals are classified into a single kingdom, Animalia. They differ from organisms belonging to other kingdoms—such as bacteria and plants—in having multicellular bodies that usually include muscles and nerves.

phylum (plural: phyla)
The animal kingdom is divided into more than 30 main phyla. One of these, Arthropoda, unites animals—including arachnids, crustaceans, and insects—with jointed legs and an exoskeleton.

class
Adults of the largest class of arthropods, Insecta, typically have three-part bodies (consisting of a head, thorax, and abdomen), six jointed legs, wings, and two antennae.

order
Members of the order Coleoptera, or beetles, have their front pair of wings modified into hard shields (elytra) that conceal the membranous hind pair when the beetle is not in flight.

family
Beetles of the family Attelabidae, called leaf-rolling weevils, have a long snout and straight antennae. Many species in this family lay their eggs in rolled-up leaves.

genus (plural: genera)
Leaf-rollers of the genus *Trachelophorus* are known as giraffe weevils—named for their extraordinary extended neck, which is longer in males than females.

species
A species, such as the striking red and black *Trachelophorus giraffa*, is usually viewed as a group of individuals that can interbreed to produce viable offspring.

Extra-long neck of males is used to push rivals from leaves in fights over females

GIRAFFE-NECKED WEEVIL
Trachelophorus giraffa

Micro-diversity
Many single-celled organisms are now known to be so distantly related to plants and animals that they are categorized in separate groups of their own. A plate from Jabez Hogg's *The Microscope* (1883) illustrates some of these microbes, including shelled forams (top and bottom), ciliates and flagellates (lower left), and wormlike parasites called gregarines (upper center). These are all shown alongside microscopic animals called rotifers that live inside transparent tubes (lower center).

GREGARINIDA, POLYCYSTINA, FORAMINIFERA, ROTIFERA, ETC.

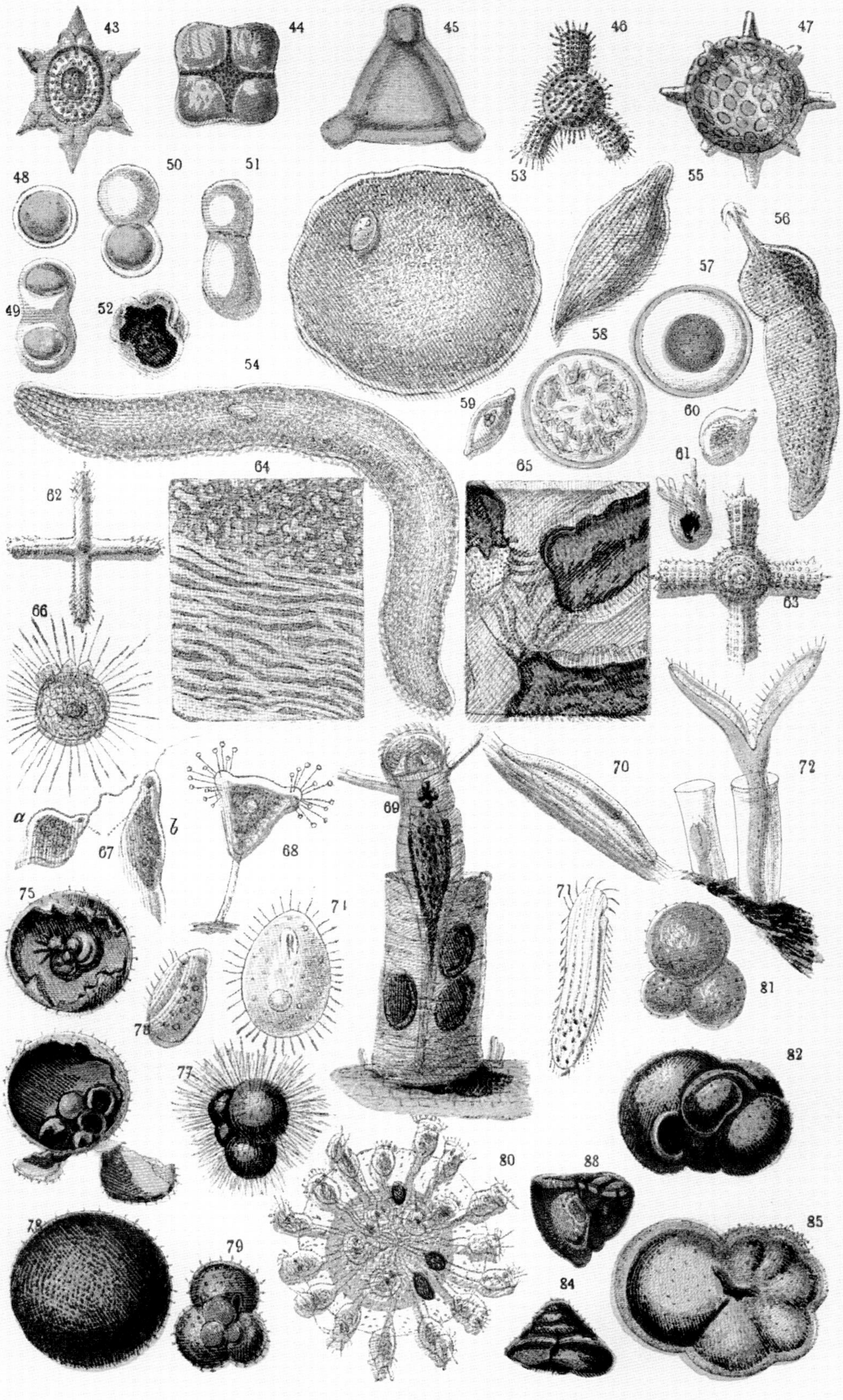

Tuffen West, del.

PLATE III.

Edmund Evans.

extreme halophiles

PHYLUM: Euryarchaeota **KINGDOM:** Archaea

Like all archaeans, extreme halophiles are microscopic, single-celled organisms lacking a nucleus. In that respect, archaeans are similar to bacteria; but they differ from bacteria in the makeup of their cell membranes and other chemical features. Their DNA indicates that archaeans are actually closer to eukaryotes (see pp.12–13).

Extreme halophiles thrive in conditions that are more than four times saltier than sea water—including on salt flats and hypersaline lakes. Their carotenoid pigments, which protect them from ultraviolet radiation, color the water red. Another pigment in their cell membrane is used to absorb light energy to help power their metabolism; however, they cannot photosynthesize to make their own food. Most need oxygen, but some grow without it.

Halobacterium
Like other halophiles, *Halobacterium* can survive only in salty environments. It lives in the Dead Sea (Jordan/Israel), Lake Magadi (Kenya), and the Great Salt Lake (US).

methanogens

PHYLUM: Euryarchaeota **KINGDOM:** Archaea

Methanogens are archaeans that respire without oxygen, instead using carbon dioxide to oxidize hydrogen gas—which releases methane as a waste product. They are very abundant in swamps and marshes, where they digest dead vegetation and produce "marsh gas" (methane, hydrogen sulfide, and carbon dioxide). They also inhabit the guts of many animals, especially herbivores. Some of these spherical or rod-shaped microbes can thrive in the most extreme conditions, including hot, dry desert soil, rocks deep in Earth's crust, and volcanic hot springs. *Methanopyrus kandleri* lives at the base of ocean vents at 176–230°F (80–110°C) and 6,500 ft (2,000 m) below sea level, while other methanogens have been found 9,800 ft (3,000 m) deep in ice cores taken from Greenland's ice sheet.

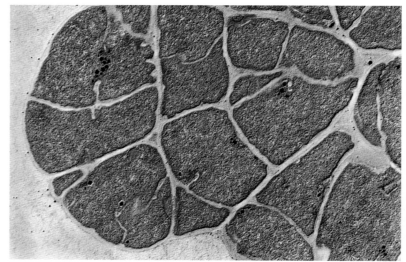

Methanosarcina
This transmission electron micrograph (TEM) shows *Methanosarcina* archaean (green with yellow cell walls). They live in a wide range of environments, including deep ocean vents, landfill sites, and the guts of animals.

hyperthermophiles

PHYLUM: Euryarchaeota, Crenarchaeota   **KINGDOM:** Archaea

Hyperthermophiles are archaeans that live anaerobically in hydrothermal vents, oil deposits, and hot springs. Typically, they grow at temperatures of 140–203°F (60–95°C). At such temperatures, the complex biological molecules of most organisms unravel or "denature" and become unworkable. But in hyperthermophiles, these molecules have chemical bonding that makes them resistant. Archaeans also have protective proteins around their DNA. Related proteins make up the chromosomes of eukaryotes—evidence that archaeans may be their ancestors. *Geogemma barossii*, from Pacific hydrothermal vents, is one of the most extreme hyperthermophiles. It can reproduce at 250°F (121°C) and remains viable at 266°F (130°C). Prior to its discovery, scientists thought a temperature of 250°F (121°C) killed all organisms.

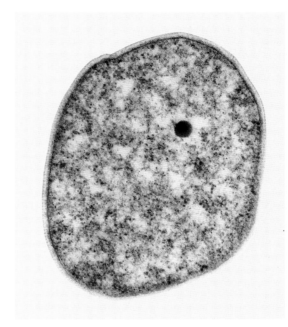

Sulfolobus
The species in this genus are strongly associated with volcanic activity, living in mudpots, hot springs, and hydrothermal vents with temperatures of 104–203°F (40–95°C). They have acid-resistant cell walls and are acidophillic, surviving only in acidic conditions with a pH range of 1–6.

ultra-small archaeans

PHYLUM: Nanoarchaea   **KINGDOM:** Archaea

The first nanoarchaeans were discovered as recently as 2002 at a hydrothermal vent off the coast of Iceland. Named *Nanoarchaeum equitans*, they were attached to a much larger thermophilic archaean. Less than half a micrometer in diameter, this—the smallest known archaean—is the size of a large virus. Other nanoarchaeans have since been found in habitats around the world, such as hot springs in the US and Russia, and hydrothermal vents in the Pacific Ocean. Their tiny dimensions limit the size of the DNA, and therefore their ability to live independent lives: most appear to depend on other microbes—living as parasites and in mutualistic partnerships with their larger host.

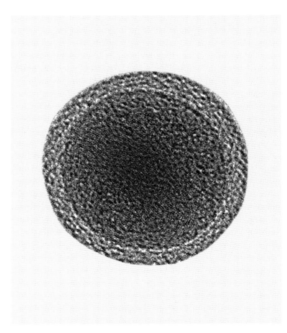

Parvarchaeum
Found in acidic water draining from mines, *Parvarchaeum* has a symbiotic relationship with other, larger archaeans.

proteobacteria

PHYLUM: Proteobacteria **KINGDOM:** Bacteria

Proteobacteria are a diverse group of Gram-negative bacteria (with an outer membrane around the cell wall, see p.33); most are rod-shaped or spiral-shaped. Several species in the group have a whiplike flagellum that they use for movement. Many proteobacteria are anaerobic—they do not need oxygen to metabolize.

Many proteobacteria are pathogens—they are responsible for digestive-system infections and diseases such as typhus. Reproduction can be rapid among pathogenic proteobacteria; for example, a parent cell of *Escherichia coli* can produce two "daughter" cells every 20 minutes. But the majority of proteobacteria are not pathogens; in fact, they play important roles in ecosystems by decomposing organic material or fixing nitrogen in soil (see p.26).

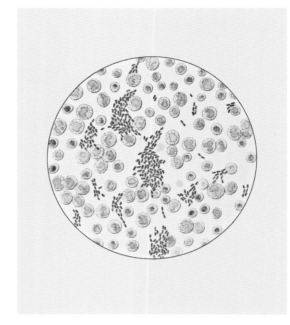

Salmonella
This proteobacteria (blue) is well known for causing food poisoning. One form of *Salmonella*, which causes typhoid fever, can pass into the blood and spread the infection around the body. Here, it has infected the human bloodstream.

firmicutes

PHYLUM: Firmicutes **KINGDOM:** Bacteria

Firmicutes are spherical or rod-shaped Gram-positive bacteria (without an outer membrane around the cell wall). Many resist hostile conditions by transforming into dormant spores. The bacterium divides within its cell wall, and one side engulfs the other to form an endospore that can resist drying out, ultraviolet radiation, high temperatures, and freezing. The endospore can remain dormant for centuries, rehydrating and metabolizing again when conditions improve.

Firmicutes use organic food, but a few kinds, such as *Heliobacterium*, also use light energy to help power their nutrition; however—unlike most true photosynthesizers—they do not generate oxygen. *Clostridium pasteurianum* belongs to the firmicutes and was the first free-living bacterium known to fix atmospheric nitrogen.

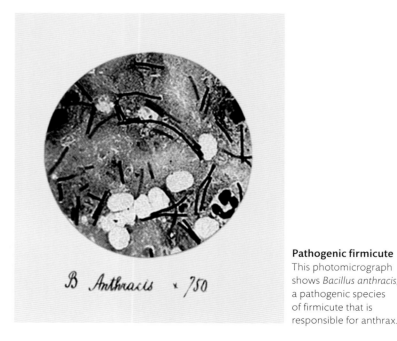

Pathogenic firmicute
This photomicrograph shows *Bacillus anthracis*, a pathogenic species of firmicute that is responsible for anthrax.

acidobacteria

PHYLUM: Acidobacteria   **KINGDOM:** Bacteria

Found in terrestrial and aquatic environments, sometimes very abundantly in soils, Acidobacteria form a very diverse group. However, their ecology and metabolism are not as well understood as that of many other groups, partly because they are hard to cultivate in the laboratory. Most members of the phylum thrive only in acidic conditions.

Acidobacteria are common in soils and other sediments that are rich in organic matter. They also thrive in freshwater and wastewater systems. Although their role is not fully understood, acidobacteria play an important part in soil ecology, by decomposing the dry matter of plants, modulating nitrogen cycling in soils, and trapping nutrients and water. When plant roots are exposed to these bacteria, they grow more vigorously.

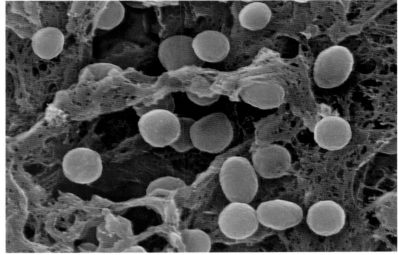

Acidobacteria
This artificially coloured scanning electron micrograph (SEM) shows cocci-shaped acidobacteria. Many members of this phylum are rod-shaped anaerobic species.

cyanobacteria

PHYLUM: Cyanobacteria   **KINGDOM:** Bacteria

Formerly called blue-green algae, cyanobacteria live in oceans, fresh water, and damp soil. They are the smallest organisms that use oxygenic photosynthesis (see pp.20–21)—they use sunlight to create food, producing oxygen in the process. Many species help fertilize the oceans by combining atmospheric nitrogen into compounds usable by other organisms. Many cyanobacteria species form hormogonia—moving filaments of connected cells. The hormogonia of *Oscillatoria* help to reorient the algae toward sunlight. Scientists believe that more than 1 billion years ago, cyanobacteria cells were engulfed by larger cells. Such cyanobacteria eventually evolved into chloroplasts, which are the light-harvesting structures found in all living plant and algal cells today.

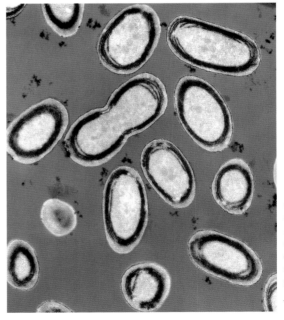

Marine cyanobacteria
Prochlorococcus is probably the most abundant and the smallest photosynthetic organism on Earth. The spherical structures are just 0.5–0.7 micrometers in diameter.

actinobacteria

PHYLUM: Actinobacteria   **KINGDOM:** Bacteria

Actinobacteria are Gram-positive (see p.33) bacteria that live in terrestrial and aquatic environments, forming colonies of branched chains of cells. Most are free-living (they do not live inside another organism), and many play a vital role in the decomposition of organic matter, such as cellulose, in soil.

Actinobacteria also have immense medicinal value. Soil-dwelling species in the genus *Streptomyces* are a source of many antibiotics; in fact, about two-thirds of all naturally derived antibiotics come from this phylum. In addition, some plants depend on actinobacteria for nitrogen fixation; *Frankia*, for example, lives on the root nodules of certain plants, supplying them with most of their nitrogen needs. A few are pathogens: *Corynebacterium*, for example, causes diphtheria, and *Mycobacterium* causes tuberculosis and leprosy.

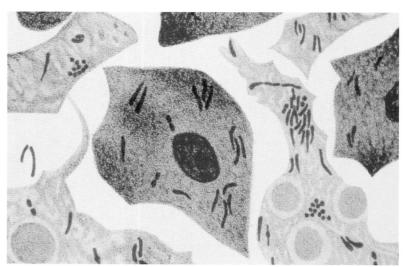

Corynebacterium diphtheriae
This illustration shows *Corynebacterium diphtheriae* (red), the pathogen responsible for diphtheria, infecting cells. It was discovered by German bacteriologists Edwin Klebs and Friedrich Loeffler in 1884.

deinococci

PHYLUM: Deinococci   **KINGDOM:** Bacteria

The earliest bacteria were probably thermophiles or hyperthermophiles, which are able to thrive at hot or very hot temperatures. Many of their modern deinococci descendants are adapted for some of the harshest conditions on Earth. Those in the genus *Thermus* live in hot springs, growing best at temperatures of 149–158°F (65–70°C) and able to survive at 176–194°F (80–90°C).

Other deinococci are highly resistant to ultraviolet light and harmful radiation. *Deinococcus radiodurans*, for instance, can survive radiation that is 1,000 times more powerful than that needed to kill a human. Scientists on the International Space Station have found that this bacterium can survive for 3 years in outer space, consistent with the idea of panspermia—that life exists throughout the universe.

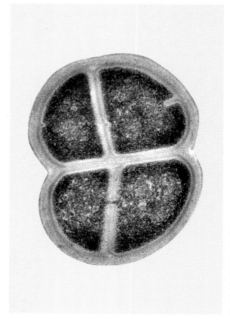

Deinococcus radiodurans
Nicknamed "Conan the Bacterium" for its hardiness, *Deinococcus radiodurans* is a polyextremophile—it is able to survive a variety of extreme conditions, including excessive heat, drought, radiation, acidity, and vacuum.

bacteroidetes

PHYLUM: Bacteroidetes   **KINGDOM:** Bacteria

Bacteroidetes are important bacteria found in the guts of humans and other animals. Gram-negative (see p.33) and rod-shaped, they form a very diverse group and may be aerobic (requiring oxygen for metabolism) or anaerobic, and motile (capable of movement) or nonmotile.

Bacteroidetes also inhabit soil and other sediments, as well as fresh and salt water. They play a vital role in the human digestive system, breaking down proteins and sugars that the body is not able to process. *Bacteroides fragilis*, for example, forms a part of the healthy microbiota of the human colon, but can cause infection if displaced into the bloodstream or surrounding tissue following surgery, disease, or trauma. *Cytophaga*, a soil-dwelling member of the phylum, breaks down cellulose from dead plants.

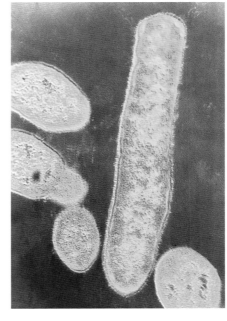

Gut bacteria
In mammals' guts, *Bacteroides* bacteria aid digestion by breaking down complex molecules into simpler ones. They dominate where the host eats mostly proteins and animal fats, whereas *Prevotella* species are more common if the diet is more carbohydrate-based.

fusobacteria

PHYLUM: Fusobacteria   **KINGDOM:** Bacteria

Fusobacteria are rod-shaped, Gram-negative bacteria adapted for life in anaerobic (oxygen-free) environments. Although their role is not fully understood, bacteria in the phylum Fusobacteria are common constituents of the healthy oral and gastrointestinal microbiota of many animals, including humans. Some are free-living in marine environments and others are pathogens.

In humans, some fusobacteria have been linked to premature labor and others have been known to cause infections responsible for ulcers, tissue necrosis, and septicemia. In the mouth, *Fusobacterium nucleatum* contributes to the buildup of calcified deposits called tartar; if unchecked, this can cause tooth decay and gum disease.

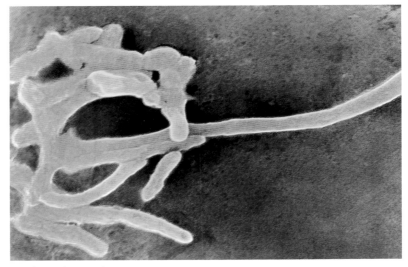

Fusobacterium nucleatum
An anaerobic Gram-negative bacterium, *Fusobacterium nucleatum* inhabits the human mouth and contributes to periodontal (gum) disease.

excavates

SUBKINGDOM: Excavata   **KINGDOM:** Protozoa

Like all organisms other than archaeans and bacteria (see pp.324–29), excavates are eukaryotic—their cells have nuclei (see pp.12–13). Excavates, which are single-celled and move using flagella, are named for their cavity, or groove, which helps with feeding in some.

The first excavate to be discovered—the freshwater genus *Euglena*—puzzled early biologists because specimens were observed to photosynthesize like plants, but could also survive without light by absorbing or capturing food. Most excavates do not photosynthesize, and live as parasites instead. The parasitic trypanosomes cause sleeping sickness and other diseases. Many excavates are anaerobic (do not need to breathe oxygen), including some that live inside termites and consume wood for them.

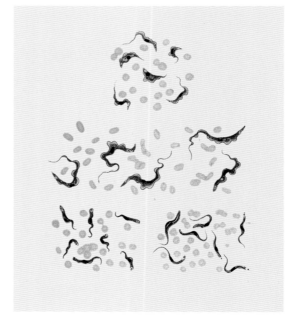

Parasitic trypanosomes
Trypanosomes (blue) are responsible for a variety of diseases, including Chagas disease and sleeping sickness (carried by tsetse flies) in humans. These examples were found in infected camels and cattle in Africa.

choanoflagellates

CLASS: Choanoflagellata   **PHYLUM:** Choanozoa   **KINGDOM:** Protozoa

Widespread in marine and fresh waters, choanoflagellates, or collar flagellates, are the nearest relatives of animals. Within the animal kingdom, sponges and some ribbon worms grow cells in their tissues that closely resemble choanoflagellate cells. About 125 species have been described.

The distinctive cells of choanoflagellates are oval in shape. They have a collar of tiny projections at one end with a beating flagellum in the middle. The flagellum creates a current that wafts food particles such as bacteria onto the collar, where they are trapped and consumed. Choanoflagellates either swim in plankton or are attached to surfaces, and they may be solitary or aggregated into colonies. Some species protect themselves with a tough outer covering, while others surround their body with mucus.

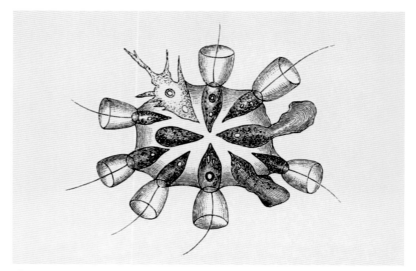

Choanoflagellate colony
This 1886 drawing of a choanoflagellate colony shows each individual with a single flagellum projecting from a conical collar.

strontium radiolarians

CLASS: Acantharea   **PHYLUM:** Radiozoa   **SUBKINGDOM:** Rhizaria   **KINGDOM:** Chromista

Radiolarians are single-celled organisms that are common in ocean plankton. They catch tiny prey by using long, needlelike pseudopods that emerge from a central cell body. Radiolarians have elaborate skeletons that feature radiating spines and often, one or more perforated spheres inside the cell. Usually, radiolarian skeletons are constructed of silica, but in the class Acantharea they are made—very unusually for living organisms—of strontium sulphate (strontium is an element similar to calcium, but heavier). Strontium sulphate dissolves after the cells die and so acantharians do not normally form fossils. Often, symbiotic algae within their cells supply them with part of their food. About 150 species have been described.

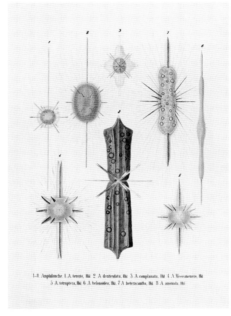

Amphilonche radiolarians
Acantharia display a bewildering array of different shapes. This illustration from Ernst Haeckel's 1862 publication *Die Radiolarien* (*The Radiolaria*) shows eight very different species from the genus *Amphilonche*.

silica radiolarians

CLASS: Polycystina   **PHYLUM:** Radiozoa   **SUBKINGDOM:** Rhizaria   **KINGDOM:** Chromista

Polycystine radiolarians have silica skeletons and represent the vast majority of radiolarians. More than 8,000 species have been described. Silica does not readily dissolve, and parts of the ocean floor are covered with "radiolarian ooze" made up of the dead skeletons of these organisms. In many polycystines the skeletons look like ornamented spheres, but in one subgroup, the Nassellaria, the skeletons are roughly cone-shaped, although there is much variation among species. Members of another subgroup, the Collodaria, form large gelatinous floating colonies with individual cells embedded in them. As with acantharians, several species of polycystine radiolarians contain symbiotic algae. There is much still to learn about radiolarians, in part because it is difficult to culture them in laboratories.

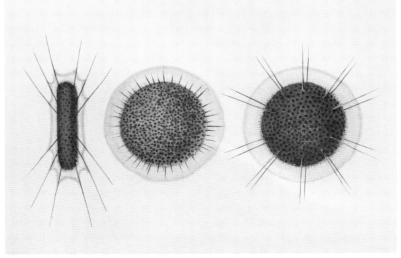

Spumellaria
This drawing from Ernst Haeckel's *Die Radiolarien* shows two species of polycystine radiolarians from the subgroup Spumellaria: *Spongotrochus brevispinus* (left and right) and *S. longispinus*.

foraminiferans

PHYLUM: Foraminifera   **SUBKINGDOM:** Rhizaria   **KINGDOM:** Chromista

Like radiolarians (see p.331), foraminiferans, or forams, are single-celled marine organisms with thin, flexible pseudopods (false feet) and shell-like skeletons. Unlike radiolarians, most foraminiferans live on the sea floor, although some species are planktonic. Forams' protective skeletons, called tests, usually develop in the form of connected chambers, with new chambers added as the cell grows. A main opening, along with other tiny perforations, allows the cell substance and pseudopods to extend into the environment. The pseudopods merge to form a constantly changing net and are used for crawling, building tests, and gathering food. The diet of forams includes dissolved organic molecules, bacteria, and invertebrate larvae. Some species harbor symbiotic algae.

The tests start out as organic material secreted by the cells but are usually hardened with calcium carbonate or cemented by sand grains and other foreign particles. Planktonic forams include *Globigerina*—upon death, their tests sink and cover much of the ocean floor. The largest forams are the xenophyophores, which grow up to 8 in (20 cm) across. They are unicellular with multiple nuclei, although only a small part of their bulk is living tissue.

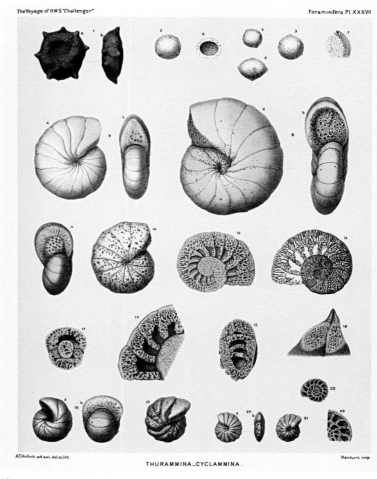

Thurammina and _Cyclammina_ forams
This illustration shows the forams collected on the expedition of HMS *Challenger* (1873–76). The species in the top row belong to the genus *Thurammina*, and the rest are *Cyclammina* species.

Planktonic foraminiferan
Globigerina sp.

Benthic foraminiferan
Baculogypsina sp.

Large xenophyophore
Xenophyophorea

Benthic foraminiferan
Peneroplis planatus

coccolithophores and relatives

PHYLUM: Haptophyta   **SUBKINGDOM:** Hacrobia   **KINGDOM:** Chromista

Haptophytes are single-celled, photosynthetic (they get their energy from light) organisms. They are mostly marine plankton, although some live in fresh water. Most haptophytes are coccolithophores—marine microalgae with hard outer coverings, called coccospheres, made of calcium carbonate. Coccolithophores are important players in the Earth's carbon cycle: following death, their fragmented skeletons sink to the ocean floor, removing carbon from the upper ocean.

Coccospheres vary in shape and are built of complex individual units called coccoliths that differ between species. During their life cycle, coccolithophores may grow different shapes of coccoliths or may even lack them entirely, making identification and classification difficult. Although transparent, coccoliths scatter light, acting like tiny mirrors dispersed in the water and making the sea look milky where population explosions, or blooms, occur. One species, *Emiliania huxleyi*, thrives in varied ocean habitats, but most other species fare best in the low-nutrient tropical oceans that are more exposed to solar radiation. Not all haptophytes are coccolithophores. *Phaeocystis*, for example, is a planktonic microalga whose blooms can produce copious foams on beaches.

Coccolithophore
The cells of *Emiliania huxleyi* are covered with oval coccoliths. It is the most widespread and abundant coccolithophore, forming a major part of the plankton in temperate, subtropical, and tropical oceans.

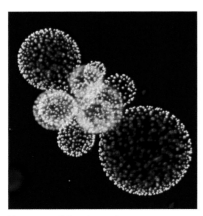

Non-coccolithophore alga
Phaeocystis sp.

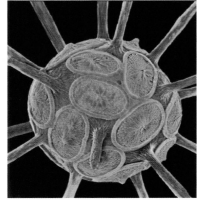

Rhabdosphaeraceaen coccolithophore
Rhabdosphaera claviger

Calyptrosphaeraceaen coccolithophore
Calyptrosphaera oblonga

dinoflagellates

PHYLUM: Dinoflagellata   **SUBKINGDOM:** Alveolata   **KINGDOM:** Chromista

Single-celled organisms, dinoflagellates are common in ocean plankton and freshwater lakes. More than 2,000 species have been described. About half the species produce food through photosynthesis, but many can also catch and consume other tiny organisms. Some live on the sea floor, while a few are parasites. Dinoflagellate cells have a groove around the middle containing a beating flagellum, while a second flagellum extends behind them. The movement of these flagella causes them to corkscrew through the water. Many dinoflagellates are protected by cellulose plates (armor) on the outside, which may be shaped into spikes or horns.

Some species of dinoflagellates, called zooxanthellae, are vital to coral reefs because they live mutualistically in the tissues of corals and other animals and produce food for them. Others have a destructive side: some free-living species may multiply excessively to produce toxic "blooms" in the water, poisoning fish and sometimes endangering humans. *Pyrocystis* and certain other dinoflagellates are responsible for producing a blue luminescence of the sea at night when stimulated by movement in the water.

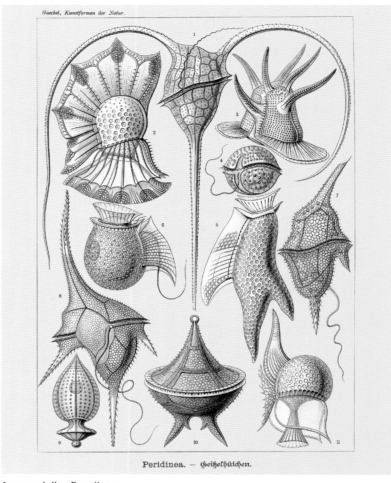

Armored dinoflagellates
This 19th-century illustration shows armored dinoflagellates, including *Tripos tripos* (previously *Ceratium tripos*, fig. 1) and *Tripos hirundinella* (previously *Ceratium hirundinella*, fig. 8), which has two flagella and two cellulose horns.

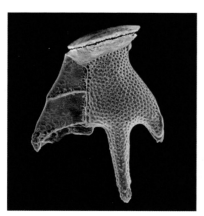

Armored marine dinoflagellate
Dinophysis sp.

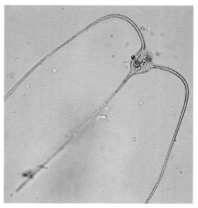

Armored marine dinoflagellate
Tripos macroceros

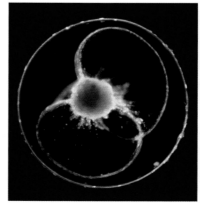

Unarmored marine dinoflagellate
Pyrocystis sp.

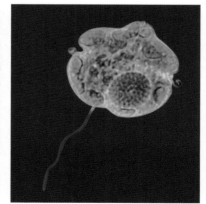

Unarmored marine dinoflagellate
Karenia sp.

ciliates

PHYLUM: Ciliophora **SUBKINGDOM:** Alveolata **KINGDOM:** Chromista

Ciliates are single-celled organisms with beating hairlike structures, called cilia, used for moving or creating feeding currents. The cilia cover their entire surface or occur in patches or rows. Some ciliates attach to the seabed or lake bottom, and others attach to floating debris in the water column. About 4,500 species have been described, all of which are widespread in marine and freshwater environments as well as in damp soil. Although single-celled, ciliates have complex structures that include the equivalent of mouths and gullets. Their shapes range from trumpet-shaped to oval to broad disks. Some of these can become colonial. Some ciliates are parasites, and many eat smaller food such as bacteria and dead particles, but others can stretch their mouths to eat prey as big as themselves, including other ciliates.

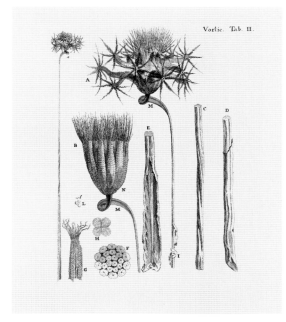

Vorticella ciliates
This illustration shows some of the 200 or so species in the genus *Vorticella*. After a free-swimming stage, these ciliates adopt a sessile lifestyle—their bell-shaped body becomes attached to a surface by a stalk.

apicomplexans

PHYLUM: Apicomplexa **SUBKINGDOM:** Alveolata **KINGDOM:** Chromista

Apicomplexans, numbering more than 5,000 species, are parasitic, single-celled organisms that include the deadly malaria parasite *Plasmodium*. They have a structure called an "apical complex," which they use to penetrate the cells of their hosts. These include all kinds of animals, from humans to worms, depending on the species of parasite involved.

Apicomplexans have complex life cycles with several stages that may live in more than one kind of host. The life cycle of *Plasmodium* involves two hosts—a human or other vertebrate and a mosquito. When inside vertebrates, the parasite targets red blood cells, multiplying until the cells burst. *Cryptosporidium* and *Toxoplasma* are other apicomplexans that cause disease in humans.

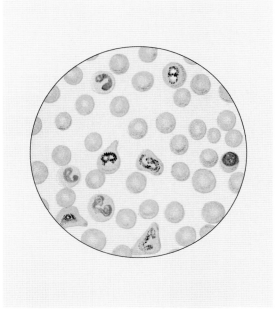

Plasmodium infecting blood cells
This microscopic view shows human red blood cells infected by *Plasmodium*. The ensuing destruction of the blood cells will lead to malaria. Five species from this genus regularly infect humans.

amoebas

CLASS: Tubulinea   **PHYLUM:** Amoebozoa   **KINGDOM:** Protozoa

An amoeba is a cell that crawls along, changing its shape and engulfing particles using temporary projections of its body—pseudopods, or "false feet." The free-living amoebas (which only occasionally invade a host) belong to a group called the Tubulinea, and are found in marine, freshwater, and soil habitats. *Amoeba proteus*, found in fresh water, is a typical species of the group. A microscope view will usually show food vacuoles (closed sacs) in its body where it has engulfed smaller single-celled organisms. *Amoeba proteus* has one nucleus, but some of its relatives have hundreds in a single cell, which can reach several millimeters across.

The types mentioned above are "naked" amoebas, but there are also "testate" species such as those in the genera *Arcella* and *Difflugia*. These amoebas build protective coverings called "tests" around the cell, either made of secretions from their own bodies or from particles available in the environment. The test has a hole in it from which the pseudopods can protrude. Amoeba tests preserve well and are common in the fossil record.

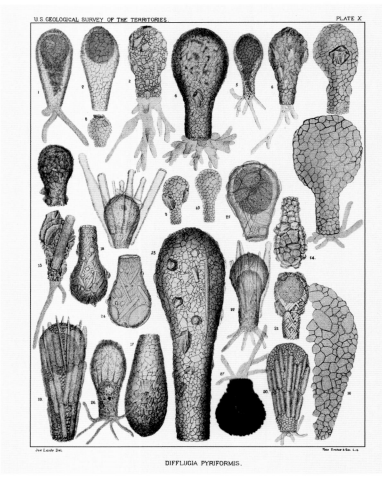

Testate amoebas
This 19th-century illustration from *Freshwater Rhizopods of North America* (1879) shows many body shapes of a testate amoeba, *Difflugia linearis*. Amoebas were formerly grouped under Rhizopoda.

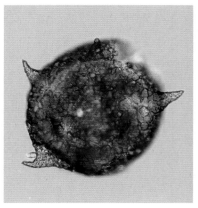

Testate amoeba
Arcella sp.

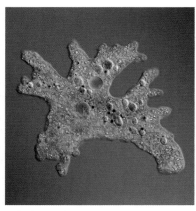

Naked amoeba
Amoeba proteus

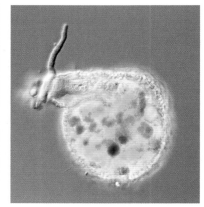

Testate amoeba
Lesquereusia sp.

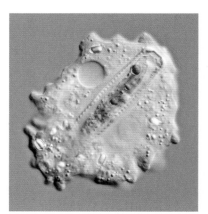

Naked amoeba
Mayorella sp.

slime molds

CLASS: Myxogastrea, Dictyostelea   **PHYLUM:** Amoebozoa   **KINGDOM:** Protozoa

Slime molds were once classed as fungi, but their 900 or so species are now classified into several unrelated groups. The best-known ones are relatives of amoebas, and begin life as single-celled amoebas. In some species, when food is short, the amoebas aggregate into a multicellular "slug." In other types, the amoebas mate and produce offspring that grow into plasmodia—extensive bodies of tissue, undivided by cell membranes, and containing many cell nuclei. It is this slimy form of plasmodium that appears on moist woodland floors. Both types of slime molds sprout capsules that release spores for dispersal. Although often found around rotting material, slime molds are not decomposers themselves—they feed on the decomposers, including bacteria, yeasts, and other fungi.

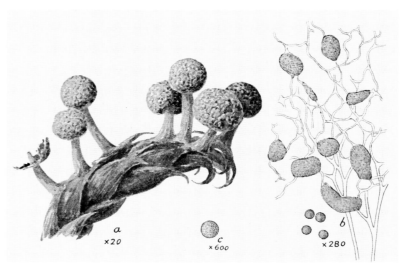

Physarum citrinum
Physarum citrinum is a distinctive plasmodial slime mold with its bright yellow stalks and spore capsules. It grows on decaying wood, often in association with mosses and liverworts.

water molds

CLASS: Oomycetes   **PHYLUM:** Oomycota   **SUBKINGDOM:** Heterokonta   **KINGDOM:** Chromista

Oomycetes are funguslike life forms—mostly parasitic—that can cause diseases in plants and aquatic animals. They are not related to true fungi—their cell walls are made of cellulose (not chitin) and they produce flagellated spores (whereas true fungi lack flagella)—but most grow in a similar way with a network of microscopic filaments, or hyphae, that absorb nutrients from living tissue or dead material. Like fungi, oomycetes cause diseases variously described as mildews, blights, rusts, molds, rots, and cankers. They are sometimes called water molds because they spread well in damp conditions and in fresh water. Some examples of the serious diseases they cause include potato blight, sudden oak death, and "white rusts" of crops such as cabbages and turnips. At least 500 species are known, and some of them are single-celled.

Downy mildew
This illustration shows downy mildew, which damages the leaves of roses and other plants (top), and potato blight. They are both caused by the water mold *Phytophthora infestans*.

centric diatoms

CLASS: Bacillariophyceae **PHYLUM:** Ochrophyta **SUBKINGDOM:** Heterokonta **KINGDOM:** Chromista

Diatoms are single-celled algae, commonly found in both marine and freshwater habitats. The cells are protected by transparent, boxlike structures of silica called frustules, which consist of two halves with one overlapping the other like the lid of a box. Often intricately patterned, these frustules are important in identifying individual species. In "centric" diatoms, the frustules are typically circular when viewed from above, although they may be square or triangular depending on the species. Centric diatoms form the main subgroup found in ocean plankton, and they play an important role as producers in the aquatic food chain. Their cells contain oil droplets that keep them from sinking. They also thrive on the underside of Antarctic sea ice.

Jewels of the sea
The frustules of centric diatoms, as shown here, can be highly patterned with pores, ribs, and spines. They are sometimes referred to as "jewels of the sea" because of their intricate structures.

pennate diatoms

SUBCLASS: Bacillariophycidae, Fragilariophycidae **CLASS:** Bacillariophyceae **PHYLUM:** Ochrophyta **KINGDOM:** Chromista

Pennate diatoms, which are thought to have evolved from centric diatoms, have an elongated shape and are more common in fresh water. They can be found both as free-floating plankton and attached to underwater surfaces. Some species grow as chains of individuals linked together. Unlike centric diatoms, they can move by gliding along surfaces, probably using threads of tissue that project through pores in their frustules. Like centric diatoms, when the cells of pennate diatoms divide during asexual reproduction, each "daughter" cell keeps one half of the original frustule and grows a smaller half to fit into it. Thus, over generations the successive cells get smaller. When the cells become smaller than half their original size, diatoms reproduce sexually and restore cell size.

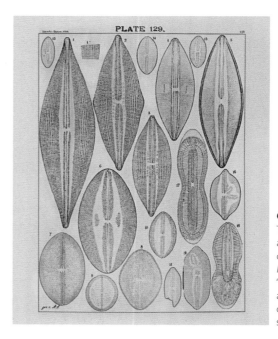

Creeping diatoms
This illustration shows a selection of pennate diatoms in the genus *Navicula*, which means "boat-shaped." They are able to creep around each other and on hard surfaces, such as microscope slides.

golden algae

CLASS: Chrysophyceae **PHYLUM:** Ochrophyta **SUBKINGDOM:** Heterokonta **KINGDOM:** Chromista

Golden algae are single-celled or colonial relatives of diatoms and brown algae. There are about 1,000 species in the group; most are found in fresh water, although some may inhabit the oceans. Like their relatives, golden algae contain a yellow-brown pigment called fucoxanthin that helps collect light for photosynthesis and masks the green color caused by the pigment chlorophyll.

Most golden algae have flagella that are used for swimming. Their cells are extremely small and can be seen only under high magnification. Some clump together in colonies, while others live in animal tissues. In non-scientific contexts, "golden algae" often refers to just one species, *Prymnesium parvum*, which can kill fish, although it is not thought to be dangerous to humans.

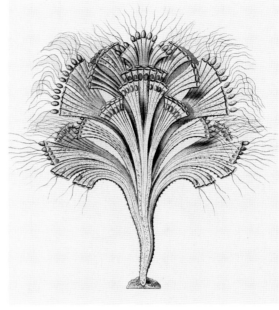

Bacteria eaters
This microscopic alga of freshwater lakes and ponds belongs to the genus *Dinobryon*, whose species get their energy from phagotrophy—engulfing bacteria—as well as photosynthesis.

brown algae

CLASS: Phaeophyceae **PHYLUM:** Ochrophyta **SUBKINGDOM:** Heterokonta **KINGDOM:** Chromista

About 1,500–2,000 species of brown algae are known. They form a large group of mostly marine multicellular algae, or seaweeds. Brown seaweeds flourish in and below the intertidal zone (where the ocean meets the land) where light penetrates, particularly in the cooler regions of the world. The structure of seaweeds is simpler than that of land plants. They attach to surfaces using a structure called a "holdfast" instead of a true nutrient-absorbing root. The largest species is the giant kelp of coastal Pacific regions, which can grow at least 150 ft (45 m) in length. Wracks are a common group of brown algae that are tolerant of exposure to air, and often cover rocky seashores in sheltered areas. Their fronds are usually branched and sometimes have gas-filled floats that keep them buoyant.

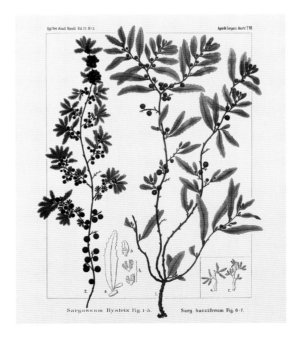

***Sargassum* seaweeds**
This illustration from *Kunstformen der Natur* (*Art Forms in Nature*) shows two species from the genus *Sargassum*. Most seaweeds in this group attach to the seabed, although some species may be free-floating.

red algae

PHYLUM: Rhodophyta **KINGDOM:** Plantae

With more than 7,300 species, red algae include a diverse range of seaweeds, as well as some unicellular species. They are usually pink or red due to phycobiliprotein, a pigment that helps them absorb more light for photosynthesis than if they contained only chlorophyll, allowing many varieties to grow in deeper waters. Red seaweeds vary in form, but most are small and delicate compared to brown seaweeds. The majority cannot survive prolonged exposure to air, so they are found either in rock pools in the intertidal zone or below the low-tide line, often growing on other seaweeds. Some temperate species have a seasonal pattern of growth, dying back in the fall and regenerating in spring. Many red seaweeds are harvested to extract agar, a substance used for gelling food and in laboratories, and some, such as laver, are collected for eating.

"Coralline" red algae, like corals (see p.353), have deposits of calcium carbonate in their tissues that give them a hard structure. This group includes the feathery coral weed (*Corallina officinalis*) and the paintweeds that look like pink stains on rocks. It also includes maerl, which grows as hard, branching nodules lying unattached on sandy sea floors, forming a habitat for other organisms.

Halymeniaceae family
This illustration of three species of red algae from the family Halymeniaceae appeared in the Swedish publication *Florideernes morphologi,* which was published in 1879.

Irish moss
Chondrus crispus

Paintweed
Lithophyllum cabiochiae

Dulse
Palmaria palmata

Coral weed
Corallina officinalis

green algae

PHYLUM: Chlorophyta, Charophyta   **KINGDOM:** Plantae

Of the many types of organisms commonly called algae, those most closely related to land plants form two groups of green algae called chlorophytes and charophytes, with about 8,000 species. Widespread in oceans, fresh water, and even on land, these green algae include single-celled phytoplankton and multicellular algae, such as green seaweeds. Green seaweeds vary in shape from lettucelike to tubular or brush-shaped, and may spread across the ocean floor. In fresh water, algae called stoneworts grow branching stems similar to land plants, while *Spirogyra* lives as tangles of filaments in ponds.

Single-celled chlorophytes play an important role in ocean plankton as photosynthesizers. Some marine species called zoochlorellae grow symbiotically in animal tissues. Other seafloor species grow as stalked cups or green globes: these giant single cells can be more than ¾ in (2 cm) across. Planktonic green algae called desmids thrive in nutrient-poor fresh waters. They are all symmetrical, divided nearly into two. On land, green algae are found in damp places such as on tree trunks. Most algae that live in lichens (see p.347) are single-celled species of this group.

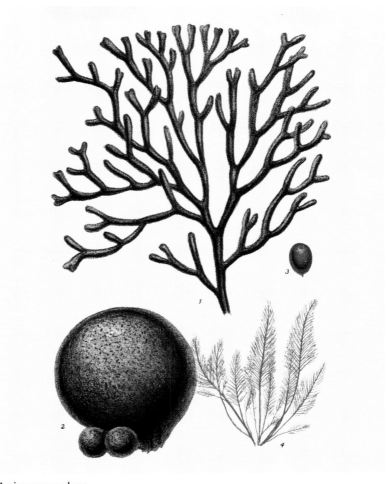

Marine green algae
This selection of marine green algae shows some contrasting forms. Velvet horn (*Codium tomentosum*, top) has a branched structure, while *Codium bursa* (bottom left) is a spongy sphere.

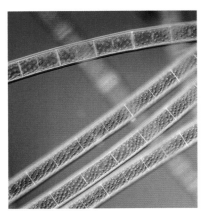

Water silk
Spirogyra sp.

Mermaid's wineglass
Acetabularia sp.

Sea lettuce
Ulva compressa

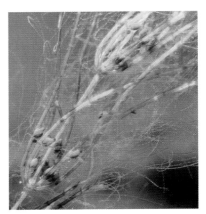

Stonewort
Chara globularis

bryophytes

PHYLUM: Anthocerotophyta, Bryophyta, Marchantiophyta   **KINGDOM:** Plantae

Most plants have an extensive system of vascular tissue that transports nutrients and water, and are referred to as vascular plants. However, the 25,000 species of hornworts (phylum Anthocerotophyta), mosses (phylum Bryophyta), and liverworts (phylum Marchantiophyta) do not have vascular tissues and are collectively called bryophytes (from the Greek *bryon*, moss, and *phyton*, plant). Bryophytes do not depend on roots for the absorption of nutrients, so they can grow in places that vascular plants cannot, such as rocks and the walls of buildings, anchored by short filaments called rhizoids. They do, however, usually require a moist environment to grow.

Bryophytes do not produce seeds or flowers; they reproduce via spores, usually housed in structures called sporangia. As with all land plants, the life cycle of bryophytes alternates between two generations: gametophytes (which produce the sex cells that participate in fertilization) and sporophytes (created by fertilization), which produce spores. Unlike other plants, the gametophyte generation of bryophytes is the longer lasting of the two.

Liverworts
Typically very small plants, liverworts have a varied range of forms when examined closely, as shown in this illustration from Ernst Haeckel's *Kunstformen der Natur* (*Art Forms in Nature*).

Hornwort
Anthoceros sp.

Red bog moss
Sphagnum capillifolium

Common haircap
Polytrichum commune

Common liverwort
Marchantia polymorpha

seedless vascular plants

CLASS: Lycopodiopsida, Polypodiopsida   **PHYLUM:** Tracheophyta   **KINGDOM:** Plantae

Vascular plants have specialized tissues, called xylem and phloem, which conduct water and nutrients. In order to reproduce, most vascular plants produce seeds. Some, however—including clubmosses (which are not true mosses), ferns, and horsetails—do not bear seeds, fruit, or flowers, but instead reproduce via spores. About 12,000 species of seedless vascular plants are known, and they vary greatly. Some are very small, but the largest tree ferns grow up to 66 ft (20 m) in height.

All plants have alternating gametophyte and sporophyte generations. In all plants, except bryophytes, the sporophyte generation is longer lasting. In seedless vascular plants, such as ferns, both generations are free-living plants. Their sporophyte generation is represented by a relatively large, spore-producing plant. In the right conditions—usually where there is moisture—these give rise to the simple and inconspicuous plants of the gametophyte generation, during which male and female reproductive structures develop. The male structures release sperm cells that swim to fertilize the female egg cells—which then form new sporophytes.

Lanceolate spleenwort
This illustration shows lanceolate spleenwort (*Asplenium obovatum*), a type of fern. Ferns have complex leaves called megaphylls, unlike horsetails and clubmosses.

Field horsetail
Equisetum arvense

Coarse tassel fern, a clubmoss
Huperzia phlegmaria

Delta maidenhair fern
Adiantum raddianum

Soft tree fern
Dicksonia antarctica

gymnosperms

CLASS: Cycadopsida, Ginkgoopsida, Gnetopsida, Pinopsida   **PHYLUM:** Tracheophyta   **KINGDOM:** Plantae

Gymnosperms form a group of more than 1,000 species of woody plants. They possess roots, stems, and leaves, as well as vascular tissues that conduct water, minerals, and the products of photosynthesis. Their seeds are not enclosed in an ovary as they are in flowering plants (angiosperms), but lie exposed on leaflike structures called megasporophylls. In conifers, which make up the biggest group of gymnosperms, collections of megasporophylls form cones. For fertilization to take place, female cones receive pollen from male cones. In yews and maidenhair trees (*Ginkgo biloba*), each cone contains a single seed.

Most conifers are evergreen and many have long, thin, needlelike leaves. The largest trees (giant sequoias) and the oldest trees (bristlecone pines) are both conifers. Slow-growing, tropical cycads form another group of gymnosperms. They have divided leaves, which makes them look superficially like palm trees. The gnetophytes make up a small and varied group, including welwitschia (*Welwitschia mirabilis*), which is native to the driest parts of the Namib Desert in southern Africa. Its ribbonlike leaves grow along the ground.

Typical conifer
The Norway spruce *(Picea abies)* has seed-bearing cones and needlelike leaves. Single female cones are concentrated at the very tips of branches (bottom and right), while the small male cones cluster around the branches (top left).

Queen sago cycad
Cycas circinalis

Ginkgo
Ginkgo biloba

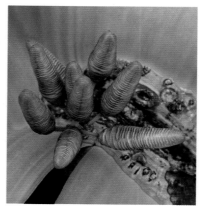

Welwitschia
Welwitschia mirabilis

Cypress
Cupressaceae

flowering plants

CLASS: Magnoliopsida   **PHYLUM:** Tracheophyta   **KINGDOM:** Plantae

With about 350,000 species, the angiosperms—or flowering plants—make up the largest group of plants. Most plants are terrestrial but many grow in fresh water and some in shallow sea water. They range in size from tiny aquatic duckweed, 0.08 in (2 mm) across, to eucalyptus trees that are 328 ft (100 m) high. What distinguishes them from other seed-bearing plants is the presence of flowers—their reproductive organs—that produce fruits containing seeds. A flower's stamens bear pollen grains containing the male reproductive cells. The ovule inside the carpel (the female part of the flower, also containing the stigma and style) contains the female reproductive cells.

The eudicotyledons, or eudicots, form the biggest group of angiosperms, containing about 175,000 species. Upon germination, their seeds develop two embryonic leaves. Eudicot flowers usually have petals and other floral parts in multiples of 4 or 5, and their leaves have branching veins. Another large group is the monocotyledons, or monocots, with more than 60,000 species. Monocot seeds produce a single embryonic leaf. Usually, their flower parts are in multiples of 3 and their leaves have parallel veins.

Coltsfoot
This illustration shows a common coltsfoot (*Tussilago farfara*), a eudicot from the family Asteraceae. It was used in traditional medicine but is now known to contain harmful toxins.

Tulip tree
Liriodendron tulipifera

Southern marsh orchid
Dactylorhiza praetermissa

Yellow pincushion
Protea leucospermum

Common hawthorn
Crataegus monogyna

pin molds and relatives

PHYLUM: Zygomycota   **KINGDOM:** Fungi

Like other fungi, pin molds grow as microscopic filaments, or hyphae, that spread to form a woolly mass called a mycelium. The mycelium releases enzymes to digest and absorb nutrients from surrounding organic matter.

Pin molds thrive on carbohydrate-rich sources, such as bread. They are named for their tiny spore-producing capsules that are borne on straight, pinlike stalks. Spores are produced asexually or sexually. The spores disperse through the air to new sources of food, where they germinate to form new mycelia. Pin molds and other fungi are decomposers: some, such as *Rhizopus* and *Mucor*, are responsible for spoilage of bread, while others—including *Pilobolus*—grow on animal dung. Their relatives include fungi that parasitize invertebrates.

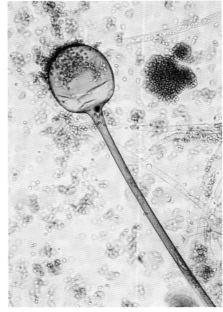

Black bread mold
The spore capsule of this common mold (*Rhizopus stolonifer*) is borne on a stalk that reaches only 0.06 in (1.5 mm) in length.

club fungi

PHYLUM: Basidiomycota   **KINGDOM:** Fungi

Club fungi, some of the most familiar fungi, are classified together because they produce their sexual spores in fruiting bodies called basidiocarps. These take the form of mushrooms, toadstools, and brackets. Inside the cap of a basidiocarp, the spores bud from microscopic structures, or basidia, and are typically released into the air through gills or pores on the cap's underside. Most of these fungi are decomposers and can digest wood, and their basidiocarps frequently sprout from tree stumps or on trunks. The hyphae of many fungi establish mycorrhizae with the roots of plants— mutualistic associations, in which nutrients are exchanged, to the benefit of both the fungus and the plant. Rusts and smuts are plant parasites, and their spore-budding basidia are released from microscopic bodies on an infected plant.

Sulfur tuft
Feeding on wood, the fruiting bodies of *Hypholoma fasciculare* sprout from dead trunks and tree stumps and release spores from gills beneath their caps.

sac fungi

PHYLUM: Ascomycota   **KINGDOM:** Fungi

The largest phylum of fungi encompasses species that grow as creeping molds, with erect mushroomlike bodies, or as single cells. They are united by having microscopic sacs called asci—the structures that produce their sexual spores. In some groups, such as *Penicillium*, the asci grow in tiny capsules similar to those of pin molds; in morels and truffles, the asci develop in much larger fruiting bodies, which release their spores when they burst open. Sac fungi include the familiar microscopic yeasts: single-celled *Saccharomyces* that are commercially used in baking and alcoholic fermentation, and *Candida*, which can cause disease. Their asci develop by division of their single cells or microscopic hyphae. Some sac fungi called dermatophytes cause skin infections, such as athlete's foot, while others are the most common fungal partners in lichens.

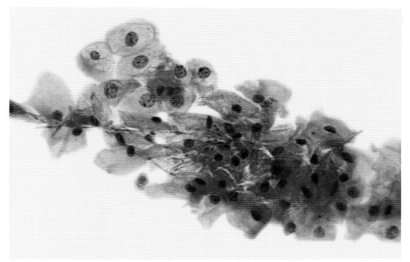

Candida yeast
Usually a benign member of human gut flora, *Candida albicans* can become pathogenic in people whose immune system has been weakened, causing the oral infection candidiasis.

lichens

PHYLUM: Basidiomycota, Ascomycota   **KINGDOM:** Fungi

Lichens are composite organisms made up of algae or cyanobacteria in a mutualistic (interdependent) relationship with fungi. Lichens are classified according to the name of the fungal partner, so all lichens belong to one of two fungal phyla. The alga or cyanobacterium makes food by photosynthesis and passes some of this onto the fungus. The fungal filaments provide protection for the partner organism as well as gathering water and minerals.

Lichens can be leaflike (foliose), branchlike (fruticose), flakelike (crustose), or powdery (leprose). They grow in most terrestrial environments, including on bare rock and tree bark, in arctic tundra, temperate zones, hot deserts, and even toxic spoil heaps. Their growth rate is very slow, but studies suggest that some can survive for thousands of years.

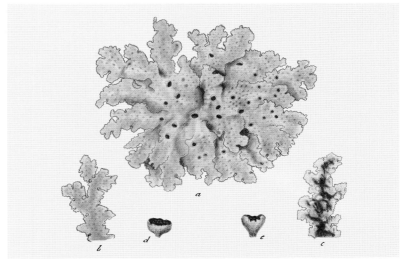

Spotted felt lichen
Lichens in the genus *Sticta* have a foliose structure and grow on rocks, tree bark, and wood. They are common in tropical forests, where those that have a mutualistic relationship with cyanobacteria can fix atmospheric nitrogen.

sponges

PHYLUM: Porifera   **KINGDOM:** Animalia

There are more than 5,000 species of sponges—the simplest multicelled animals alive today. They live in water, anchored to one spot on a rock or soft sediment, mostly in oceans, but a few favor fresh water. Many large and colorful species live on coral reefs. Although sponges contain various types of cells, they do not have the organs or specialized tissues that other animals possess. They also lack a nervous system.

The simplest sponges are small and shaped like a hollow vase. They take in water through small pores in their sides and, after sieving out food particles, eject it through a larger aperture at the top called the osculum. Thousands of cells with tiny beating hairs, or flagella, maintain water flow. Most sponges have a more complicated arrangement of chambers and internal channels, but the principle is the same. Sponges require clear water to thrive because sediment blocks their pores. Their bodies are supported in a variety of ways. Calcareous sponges have tiny skeletal units, or spicules, made of calcium carbonate. Glass sponges have spicules made of silica, and demosponges are supported by a fibrous protein called spongin.

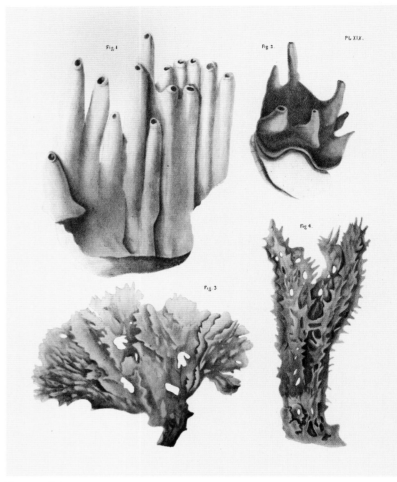

Sponge shapes
Sponges come in a range of shapes. *Neopetrosia subtriangularis* (top left) is tubular, *N. carbonaria* (top right) is cup-shaped, and *Clathria virgultosa* and *Mycale angulosa* (bottom) are dendritic, or tree-shaped.

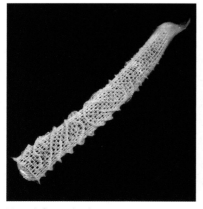

Venus's flower basket
Euplectella aspergillum

Giant barrel sponge
Xestospongia muta

Yellow tube sponge
Aplysina fistularis

Burrowing sponge
Cliona delitrix

comb jellies

PHYLUM: Ctenophora   **KINGDOM:** Animalia

Comb jellies are almost transparent animals that live in marine waters worldwide. About 100–150 species are known. Although they look similar to jellyfish, they are not closely related. Comb jellies are often round or oval, although some species may be long and strap-shaped. They do not have a brain, but they possess a simple nervous system.

Comb jellies have eight comblike rows of beating hairs, or cilia, on the outer surface of their body—"ctenophore" means "comb-bearer." The cilia are used to propel the animal through water; comb jellies are the largest animals to use their cilia for swimming. They are predators, feeding mostly on small plankton, although some species eat larger copepods, krill, jellyfish, and other comb jellies. Most ctenophores have two branched, retractable tentacles with sticky structures called colloblasts that are used to catch prey. Once swallowed, the prey is liquefied by enzymes, and the liquid is pushed by the cilia through the animal's canal system for digestion. Many comb jellies are bioluminescent—they can produce their own light, which is usually blue or green.

Varied forms
This illustration of comb jellies in German biologist Ernst Haeckel's 1904 publication *Kunstformen der Natur* (*Art Forms in Nature*) shows a range of body forms.

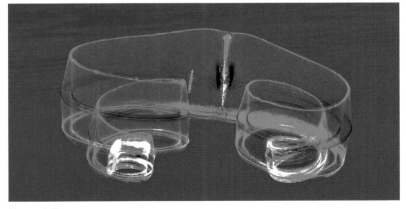

Venus girdle
Cestum veneris

Spotted comb jelly
Leucothea pulchra

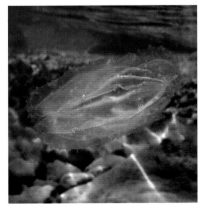

Sea walnut
Mnemiopsis leidyi

hydrozoans

CLASS: Hydrozoa **PHYLUM:** Cnidaria **KINGDOM:** Animalia

Hydrozoans belong to the same phylum as jellyfish, sea anemones, and corals—Cnidaria. Cnidarians are marine animals noted for the stinging cells (cnidae) on their body surfaces. The guts of cnidarians have a single opening that serves as both mouth and anus. Stinging tentacles around this opening reach out to catch prey. Upright, anchored forms are called polyps, whereas upside-down swimming forms are called medusae.

Hydrozoan life cycles typically alternate between two generations. A polyp generation reproduces asexually by budding to form medusae. The medusa generation produces sex cells that fertilize to form new polyps. Most hydrozoans are colonial and marine. One subgroup, the hydroids, look like seaweeds and grow as branched colonies on the sea floor. The individual polyps share food through channels in the branches, and the colony includes specialized polyps that only reproduce. Many floating, colonial hydrozoans also exist, notably the siphonophores, which include the Portuguese man o'war. In this species, one polyp becomes a huge gas-filled float, which is attached to trailing tentacles made up of stinging polyps. Other siphonophores have polyps that are specialized in swimming.

Porpid hydrozoans
This illustration from *Kunstformen der Natur* (*Art Forms in Nature*) shows siphonophores that form colonies on the ocean surface.

Portuguese man o'war
Physalia physalis

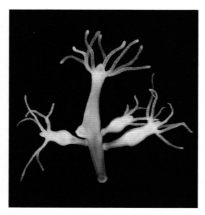

Freshwater hydrozoan
Hydra vulgaris

Sea fan hydrozoan
Solanderia ericopsis

Oaten pipes hydrozoan
Tubularia indivisa

jellyfish

CLASS: Cubozoa, Scyphozoa, Staurozoa   **PHYLUM:** Cnidaria   **KINGDOM:** Animalia

The "true" jellyfish of the class Scyphozoa are marine animals that swim slowly by contracting muscles in their saucer-shaped bodies (called bells). About 200 species have been described, the largest of which can reach 6 ft (2 m) across, with stinging tentacles that can be 98 ft (30 m) in length or more. The bells are stiffened by a jellylike layer called mesoglea. Jellyfish also feature four or eight frilly "arms" hanging down from a central mouth. Most species capture fish and other large prey, and some are dangerous to humans. One subgroup, the rhizostomes, have "micro-mouths" on their arms and ingest plankton that the arms trap in mucus. Jellyfish are normally either male or female. The eggs usually grow into tiny polyps on the sea floor and in turn give rise to miniature medusae.

Box jellyfish (class Cubozoa) live in warm waters. They are usually small but can cause excruciating pain and even death to humans. They swim faster than other jellyfish and have small eyes that help them to hunt. The largest cubozoan is the sea wasp. Stalked jellyfish (class Staurozoa) include small, trumpet-shaped animals that are attached to one place, and are harmless to humans.

True jellyfish
This selection of true jellyfish in the genera *Aurelia* (moon jellyfish), *Drymonema*, *Floresca*, and *Pelagia* shows a wide range of jellyfish body forms.

Sea wasp
Chironex fleckeri

Pacific sea nettle
Chrysaora fuscescens

Lion's mane jellyfish
Cyanea capillata

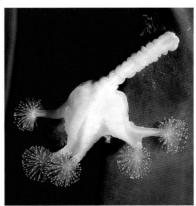

Stalked jellyfish
Lucernaria quadricornis

sea anemones

ORDER: Actiniaria, Ceriantharia   **CLASS:** Anthozoa   **PHYLUM:** Cnidaria   **KINGDOM:** Animalia

More than 1,000 species of sea anemones are known. A sea anemone's body, or polyp, is typically cup-shaped and is usually attached to a rocky or coral surface, though occasionally to soft sediment. Although most are fixed to one place, some float in the water. Atop the polyp's columnar trunk is an oral disk, which has a slitlike mouth surrounded by a ring of tentacles. The tentacles are armed with cnidae, which can be extended to trap passing prey—plankton, shrimp, or small fish—and drawn back into the body. Some tropical species can reach 3 ft (0.9 m) across.

Many species of sea anemones receive additional nutrition from mutualistic relationships with single-celled dinoflagellates (zooxanthellae) living within their cells. To reproduce sexually, they release sperm and eggs into the water. Fertilized eggs develop into free-floating larvae, which later develop into polyps. Sea anemones can also reproduce asexually, breaking into pieces that regenerate as new polyps. Compared to sexual reproduction, asexual reproduction allows for larger numbers of individual offspring to be produced more quickly.

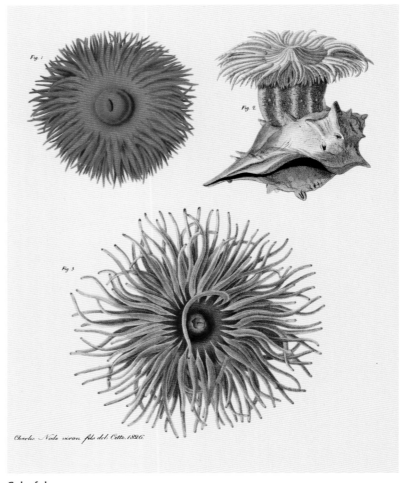

Colorful sea anemones
Hermit crab anemones (top right) often attach to shells inhabited by hermit crabs. The crab is protected by the sea anemone's stinging tentacles, and the latter eats food particles stirred up by the crab's movements.

Tube anemone
Cerianthus membranaceus

Bubble-tip anemone
Entacmaea quadricolor

Plumose anemone
Metridium senile

McPeak anemone
Urticina mcpeaki

stony corals

ORDER: Scleractinia **CLASS:** Anthozoa **PHYLUM:** Cnidaria **KINGDOM:** Animalia

There are more than 1,600 species of stony corals. Also known as hard corals, they have a rigid skeleton unlike soft corals and are the main creators of reefs, such as Australia's Great Barrier Reef. A stony coral is made up of small individual polyps that multiply by budding and remain connected, resulting in a continuous tissue of thousands of individuals. The exception is the solitary stony corals, which do not bud but gradually increase in size as they deposit more calcium carbonate. There are some similarities between stony corals and sea anemones, both of which are members of the class Anthozoa. In both, the top of each cylindrical polyp has an oral disk with a mouth, tentacles, and stinging cells.

In stony corals, the polyps build a permanent and growing support by laying down a skeleton made of calcium carbonate below their body. Common colony shapes include columnar, brain-shaped, leaflike, platelike, encrusting, and massive. For protection, a coral polyp can pull itself back into the stony cup, or corallite, in which it sits. Although corals are animals, many shallow-water species contain symbiotic dinoflagellates (zooxanthellae) that contribute to their food supply (see p.334).

Lettuce coral
Named for its spiraling plates that resemble a growing lettuce, the lettuce coral (*Pectinia lactuca*) grows in colonies. The plates may be up to 3 ft (0.9 m) in diameter.

Grooved brain coral
Diploria labyrinthiformis

Table coral
Acropora sp.

Torch coral
Euphyllia glabrescens

Sun coral
Tubastraea sp.

flatworms

PHYLUM: Platyhelminthes   **KINGDOM:** Animalia

More than 20,000 species of flatworms are known. Some are parasites of other animals, while others live in the ocean or in damp terrestrial environments. Unlike cnidarians and ctenophores (see pp.349–53), flatworms have a brain and other organs. They use tiny beating hairs called cilia or their muscular system to crawl along surfaces. They have thin, flattened bodies that can absorb oxygen from water without the need for gills. Free-living (nonparasitic) flatworms have a tubelike mouth on their underside and are mostly carnivores—some species prey on sea squirts or corals. Tapeworms and flukes are parasitic flatworms that can cause serious diseases. Adult tapeworms live in the guts of vertebrates, including humans, and flukes live in body tissues.

Bilateral symmetry
Although variable in form, flatworms are among the simplest animals to show bilateral symmetry—unlike other simple animals such as sponges and jellyfish, they have a left and right side—and have a head and a tail.

rotifers

PHYLUM: Rotifera   **KINGDOM:** Animalia

Rotifers, numbering about 2,200 species, are microscopic animals abundant in freshwater ecosystems. They have a structure called a corona, which bears circles of beating hairs (cilia) that are used to trap food, for swimming, or both. The corona's shape varies between species. Rotifers' mouths contain hard parts (trophi) that crush food or pierce prey. Many species attach themselves to underwater surfaces using adhesive produced in their feet, and a few species form colonies. Rotifers can survive dry conditions by shrinking and losing body water or producing drought-resistant eggs that hatch only when conditions improve. In one group, the bdelloid rotifers, only females are known. Acanthocephalans, once classified as a separate phylum, are highly modified parasitic rotifers that can be up to 25 in (65 cm) long.

Floscularia ringens
This sessile (fixed in one place) freshwater rotifer grows up to 0.06 in (1.5 mm) in length inside a protective tube it builds from tiny particles of detritus. The tube, or holdfast, attaches to the leaf of an aquatic plant.

bryozoans

PHYLUM: Bryozoa   **KINGDOM:** Animalia

Bryozoans, or moss animals, are aquatic organisms that live mostly in colonies of interconnected, genetically identical individuals called zooids. Each zooid is only about 0.5 mm across, but a colony may contain millions of individuals. Although the zooid colonies resemble colonies of corals, the zooids themselves are more anatomically complex. They filter-feed with an organ called a lophophore, which is a "crown" of hollow, ciliated tentacles, and they have a digestive system with two openings—a mouth and an anus. Bryozoan colonies grow as mats on seaweed fronds or form branched structures up to an inch or so high on undersea surfaces. Each zooid is protected by a tough or chalky box. All colonies contain autozooids, which are responsible for feeding and excretion, as well as other zooids specialized for defense, reproduction, or storing nutrients.

Marine bryozoans are preyed upon by sea slugs, fish, crustaceans, and starfish, while freshwater species are eaten by snails and insect larvae. More than 4,000 living species of bryozoans have been described, and extinct species feature prominently in the fossil record.

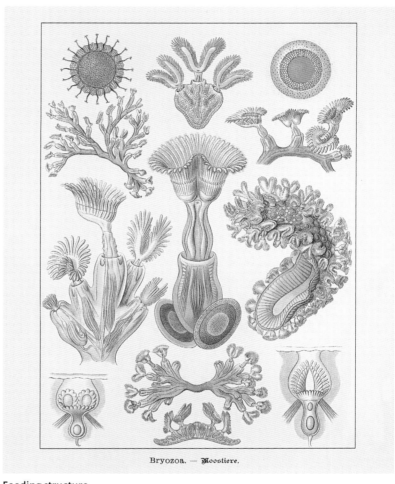

Feeding structure
The bryozoans shown here have their lophophores extended. These structures surround the animal's mouth and, unlike the tentacles of many other aquatic animals, they are hollow.

Freshwater bryozoan
Cristatella mucedo

"False coral"
Myriapora truncata

brachiopods

PHYLUM: Brachiopoda   **KINGDOM:** Animalia

Brachiopods, or lamp shells, are two-shelled marine invertebrates that resemble bivalve mollusks (see p.358) externally but have a very different body structure. Their shells are of unequal sizes and represent the top and bottom of the animal, not the left and right sides as in the case of bivalves. Brachiopods usually attach to the seafloor via a muscular stalk. The main body lies between the shells, toward the back and near the stalk, while the front space is occupied by a coiled, U-shaped filtering organ with tentacles called a lophophore (see p.355). The animal opens the shells slightly to draw water through the lophophore and extract tiny particles of food. Most brachiopods are small; the largest have shells up to 4 in (10 cm) long. There are about 330 living species, with many more in the fossil record.

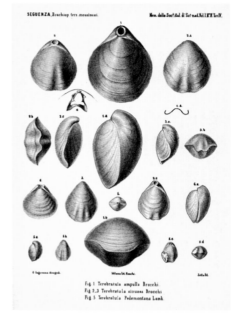

Asymmetrical brachiopods
This illustration shows the asymmetry of the brachiopods in the genus *Terebratula*. The size and shape of the upper and lower shells are clearly different.

ribbon worms

PHYLUM: Nemertea   **KINGDOM:** Animalia

Comprising about 900 species, ribbon worms are predatory worms that mostly crawl on the ocean floor, although some can swim. Most ribbon worms are long, slow-moving, narrow-bodied creatures with soft, mucus-covered skin, and some species are bright yellow, orange, red, or green in color. They have a unique tubelike proboscis, which is sometimes equipped with a spine that they shoot out from the head or mouth to grab prey such as annelid worms, bivalves, and crustaceans. They then subdue the prey with toxins. Ribbon worms are also scavengers. The bootlace worm (*Lineus longissimus*) is probably the world's longest animal, growing to an incredible 100 ft (30 m) or more. It feeds partly by absorbing dissolved organic substances through its skin.

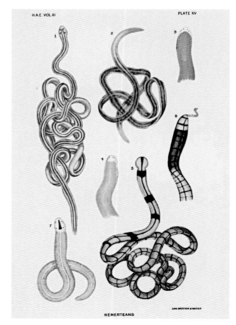

Ribbon worms
This illustration of ribbon worms from the genera *Nemertopsis*, *Paranemertes*, *Lineus*, *Carinella*, and *Tetrastemma* is dated to 1846. All ribbon worms are characterized by a very long, thin shape.

chitons

CLASS: Polyplacophora   **PHYLUM:** Mollusca   **KINGDOM:** Animalia

Chitons are shelled, marine mollusks that live on rock surfaces and graze algae. Most mollusks have a tonguelike structure called a radula, which they use to scrape food. A fleshy "mantle" covers a part or whole of the animal's upper surface. Beneath the overhanging mantle are pairs of gills. Mollusks crawl along using an adhesive, muscular "foot" on their underside. Early species probably had only one shell, but chitons have eight shell plates that overlap, allowing them to bend and curl up into a ball.

The largest chitons can reach 12 in (30 cm) long, but the majority are much smaller. Like the eggs of most other mollusks, chiton eggs hatch into tiny swimming larvae called trochophores, which mature into the adult form. Most of the 940 species live in shallow waters.

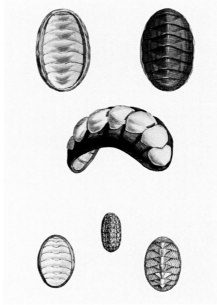

Many plates
Chitons form the class Polyplacophora, which means "bearing many plates." They have eight separate shell plates, visible in these examples.

aplacophorans

CLASS: Caudofoveata, Solenogastres   **PHYLUM:** Mollusca   **KINGDOM:** Animalia

Aplacophorans are wormlike mollusks that live on the sea floor. They have no shells, although they share some general features with mollusks such as usually having a radula. Aplacophorans range in length from about 0.04–12 in (1 mm to 30 cm). Many have short hairs or scales covering their body.

Members of one subgroup, the Caudofoveata, extract food particles from mud and live head-down in ocean sediment with gills at the rear that are exposed to the water. The other subgroup, Solenogastres, includes carnivores that feed on animals such as hydroids. The exact relationships of aplacophorans are uncertain, but it is thought they may have evolved from chitonlike shelled ancestors. More than 300 species have been described.

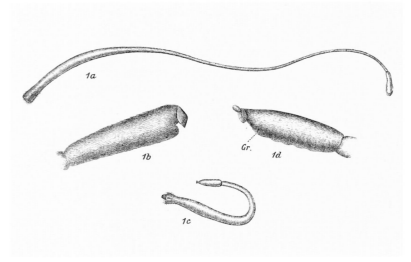

Shell-less mollusk
This illustration shows details of the head region of two species of aplacophoran mollusks in the subgroup Caudofoveata: *Chaetoderma productum* (1a, b) and *C. nitidulum* (1c, d).

bivalves

CLASS: Bivalvia   **PHYLUM:** Mollusca   **KINGDOM:** Animalia

There are about 9,200 species of bivalves. This group of aquatic mollusks includes cockles, mussels, clams, oysters, and scallops. Most bivalves are marine, but more than 1,000 species live in fresh water. All species have two shells, hinged together, that enclose the soft body of the mollusk inside. Anatomically, the shells, or valves, represent the left and right sides of the body. Strong muscles clamp the valves shut for protection against dehydration or predators. Bivalves have a foot, but it is often tongue-shaped and used to burrow rather than for crawling. They have no obvious head and a small brain. Beneath their shells, these mollusks have large, flattened gills that are also used for filtering food particles.

Bivalves are mainly sedentary animals that may live in a range of different environments above or below the water, such as buried in sand or mud, or attached to hard surfaces. Burrowing species have a pair of retractable tubes called siphons, which they extend into the water. While one siphon sucks in water and food particles, the other expels water and waste. Shipworms are long-bodied bivalves that burrow into submerged wood, and some burrowing species bore into stone.

Scallops
The distinctively shaped ribbed shells of scallops can be seen in this illustration. Scallops are capable of swimming by opening and closing their shells to generate a kind of jet propulsion.

Giant clam
Tridacna gigas

Flame scallop
Ctenoides scaber

Blue mussels
Mytilus edulis

Common cockle
Cerastoderma edule

gastropods

CLASS: Gastropoda   **PHYLUM:** Mollusca   **KINGDOM:** Animalia

Snails and slugs, known as gastropods, make up the largest class of mollusks—with more than 65,000 species—and they live in the ocean, in fresh water, and on land. All gastropods have a well-defined head, usually with eyes and tentacles. Most species crawl, much as chitons do. Their upper body becomes twisted during development, so their internal anatomy is asymmetrical. A majority of species can pull their body into a spiral-shaped shell for protection. Many gastropods have a horny operculum, or "door," which seals the opening to their shell when they retreat inside.

A majority of gastropods use a radula—a rasping, tonguelike structure—for feeding; some sea snails, such as whelks, have their radula at the end of a proboscis, like a drill. These snails are carnivores, and some of them can drill through other shells to reach the flesh of their prey. Slugs are gastropods without shells, and they use alternative forms of protection. Some sea slugs are often brightly colored as a warning signal to predators that they are poisonous or venomous. Shell-less gastropods called sea butterflies can swim by flapping finlike projections called parapodia.

Land snails
This illustration depicts relatively large terrestrial snails native to Europe, including garden snail (*Cornu aspersum*, top), Roman snail (*Helix pomatia*, center), and grove snail (*Cepaea nemoralis*, bottom left).

Giant African land snail
Achatina fulica

Variable neon sea slug
Nembrotha kubaryana

Mottled sea hare
Aplysia fasciata

octopuses, squid, and cuttlefish

CLASS: Cephalopoda   **PHYLUM:** Mollusca   **KINGDOM:** Animalia

Squid, octopuses, nautiluses, and cuttlefish are cephalopods, a class of intelligent, predatory mollusks. About 800 species have been described and all live in marine environments. Cephalopod means "head-foot" in Greek, in reference to the way the animal's head connects to its arms. The original molluskan foot evolved into the front of the body, which is equipped with grasping arms.

Nautiluses are the most primitive living cephalopods, with an external snaillike shell and 60–90 suckerless arms, or cirri. Squid and cuttlefish have eight arms and two longer tentacles, all with suckers, and octopuses have eight suckered arms and no tentacles. Squid are fast-moving hunters equipped with fins; they propel themselves by jet propulsion, forcing sea water rapidly out of a tube called a siphon. Octopuses also have a jet-propulsion ability but they usually move more slowly. Cephalopods tear food with a hard beak, and have large eyes with good color vision. They can quickly change skin color to communicate or camouflage themselves. Squid and cuttlefish have an internal rodlike shell, which octopuses lack. Cephalopods' eggs hatch into miniature adults rather than larvae as in other mollusks.

Octopus and cuttlefish
At the top of the illustration is the common octopus (*Octopus vulgaris*), which is found in all the world's oceans. The pink cuttlefish (*Sepia orbignyana*, bottom) can be found at depths of up to 1,500 ft (450 m).

Caribbean reef squid
Sepioteuthis sepioidea

Coconut octopus
Amphioctopus marginatus

Chambered nautilus
Nautilus pompilius

Giant cuttlefish
Sepia apama

annelid worms

PHYLUM: Annelida **KINGDOM:** Animalia

There are more than 22,000 species of segmented worms, or annelids. The phylum includes earthworms, leeches, and marine worms such as ragworms. The segments of their bodies are often visible as rings on the surface. Annelids lack a hard skeleton and keep their shape through a combination of muscle action and internal hydrostatic (fluid) pressure.

The polychaete worms make up the most diverse group. In polychaetes, meaning "many bristles," each body segment bears bristles, or chaetae. They come in many shapes; the sea mouse is rounded, unlike the usual worm shape. The sides of many worms have leglike flaps called parapodia that bear the bristly chaetae. Some are swimmers, but most live on the ocean floor. Polychaetes include active predators such as ragworms, with tough jaws; mud-eating lugworms; and tube worms, which create a permanent tube around their body that they never leave. Tube worms have tentacles or fans, which they spread to obtain food particles from sea water or mud. Oligochaetes, meaning "few bristles," include earthworms and leeches. Leeches have suckers at both ends that aid movement. Most leeches are bloodsucking parasites, although some are predators.

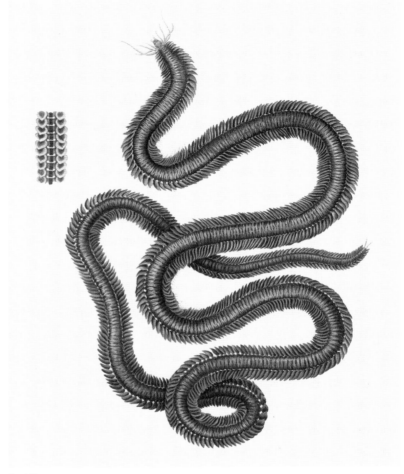

Paddle worm
Phyllodoce lamelligera is a type of polychaete called a paddle worm. It has several hundred segments, each with paddle-shaped, swimming parapodia.

Blue earthworm
Perionyx sp.

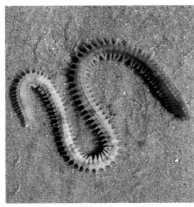

Slender ragworm
Nereis pelagica

Peacock worm
Sabella pavonina

Leech
Hirudinea

nematodes

PHYLUM: Nematoda **KINGDOM:** Animalia

Nematodes, or roundworms, are found in virtually all habitats on Earth, and more than 25,000 species are known. They are typically no more than a few millimeters long, although parasitic species can be much larger. Nematodes abound in soil and marine sediments, where they feed on bacteria or tiny animals, often piercing their prey before pumping out its fluids.

Nematodes have long, slender bodies that taper at the ends. Like insects, they molt their tough outer "skin" (cuticle). They can reproduce as separate sexes or as hermaphrodites. Many species are major pests of crop plants. Animal parasites include some species that cause diseases, such as elephantiasis, in humans. The largest known nematode is more than 25 ft (8 m) long and parasitizes sperm whales.

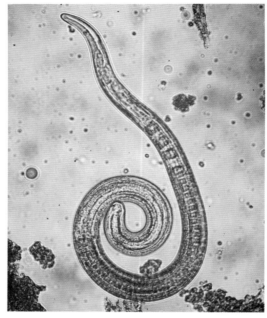

Trichinella
Nematodes in the genus *Trichinella* parasitize humans and domestic animals. If their larvae migrate to muscle tissue, they can cause the disease trichinosis.

arrow worms

PHYLUM: Chaetognatha **KINGDOM:** Animalia

Arrow worms are dart-shaped predators that exist in large numbers in plankton and on the sea bed, forming an important part of the marine food chain. More than 100 species have been discovered, growing up to 5 in (12 cm) long. They are hermaphrodites—individuals have both male and female reproductive organs—and the eggs develop directly into miniature adults.

On detecting prey, such as a copepod or fish larva, an arrow worm darts forward with flicks of its tail and grasps the prey with tough mouth bristles, sometimes paralyzing it by injecting venom. A few species follow their prey by rising to the surface at night to feed and sinking down during the day. Arrow worms in turn are preyed upon by many other animals, from jellyfish to manta rays.

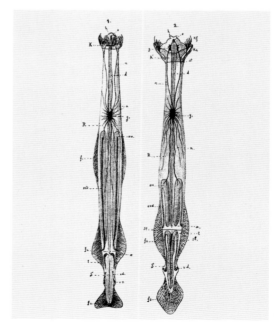

Bristle-jawed worms
Arrow worms have a stiff but flexible, transparent body equipped with fins. Their mouth has bristles—the name Chaetognatha means "bristle jaws."

gastrotrichs

PHYLUM: Gastrotricha **KINGDOM:** Animalia

Gastrotrichs are tiny wormlike animals usually less than 0.04 in (1 mm) long, and live in sea and fresh water. They live between sediment grains or on surfaces. Gastrotrichs feed by sucking up tiny prey and dead organic material. Like flatworms (see p.354), they mainly use cilia on their flat undersides to glide over surfaces, although some can loop along like inchworms, and there are a few planktonic swimming species. Many gastrotrichs have spines, scales, or bristles covering their bodies, as well as adhesive tubes used for gripping and releasing sand grains. They are either hermaphrodites (individuals with both male and female sex organs), or they exist as female-only populations. The eggs hatch into miniature adults without going through a larval stage. Nearly 800 species have been described.

Chaetonotus simrothi
This image of *Chaetonotus simrothi* shows the typical unsegmented body of a gastrotrich. The body is covered with a spiny cuticle, and two adhesive tubes are visible at the rear end (right).

kinorhynchs

PHYLUM: Kinorhyncha **KINGDOM:** Animalia

Kinorhynchs, popularly called mud dragons, are tiny marine animals that live mainly in muddy sediments and extract food from the mud that they swallow. About 270 species are known. Usually less than 0.04 in (1 mm) long, their bodies are divided into segments, or zonites, covered by protective plates, making them look a little like microscopic insect larvae. Kinorhynchs can withdraw their spiny head region into their body for protection. For movement, they extend their head region forward and anchor it in the mud, while drawing up the body behind. Kinorhynchs have movable bristles down their sides and, typically, longer spines at the tail end. Like arthropods and nematodes, kinorhynchs molt (shed) their outer "skin" (cuticle) as they grow. There are separate sexes, but males and females look similar.

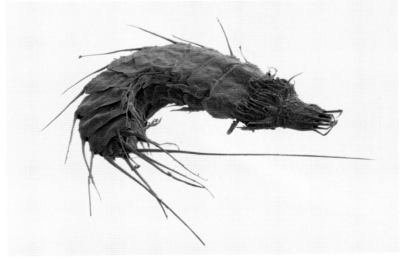

***Antygomonas* kinorhynch**
An external cuticle protects the body of a kinorhynch, such as this *Antygomonas* species. The animal's body is divided into spine-bearing segments, and its head (right) is fully retractable.

tardigrades

PHYLUM: Tardigrada   **KINGDOM:** Animalia

Also called water bears, tardigrades are tiny crawling animals, usually less than 0.04 in (1 mm) long. They live on land, in damp habitats such as the surfaces of mosses, as well as in fresh water and on the ocean floor. They have a tubelike mouth equipped with pointed structures called stylets, which they use to puncture the cells of plants and microscopic animals before sucking up the liquid contents. Like insects, they molt (shed) their tough outer "skin" (cuticle) as they grow. Males and females are separate, although some populations are female-only. Tardigrades are notable for their ability to survive unfavorable conditions by losing nearly all of their body water and becoming a shriveled form called a tun. More than 1,300 species have been described.

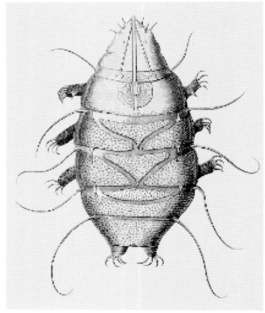

Echiniscus
Like other tardigrades, species in the genus *Echiniscus* have a segmented, barrel-shaped body and four pairs of stubby, unjointed legs with tiny claws.

velvet worms

PHYLUM: Onychophora   **KINGDOM:** Animalia

There are more than 220 species of velvet worms—caterpillarlike predators that live in moist tropical forests. Like annelids (see p.361), the body of a velvet worm is divided into segments; it has two antennae on the head, and its internal organs lie in internal body cavities that are bathed in blood. The numerous legs of the velvet worms lack joints and are stubby, and the animal has no tough exoskeleton outside its soft skin.

Velvet worms dry out easily, so they can live only in damp places. They are nocturnal ambush predators that feed on other invertebrates, sometimes as large as themselves, using a sticky slime that they squirt to immobilize their prey.

Many-legged worm
The Cape velvet worm (*Peripatopsis capensis*) has 17 pairs of legs. Other members of its family (Peripatopsidae) have 13–25 pairs.

sea spiders

CLASS: Pycnogonida   **SUBPHYLUM:** Chelicerata   **PHYLUM:** Arthropoda

More than 1,300 species of sea spiders are known. They are found scuttling along sea floors throughout the world's oceans. The central body of sea spiders is tiny compared to their limbs, and their digestive system extends into their legs. Sea spiders are usually small, but the largest species have a leg span of up to 30 in (75 cm). Most species have four pairs of walking legs, but some have five or six pairs. Sea spiders are carnivores that feed on stationary animals, such as sponges and corals. Their feeding appendages, or chelifores, may be related to the clawlike mouthparts (chelicerae) of arachnids, so they are placed with them in the subphylum Chelicerata. In addition to the chelifores, their mouths are equipped with a slightly mobile, tubular proboscis that sucks up food.

Colossendeis robusta
This sea spider species feeds mostly on hydrozoans and is found on the continental shelf around Antarctica. It has been recorded at depths of 11,811 ft (3,600 m).

horseshoe crabs

CLASS: Merostomata   **SUBPHYLUM:** Chelicerata   **PHYLUM:** Arthropoda

Horseshoe crabs are more closely related to arachnids than they are to true crabs (see p.371). There are just four species of these marine animals, and they share a number of features: a strong, horseshoe-shaped shield called a carapace, which protects the front of their body; and a long, movable spine at the rear, which they use to turn themselves upright when necessary. Horseshoe crabs have five pairs of legs and, like most arthropods, compound eyes—eyes made up of many facets. Horseshoe crabs live on the sea floor, where they eat a wide range of food, both live prey and scavenged dead fish. In the breeding season, they congregate by the shoreline in large groups to mate. Afterward, the females bury their eggs in the sand to hatch.

Atlantic horseshoe crab
With its smooth, tough carapace and tail-spine (telson), the Atlantic horseshoe crab (*Limulus polyphemus*) is typical of the class Merostomata. It lives along the Atlantic coast of the United States.

spiders, scorpions, and relatives

CLASS: Arachnida **SUBPHYLUM:** Chelicerata **PHYLUM:** Arthropoda

Spiders, scorpions, mites, and ticks are some familiar members of a large group of air-breathing arthropods called arachnids, which contains about 65,000 species. Arachnids usually have four pairs of legs and a two-part body: a combined head and leg-bearing section (cephalothorax) and an abdomen. Unlike most arthropods, they lack antennae.

Spiders (order Araneae) inject paralyzing venom into prey through their fangs, whereas scorpions (order Scorpiones) use the sting in their tails to subdue prey. Spiders' ability to make silk has many uses: constructing webs, lining burrows, tying down prey, and making cocoons to protect their eggs. Silk is also used, especially by young spiders, to create long, gossamer threads that get caught by the wind and carry the airborne animals over long distances. Many spiders are active hunters or ambush predators. Pseudoscorpions (order Pseudoscorpiones), or false scorpions, are similar to scorpions in shape, but lack a tail sting. They form a distinct group.

Ticks and mites (order Acari) form a large group of small or microscopic animals. Many species live off larger animals, feeding from their skin or sucking blood. Others are miniature predators, with pincerlike mouthparts.

Orb web spiders
These orb web spiders (family Araneidae) were illustrated by Carl Wilhelm Hahn in the 1820s. The spider in its web is a female *Argiope lobata*, one of the largest spiders in Europe.

Mexican redknee tarantula
Brachypelma smithi

Jumping spider
Carrhotus xanthogramma

House pseudoscorpion
Chelifer cancroides

Sheep tick
Ixodes ricinus

centipedes and millipedes

SUBPHYLUM: Myriapoda   **PHYLUM:** Arthropoda

Myriapods are many-legged terrestrial arthropods with segmented bodies. More than 16,000 species have been described, including millipedes, centipedes, and some smaller groups such as the symphylans. Millipedes usually live among leaf litter or fallen logs. They mainly eat dead or living vegetable material, although a few are predators. One millipede species has up to 750 legs, but some myriapods have fewer than ten. Each body segment of a millipede has two pairs of legs, unlike the single pair in centipedes. Millipedes can be cylindrical or flattened; members of one short-bodied family resemble woodlice (see p.369) and can curl up in a ball for protection.

Centipedes are predators that attack their prey using venom-bearing claws behind their heads. They have one pair of legs on each body segment, and the species in the family Scutigeridae have very long legs. Centipedes are mainly nocturnal and usually flattened in shape. The largest species can catch and eat small reptiles, mammals, and birds. Members of one group of myriapods, the symphylans (order Symphyla), or pseudocentipedes, are small, nonvenomous, and live in soil, where they feed on decaying vegetation.

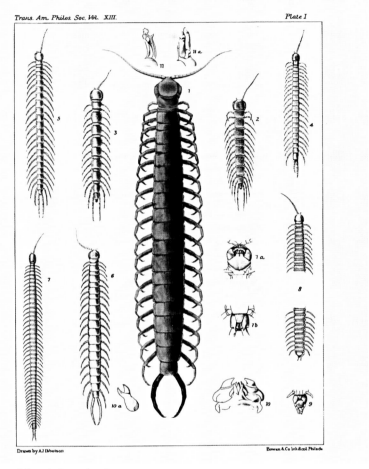

Giant red-headed centipede
Growing up to 8 in (20 cm) long, the bright colors of the venomous giant red-headed centipede (*Scolopendra heros*) serve to warn off predators.

Yellow-banded millipede
Anadenobolus monilicornis

Shocking pink dragon millipede
Desmoxytes purpurosea

Megarian banded centipede
Scolopendra cingulata

Garden centipede
Scutigerella immaculata

water fleas and relatives

CLASS: Branchiopoda   **SUBPHYLUM:** Crustacea   **PHYLUM:** Arthropoda

A diverse, mainly freshwater group, water fleas—or branchiopods—comprise about 800 species of small crustaceans, the subphylum that includes crabs and shrimp of all kinds. They are so named because many of the animals' appendages bear gills—branchiopod means "gill-feet." Some branchiopods thrive in salt lakes and a few others are found in the sea. A group called water fleas is very common in ponds. The longer bodied fairy shrimp may grow to several inches and are notable for swimming "upside down."

Branchiopods catch food mainly using tiny hairs on their limbs as they swim. Tadpole shrimp are branchiopods that grub around at the bottom of ponds. Many branchiopods produce eggs that can survive droughts and are blown to new locations on the wind.

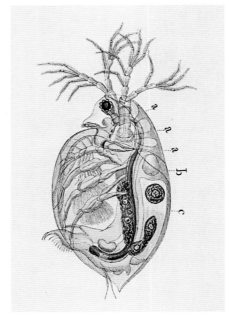

Daphnia
The water flea *Daphnia pulex* was the first branchiopod to be studied. Since most of its outer covering is transparent, its internal organs are easily visible, as shown in this 1798 illustration by George Shaw.

copepods

SUBCLASS: Copepoda   **CLASS:** Maxillopoda   **SUBPHYLUM:** Crustacea   **PHYLUM:** Arthropoda

Like branchiopods, copepods are crustaceans. About 13,000 species are known. Free-swimming copepods exist in large numbers in plankton and are a vital part of the marine food chain, but about 2,800 species live in fresh water. Typically just 0.04–0.08 in (1–2 mm) long, most copepods have distinctive, long antennae and feed by filtering tiny phytoplankton from the water with sievelike mouthparts, as well as by preying on other small animals. Many copepods undergo daily vertical migration, retreating to deeper waters during the day to avoid predators and rising to the surface at night. Some live on the sea floor, and others are parasites of fish and invertebrates. The body shape of parasitic copepods is often very different from that of their free-living relatives, and they can grow to as much as 12 in (30 cm) long.

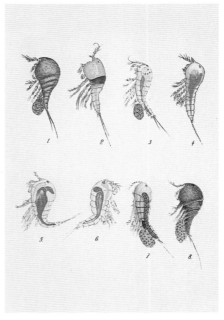

Harpacticoid copepods
This illustration from 1921 shows a selection of copepods from the order Harpacticoida. This group is distinct from other copepods because of having only a very short first pair of antennae.

barnacles

SUBCLASS: Thecostraca **CLASS:** Maxillopoda **SUBPHYLUM:** Crustacea **PHYLUM:** Arthropoda

Barnacles live fixed to one spot and protect themselves by secreting chalky plates around their bodies. About 1,000 species are known. Barnacles mainly come in two shapes: acorn barnacles, which have a roughly conical body that cements directly to rocks or other surfaces; and goose-necked barnacles, which have a larger, flattened body sitting on a fleshy stalk. Rhizocephalans are barnacles that parasitize mostly decapods. Females attach themselves to crabs, spreading tendrils throughout the host's insides to absorb nourishment, rendering the crab infertile.

Barnacles living inside chalky plates capture food particles from sea water by extending their cirri (appendages that filter food). Although adult barnacles cannot move around, their larvae can swim and select new sites to settle on.

Acorn barnacles
Variable in shape, acorn barnacles are barrel-shaped or conical crustaceans that live in the intertidal zone. This illustration depicts one species, *Megabalanus tintinnabulum*, which is one of the larger barnacles, growing up to 2 in (5 cm) tall.

isopods

ORDER: Isopoda **CLASS:** Malacostraca **SUBPHYLUM:** Crustacea **PHYLUM:** Arthropoda

This order includes some 10,000 species. About half are land-living woodlice and half are aquatic—mostly marine—species. Typically, isopods have flattened bodies protected by a series of horny plates. Most are small, but giant deep-ocean species can reach up to 20 in (50 cm) long. Sea slaters, which dwell on shorelines, can survive above or (temporarily) below water, but most marine isopods are fully aquatic, burrowing or crawling on the sea floor. Female isopods have brood pouches on their underside that protect their eggs, which hatch into miniature adults, unlike the free-swimming larvae of other crustaceans. Isopods usually feed on algae or dead material, except one subgroup, which parasitizes other crustaceans. The gribble worms (family Limnoriidae) eat wood.

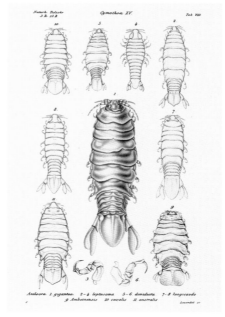

Parasitic sea slaters
Isopods from the genus *Anilocra* are fish parasites. They quickly latch onto passing fish, biting the host and sucking blood and other fluids until the fish dies. Female *Anilocra gigantea* (center) may grow to 4 in (10 cm) in length.

lobsters, shrimp, and relatives

ORDER: Decapoda **CLASS:** Malacostraca **SUBPHYLUM:** Crustacea **PHYLUM:** Arthropoda

Decapods have a variable number of appendages (up to 38), of which 10 are considered to be legs. The first pair of legs is sometimes modified into heavy claws. The order includes most shrimp, as well as prawns, crayfish, lobsters, and crabs. Crabs belong to two infraorders, Anomura and Brachyura (see p.371). About 3,000 species of shrimp have been described, and a majority are omnivorous. Most are marine but some live in fresh water. Shrimp have a long, thin abdomen, and their head and thorax are fused to form a cephalothorax, with two long antennae. They move mostly by swimming. Many groups of crustaceans that are commonly called shrimp, including mantis shrimp, lie outside the Decapoda.

The term "lobster" is used for various crawling, long-bodied decapods. The true lobsters (infraorder Astacidea) have pincerlike claws and crawl along the ocean floor, usually at night, in search of living or dead prey. The spiny lobsters (infraorder Achelata) share some similarities with the true lobsters but lack claws; instead, they are protected by sharp spines on their antennae. The flat-bodied slipper lobsters also belong to this infraorder.

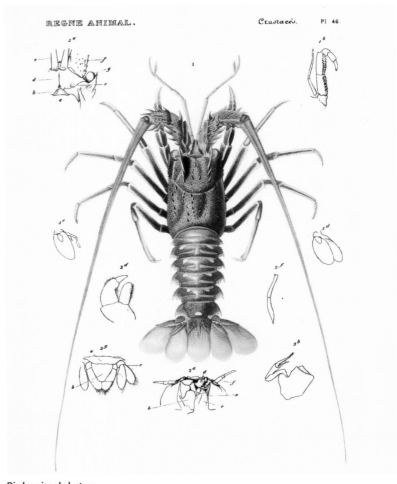

Pink spiny lobster
Feeding on small crabs and worms by night and hiding in rock crevices during the day, the pink spiny lobster (*Palinurus mauritanicus*) is named for its forward-pointing spines.

Harlequin shrimp
Hymenocera picta

Pacific cleaner shrimp
Lysmata amboinensis

Blue crayfish
Procambarus alleni

California spiny lobster
Panulirus interruptus

crabs

INFRAORDER: Anomura, Brachyura   **ORDER:** Decapoda   **CLASS:** Malacostraca   **SUBPHYLUM:** Crustacea

In contrast to other decapods, the true crabs (infraorder Brachyura) have a tiny abdomen that is tucked under a fused head and thorax (cephalothorax), all protected by a hard covering called the carapace. This short, squat, wide body gives them good balance and enables them to walk or run much faster than lobsters. The anatomy varies, from the sturdy species of rocky shores to spider crabs and arrow crabs, whose long legs allow them to spread their weight on soft sediments. Some true crabs have paddlelike hind legs, allowing them to swim and even catch fish. The shore-dwelling fiddler crabs and ghost crabs forage out of water, watching for danger with their long-stalked eyes. Other true crabs live in fresh water or on land, though the latter must return to the water to breed.

The hermit crabs (infraorder Anomura) carry around empty mollusk shells for protection, into which they insert their soft, curved abdomens. This group also includes the large, land-living robber crab and other species that look like true crabs. Anomurans have 10 legs but appear to have only four pairs of legs, since the hind pair is much reduced in size.

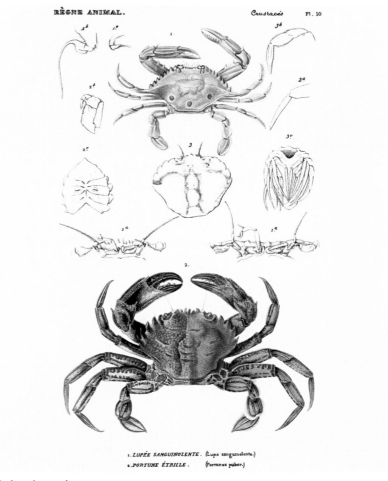

Swimming crabs
The paddlelike hind legs of the three-spot swimming crab (*Portunus sanguinolentus*, top) and the velvet swimming crab (*Necora puber*, bottom), help them to swim. The left side of fig. 2 (bottom) shows the body after molting.

Arrow crab
Stenorhynchus seticornis

West African fiddler crab
Afruca tangeri

Sally Lightfoot crab
Grapsus grapsus

White-spotted hermit crab
Dardanus megistos

springtails

CLASS: Collembola **SUBPHYLUM:** Hexapoda **PHYLUM:** Arthropoda

Once considered to be insects, springtails are now classified in a separate group that contains about 3,600 species. Like insects, these small, wingless arthropods have six legs. However, they lack the spiracles and tracheae (breathing holes and tubes) of true insects. Springtails live mainly in or around damp soil, where they can be very common.

Some species are almost globular, while others have long, thin bodies. Springtails are named for their most distinctive feature, a forked structure called a furca, which is held against their underside by a catch. If a springtail is threatened by a predator, it releases the catch, forcing the furca to the ground very fast and sending it tumbling into the air, landing some distance away—and hence evading predation.

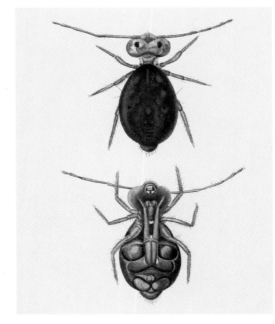

Globular springtail
This illustration shows the top and underside of a globular springtail from the family Sminthuridae. The fork-shaped furca can be seen in the view from below.

silverfish

ORDER: Zygentoma **CLASS:** Insecta **SUBPHYLUM:** Hexapoda **PHYLUM:** Arthropoda

Silverfish—about 550 species—live in damp terrestrial habitats worldwide, under rocks, in caves, and often in human habitations. In homes, they feed mostly on sugars and starches in cereals, fabrics, and even the glue in the bindings of books. Although wingless, they are true insects, with a body divided into a head, a thorax (midsection) with three pairs of legs, and an abdomen. They have compound eyes (made up of several tiny lenses) and, like other insects, breathe using holes down their sides called spiracles, which lead to air-filled tubes (tracheae) that branch throughout their bodies. Silverfish have a flattened, streamlined body covered with shiny scales and three long tail filaments. Unlike other insects, however, they continue to molt (shed) their exoskeletons as adults, passing through several mature instars (phases between two moltings).

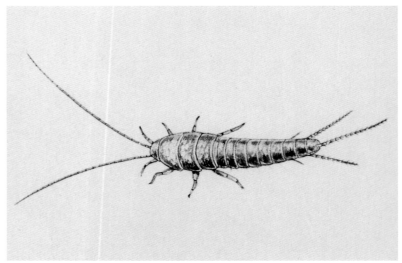

Common silverfish
This nocturnal species is commonly found in homes, where it feeds on hair, carpet fibers, paper, coffee, and even dandruff. It requires a humidity level of 75 percent or more to thrive.

mayflies

ORDER: Ephemeroptera   **CLASS:** Insecta   **SUBPHYLUM:** Hexapoda   **PHYLUM:** Arthropoda

Mayflies live most of their life in fresh water as nymphs, breathing using gills and feeding on algae, diatoms, or the larvae of other insects. Nymphs molt 40–50 times—far more than other insects. It may be years before they metamorphose into flying insects. About 3,000 species of mayflies are known.

Mayflies are "primitive" (near the base of the insect family tree), and one sign of this is that they cannot fold their wings flat over their bodies, as most other insects do; instead, they hold them vertically behind the body. Their hindwings are much smaller than the forewings and are completely absent in some species. Mayflies are weak fliers and the adults do not feed, living only long enough to mate and lay eggs—which in some species is just a few hours. Adults and nymphs typically have three long "tails" at the end of their body.

Ephemera danica
During summer, male mayflies of this species—sometimes called the green drake—fly in large numbers near slow-moving rivers and perform an aerial "nuptial dance" before mating with females.

lacewings

ORDER: Neuroptera   **CLASS:** Insecta   **SUBPHYLUM:** Hexapoda   **PHYLUM:** Arthropoda

Lacewings and their relatives in the order Neuroptera make up one of the oldest groups of endopterygotes—insects whose wings develop out of sight within their growing larvae. More than 4,000 species are known. Butterflies, moths, flies, bees, and beetles are also endopterygotes.

Lacewing larvae are wingless predators that eat small insects. They have grooved biting jaws through which they suck up the juices of their prey. Larvae of the antlion family hide in pits with only their jaws showing, catching any insects that fall in. The larvae enter a pupal stage, within which the adult body is formed. The flying adults have delicate, netlike wings, which they hold over their slender bodies when at rest. Some species are predatory, while others consume only nectar, and some do not feed at all.

Antlion
This 18th-century illustration depicts an adult antlion from the genus *Myremeleon*. Antlions are close relatives of lacewings. Their larvae are called "doodlebugs" in North America because of the tracks they leave in the sand.

dragonflies and damselflies

ORDER: Odonata **CLASS:** Insecta **SUBPHYLUM:** Hexapoda **PHYLUM:** Arthropoda

Dragonflies are long-bodied, often brightly colored predatory insects. They have two pairs of wings that allow them to fly forward and backward, and to hover. This, combined with two large eyes, enables them to catch insects in the air or snatch them off the ground with their forward-pointing legs. Some dragonflies pursue prey in flight, whereas others are ambush predators, perching on vegetation until their prey flies by. Males often defend territories from other males. Some species migrate long distances. Most of a dragonfly's life is spent as a nymph in fresh water. The nymphs are also predatory and have a hinged structure under their head, called a mask, which they shoot out to catch prey. Nymphs' diet includes other invertebrates, young amphibians, and even small fish.

Damselflies are also members of the order Odonata, which includes about 5,000 species. Damselflies have a more delicate body with a long, thin abdomen. Although they are less agile fliers than dragonflies, their lifestyles are generally similar. Damselflies fold their sometimes colorful wings behind their back at rest, as mayflies do.

Darners
The green darner (*Anax junius*, top) and swamp darner (*Epiaeschna heros*) are two members of the dragonfly family Aeshnidae. Named for their resemblance to darning needles, they are predators of small flying insects.

Sahara bluetail
Ischnura saharensis

Darner
Aeshnidae

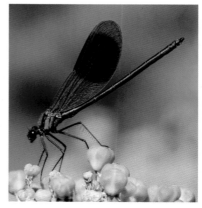

Banded demoiselle
Calopteryx splendens

Hine's emerald dragonfly
Somatochlora hineana

grasshoppers, crickets, and bush crickets

ORDER: Orthoptera   **CLASS:** Insecta   **SUBPHYLUM:** Hexapoda   **PHYLUM:** Arthropoda

Collectively, grasshoppers, crickets, and bush crickets are called orthopterans. There are more than 20,000 species in the group. They are easily recognized by their large hind legs, which enable them to jump long distances. Orthopterans share the ability to produce sound by "stridulation"—rubbing parts of their body against each other—although the sound varies between species. Adults have two pairs of wings, but they generally use their wings simply to lengthen their leaps. With the exception of locusts, most species do not fly far. Their hindwings may be brightly colored to startle predators.

True grasshoppers, including locusts, have short antennae, and many specialize in eating grass. Locusts are highly gregarious, sometimes swarming in millions; they can become agricultural pests, as they voraciously eat crops. Bush crickets, or katydids, have long antennae and may eat insects as well as plants. True crickets have long antennae and wider, flatter bodies than other orthopterans. Mole crickets live much of their life underground in burrows that they dig with their spadelike front legs. There, they mate, raise young, and search for roots, insect larvae, and worms to eat. Young orthopterans resemble small, wingless adults.

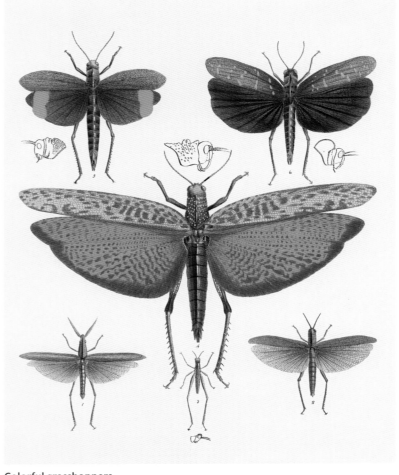

Colorful grasshoppers
This illustration shows a range of grasshoppers with brightly colored wings. The males of some species display these vibrant colors to help attract females.

Large marsh grasshopper
Stethophyma grossum

Mount Arthur giant weta
Deinacrida tibiospina

Bush cricket
Saga pedo

stick and leaf insects

ORDER: Phasmida   **CLASS:** Insecta   **SUBPHYLUM:** Hexapoda   **PHYLUM:** Arthropoda

Also known as phasmids, stick insects and leaf insects are mainly a tropical group of more than 3,000 species. The longest insect in the world, *Phryganistria chinensis*—measuring up to 24 in (62.4 cm) long—is a phasmid. Some leaf insects have a cylindrical, sticklike body, while others are flattened and may look like fallen leaves, complete with apparently decaying and "nibbled" regions. Stick insects undergo incomplete metamorphosis. Their eggs, usually dropped onto the ground, may take months to hatch. The nymphs look like miniature adults and develop through a series of stages, or instars. There is no pupal stage.

Mostly nocturnal, phasmids remain motionless among vegetation by day and are excellent at camouflage. They become active at night, feeding mostly on the leaves of trees and shrubs. Many stick insects are wingless, but others have wings and can fly. Some winged species startle predators by flashing their brightly colored hindwings. Other phasmids have defensive spines on their legs. Many species are parthenogenetic—females can produce eggs that hatch without fertilization by males.

Indian stick insect
This Indian stick insect (*Carausius morusus*) was illustrated by entomologist John Obadiah Westwood in *The Cabinet of Oriental Entomology* (1848). He described the defensive spines on the insect's legs.

Leaf insect
Phyllium philippinicum

Lord Howe Island stick insect (nymph)
Dryococelus australis

Golden-eyed stick insect
Peruphasma schultei

Margin-winged stick insect
Ctenomorpha marginipennis

mantises

ORDER: Mantodea   **CLASS:** Insecta   **SUBPHYLUM:** Hexapoda   **PHYLUM:** Arthropoda

Mantises form a large group—about 2,400 species—of long-bodied predatory insects. They have a triangular head and bulging, widely spaced eyes. They can turn their head around to look behind them, which allows them to locate and judge the distance of prey with great accuracy. Most mantises are ambush predators that wait, with their spiny front legs folded, for prey to come close before pouncing. They eat insects of all kinds, and the larger species also eat small vertebrates.

Males often fly at night, when there is less danger from predatory birds, to seek females. Mantises are excellent at camouflage—most species are green or brown, which matches their environment of sticks, barks, and leaves. Flower mantises imitate flowers such as orchids, and may be bright pink, which camouflages them and helps to lure prey.

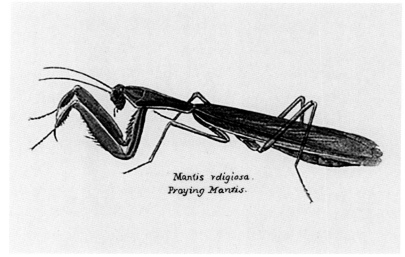

Praying mantis
Named for its prominent front legs, which are bent at an angle suggestive of a prayer position, the praying mantis preys on moths, flies, grasshoppers, and other invertebrates.

earwigs

ORDER: Dermaptera   **CLASS:** Insecta   **SUBPHYLUM:** Hexapoda   **PHYLUM:** Arthropoda

About 1,900 species of earwigs have been described. They are nocturnal insects with flattened bodies that allow them to squeeze into tight crevices, such as under tree bark. They have pincerlike forceps called cerci at their tail end, which are used for defense and by the male during mating. Most earwigs are scavengers, but some are carnivores or omnivores, eating dead material, plants (including flower petals), and smaller insects. By day, they hide in dark crevices. There is no larval or pupal stage in their life cycle, so the young look like small adults. The forewings are short, with tough covers protecting the folded hindwings that lie beneath the covers, but many earwigs are wingless or rarely fly. Unusually for insects, they show maternal care, with the females protecting their eggs and feeding the newly hatched young.

Earwig cerci
This illustration of three species of earwigs from a 1907 issue of the *Journal of the Entomological Society of London* shows their distinctive pincers, or cerci.

cockroaches

ORDER: Blattodea **CLASS:** Insecta **SUBPHYLUM:** Hexapoda **PHYLUM:** Arthropoda

Insects with long antennae, cockroaches have a flattened body and chewing mouthparts. They are often large, growing up to 4 in (10 cm), and usually nocturnal. They are most common in warmer regions of the world and eat a wide variety of foods. About 4,600 species are known.

The best-known cockroach species are the fast-running, brownish pests found in the warmth of buildings worldwide. If plentiful food is available, they congregate in large numbers and taint food with an unpleasant odor. Less well known are thousands of wild species that live in leaf litter, decaying wood, and other concealed habitats. In some species both sexes are winged, in others only the males are winged, and in some both sexes are wingless. In winged species, tough forewings protect the delicate hindwings underneath.

Cuban burrowing cockroach
This flightless nocturnal species (*Byrsotria fumigata*) grows up to 2 in (5 cm) long. It burrows in soil and consumes leaf litter in humid forests.

lice

ORDER: Psocodea **CLASS:** Insecta **SUBPHYLUM:** Hexapoda **PHYLUM:** Arthropoda

Numbering more than 10,000 species, lice are small insects that include the winged bark and book lice—which feed on fungi, algae, and starchy materials—and some that are parasites. Parasitic lice are wingless and live on birds or mammals. Biting lice mainly parasitize birds, while blood-sucking lice live on mammals. Most of these parasites spend their entire lives on their hosts, many restricted to specific host species.

Parasitic lice have a flattened body with hooked claws that grasp hairs or feathers. Some have wide bodies, while others are elongated. Lice glue their eggs (nits) onto hairs so that they remain on the host until hatching. Two species of human lice irritate the skin and can spread diseases such as typhus.

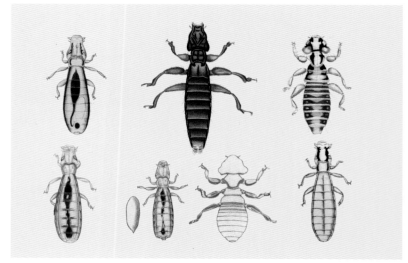

Biting lice
This selection of biting lice shows their shortened front legs, which are used to move food to the mouth. They have chewing mouthparts for eating dead skin and feathers.

true bugs

ORDER: Hemiptera **CLASS:** Insecta **SUBPHYLUM:** Hexapoda **PHYLUM:** Arthropoda

Many insects are loosely called "bugs," but entomologists define true bugs as members of the extremely diverse order Hemiptera. There are about 75,000 species of true bugs. Terrestrial leafhoppers, shieldbugs, stinkbugs, cicadas, and aphids are all true bugs. So, too, are aquatic pond skaters and backswimmers.

Most true bugs are small or very small, but the largest grow up to 6 in (15 cm). Many can fly, and some species have large-winged, short-winged, and wingless forms—this is an example of polymorphism. Leafhoppers and cicadas are great jumpers. All hemipterans have mouthparts in the form of a long sucking tube called the rostrum, which is equipped with piercing stylets (needlelike appendages). The rostrum is folded under the body when not in use. Most true bugs feed on plant sap, although some—including assassin bugs and water striders—suck the body fluids of animal prey. Aphids and their relatives, including scale insects and whiteflies, are serious pests of cultivated plants. The females are often parthenogenetic—they can reproduce without the need for males—and multiply rapidly.

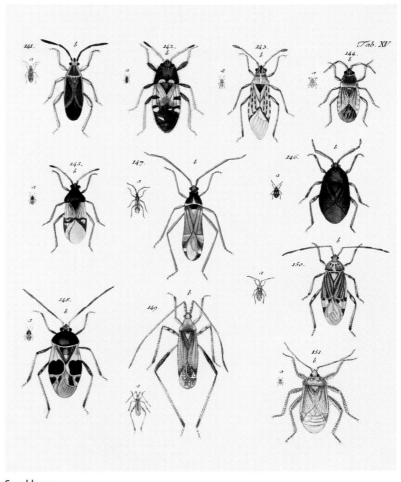

Seed bugs
The bugs shown here are collectively known as seed bugs or milkweed bugs. The bright colors of some act as a warning to predators that they taste bad.

Red-banded leafhopper
Graphocephala coccinea

Squash bug
Coreidae

Periodical cicada
Magicicada sp.

Water strider
Gerris sp.

ants, bees, and wasps

ORDER: Hymenoptera **CLASS:** Insecta **SUBPHYLUM:** Hexapoda **PHYLUM:** Arthropoda

This is one of the most diverse orders of insects, collectively known as hymenopterans. There are about 115,000 species. Typically, the species in this group have two pairs of transparent wings, with the forewings being larger than the hindwings. The wings are membranous. Hymenopterans usually have biting mouthparts, but many bees have long, nectar-drinking tongues instead.

Ants, bees, and wasps have a narrow constriction, or "waist," between the first and second segments of the abdomen. This gives their body flexibility, which helps the parasitic species to aim their needlelike egg-laying organ (ovipositor) into their insect or spider prey. In these species, the larvae begin to consume their host once they have hatched. Some groups of hymenopterans, including sawflies, are not "waisted," however, and the female's sawlike ovipositor cuts into the plant tissue that the larvae eat. In bees, typical wasps, and some ants, the ovipositor is modified into a sting. Ants, as well as many bees and wasps, live in social colonies headed by a queen, helped by sterile workers who feed and raise the grublike larvae of the next generation.

Bee variety
From top to bottom this plate shows four very different bees from the genera *Bombus* (bumblebees), *Xylocopa* (large carpenter bees), *Acanthopus*, and *Melecta* (mourning bees).

Sawfly
Symphyta

Honeybee
Apis sp.

Weaver ant
Oecophylla sp.

Hornet
Vespa sp.

beetles

ORDER: Coleoptera **CLASS:** Insecta **SUBPHYLUM:** Hexapoda **PHYLUM:** Arthropoda

Beetles are one of the most species-rich groups of insects, with some 400,000 species described, and it is certain that there are many more yet to be discovered. Beetles range from numerous minute species to some of the heaviest insects. The success of this group derives from their strong, compact body plan, which features hardened forewings, or elytra, that fit snugly over the body and protect the usually functional hindwings underneath.

Beetles practice complete metamorphosis, with a pupal stage between the larval and adult stages. Both adults and larvae consume a wide range of foodstuffs, including wood, plant leaves, roots, fungi, carrion, and dung. Some species are predatory. Aquatic beetles use a variety of techniques to retain air beneath the water's surface. For example, diving beetles (family Dytiscidae) hold air between their abdomen and elytra when they dive. Some beetles migrate in search of food; for instance, mountain pine beetles (*Dendroctonus ponderosae*) can fly up to 70 miles (110 km) in a day.

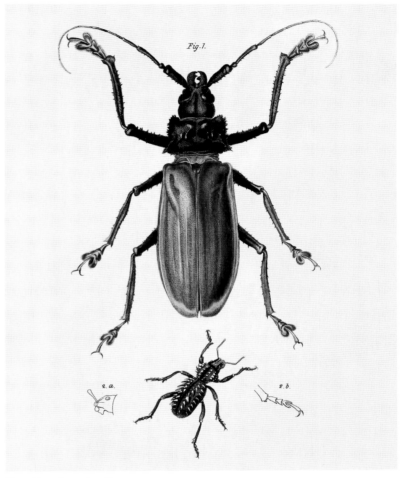

Hopei longhorn beetle
Rhaphipodus hopei is a forest-dwelling longhorn beetle native to Southeast Asia. Like other longhorns, it has very long antennae.

Ground beetle
Calosoma sycophanta

Seven-spotted ladybug
Coccinella septempunctata

Diving beetle
Noterus sp.

European stag beetle (male)
Lucanus cervus

thrips

ORDER: Thysanoptera   **CLASS:** Insecta   **SUBPHYLUM:** Hexapoda   **PHYLUM:** Arthropoda

Also called thunderbugs or thunderflies, thrips are fairly common, tiny, slender-bodied insects. More than 6,000 species have been described. Adult thrips may be spotted on human skin on warm, humid days. Most species feed on plants, piercing leaves and flower petals with their sucking mouthparts to consume the cells beneath. They can occur in large numbers, and some species are serious pests of crop plants—such as tomatoes and peas—damaging them either directly or by spreading plant viruses. Thrips may also eat pollen and fungal spores, and some larger species are predators of other insects. Members of this order have four wings fringed with long hairs (Thysanoptera means "fringe-winged"), but many species and individuals are wingless as adults.

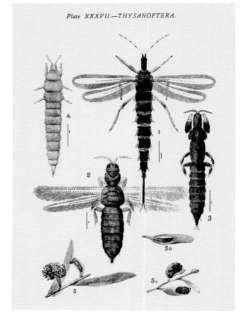

Thrips and galls
This illustration shows a selection of Australian thrip species and the galls (abnormal outgrowths) produced by plants as a reaction to the insects feeding on their cells.

caddisflies

ORDER: Trichoptera   **CLASS:** Insecta   **SUBPHYLUM:** Hexapoda   **PHYLUM:** Arthropoda

Caddisflies form a group of about 14,500 species whose aquatic, gill-breathing larvae are an important part of freshwater ecosystems. Caddisfly larvae have evolved a variety of lifestyles based on their ability to produce silk. Many use the silk to help construct protective cases for themselves using vegetation, sand grains, or mollusk shells. Other species build silken nets to trap food particles, or create retreats from which they can rush out to catch small prey. The mainly nocturnal adults often look like small, brownish moths and some do not feed at all, although others can sip nectar. Adults' wings are covered with small hairs (Trichoptera means "hair-wing"). The larvae of a few New Zealand species develop in intertidal rocks pools and are among the very few marine insects.

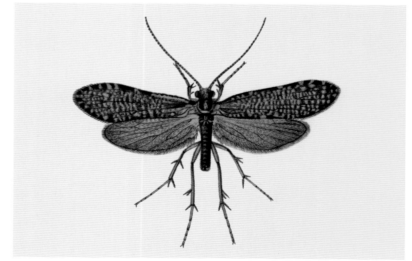

Chimarra marginata
The adults of this species begin to emerge from European rivers in June, and are on the wing until October.

moths and butterflies

ORDER: Lepidoptera   **CLASS:** Insecta   **SUBPHYLUM:** Hexapoda   **PHYLUM:** Arthropoda

More than 175,000 species of lepidopterans have been described. Butterflies and moths have very similar lifestyles, but there are some anatomical and behavioral differences between the two. Most butterflies fly by day, whereas most moths are nocturnal. Butterflies usually have club-ended antennae, and they typically hold their wings vertically above the body when at rest. Moths have straight or feathered antennae, and often—but not always—fold their wings over their abdomen. In most moths, the fore- and hindwings are coupled together with a hooklike structure called a frenulum. All lepidopterans have scales, which can be brightly colored to give them their patterns and markings. If the adults feed (in some species they do not), they consume nectar or other similar fluids via a long tube called a proboscis. In the lepidopteran family tree, moths occupy the trunk and most of the branches, with butterflies representing just one large branch. Butterflies and moths undergo complete metamorphosis. Caterpillars have chewing mouthparts and nearly all eat plants, often specializing in particular species. They have three pairs of jointed legs and usually five pairs of stubby limbs, called prolegs, on the abdomen.

Life cycle of a moth
This 17th-century depiction of the four-stage life cycle of an emperor moth shows the eggs, a caterpillar (larval stage), the pupa, and a flying adult.

Pine hawkmoth
Sphinx pinastri

Common brimstone
Gonopteryxs rhamni

Lime butterfly
Papilio demoleus

Metalmark
Family Riodinidae

true flies

ORDER: Diptera **CLASS:** Insecta **SUBPHYLUM:** Hexapoda **PHYLUM:** Arthropoda

The "true" flies make up a large group of flying insects and have a single pair of wings. Instead of hindwings, they have club-shaped sensory organs, called halteres, which sense movement and help the insect to balance as it performs complex aerial maneuvers. More than 125,000 species of flies have been described, including major groups such as blowflies, hoverflies, craneflies, and mosquitoes. Most adults eat liquid food, but their mouthparts vary depending on whether they suck up liquids, feed on pollen, prey on other insects, or pierce flesh to consume blood, as mosquitoes do.

Pollen- and nectar-feeding species are important pollinators of plants. On the other hand, a few mosquitoes transmit the parasitic microorganism *Plasmodium*, which is responsible for malaria in humans. While many flies are dull in color, others, such as hoverflies, mimic bees and wasps in appearance and have black and yellow stripes. Fly larvae, which do not have legs, are more diverse than the adults. They vary from the bristly, water-living larvae of mosquitoes to almost featureless maggots. The diet of the larvae is often very different from that of adults.

March fly
Flies in the genus *Bibio* are called March flies, or St. Mark's flies. They feed on dead and decaying vegetation, exposed plant roots, and nectar.

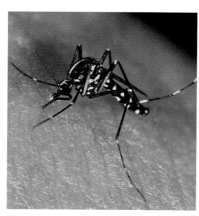

Mosquito
Culicidae

Hoverfly
Myathropa florea

Beefly
Bombyliidae

Blowfly
Calliphoridae

fleas

ORDER: Siphonaptera **CLASS:** Insecta **SUBPHYLUM:** Hexapoda **PHYLUM:** Arthropoda

Fleas are small, specialist bloodsuckers of mammals and birds. About 2,600 species are known. Although wingless, fleas can leap great distances—about 12 in (30 cm), which is over 200 times their body length. Their narrow bodies help them move among the hairs or feathers of their hosts.

Unlike lice, fleas do not remain on their hosts all the time. They often make use of animal nests and dens, where the legless, maggotlike larvae feed on organic debris before transforming into adults. Individual species of fleas tend to specialize in particular hosts but there is some flexibility—cat fleas can live on dogs, for example. Fleas can spread serious diseases in humans, including Bubonic plague, while a skin-burrowing species known as the jigger (*Tunga penetrans*) is a major problem in tropical regions.

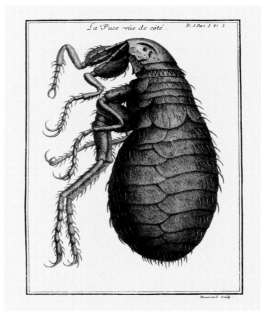

Human flea
Despite its name, the human flea (*Pulex irritans*) can be found on a range of mammals, including domestic dogs and cats. It can be a carrier of disease.

scorpionflies

ORDER: Mecoptera **CLASS:** Insecta **SUBPHYLUM:** Hexapoda **PHYLUM:** Arthropoda

Scorpionflies form a group of about 600 species that are found mainly in damp habitats. Their heads bear a distinctive, downward-pointing, beaklike projection with jaws set at the end. The males of the largest family (Panorpidae)—which gives the group its name—have tails curved up over their bodies, like scorpions, although these are used for mating and are not stingers. The adults of this family feed mainly on dead or dying insects, while their caterpillarlike larvae scavenge on the ground. Another family of mecopterans, the hangingflies, dangle from branches and catch insects with their long hindlegs. The family of snow scorpionflies, or snow fleas, comprises small wingless insects that can jump like fleas; they live among mosses in colder regions.

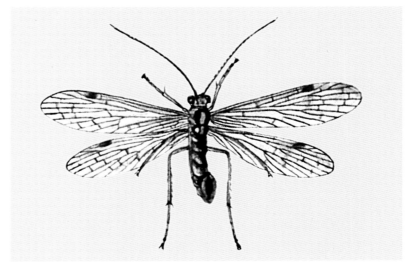

German scorpionfly
Like other species in the order Mecoptera, the German scorpionfly (*Panorpa germanica*) is completely harmless, feeding on rotting fruit, plants, and dead insects.

sea urchins

CLASS: Echinoidea **PHYLUM:** Echinodermata **KINGDOM:** Animalia

About 940 species of sea urchins are known. They belong to a phylum of exclusively marine animals called echinoderms, whose bodies have a five-sided, or pentaradial, symmetry. Sea urchins have skeletons made of small units called ossicles, some of which are fused into a round, or almost round, case called a test. They are usually rigid and have five bands of of flexible projections called tube feet running down their sides. The tube feet are used for clinging, climbing, and sometimes walking. A sea urchin's downward-facing mouth has chisel-like teeth. Many species graze algae from rocks, and others extract food from the ocean floor.

The bodies of sea urchins are covered in long, mobile spines. Some sea urchins can inject venom with their spines. Slate pencil urchins have extra-large, thick spines. Typical sea urchins exhibit pentaradial symmetry, but some species have lost that symmetry and have distinct "front" and "back" ends. These include burrowing species called heart urchins, or sea potatoes, which tunnel slowly through mud, extracting food from sediment and water. Sand dollars have a flattened shape and a dense covering of short spines. They live buried or half-buried in sand, where they filter-feed.

Sputnik sea urchin
This 19th-century illustration of several species of sea urchins includes the Sputnik sea urchin (*Phyllacanthus imperialis*, upper center), which lives in crevices in tropical coral reefs.

Red slate pencil urchin
Heterocentrotus mamillatus

Sand dollar
Clypeasteroida

Diademed sea urchin
Diadema sp.

European edible sea urchin
Echinus esculentus

sea cucumbers

CLASS: Holothuroidea   **PHYLUM:** Echinodermata   **KINGDOM:** Animalia

Sea cucumbers make up a diverse class of echinoderms, numbering about 1,700 species. They resemble sea urchins that have been stretched vertically and laid on their side, with the mouth now at the front end. Larger species such as the snake sea cucumber (*Synapta maculata*) can grow to more than 8 ft (2.5 m) long. Typical sea cucumbers look more like pickles, with rough skin dotted with protuberances. They have tube feet on the underside that can walk and grip. Branched tentacles surround their mouth and help them filter-feed or collect and swallow sediment. Sea cucumbers vary greatly in color, and the colors often contrast with the surroundings, presumably to warn predators. Some species have toxins in their bodies, so they do not need camouflage.

Some species of sea cucumbers, if disturbed, can eject some of their internal organs and grow a new set. Sea pigs have large, leglike podia (tube feet) and move like herds of cattle across the ocean bottom as they feed on muddy sediment. Sea pigs and deep-water sea cucumbers have fins on their bodies, which enable them to swim, and they descend to the ocean bottom only to eat.

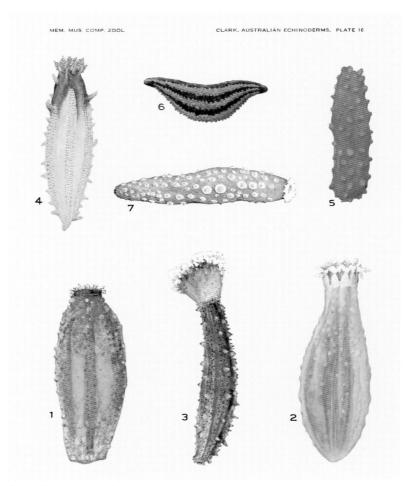

Australian sea cucumbers
The tentacles are visible around the mouths of some of these Australian sea cucumbers, including four species from the genus *Colochirus* (top left, and bottom row).

Pineapple sea cucumber
Thelenota ananas

Yellow sea cucumber
Colochirus robustus

Leopard sea cucumber
Bohadschia argus

sea lilies and feather stars

CLASS: Crinoidea   **PHYLUM:** Echinodermata   **KINGDOM:** Animalia

Most sea lilies are found in deeper waters than starfish and sea urchins. They attach to the sea floor using a stalk. Their arms, covered with thousands of flexible projections called tube feet, trap food particles and move them toward the mouth in the center. Like other echinoderms, sea lilies have a skeleton made of chalky units called ossicles and exhibit five-fold body symmetry. They have a water-vascular system—a hydraulic system of water-filled tubes—that operates the tube feet.

Feather stars lack the stalk of sea lilies but feed in the same way. These often colorful, night-feeding animals perch on surfaces such as corals to catch food. They can swim slowly by beating their arms. Together, there are about 600 species of sea lilies and feather stars.

Feather star
This print depicts a ventral (underside) view of the 37-armed feather star *Comatella stelligera*. It was illustrated during the Dutch Siboga expedition to Indonesia in 1899–1900.

brittle stars

CLASS: Ophiuroidea   **PHYLUM:** Echinodermata   **KINGDOM:** Animalia

There are 1,600 species of brittle stars. They typically have five arms and resemble starfish, but their bodies are structured differently. They use their suckerless podia (tube feet) for feeding, probing, and sensing rather than for walking. Instead, they "walk" along the sea floor with sweeping motions of their arms, each one strengthened by ossicles. Brittle stars filter small food particles from the water, often at night, by raising their arms and using their podia, spines, and mucus secretions as traps. Some species burrow in mud for food, and others can catch larger prey such as small fish and shrimp. Although most brittle stars are only an inch or so in diameter, a subgroup called basket stars can grow up to 3.3 ft (1 m) across.

Brittle star
This illustration shows the central part of the body of a *Macrophiothrix capillaris*, without its five long arms. Its free-swimming larvae sink to the sea floor, where they grow into adults such as this. They can be found at depths of up to 2,460 ft (750 m).

starfish

CLASS: Asteroidea **PHYLUM:** Echinodermata **KINGDOM:** Animalia

Starfish, or sea stars, are predators that live on the ocean floor. The largest species grow up to 3.3 ft (1 m) across. There are about 1,800 species; most have five arms but some have more. The mouth of a starfish is on the underside and can often stretch to swallow relatively large but slow-moving prey. Some starfish can extend their stomachs over prey, digesting it externally. Typical starfish have flexible arms with thousands of suckered tube feet that walk, grip, and pull prey apart. Other species have stiffer arms and burrow in sediment to seek buried prey or feed on smaller particles. The tube feet of burrowing species lack suckers and are used for digging. Most starfish release eggs and sperm into the sea, which fertilize to form swimming larvae.

North American starfish
This selection of North American starfish includes ochre starfish (*Pisaster ochraceus*, top left), which comes in orange and purple forms; common starfish (*Asterias rubens*, top right); and chocolate chip starfish (*Nidorellia armata*, bottom left).

acorn worms

CLASS: Enteropneusta **PHYLUM:** Hemichordata **KINGDOM:** Animalia

Acorn worms are slow-moving marine animals with a distinctive bulbous "nose" (proboscis) that resembles an acorn in its cup. They use the cilia on their bodies and mucus secretions to filter food particles from water or mud. A majority of the 110 species dig burrows in the soft sediment where they live, and some deep-sea species crawl on the sea floor. In most species, fertilized eggs hatch into swimming larvae. Adult acorn worms range from an inch or so to more than $6\frac{1}{2}$ ft (2 m) in length. The phylum Hemichordata contains another class, the pterobranchs, which are small tube-building worms. "Hemichordata" indicates a possible evolutionary connection with chordates (see pp.390–99), although echinoderms are now thought to be the closest living relatives of hemichordates.

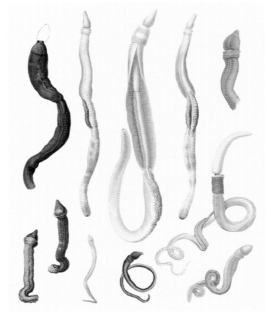

Balanoglossus
These illustrations show species from the genus *Balanoglossus*. They are burrowing marine acorn worms and feed by wafting their cilia to direct a current of water, carrying organic particles, into their mouth.

lancelets

SUBPHYLUM: Cephalochordata   **PHYLUM:** Chordata   **KINGDOM:** Animalia

There are only about 30 species of lancelets—small, fishlike creatures that live half-buried in sandy sediments on shallow sea floors. Lancelets are invertebrates but they belong to the same phylum, Chordata, as vertebrates. The body of a chordate is strengthened by a stiffening rod called a notochord, at least in the larval or embryonic stages. In vertebrates, the notochord is present in embryos but it develops into the backbone in adults. In lancelets, the notochord is a stiff supporting rod running through the back. Although lancelets can swim, mostly they live a benthic (bottom-dwelling) lifestyle, sieving plankton from water. If disturbed, lancelets swim away in a fishlike manner by contracting muscles along their sides. They do not have a true skeleton and exchange gases entirely through their skin.

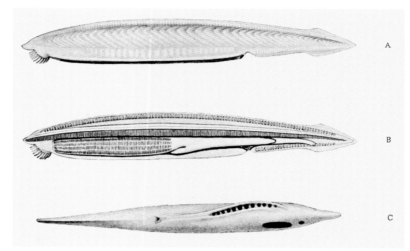

Lancelet
Three sketches of a lancelet show the small filaments called cirri surrounding its mouth and the narrow caudal (tail) fin (top); the straight intestine running from the mouth to the anus in a dissected adult (center); and a larva (bottom).

hagfish

CLASS: Myxini   **SUBPHYLUM:** Vertebrata   **PHYLUM:** Chordata   **KINGDOM:** Animalia

Hagfish are not related to eels, although they look similar. They are living representatives of the ancient jawless fish that swam in the world's oceans before fish with jaws evolved about 400 million years ago. There are about 80 species of hagfish and all live on the ocean floor. They feed on invertebrates, such as polychaete worms, as well as the carcasses of large animals such as whales, which they eat from the inside out. Hagfish have rudimentary eyes and no true fins. Their mouth contains a tongue equipped with horny teeth and is surrounded by slender sensory organs called barbels. Lines of pores run along the body; these discharge thick slime, which is used for defense. Their skull is made of cartilage, and the body is stiffened by a flexible notochord rather than a vertebral column.

Hagfish
Barbels surround the mouth of this Atlantic hagfish (*Myxine glutinosa*), which has many mucus-secreting pores along the length of its body.

sea squirts and relatives

SUBPHYLUM: Urochordata **PHYLUM:** Chordata **KINGDOM:** Animalia

About 3,050 species of tunicates are known. They are divided into three classes: Ascidiacea (ascidians, including the immobile sea squirts); Thaliacea (free-floating species, including salps); and Appendicularia (planktonic larvaceans). Sea squirts live mostly in shallow water. A notochord is present in the larvae of sea squirts and salps, but it persists in adult larvaceans. The notochord is also present in vertebrate embryos; this similarity with them puts tunicates in the same phylum as vertebrates, Chordata. The adults attach themselves to surfaces and filter food from water like sponges. Unlike the sponges, their body is more complex and is covered by a tough protective layer called the tunic. Sea squirts sieve food using two tubelike siphons; one draws in sea water and the other expels the waste.

Some small individual sea squirts, or zooids, merge together to form colonies up to several feet in diameter. Unlike ascidians that are fixed to one place, salps are free-floating and move using a system of jet propulsion—as the water is drawn in and ejected through the body, the resulting thrust propels the animal forward. They can swim alone or in the form of long, stringy colonies up to 10 ft (3 m) long.

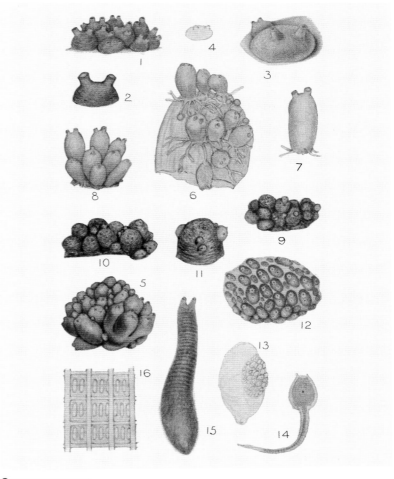

Orange sea grapes
This 20th-century illustration shows British tunicates, including orange sea grapes (*Stolonica socialis*, fig. 6), a colonial species of sea squirts. Fig. 14 (bottom right) shows a free-swimming, tadpolelike tunicate larva.

Blue sea squirts
Rhopalaea sp.

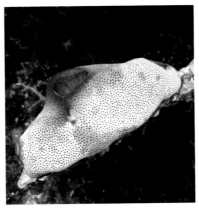

Green barrel sea squirt
Didemnum molle

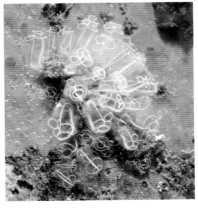

Light-bulb sea squirt
Clavelina lepadiformis

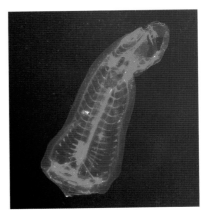

Free-floating salp
Salpida

lampreys

CLASS: Petromyzonti **SUBPHYLUM:** Vertebrata **PHYLUM:** Chordata **KINGDOM:** Animalia

Biologists have described 38 species of lampreys that, along with hagfish, are the only living fish that lack jaws. They have a rudimentary vertebral column. All species spawn in fresh water, but about one-quarter of them move to the oceans as adults. The larvae, called ammocoetes, live in river sediments, feeding by filtering the water for food particles, until they transform into adults that look like eels. In the sea lamprey and some other species, young adults head downstream to the ocean, where they live as parasites of fish. They attach themselves to their hosts with the rasping teeth on their suckerlike mouths. Later, they swim back to fresh waters to breed. Some adults do not feed at all; instead, they live off the reserves acquired from when they were larvae until they breed and die.

Vampire fish
This sea lamprey (*Petromyzon marinus*) is a parasitic species that grips its prey with a suction-cup mouth, rasping the victim's flesh with sharp, tough teeth.

chimaeras

CLASS: Holocephali **SUBPHYLUM:** Vertebrata **PHYLUM:** Chordata **KINGDOM:** Animalia

Comprising about 53 species, chimaeras live close to the ocean floor, mainly in deep waters. Along with sharks and rays, they have a cartilaginous skeleton. Chimaeras have a large head, a dorsal fin spine, and platelike teeth. The largest species can be more than 5 ft (1.5 m) in length. Chimaeras swim slowly and feed mainly on invertebrates. As in sharks, fertilization is internal, and the females lay large eggs.

There are three families. Short-nosed chimaeras, or ratfish, have a long, tapering tail. Plough-nosed chimaeras, or elephant-fish, have remarkable protuberances on their snouts that may be used to dig up prey. Long-nosed chimaeras, or spookfish, have a fairly long snout with numerous sensory nerve endings, which they use to find prey.

Rabbit fish
This illustration shows a species of ratfish, *Chimaera monstrosa*. The species is also called a rabbit fish because of its characteristic large eyes and short, overhanging snout.

sharks and rays

CLASS: Elasmobranchii　**SUBPHYLUM:** Vertebrata　**PHYLUM:** Chordata　**KINGDOM:** Animalia

Unlike bony fish (see pp.394–95) that have bones, sharks and rays have a skeleton made of cartilage. There are almost 500 species of sharks and more than 600 species of rays. Some shark species have replaceable, razor-sharp teeth. Sharks have highly developed senses, including the ability to detect tiny electric currents given off by prey. They reproduce using internal fertilization, unlike bony fish. Females either give birth to live young or lay large egg capsules. Some sharks are fast, streamlined in shape, and apex predators (at the top of the food chain). However, a few large species, such as the whale shark (*Rhincodon typus*) and basking shark (*Cetorhinus maximus*), are slow-moving and feed on plankton. Other species feed on mollusks and crustaceans on the ocean floor.

Rays are flat-bodied fish that usually live close to the seafloor, often half-concealed in sand or mud, and use their pectoral fins (on the sides) to propel themselves through water. They have gill slits underneath the body, as well as spiracles (holes) behind their eyes that take in oxygen from the water. Sawfish, guitarfish, and banjofish are longer bodied fish that look like flattened sharks. The long "saw" of the sawfish contains electroreceptors used to detect prey.

Thorny skate
The thorny skate (*Amblyraja radiata*), a species of ray, has a mixture of large and small defensive thorns on its upperside. The claspers of this male can be seen as fingerlike appendages on either side of the tail.

Bull shark
Carcharhinus leucas

Largetooth sawfish
Pristis pristis

Reticulate whipray
Himantura uarnak

lobe-finned fishes

CLASS: Coelacanthi, Dipnoi **SUBPHYLUM:** Vertebrata **PHYLUM:** Chordata **KINGDOM:** Animalia

Lobe-finned fishes and ray-finned fishes are both classes of bony fish, but the two groups are very different. The former have fleshy pectoral and pelvic fins that contain small, sturdy bones equivalent to those in human arms and legs. The fishes from the latter group have flimsy rays of bones supporting their fins. All land vertebrates are believed to have evolved from sarcopterygian ancestors.

Just eight sarcopterygian species are alive today: two marine coelacanths and six freshwater lungfishes. Coelacanths were thought to be extinct until a living specimen was identified in a fisherman's catch off South Africa in 1938. They grow up to 6.5 ft (2 m) long, and their skin is protected by thick, bony scales. Coelacanths are slow-swimming, deep-sea fishes that favor caves. Their diet mainly consists of cephalopods and small fish. Females produce large eggs, which they hatch in their bodies, giving birth to live young. Lungfishes live in swamps and slow-moving rivers. Most have two lungs—an Australian species has one—and can gulp air at the surface. In the dry season, they go into a resting, inactive state called estivation until the rainy season.

South American lungfish
South American lungfishes (*Lepidosiren paradoxa*) have threadlike fins and are at home on mud when water levels fall. When in water, they have to come to the surface to breathe.

West Indian Ocean coelacanth
Latimeria chalumnae

West African lungfish
Protopterus annectens

South American lungfish
Lepidosiren paradoxa

ray-finned fishes

CLASS: Actinopterygii **SUBPHYLUM:** Vertebrata **PHYLUM:** Chordata **KINGDOM:** Animalia

Ray-finned fishes comprise the largest class of vertebrate animals, with about 30,000 marine and freshwater species. In contrast to lobe-finned fishes, their fins are supported by rays of bony spines.

Actinopterygians vary in size, anatomy, and lifestyle. Oarfishes grow up to 36 ft (11 m) in length while *Paedocypris* is only 0.3 in (8 mm) in length, and ocean sunfish can weigh up to 5,000 lb (2.3 tonnes). Eels have a long, snakelike body and swim with a sinusoidal (wavy) motion, and some eels have more than 100 vertebrae in their flexible spine. Long, eel-like dragonfishes hunt in the dark ocean depths by luring prey using photophores (luminous spots) that run down their flanks. Many species, including herrings, have a streamlined, torpedo-shaped body that enables fast, darting movements. Sailfishes are even faster, reaching a speed of 70 mph (110 kph) as they leap out of the water. Some salmon have fascinating life cycles: they spend most of their life in the ocean but move to fresh water to spawn. The journey to fresh water can be several thousand miles long, and the fish have to swim against fast currents, which they negotiate by jumping up waterfalls and rapids as they progress upstream.

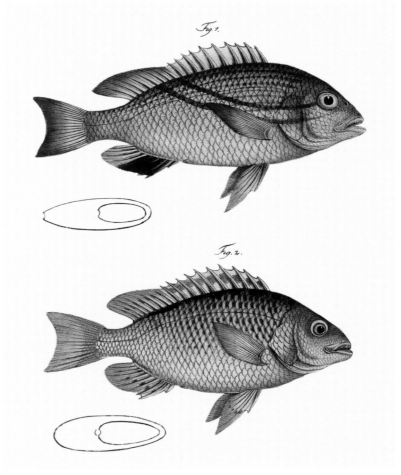

Splendid perch
This illustration shows the Japanese splendid perch, *Callanthias japonicus*. The splendid perch is one of many groups within Perciformes, the largest order of vertebrates.

Leopold's angelfish
Pterophyllum leopoldi

Japanese anchovies
Engraulis japonicus

Yellow seahorse
Hippocampus kuda

Ocean sunfish
Mola mola

amphibians

CLASS: Amphibia   **SUBPHYLUM:** Vertebrata   **PHYLUM:** Chordata   **KINGDOM:** Animalia

This group of about 8,000 species includes salamanders and newts (order Caudata), frogs and toads (anurans, order Anura), and caecilians (order Gymnophiona). Amphibians have moist, water-permeable skin. Their typical life cycle involves an aquatic, gill-breathing juvenile stage, followed by terrestrial adulthood. Some species are fully aquatic, while others are entirely terrestrial.

Frogs and toads make up the largest group in this class. Those with rougher skin, a heavy body, and shorter legs are usually called toads, but the distinction is not exact. Adult anurans are carnivores. Most anurans have bodies built for jumping, with long hindlegs and a short, inflexible spine. Frogs and toads have large heads and mouths that, combined with good binocular vision, allow them to catch and devour relatively large prey. Many species produce toxins in their skin to deter predators, advertising it with bright colors. Salamanders and newts are predators, feeding mainly on invertebrates. Caecilians are limbless amphibians that resemble snakes or worms. Their eyesight is rudimentary, but they can detect prey—such as earthworms—using retractable tentacles that transmit smells to their nose.

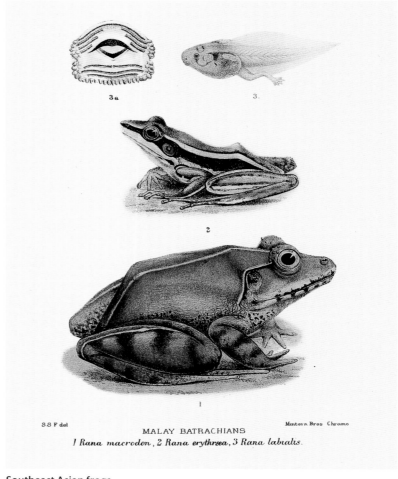

Southeast Asian frogs
From top to bottom, this illustration depicts the tadpole of a white-lipped frog (*Chalcorana labialis*), the common green frog (*Hylarana erythraea*), and Blyth's river frog (*Limnonectes macrodon*). All three species are found in Southeast Asia.

Harlequin poison frog
Oophaga histrionica

Northern crested newt, larva
Triturus cristatus

Fire salamander
Salamandra salamandra

Koh Tao Island caecilian
Ichthyophis kohtaoensis

reptiles

CLASS: Reptilia **SUBPHYLUM:** Vertebrata **PHYLUM:** Chordata **KINGDOM:** Animalia

There are about 10,000 species of reptiles. All are cold-blooded, so they regulate their body temperature by moving between warmer and cooler places. Lizards (order Squamata) are the most diverse group, and include geckos, skinks, and iguanas. In contrast to amphibians, lizards have waterproof skin typically covered with scales. Most lay eggs on land, while others give birth to live young. The eggs are surrounded by waterproof membranes that prevent them from drying out. The majority of lizards are predators. Snakes (also part of the order Squamata) can make the sides of their jaws move apart to swallow large prey whole.

Most species of turtles (order Testudines) live in fresh water, but there are many land-living types (tortoises), as well as seven species of marine turtles. Both the upper shell (carapace) and lower shell (plastron) of turtles and tortoises are made of interlocking flat bones. All turtles lay their eggs on land, which, in the case of marine turtles, can mean long migrations. Crocodilians (order Crocodilia) are semiaquatic predators and scavengers that spend most of their time in water. Strictly, the class Reptilia includes the feathered descendants of the long-extinct dinosaurs, the birds (see p.398).

Arboreal lizard
A South American species, the green iguana (*Iguana iguana*) is an arboreal, or tree-dwelling, member of the order Squamata.

Nile crocodile
Crocodylus niloticus

Leopard gecko
Eublepharis macularius

Rein snake
Gonyosoma frenatum

Hawksbill turtle
Eretmochelys imbricata

birds

CLASS: Aves   **SUBPHYLUM:** Vertebrata   **PHYLUM:** Chordata   **KINGDOM:** Animalia

Birds form a diverse group of warm-blooded, feathered vertebrates, and more than 11,000 species are known. The perching birds (passerines), which include many songbirds, form the largest group. They are characterized by a three-toes-forward, one-toe-back arrangement of the feet, which is ideal for grasping branches. In contrast, woodpeckers usually have a zygodactyl toe arrangement, with two toes pointing forward and two backward. This, together with stiff tail feathers that are used as a prop, allows woodpeckers to climb trees. Diurnal (active during the day) birds of prey have excellent eyesight and use their feet to catch prey, which they then dismember with their hooked bills. One bird of prey, the peregrine falcon (*Falco peregrinus*), can travel at speeds of 186 mph (298 km/h), making it the fastest living animal. While most birds can fly, there are several flightless groups, including penguins, which are marine except when breeding and are adapted to fast underwater pursuits of prey. Some birds have a broad dietary range, while others have a very specialized diet. Most hummingbirds, for example, probe for nectar in flowers. Their long, curved bills match the size and shape of their favored flowers.

Endangered passerines
Regent honeyeaters (*Anthochaera phrygia*) are critically endangered passerines of eastern Australian forests. Bush fires in 2020 destroyed much of their already limited habitat.

Red-tailed hawk
Buteo jamaicensis

Emperor penguin
Aptenodytes forsteri

Northern cardinal (male)
Cardinalis cardinalis

Pileated woodpecker
Dryocopus pileatus

mammals

CLASS: Mammalia **SUBPHYLUM:** Vertebrata **PHYLUM:** Chordata **KINGDOM:** Animalia

Mammals make up a group of vertebrates in which mothers suckle their young with milk. The majority of mammals are covered with fur or hair. Most of the 6,500 species are terrestrial, but some are aquatic and others are capable of flight. The primates, a group of about 450, include our own species and other great apes (bonobos, gorillas, chimpanzees, and orangutans). Most primates are forest-dwelling and confined to tropical regions. Some have a grasping tail, which acts as a fifth limb as they move deftly through the tree canopy. Some great apes are accomplished toolmakers. Bats are the only mammals capable of true flight—some use vision and smell to find food, while others orient themselves by echolocation (using sound) as they hunt insects by night. Cetaceans (whales and dolphins) also include many species that use echolocation; dolphins have a waxy lobe in the head that focuses sounds into beams and works as part of their sonar system. Hoofed animals, including horses, cattle, giraffes, and deer, are herbivorous grazers and browsers. In contrast, many mammals are carnivores, including pack-hunting wolves and the mostly lone hunters of the cat family, Felidae.

Rainforest monkeys
Native to a small area of Colombian rainforest, white-footed tamarins (*Saguinus leucopus*) are small monkeys, with relatively long tails. Like other tamarins and marmosets, these monkeys typically give birth to twins.

African bush elephant
Loxodonta africana

Giraffe
Giraffa camelopardalis

Harvest mouse
Micromys minutus

Tiger
Panthera tigris

glossary

ABDOMEN Rear part of an animal's body behind the chest cavity or thorax, and typically containing organs of the digestive and reproductive systems.

ACROSOME Fluid-filled vesicle (sac) in the head of a sperm that contains digestive enzymes, which break through the coat of an egg during fertilization.

ACTIN FILAMENT Threadlike protein structure in a cell, and part of the supporting cytoskeleton. Actin filaments are among the protein filaments found in muscles. *See also* cytoskeleton.

ACTIVE TRANSPORT Process by which organisms use energy to move substances across a cell membrane, from low to high concentration.

ADIPOSE TISSUE Animal tissue that contains accumulations of fat.

AEROBIC Describing a process, such as respiration, that uses oxygen, or an environment that contains high oxygen levels.

ALGA (pl. **ALGAE**) Organism, such as a diatom or a seaweed, that photosynthesizes, but is not a plant.

ALTERNATION OF GENERATIONS Life cycle in plants that alternates between a stage that produces sex cells (sperm or pollen nucleus and eggs) and a stage that produces spores. Some plants, such as mosses and ferns, release their spores, but in seed plants, the spores are retained and develop into the stage that generates sex cells within a cone or flower.

ALVEOLUS (pl. **ALVEOLI**) (i) Microscopic air-filled sac in the lungs of mammals and site of gaseous exchange between air and blood; (ii) Fluid-filled vesicle that makes up the stiff layer, or pellicle, beneath the cell membrane of some microbes, such as dinoflagellates.

ANAEROBIC Describing a process, such as respiration, that does not use oxygen, or an environment that contains low oxygen levels.

ANNULUS (pl. **ANNULI**) Ring or band—such as might occur in the body of a wormlike animal, or in the stem of a mushroom.

ANTENNA (pl. **ANTENNAE**) One of a pair of typically elongated sensory organs on the head of an invertebrate animal.

ANTHER Male sexual part of a flower that produces pollen.

ANTIBIOTIC Substance produced by a microorganism, such as a bacterium or fungus, that suppresses or kills another competing microbe. Many antibiotics, such as penicillin, are used in medicine for treating bacterial infections.

ANTIBODY Protein secreted by cells of the immune system, called lymphocytes, used for disabling or destroying harmful foreign particles, such as pathogens.

ARACHNID Member of a group of arthropods usually with four pairs of jointed legs and lacking antennae, such as spiders, scorpions, and mites.

ARBUSCULE Swelling produced from a fungal filament that develops inside plant root cells in some kinds of plant-root associations called mycorrhizae.

ARCHAEAN Member of a group of single-celled organisms (Archaea), superficially resembling bacteria in lacking a nucleus, but differing from them in their chemical makeup. Many members of the Archaea are adapted to thrive in extreme conditions, such as high temperature or salinity.

ARTERIOLE Tiny blood vessel that delivers blood from arteries to capillaries.

ASCUS (pl. **ASCI**) Saclike spore-producing structure in certain kinds of fungi, such as *Aspergilllus*.

ASEXUAL Reproduction that does not mix genes from different cells, and so produces offspring that are genetically identical clones.

AUTOTROPH Organism, such as a plant, that makes its own food (such as glucose) from simple inorganic materials—including carbon dioxide and water.

AXOPOD Stiffened spikelike pseudopod, or extension of a cell, used by single-celled organisms called radiolarians to catch prey. *See also* pseudopod.

BACILLUS (pl. **BACILLI**) Any bacterium that is rod shaped.

BACTERIUM (pl. **BACTERIA**) Member of a group of single-celled organisms (Bacteria) that lack a nucleus and other internal compartments, such as mitochondria and chloroplasts. Bacteria are chemically different from superficially similar organisms called archaeans.

BASIDIUM (pl. **BASIDIA**) Clublike spore-producing structure in certain kinds of fungi, such as common field mushrooms.

BENTHIC Describing organisms that live at the bottom of aquatic habitats, such as the ocean, ponds, and rivers.

BIOFILM Thin layer of microbes living attached to a surface, such as a rock or inner lining of an animal's gut.

BIOLUMINESCENCE Production of light by an organism. The light is generated by a chemical reaction catalyzed (boosted) by an enzyme.

BIOREMEDIATION Human use of microorganisms that can metabolize substances that are harmful to other organisms, in order to reduce levels of pollutants.

BRYOPHYTE A simple plant, including liverworts and mosses, that lacks the supporting and vascular tissues (xylem and phloem) found in ferns and seed plants. *See also* phloem, vascular, xylem.

BUDDING (i) Formation of a growing shoot in a plant or alga; (ii) Formation of a swelling on an animal, such as a hydra, that will detach from the parent body and develop into an independent individual; (iii) Production of new yeast cells by division from parental ones.

CALCAREOUS Describing a chalky mineral, made of calcium carbonate.

CAMBRIAN EXPLOSION Prehistoric event about 540 million years ago when most phyla, or major groups, of animals alive today originated by diversifying from their ancestors.

CAMBIUM Ring of tissue in plant stems containing dividing cells, which leads to thickening of the stem during growth.

CAPILLARY A microscopic type of blood vessel, with a thin wall only one cell thick, and the site of exchange of material between blood and surrounding tissues.

CARAPACE The hardened, shell-like covering on the back of an animal, such as many crustaceans.

CARBOHYDRATE Class of organic substance containing carbon, hydrogen, and oxygen. Carbohydrates include sugar and starch, which release energy, and cellulose, which supports body structures.

CARNIVORE Organism that eats live animals.

CARNIVOROUS PLANT Plant adapted to obtain at least some of it nutrients, especially nitrogen, from animals, by trapping and digesting their bodies.

CAROTENOID Member of a group of yellow, orange, or red pigments. Carotenoids are used for multiple purposes by different organisms, including screening light, and helping to absorb light energy for photosynthesis.

CARPEL Female sexual part of a flower that produces eggs. Also called a pistil.

CARTILAGE Tough, rubbery kind of connective tissue, and a component of vertebrate skeletons.

CELLULOSE Tough, fibrous carbohydrate that makes up cell walls of plants and some other kinds of organisms.

CELL WALL Structural layer around the outside of the cell membrane in some kinds of organisms, such as bacteria, plants, fungi, and most algae.

CERIANTHID Anemone that lives inside a tube made from mucus and accumulated sediment.

CGI (COMPUTER GENERATED IMAGERY) Images produced using computer software.

CHAETA (pl. CHAETAE) Hairlike structure that develops from the surface of an invertebrate, such as an insect or a worm.

CHELICERA (pl. CHELICERAE) Pincerlike or fanglike mouthparts of arachnids (mites, spiders, and relatives), used for collecting food.

CHEMOSYNTHESIS Process by which some kinds of bacteria and archaean can use energy in minerals to convert simple materials—including carbon dioxide, water, and minerals—into food (organic materials), such as sugar.

CHITIN Tough, fibrous, nitrogen-containing carbohydrate that makes up cell walls of fungi and exoskeletons of arthropods.

CHLOROPHYLL Green pigment used in plants and other photosynthetic organisms to absorb light energy for making food.

CHLOROPLAST Compartment, or organelle, inside a cell that contains chlorophyll and is used for photosynthesis.

CHROMOSOME Threadlike structure that carries most of a cell's DNA. Chromosomes appear during cell division in cells of eukaryotes (such as animals and plants), but are absent from cells of prokaryotes (bacteria and archaeans).

CILIUM (pl. CILIA) Short, beating, hairlike structure attached to a cell and used for locomotion or to generate currents. Many cilia typically cover all or part of the cell's surface.

CLEISTOTHECIUM (pl. CLEISTOTHECIA) Spore capsule or certain kinds of sac fungi, such as *Aspergillus*.

CLONE Genetically identical individual.

CLYPEUS Region on the front of an insect's head above the labrum (upper "lip").

CNIDA (pl. CNIDAE) Specialized cell found in cnidarians (jellyfish, corals, and relatives) that contains an explosive vesicle carrying a thread that may deliver a sting in defense or to disable prey.

CNIDOCIL Hairlike trigger used to discharge a cnida.

COCCOBACILLUS (pl. COCCOBACILLI) Short, rod-shaped bacterium intermediate in shape between a coccus and a bacillus.

COCCUS (pl. COCCI) Any bacterium that has a spherical shape.

COLLAGEN Type of fibrous protein that is used for supporting and strengthening tissues of animals.

COMPETITION Ecological relationship between two organisms arising from shared use of a common resource, such as food, to the detriment of both.

COMPOUND EYE Eye made up of multiple tiny facets, or ommatidia, each with its own lens, and present in many kinds of arthropods, including insects and crustaceans.

CONIDIUM (pl. CONIDIA) Type of asexual spore produced by fungi.

CONIDIOPHORE Reproductive stalk that produces conidia in fungi.

CONJUGATION Sexual process in some organisms, such as bacteria and fungi, where cells fuse together to transfer or exchange DNA.

CONNECTIVE TISSUE Animal tissue that develops to fill the spaces between other kinds of tissue. Examples include cartilage, bone, adipose tissue, and dense fibrous tissue.

CORONA RADIATA Outer layer of follicle cells around the unfertilized egg of some kinds of animals, including those of mammals. *See also* follicle.

COTYLEDON A leaflike structure that acts as a food store or unfurls shortly after germination to fuel a seed's growth.

CRISTA (pl. CRISTAE) Inner membrane fold of a mitochondrion, used for increasing the surface area of membrane-bound chemical reactions involved in aerobic respiration.

CRUSTACEAN Member of a group of arthropods—typically with multiple pairs of jointed legs, two pairs of antennae, and sometimes a hard carapace—such as water fleas, shrimp, and crabs.

CRYPTOBIOSIS A dormant state achieved by some kinds of organisms, such as tardigrades, in which body functions slow to help survive extreme conditions.

CUTICLE (i) Waxy protective layer on the surface epidermis of plant leaves and stems; (ii) Outer protective layer on the surface epidermis of some invertebrates, and hardened into a supporting exoskeleton in arthropods, such as insects.

CYANOBACTERIUM Type of bacterium that photosynthesizes to make food.

CYTOPLASM Thin, jellylike material inside living cells and between organelles, such as the nucleus, mitochondria, and chloroplasts.

CYTOSKELETON Supporting framework of microtubules, actin filaments, and intermediate filaments, made from protein inside a cell.

CYTOSTOME Mouthlike cavity in the side of some kinds of single-celled organisms, such as *Paramecium* and *Euglena*.

DECOMPOSER Organism that feeds on dead and waste organic matter and, in doing so, helps to break it down into simpler materials, thereby helping to recycle nutrients; this process is called decomposition. Most fungi and many bacteria are decomposers.

DENITRIFICATION Process performed by certain kinds of soil bacteria that converts nitrate into atmospheric gaseous nitrogen, and therefore reduces fertility of soils.

DERMIS Thickest tissue layer of vertebrate skin, beneath the thinner surface epidermis.

DESMID Kind of alga, mostly single-celled, but with their symmetrical cells divided into halves, or "demi-cells," that are mirror images of one another.

DETRITIVORE Animal that feeds on detritus: particles of dead and waste organic matter.

DIATOM Kind of single-celled or colonial alga, with cell wall hardened with silica.

DICOTYLEDON (or DICOT) Flowering plant having two seed leaves (cotyledons) in its seeds.

DIFFUSION Movement of particles from a high to low concentration, resulting in their gradual dispersion.

DINOFLAGELLATE Kind of single-celled alga with beating flagella and supported by a stiff surface layer, or pellicle, that is often hardened to form armorlike plates. Some dinoflagellates live in tissues of corals and other animals, where they help supplement the host's food by their photosynthesis.

DIPLOCOCCUS (pl. DIPLOCOCCI) Bacterium that occurs in pairs of cocci (spherical-shaped cells).

DNA (DEOXYRIBONUCLEIC ACID) Genetic material that encodes inherited characteristics in all cellular organisms and some viruses.

DOMAIN One of the three most basic divisions in the tree of life: Bacteria, Archaea (both with simple cells lacking complex internal structures), and Eukaryota (with complex cells that contain structures including nucleus, mitochondria, and often chloroplasts).

ECTODERM Outer layer of cells that forms in an early animal embryo, and which in cnidarians (anemones, jellyfish, and relatives) is retained into adulthood. In other animals, the ectoderm develops into multiple tissues, such as epithelium and nervous tissue.

ECTOMYCORRHIZA—see mycorrhiza.

ECTOPLASM Outer, stiffer layer of cytoplasm used by amoebas for producing their creeping pseudopods.

EGG (i) (or OVUM) Female sex cell, typically with a store of nutrients (yolk) and lacking any means of movement. An ovum develops into an embryo after it is fertilized; (ii) Shelled egg of an animal containing a developing embryo.

ELYTRON (pl. ELYTRA) One of two modified front wings of a beetle, hardened into protective shields that completely enclose the membranous hindwings when not in flight.

EMBRYO Very young plant or animal that develops from a fertilized egg.

ENDOMYCORRHIZA—see mycorrhiza.

ENDOSKELETON Skeleton that develops within the body of an animal and grows with it, such as the bony skeleton of a vertebrate.

ENDOTHELIUM Tissue that is a single layer of cells forming the wall of a blood capillary, or making up the lining of a bigger blood vessel.

ENTEROCYTE Cell making up the lining of the intestine of humans.

ENZYME Protein that works by catalyzing (boosting) chemical reactions in living organisms. Each reaction needs a different kind of enzyme.

EPIDERMIS Tissue that makes up the surface layer of cells of a plant or animal.

EPIPODIUM Opening in the shell of some kinds of amoebas, through which they can extend their pseudopods.

EPITHELIUM Animal tissue that lines a body surface or internal organs.

ESOPHAGUS Food pipe leading to stomach in the digestive system of a vertebrate.

EUDICOTYLEDON (or EUDICOT) A "true" dicotyledon: member of a group of flowering plants that contains the vast majority of dicotyledons considered to be descended from a common ancestor, but excluding certain "basal dicots," such as magnolias and water lilies, now known to be not directly related to them.

EUGLENID Kind of single-celled alga, typically with a beating flagellum. Some kinds of euglenids can switch between making food by photosynthesis and consuming it from their surroundings.

EUKARYOTE Organism made up of complex cells that have internal organelles, including nucleus, mitochondria, and often chloroplasts. Animals, plants, fungi, protozoans, and chromistans (including most algae) are all eukaryotes.

EXTREMOPHILE Organism adapted to living in extreme conditions, such as high temperature or acidity, and cannot thrive elsewhere.

FACULTATIVE ANAEROBE Organism that can respire aerobically or anaerobically. See also aerobic, anaerobic.

FALSE COLOR Artificial color added to an electron micrograph, after it has been produced, to highlight features.

FERMENTATION Alternative term for anaerobic respiration.

FERTILIZATION Fusion of sperm with an egg (ovum) to produce a fertilized egg (called a zygote) that can develop into an embryo.

FILTER FEEDER Animal that collects particles of food suspended in water by straining them through a mesh or similar structure.

FLAGELLUM (pl. FLAGELLA) Long, hairlike structure attached to a cell used for locomotion or to collect food. Flagella of bacteria rotate like a propeller; flagella of eukaryotes beat like a whip.

FOLLICLE Cluster of supporting cells around an unfertilized egg in an animal's ovary.

FORAMINIFERAN Kind of amoebalike single-celled organism that is encased in a microscopic shell.

FOSSIL Mineralized remains or impression of an organism that lived in the prehistoric past.

FRUIT Ripe female part of a flowering plant that contains seeds.

FRUITING BODY Mushroomlike, spore-producing structure of a fungus.

FRUSTULE Cell wall of a diatom alga.

FUNGUS (pl. FUNGI) Member of a group of organisms that grow as a network of filaments called hyphae and reproduce by spores. Fungi include molds, mushrooms and toadstools, brackets, and single-celled yeasts.

GANGLION (pl. GANGLIA) Mass of nervous tissue that helps to control the body of an animal.

GENETIC Describing a characteristic that is determined by DNA, and therefore inherited from one generation to the next.

GERMINATION Development of a seed into an infant plant, with functional roots and leaves.

GILL Structure, typically feathery in shape, used to absorb oxygen in aquatic animals.

GLUCOSE The commonest kind of sugar, or soluble carbohydrate, used for respiration inside living cells.

GOLGI BODY Stack of vesicles inside a cell, used for sorting and secreting substances.

GRAM STAIN Technique, devised by Hans Christian Gram, to distinguish two groups of bacteria. Gram-positive bacteria have thick walls, and stain purple; Gram-negative bacteria have thinner walls overlaid by a second cell membrane, and stain pink.

HABITAT Place in which an organism or species naturally lives.

HALTERE One of a pair of knoblike structures derived from the hindwings of a true fly (order: Diptera) and used to help with flight control.

HAUSTORIUM (pl. HAUSTORIA) Rootlike structures in fungi and some parasitic plants, used to help absorb nutrients.

HEME Chemical group, containing iron, that is the oxygen-carrying part of the hemoglobin, found in red blood cells.

HEMOGLOBIN Red blood protein used for binding and carrying oxygen in the bloodstream.

HERBIVORE Organism that eats live plants or algae.

HETEROTROPH Organism that consumes or absorbs food instead of making it by photosynthesis or chemosynthesis.

HOLOTRICHOUS Describing a single-celled organism that is completely covered in cilia.

HOST Organism that carries or contains another organism, such as a parasite.

HYPHA (pl. HYPHAE) Filament of a fungus that digests and absorbs surrounding food.

HYPOPHARYNX Tonguelike structure that is a component of insect mouthparts.

INTERMEDIATE FILAMENT Protein filament in a cell, and part of the supporting cytoskeleton.

INVERTEBRATE Any animal without a backbone.

JEJUNUM First part of the mammalian small intestine.

KERATIN Type of tough, fibrous protein that makes up hair, feathers, claws, horn, and surface skin of vertebrates.

LABIUM Lower "lip" component of mouthparts of an insect.

LABRUM Upper "lip" component of mouthparts of an insect.

LARVA (pl. LARVAE) Juvenile form of an animal that is usually very different in form from the adult and must therefore undergo metamorphosis during its development.

LEGHEMOGLOBIN Form of hemoglobin found in plant root nodules, where it binds to oxygen, preventing it from interfering with nitrogen-fixing bacteria that the plant relies on.

LICHEN Composite organism made up of a fungus and alga, living together in a mutually beneficial partnership.

LIGNIN Tough, fibrous material and the principal chemical component of wood.

LIMESTONE Sedimentary rock formed by compaction of calcareous (chalky) particles, often including shells of foraminiferans or coccolithophores.

LM (LIGHT MICROGRAPH) Image produced by photographing through a light microscope.

LUCIFERASE Enzyme that drives the chemical reaction involving the conversion of luciferin to produce light in bioluminescence.

LUCIFERIN Light-releasing protein found in bioluminescent organisms.

LYSOSOME Vesicle, or sac, inside a cell that contains digestive enzymes.

MANDIBLE (i) One of a pair of structures that make up the mouthparts of arthropods, including insects, and are typically involved in biting or piercing; (ii) Bone that makes up the jaws of a vertebrate.

MASTAX Muscular throat of a rotifer, used to help grind up food.

MAXILLA (pl. MAXILLAE) One of the paired structures that make up the mouthparts of arthropods, and usually used to help process or manipulate food.

MEIOFAUNA Community of tiny animals living between particles in soil, mud, or sand, and ranging in size from 1 mm to 0.05 mm.

MEIOSIS Division of nucleus in a type of cell division whereby the chromosome number is halved, such as in the formation of sex cells (sperm and eggs) in animals, or spores in plants.

MEMBRANELLAR BAND Band of cilia found in some kinds of single-celled organisms, such as *Stentor*.

MESOGLEA Jellylike layer in the body wall of cnidarians (anemones, jellyfish, and relatives).

MESOPHYLL Tissue layers in a plant leaf, between upper and lower epidermis, used for photosynthesis and gaseous exchange.

METABOLISM All the chemical reactions taking place inside a living organism.

METAMORPHOSIS Transformation of body form during development, such as a caterpillar metamorphosing into a butterfly, or a tadpole into a frog.

METHANOGEN Microbe, typically an archaean, that generates methane gas by its metabolism.

MICROAEROPHILE Organism that thrives at low oxygen concentrations.

MICROBIOTA Community of microorganisms living in one place.

MICROGRAPH Photograph taken using a microscope.

MICROHABITAT Habitat of small organism confined to a small area or space.

MICROMETER Unit of measuring length: a millionth of a meter.

MICROSCOPY Technique involving the use of a microscope.

MICROTRICHIUM (pl. MICROTRICHIA) Minute sensory hair on the cuticle of an insect.

MICROTRIX (pl. MICROTRICHES) Microscopic hairlike projection of a tapeworm. A coating of microtriches increases the surface area for the parasite to absorb food.

MICROTUBULE Hollow protein tube in a cell, and part of the supporting cytoskeleton. *See also* cytoskeleton.

MICROVILLUS (pl. MICROVILLI) One of many short, hairlike projections on a cell, used for increasing surface area for absorbing food. Cells lining the intestine have microvilli for this purpose.

MIDPIECE Middle "neck" part of a sperm cell, containing an energy-releasing mitochondrion that powers swimming.

MITOCHONDRION (pl. MITOCHONDRIA) Organelle inside the cell of a eukaryote that is the main site of aerobic respiration. *See also* aerobic, eukaryote, organelle.

MITOSIS Process in which the nucleus of a eukaryotic cell divides to separate replicated chromosomes during cell division. Mitosis produces cells with the same number of chromosomes and identical DNA, and happens during growth of a multicelled body and in sexual reproduction of single-celled organisms.

MONOCOTYLEDON (or MONOCOT) Flowering plant having one seed leaf (cotyledon) in its seeds.

MUCILAGE Slimy substance produced by plants.

MUCUS Slimy substance produced by animals, usually to help with defensive functions.

MUTUALISM Ecological relationship between two organisms of different species, in which both benefit.

MYCELIUM Nutrient-absorbing network of hyphae (filaments) in a fungus.

MYCORRHIZA Association between the roots of a plant and the hyphae of certain kinds of fungi, for mutual benefit in the exchange of nutrients. In ectomycorrhizae, the fungal hyphae grow between root cells; in endomycorrhizae (or arbuscular), the hyphae penetrate the root cells.

NEMATOCYST A stinging cell of a cnidarian (anemones, jellyfish, and relatives). *See also* cnida.

NERVE FIBER Filamentous extension of a nerve cell, used for transmitting an electrical impulse over a distance.

NEURONE Nerve cell that makes up the nervous system of an animal.

NEUTROPHIL White blood cell that is part of the immune system. It engulfs foreign particles by phagocytosis.

NITRIFICATION Process performed by certain kinds of soil bacteria that converts nitrogen compounds released from decomposition into nitrate—a form that can be absorbed and used by plants.

NITROGEN FIXATION Process performed by certain kinds of bacteria and a few other microbes that helps convert atmospheric gaseous nitrogen into an organic form—such as amino acids. Some nitrogen-fixing bacteria live free in the soil and others live in a mutually beneficial partnerships with some kinds of plants, inside their root nodules.

NUCLEUS Organelle inside a cell of a eukaryote that contains DNA.

NYMPH Juvenile form of an insect that differs from the adult flying form mainly in having undeveloped wing buds.

OBLIGATE AEROBE Organism that must respire aerobically (with oxygen) to survive.

ODONTOPHORE Stiff, tonguelike projection that carries the radula in the mouth of a mollusk.

OOTHECA (pl. OOTHECAE) Egg case of some kinds of insects, such as cockroaches and praying mantises. Each ootheca contains multiple eggs.

OPISTHOSOMA Abdomen of an arachnid, such as a spider.

ORGAN Structure in the body—made of multiple kinds of tissues—that performs a particular set of functions, such as a heart.

ORGANELLE Structure inside a cell, such as a nucleus or mitochondrion, involved in performing a particular set of functions.

ORGANISM Living thing.

OSMOSIS Movement of water across a membrane from a region of low solute concentration to high solute concentration.

OSSICLE (i) Hard, bonelike structure that makes up the skeleton in the skin of an echinoderm (starfish, sea urchin, and relatives); (ii) Tiny bone in vertebrates.

OVARY Reproductive organ in a plant or animal, used to produce eggs (ova).

OVIPOSITOR Egg-laying tube of many kinds of female animals, including insects.

PALISADE Part of the mesophyll tissue of a leaf, packed with many chloroplasts and used mainly for photosynthesis.

PALP Fingerlike projection in the mouthparts of an invertebrate, used to manipulate food or for other functions.

PAPILLA (pl. PAPILLAE) Any small projection on the body of a plant or animal.

PARASITISM Ecological relationship between two organisms of different species, a parasite and a host, where the parasite benefits and the host is harmed. Parasites typically live in or on their host, using them as a source of food.

PARTHENOGENESIS Asexual production of eggs that develop without fertilization, restricted to certain kinds of animals such as aphids and water fleas.

PATHOGEN Disease-causing organism.

PEDIPALP The palp of an arachnid.

PELLICLE Stiff, supporting layer under the cell membrane of some kinds of single-celled organisms, such as *Euglena* and dinoflagellates.

PEPTIDOGLYCAN Tough material that makes up the cell walls of bacteria.

PERISTALSIS Waves of muscular contraction in the wall of the gut used to move the contents of the digestive system.

PHAGE (or BACTERIOPHAGE) Virus that infects bacteria.

PHAGOCYTOSIS Behavior exhibited by some kinds of cells, such as single-celled amoebas and some white blood cells, whereby they engulf and digest particles.

PHASE CONTRAST Technique used in microscopy that accentuates the contrast of a transparent specimen, making it easier to distinguish structural detail.

PHLOEM Transport pipe, or vessel, in plants used to carry soluble food—mainly sugar—produced by photosynthesis.

PHOTORECEPTOR Sensory structure that is stimulated by light.

PHOTOSYNTHESIS Process by which some kinds of organisms, mainly plants and algae, can use light energy to convert simple materials—including carbon dioxide, water, and minerals—into food, such as sugar.

PHYTOPLANKTON Part of plankton, including algae, that can photosynthesize to make food.

PILUS (pl. PILI) Straight, bristlelike structure on a bacterium, used for attaching to surfaces, as well as other bacteria to allow exchange of DNA.

PLANKTON Community of usually small organisms that drift in open water.

PLASMODIUM (pl. PLASMODIA) *See* syncytium.

PLATELET Blood cell fragment involved in blood clotting.

POLLEN Reproductive grains produced by male parts of a plant's cone or flower. Each pollen grain contains a male nucleus that can fertilize an egg after pollination.

POLLEN TUBE Tube that grows from a pollen grain that germinates on the female part of a plant's cone or flower. The tube delivers the male nucleus to the egg.

PROBOSCIS Long nose, or noselike organ, such as the coiled mouthpart of a nectar-drinking butterfly.

PROKARYOTE Organism made up of simple cells that have no nucleus or other internal organelles. Bacteria and archaeans are prokaryotes.

PROSOMA First body part (consisting of fused head and thorax) of an arachnid, such as a spider.

PROTEIN Complex organic substance containing carbon, hydrogen, oxygen, nitrogen, and often sulfur. Proteins are the most diverse kinds of molecules found in living things, and their different types, including enzymes, antibodies, carriers, and signalers, perform many functions.

PROTOZOAN Complex, single-celled organism, such as an amoeba or *Paramecium*, that consumes organic food, rather than making it by photosynthesis. Organisms classified in the formal kingdom Protozoa include amoebas, slime molds, and their close relatives.

PROTRACTOR MUSCLE Muscle that extends part of the body, typically a limb.

PSEUDOPOD Extension of cellular cytoplasm, produced by cells such as amoebas or certain white blood cells and used for locomotion and phagocytosis.

PSEUDOSCORPION Tiny arachnid, shaped like a scorpion, but lacking a tail sting and classified in a separate group.

PTYCHOCYST Kind of cnida found in tube anemones, used for assembling the supporting tube.

PUPA (pl. PUPAE) Intermediate, typically encased and motionless, stage in the life cycle of certain kinds of insects, such as beetles and butterflies, in which a larva undergoes complete metamorphosis into an adult.

RADICLE First root produced by a germinating plant seedling.

RADIOLARIAN Kind of amoebalike, single-celled organism that has long spines (axopods) made from hard silica or strontium mineral, and filamentous pseudopods.

RADIOLE Featherlike tentacle of certain kinds of marine worms, used to collect oxygen and particles of food.

RADULA Rasping surface of the tonguelike organ of a mollusk, used for gathering food.

RAMUS (pl. RAMI) Long, spiny structure found in some kinds of arthropods. Taillike caudal rami are found in many kinds of crustaceans.

RECEPTOR Molecule or cell that detects a stimulus and triggers a response.

REFRACTION Bending of light rays, which happens when the density of the medium through which it is travelling changes.

RESPIRATION Chemical process in an organism by which it releases usable energy from food molecules.

RETRACTOR MUSCLE Muscle that bends or retracts a part of the body.

RHIZINE Rootlike structure of a lichen, used to anchor it to a surface.

RHIZOID Rootlike structure of a moss or liverwort, used to anchor it to a surface.

RIBOSOME. Granule inside a cell that makes protein, using information from the cell's DNA.

RNA (ribonucleic acid) Genetic material in cellular organisms used to enable the instructions in DNA to be "translated" into proteins. Some kinds of viruses (such as coronaviruses) contain only RNA and not DNA.

ROOT NODULE Swelling that develops in roots of certain types of plants (such as legumes), in response to infection by beneficial nitrogen-fixing bacteria.

SCAPE Segment at the base of an insect's antenna.

SCINTILLON Vesicle containing chemical used for bioluminescence, found in some kinds of dinoflagellates.

SCLERITE Armorlike plate that makes up part of the exoskeleton of an arthropod.

SCOLEX Head of a tapeworm, with suckers and hooks used to attach to the host's gut wall.

SEDIMENTARY ROCK Rock, such as limestone, formed by accumulated sediment being compacted and cemented together.

SEGMENT (i) One of a number of repeated units of the body of an animal, such as the segments on an earthworm or centipede; (ii) One of the sections of the leg of an arthropod, between the joints.

SEM (SCANNING ELECTRON MICROGRAPH) Image produced by using a scanning electron microscope. The specimen illustrated appears in three dimensions.

SENSILLUM (pl. SENSILLA) Small sensory structure on the exoskeleton of an invertebrate animal.

SETA (pl. SETAE) Hairlike structure, often with sensory function, of an invertebrate.

SILK Strong, fibrous protein produced by certain kinds of invertebrates, such as spiders and some caterpillars.

SORUS (pl. SORI) Cluster of spore-producing capsules in a fern. In most ferns, sori develop on the underside of the leaf, beneath an umbrellalike covering called an indusium.

SPECIES Type of organism, usually regarded as consisting of individuals that can potentially interbreed to produce viable, fertile offspring.

SPERM (or SPERMATOZOON, pl. SPERMATOZOA) Male sex cell, with at least one beating hairlike flagellum, which provides propulsion when swimming toward the egg.

SPIRACLE Breathing hole in the body wall of insects and some other related arthropods.

SPIRILLUM (pl. SPIRILLA) Spiral-shaped bacterium.

SPIROCHETE Corkscrew-shaped bacterium.

SPIROCYST A cell of a cnidarian (anemones, jellyfish, and relatives) that discharges threads to entangle prey. *See also* cnida.

SPORE Reproductive single cell that develops into an adult organism without fertilization, and is used for dispersing organisms such as fungi, ferns, and mosses.

STAPHYLOBACILLUS (pl. STAPHYLOBACILLI) Bacterium that occurs in clusters of bacilli (rod-shaped cells).

STAPHYLOCOCCUS (pl. STAPHYLOCOCCI) Bacterium that occurs in clusters of cocci (spherical-shaped cells).

STIGMA (pl. STIGMATA) "Eyespot" of pigment that gives light-direction information in some microorganisms.

STOMA (pl. STOMATA) Pore in the surface epidermis of a plant leaf or stem that allows gaseous exchange and transpiration. Each stoma is typically bordered by two guard cells.

STREPTOBACILLUS (pl. STREPTOBACILLI) Bacterium that occurs in chains of bacilli (rod-shaped cells).

STREPTOCOCCUS (pl. STREPTOCOCCI) Bacterium that occurs in chains of cocci (spherical-shaped cells).

STYLET Cutting blade in the piercing mouthparts of some insects, such as aphids and mosquitoes.

SUSPENSION FEEDER Animal that collects particles of food suspended in water.

SYMBIOSIS Any intimate ecological relationship, such as parasitism and mutualism, between two species.

SYNCYTIUM (pl. SYNCYTIA) Type of organization found in limited kinds of organisms where the body is not divided into discrete cells by membranes or walls, but instead their many nuclei are scattered through a sprawling mass of common cytoplasm. The syncytium of a slime mold is called a plasmodium (pl. plasmodia).

TARSUS Foot of an animal. In arthropods, the tarsus is the terminal segment at the end of the leg.

TEM (TRANSMISSION ELECTRON MICROGRAPH) Image produced by using a transmission electron microscope. The specimen used is a thin section, so it appears in two dimensions.

THIGMOTAXIC Describing defensive behavior in an animal that tends to move toward a confined space so its body is in contact with surfaces.

THORAX Middle part of an animal's body, between the head and abdomen.

THYLAKOID Flattened sac within a chloroplast, carrying the light-absorbing pigment chlorophyll, which carries out photosynthesis.

TISSUE Aggregation of cells that work together to perform a function or set of functions. The epidermis of a plant's leaf and the blood of an animal are examples of tissues.

TOXICYST Kind of trichocyst in some kinds of predatory single-celled organisms that discharges poisonous threads, which catch prey.

TRACHEA (i) One of many breathing tubes in the body of an insect and related animals that carries oxygen from spiracles (pores) at the body surface to respiring tissues deep inside; (ii) Windpipe of an air-breathing vertebrate.

TRACHEOLE Microscopic branch arising from a trachea in an insect or related animal that penetrates cells of tissues, to bring oxygen to the site of respiration.

TRANSPIRATION Loss of water from the surface of a plant's leaf, or other surfaces, caused by evaporation from tissues.

TRICHOCYST Structure in some kinds of single-celled organisms, such as *Paramecium*, that discharge threads for the purposes such as catching prey or attachment to surfaces.

TUN Dehydrated husk of tiny animals called tardigrades, in which they enter a dormant state. *See also* cryptobiosis.

VACUOLE A fluid-filled sac inside a cell. Plants cells typically contain a single large sap vacuole that helps to keep plant tissues firm. Some single-celled organisms and animal cells contain food vacuoles that are sites in which they digest particles of food.

VASCULAR BUNDLE Bundle of transport vessels—including xylem and phloem—running through the roots, stem, and leaf veins of a plant.

VASCULAR TISSUE Tissue made up of transport vessels, such as xylem and phloem in plants, or blood vessels in animals.

VENOM Poison produced by organisms for defense or to overpower prey, and injected into their target, such as through a sting or fangs.

VENULE Tiny blood vessel that delivers blood from capillaries to veins.

VERTEBRATE Animal with a backbone.

VESICLE Small, fluid-filled sac inside a cell. Smaller than a vacuole.

VIBRIO Bacterium that is comma-shaped.

VIRUS Tiny infectious particle, typically consisting of a protein coat containing genetic material (DNA or RNA). Viruses are not cellular organisms, but must invade cells of organisms in order to multiply. Since viruses are not capable of independent replication, many experts argue that they do not qualify as "living."

VISCOSITY A measure of a fluid's resistance to flow. Thick, poorly flowing fluids, such as molasses, have higher viscosity than thin fluids that flow easily, such as water.

XENOSOME Particle of debris that is incorporated into the shell or casing of some kinds of single-celled organisms.

XYLEM Transport pipe, or vessel, in plants used to carry water and minerals.

ZOOCHLORELLA (pl. ZOOCHLORELLAE) Green alga that lives inside the tissues of another organism, such as *Hydra* or *Paramecium*, with mutual benefit.

ZOOPLANKTON Part of the plankton, including small animals and animal-like microorganisms, that consume food rather than photosynthesize.

ZOOXANTHELLA (pl. ZOOXANTHELLAE) Yellow alga, typically a dinoflagellate, that lives inside the tissues of another organism, such as coral, sea slug, or giant clam, with mutual benefit.

index

A

A25 phage virus 220–21
abdomens
    ants 125, 210, 313
    arachnids 59, 64, 212
    damselflies 271
    flies 85, 300
    lice 306
    mosquitos 302
    pseudoscorpions 56
acantharian radiolarians 331
Acanthaspis sp. (ant-snatching assassin bug) 194
Acanthognathus teledectus (trap-jawed ant) 210
Acheronia sp. (death's-head hawkmoth) 84
Acheta domesticus (house cricket) 106–07
Achnanthidium sp. (diatom) 178
acidobacteria 327
acidophiles 278
Acidovorax 76–77
acorn worms 389
actin 133, 141, 177
actinobacteria 328
Actinosphaerium eichhorni 290
active suspension feeding 44
adipose tissue 14
Aegimia elongata (bush cricket) 194
aerobic respiration 75, 78, 80–81
aeroplankton 168
aeropyles 259
aerotolerant anaerobes 80
Agabus sp. (diving beetle) 140–41
Aiptasia (glass anemone) 319
air plants 208
air sacs 88, 96, 106, 146–47
Alcea sp. (hollyhock) 237
alcohol 78
algae 20, 66–67, 290–91
    see also diatoms; lichens
    algal blooms 227
    cells 254
    classifications 339–41
    colonies 252–53
    and coral partnerships 319
    as food 38, 130, 132, 134, 151
    reproduction 226–27
alkaloids 190, 191
Allium cepa (onion) 262
alveoli 88, 184
Amanita muscaria (fly agaric) 314–15
amoebas 12, 38–39, 132–33, 244
    classifications 336–37
Amoeba proteus (common amoeba) 38, 132–33

amphibians 195, 257, 396
Amphiprora alata (diatom) 179
amphiesma 184
Amphiprion percula (orange clownfish) 256–57
ampulla 148, 196
anaerobic respiration 30, 78, 80–81
anemones 42, 138, 319, 352
angels trumpet 236
angiosperms 262, 345
animals, classifications of 348–99
    arthropods 365–85
    chordates 390–99
anisochela spicules 187
annelids 361
annuli 102, 260–61
Anopheles (mosquito) 302–03
antennae 102–03, 154
    ants 52, 125
    aquatic life 146, 151, 272, 288
    water fleas 142, 143, 242
    flies 112, 246
    mosquitoes 302
    springtails 161, 297
    thrips 169
    velvet worms 156
antheridiophores 266
anthers 235
Anthophila (bees) 239
Anthracothorax prevostii (green-breasted mango) 96–97
antibiotics 32, 36
antibodies 217, 308
anticoagulants 302
antigens 216–17
ant-mimic jumping spiders 154–55
ants 380
    feeding 48, 52
    parasites 172, 312–13
    sensory systems 124–25
    stingers 210–11
ant-snatching assassin bugs 194
aphids 24, 51, 52–53, 243, 300
Aphodius prodromus (dung beetles) 172–73
apicomplexans 335
Apis mellifera (European honeybee) 49, 172
aplacophorans 357
Apocarchesium sp. 291
apples 78
appositional eyes 112
Aquaspirillum 28
aquatic life
    see also algae; crustaceans; fish; swimming
    copepods 288–89

cyanobacteria 18
diving beetles 140–41, 292
feeding 44–45
freshwater life 152–53, 292–93
larvae and nymphs 87, 272, 285
marine plankton 286–87
pond life 290–91
Arabidopsis thaliana (thale cress) 209
arachnids 57, 58–59, 102, 366
    see also mites; spiders
    legs 158, 190
    pseudoscorpions 56–57, 172
Araneus sp. (orb-weaver spider) 64
Araucaria bidwillii (bunya pine) 262
arbuscules 314
archaeans 12, 13, 29, 68, 80
    extreme environments 278–79
    ultra-small 325
Argiope sp. (banded spider) 59
Argonauta argo (greater argonaut octopus) 120–21
arrow worms 362
Artemia sp. (brine shrimp) 144–45
arteries 92, 93
arteriole 60
arthropods 102, 297
    see also arachnids; crustaceans; insects
    classifications of 365–85
    exoskeletons 193, 200, 270
    legs 156–57, 158–59
    springtails 160–61, 188–89, 297, 299
    asexual reproduction 226–27
    fungi 36, 230, 231
    self-fertilization 235
    simple plants 266
    single-cell organisms 30, 129
    Volvox algae 252
    water fleas 242, 243
Asian praying mantises 47
Aspergillus 216–17
assassin bugs 194
Asterias rubens (five-armed starfish) 148–49
Atlantic ctenophores 136–37
ATP (chemical energy) 73
Atropacarus sp. (box mite) 296–97
Aulacodiscus oreganus (diatom) 178
Augeneriella dubia (polychaete) 299
Aulorhynchus flavidus (tube-snout fish) 200–01
Automeris io (io moth) 194
Automeris naranja (saturnid moth) 214
autotrophs 32
axons 122
axopods 143, 182

B

babies 248–49, 272
bacilli (shape) 28–29, 76, 277, 308
Bacillus bacteria 28–29, 81, 277, 308
backswimmers 152–53
bacteria 11, 12, 13, 28–29, 276–77
    see also cyanobacteria
    classifications of 326–29
    E. coli 11, 28, 30–31, 40
    energy sources of 76–77
    extremophiles 278
    flagellum 129
    in the gut 308–09
    multicellular 224–25
    nitrogen-fixing 26–27
    nourishment for 32–33
    phage viruses 220–21
    as prey 40
    responses to oxygen 80–81
    Rhizobium 26
    sensory receptors 100
bacteroidetes 329
ballistic movement 160–61
bamboo 206
banded spiders 59
banded sugar ants 124–25
barbs 42, 170, 212
barbules 170, 171
bark spiders 194
barnacles 45, 369
basil 208
batteries 42
Bdelloidea (rotifers) 299
Bdellovibrio bacteriovorus 40
bees 238–39, 380
    mouthparts 49
    parasites of 172
    stingers 210
beetles 381
    aquatic 87, 150–51, 292
    body structure 102, 140–41, 158
    exoskeleton 192–93
    mouthparts 46–47, 49
    wings 165
    iridescence 116
    parasites of 172–73
Bifidobacterium sp. 29, 80–81, 308
bills 97
Bilobella (springtail) 188–89
binary fission 130, 179
binocular vision 111
bioluminescence 72–73, 118–19, 199
    see also iridescence
Biorhiza pallida (oak apple gall wasp) 300, 301
birds 96–97, 170–71, 398

birds of paradise 247
bivalves 358
black soldier flies 112
bladders 147, 148
Blastophaga psenes (fig wasp) 241
blastula 256
blood 15
    red blood cells 14, 88, 90–91, 93, 310–11
    white blood cells 90, 91, 133, 217
blood suckers 50–51, 302
blue butterflies 10–11
blue-green algae 327
    see also cyanobacteria
bog moss 204–05
bone 14, 108, 200–01, 257
    ossicles 108, 196
bottle flies 48
Bouteloua gracilis (mosquito grass) 236
box mites 296–97
brachiopods 356
Brachypelma smithi (red-knee tarantula) 212–13
bracts 240, 262
brain 122–23, 125, 312–13
branched green algae 291
branchiopods 145, 368
breathing see respiration
brewer's yeast 78–79
brewing 78
brine shrimp 144–45
brittle stars 148, 287, 388
broadclub cuttlefish 120
brochosomes 190, 191
bronchi 88
bronchioles 88
brood chambers 120, 242, 243
brown algae 339
Brucella abortus 29
Brugmansias sp. (angels trumpet) 236
bryophytes 342
bryozoans 355
buds 226, 227
bugs see true bugs
bunya pine 262
buoyancy 146–47, 151, 152, 182
bush cricket 194
butterflies 10–11, 383
    caterpillars 64, 212–13, 214–15, 258
    mouthparts 49, 51
    sensory systems 104, 114–15
    wings 168

C

caddisflies 382
Caenorhabditis elegans (nematode) 294–95, 299
Caerostris sp. (bark spider) 194
calcium carbonate 181
Californian poppies 236
Calliphora vicina (bottle fly) 48
Caloplaca (lichen) 317
Cambrian Explosion 19
camouflage 194–95, 215, 297
    color changes 120–21
Camponotus consobrinus (banded sugar ant) 124–25
Camponotus gigas (giant forest ant) 48
cancer cells 33
Cannabis sativa 208, 209
Cape sundew 24–25
capillaries 88, 90, 92, 93
capillitium 244
Carabus rutilans (Reddish ground beetle) 46–47
carapaces 143, 242, 296
cardinal tetra 116–17
Carex sp. (sedge) 262, 263
carnivorous plants 24–25
carotenoids 100, 268
carrot 263
cartilage 200, 201
Caryophyllaceae (dianthus) 236
catapult movement 160–61
caterpillars 64, 212–13, 214–15, 258
cat fleas 50–51
Cavia porcella (guinea pig) 108–09
Cecidomyiidae sp. (gall midge) 300
cells 13
    bioluminescence 122–23
    colonies 252–53
    division of 254–55, 256–57
    nervous systems 122–23
    skeletons 176–77
    viruses 220–21
    walls 79, 179, 204–05, 261
centipedes 102, 297, 367
centric diatoms 178–79, 338
centriole 177
cephalopods 120–21
cephalothoraxes 59
cerci 272
cerebellum 122–23
cerotegument 191
Cestum veneris (Venus's girdle) 136
chaetae 57, 87, 188, 304
chameleons 120
Chaoborus sp. (phantom midge) 146–47
Chaos carolinense (amoeba) 132
chelae spicules 186
chelicerae 56, 58, 59, 249, 296
    classifications of 365–66
chemical energy 73
chicory 236
chimaeras 392
chitin 85, 87, 193, 216
chitons 357
Chlamydomonas (algae) 252
Chlorella (algae) 67
chlorophyll 20, 204, 316
chlorophytes 290
chloroplasts 20, 94, 118, 205
    algae 66, 100, 252
choanoflagellates 330
chordates 390–99
chorions 258, 259
Chorotypus (grasshopper) 195
Christmas cacti 236
chromatophores 120–21, 147
chromistae 12, 333, 338–39
Chroococcus turgidus 291
cicadas 48, 163
Cichorium intybus (chicory) 236
Cicindela campestris (green tiger beetle) 49
cilia 128–29, 133
    of ctenophores 136–37
    in the ear 108
    of gills 82
    of protozoans 41
    of rotifers 134–35
    of sea urchins 268
    of stentors 44–45
ciliates 44–45, 335
cingula 184
circulatory systems 88, 92–93
Cladonia rangiferina (reindeer lichen) 282–83
class (classification) 322
classification 322–23
claws
    arachnids 155, 191
    lice 306
    midge larvae 285
    velvet worms 157
cleavage 254
cleistothecia 35
Clematis sp. 206–07
climbing 154–55
clones 129, 226, 256
clovers 26–27
clusters (bacteria) 29
clypeus 46
cnidarians 42–43, 138–39, 350–53
coccobacillus bacteria 29
coccus bacteria 28
cochleas 108
cockroaches 258, 272, 378
Codosiga umbellata 45
Coix lacryma (Job's tears) 262
cold environments 278, 282–83
collagen 14, 93, 177
color
    changes 120–21
    iridescence 116–17, 137
    production of 114–15, 266
colpi 236–37
comb jellies 136–37, 349
comb plates 136
commensal bacteria 308
complex shapes (bacteria) 28–29
complex shapes (sponges) 186–87
compound eyes 112–13, 167, 246, 288, 306
conidia 36, 230–31
conidiophores 36, 231
conifers 237, 262–63
conjugation 30, 130, 230
contamination 77
Convallaria majalis (lily of the valley) 94–95
copepods 286, 288–89, 299, 368
Coprinellus micaceus (glistening inkcap) 35
Coptotermes niger (termite) 68
corals 42, 138, 319, 353
corbiculae 239
cord mosses 232–33
Coreus sp. (leather bug) 48
corona (cillum) 134, 135
coronavirus 11, 222–23
cortex cells 202, 206
Corti 108
Coscinodiscus sp. (diatom) 178
cotyledons 264, 265
courtship rituals 247
COVID-19 11, 222–23
crabs 194, 370
    see also horseshoe crabs
crab spiders 194
creeping fingerworts 266–67
crickets 106–07, 158–59, 163, 194, 375
cristae 75
Crossaster papposus (common sunstar) 196–97
crustaceans 368–71
    body structure 102, 142, 158
    copepods 286, 288–89, 299
    food webs 268, 292
    swimming 144–45
    water fleas 142–43, 243
cryophiles 278
crypsis 194–95
cryptobiosis 281
ctenes (comb plates) 136
Ctenocephalides felis (cat flea) 50–51
ctenophores 136–37, 349
Culex sp. (mosquito) 164–65
cuticle
    ears 106
    exoskeletons 63, 85, 159, 188, 190, 193, 270–71
    eyes 110, 112
    hair 302
    leaves 266
    wings 165
cuttlefish 120, 360
cyanobacteria 18–19, 20, 151, 282, 286, 291, 327
Cycas revoluta (sago palm) 262
Cyclops (copepods) 288–89
Cypris sp. (seed shrimp) 292
cytokinesis 254

cytoplasm 22, 32, 38, 78, 118, 180, 201, 254, 255
   cytoplasmic streaming 78, 133
   endoplasm 132
cytoskeleton 177
cytostome 40, 41

D

damselflies 86–87, 168, 270–71, 374
Daphnia sp. (water flea) 142–43, 242–43
Daucus carota (carrot) 263
death's head hawkmoth 84
decomposers 35, 36, 297
   see also detritivores
decontamination 77
defenses
   see also spines (defenses)
   caterpillars 214–15
   chemical 190–91
   immune system 216–17
   stingers 42, 210–11
deinococcus 328
Delta latriellei (orange potter wasp) 248
Demodex sp. (follicle mite) 305
dendrites 104, 122, 308
Dennyus hirundinis (feather louse) 306
denticles 198–99
dentine 198
Derocheilocaris typica (mystacocarid) 299
desmid algae 226–27, 290, 291
detritivores 36, 294, 296–97
   see also decomposers
deutonymphs 172
dianthus 236
Diasemopsis meigenii (stalk-eyed-flies) 246–47
diatoms 18–19, 39, 178–79, 284, 286, 290, 338
Didinium nastum 40–41
Difflugia sp. (shelled amoeba) 38–39, 290
diffusion 82, 84–85, 88
digestion 60–61
   arachnids 58–59
   carnivorous plants 24
   rotifers 134
   snails 54
dinoflagellates 118–19, 184–85, 286, 319, 334
diplococcus bacteria 28
diploid cells 260
distance judgment 110–11
diving beetles 140–41, 292
DNA (deoxyribonucleic acid) 12, 228, 254–55
   of tardigrades 281
   of viruses 223
dorsal shields 58
dragonflies 112, 374

Drosera capensis (Cape sundew) 24–25
Drosophila melanogaster (fruit fly) 84–85
Dryopteris filix-mas (fern) 260
duckweeds 20
dung beetles 172–73
Dytiscus sp. (diving beetle) 292

E

ears 106, 108–09, 377
earwigs 168
Echiniscus granulatus (tardigrade) 280–81, 298
Echinocardium cordatum (sea potato) 268–69
echinocytes 91
echinoderms 148–49, 196–97, 386–89
echinopluteus 268
echolocation 106
E. coli 11, 28, 30–31, 40
ectomycorrhizae 314
ectoplasm 132, 133
Ectopleura larynx 138–39
Ecuadorian orchids 15
eggs 248–49
   clownfish 256–57
   human (ovum) 228–29
   insects 258–59, 306, 307
   ovipositors 102, 194, 210, 300
   tapeworms 62
   water fleas 242, 243
einkorn 265
electrical energy 73
electrical signals 122, 141
Elphidium (foram) 180
elytra 165, 193
embryos 235, 254, 256–57, 258, 265
emmer 265
enamel 198
endites 144
endomycorrhizae 314
endoplasm 132, 133
endoskeletons 196
energy consumption 72–73, 96–97
energy release 75
Enterococcus faecalis 28
enterocyte cells 60
enzymes
   digestive 25, 54, 68
   light producing 119
   nitrogenase 26
   in sexual reproduction 228
   urease 32–33
   in viruses 221
Eobrachychthonius sp. (oribatid soil mites) 190–91
epidermis 94, 135, 202–03, 206, 305
epidodium 38
epigeal germination 265
epithecas 178, 184
epithelial cells 14, 87, 138

Erysiphe adunca (powdery mildew fungus) 34–35
Escherichia coli (*E. coli*) 11, 28, 30–31, 40
Eschscholzia californica (Californian poppy) 236
ethanol 78
eudicot seeds 262–63, 264–65
Euglena sp. 11, 67, 100–01
euglenids 66, 67
eukaryotes 12, 13, 80, 129
   classification of 330–31, 334–35
Euplectella aspergillum (glass sponge) 187
Eurasian pygmy shrews 97
euraster spicules 187
European honeybees 49, 172
excavates 330
excretion 93
   see also honeydew
exine 236
exoskeletons 140, 192–93
   arachnids 155
   crickets 107
   dorsal shields 58
   molting of 270–71
   water fleas 242
   wings 165
extreme halophiles 324
extremophiles 278–79
eyes
   ants 124
   aquatic insects 146, 150
   arachnids 110–11, 293
   compound 112–13, 167, 246, 288, 306
   in embryos 257
   fish 200
   mayflies 273
   springtails 189
   stalk-eyed flies 246–47
eyespots 252

F

facets (eyes) 112, 113, 150
facultative microbes 40, 80
false feet (pseudopods) 38, 132–33, 180–81
families (classification) 322
fanworms 82
Favulina (foram) 180–81
feather duster worms 82–83
feather lice 306
feathers 96, 170–71, 306
feather stars 388
feelers *see* antennae
feet
   arthropods 158, 159, 154–55
   echinoderms 148–49
   pseudopods 38, 132–33, 180–81
   rotifers 135

female reproductive parts (plants) 235
femurs 159, 162, 163
fennel 262
fermentation 30, 78–79
ferns 260–61
fertilization 228–29, 242, 256
   in plants 232, 235
fibroblast 177
fibrous tissue 14
Ficus sp. (fig) 240–41
fig wasps 241
filament bacteria 28, 308
filamentous fungi 346
filamentous molds 346
filter feeding 44, 45, 291
fire ants 190
fireflies 72–73
firmicutes 308, 326
fish 394–95
   embryo development 256–57
   iridescence 116–17
   movement 142
   skeletal systems 200–01
flagella 129
   in algae 66, 100, 101, 184, 252
   antennae 102
   in bacteria 32
   in stentors 45
Flammulina velutipes (velvet shank mushroom) 230
flat-faced longhorn beetles 102–03
flatworms 354
fleas 50–51, 142–43, 161, 385
   see also water fleas
flies 384
   body structure 48, 112–13
   breeding 246–47, 300
   flight 166–67
   as prey 153
   respiration 84–85
flight 166–67, 168–69
   feathers 170–71
floating ferns 208
flowering plants 234–35, 239
   pollen 236–37
flower mantises 194
fly agaric fungi 314–15
flying *see* flight
Foeniculum vulgare (fennel) 262
follicle mites 305
follicles 170, 202, 228, 305
Folsomia candida (springtail) 299
food chains 190, 292, 297
food consumption
   in bacteria 224
   in circulatory systems 93
   digestion 60–61
   snails 54–55
food webs *see* food chains
forams (foraminiferans) 180–81, 286, 332
forewings 168, 169

fornix 143
fossils 180, 182, 201
freshwater life 292–93
friction 142–43
"friendly" bacteria 81, 308
frogs 190
fronds 260, 29
Frontipoda (mites) 293
fruit flies 84–85
frustules 39, 179
Funaria hygrometrica (common cord moss) 232–33
fungi 12
  see also molds; yeasts
  absorbing food 34–35
  catapulting 160, 161
  classifications of 346–47
  habitats 297, 314–15
  lichens 282, 316–17
  parasites 312–13
  Penicillium 36–37
  reproduction 230–31
furca 160
fusobacteria 329

G

gall midges 300
galls 241, 300–01
Gamasellus sp. (predatory mite) 58–59
gametes 232, 235, 263, 310
gametocytes 310, 311
gametophytes 232
ganglion 125
gas exchange 82, 88
Gasteracantha sp. (orb-weaver spider) 64
gastropods 359
gastrotriches 299, 363
gemma cups 266
gem soldier flies 112–13
genera (classification) 322
genes 228
Geobacter metallireducens 77
Geranium sp. (geranium) 208, 236
germination 23, 232, 244, 264–65
Gerris lacustris (pond skaters) 151
giant forest ants 48
Giardia 80
gills 87
  feather duster worms 82–83
  fungal 230
  larvae 284, 285
  nymphs 273
  skeletal systems 200
Ginkgo biloba (gingko) 262
giraffe-necked weevil 322
glass anemone 319
glass sponges 187
glassworms (phantom midge larvae) 146–47
gliding 168

glistening inkcaps 35
glucose 67, 75
Glycine max (soya bean) 209
goblet cells 60, 61
golden algae 339
golden poison frogs 190
gonophores 139
Gram straining 32
grasses 236, 263, 265
grasshoppers 375
  body structure 47, 163, 271
  camouflage 195
  nervous systems 125
  as prey 59
  reproduction 258
green algae 341
green-breasted mango 96–97
greenhouse gases 68
green tiger beetles 49
grey matter 122
Gryllotalpa gryllotalpa (mole cricket) 158–59
guanine 116, 117
guinea pigs 108–09
gummy sharks 198–99
gut bacteria 308–09
gut microbes 68
gymnosperms 262–63, 344
Gyrinus sp. (whirligig beetles) 150–51

H

Haeckel, Ernst 134, 183, 268
Haematococcus pluvialis 290
hagfish 390
hairs
  see also cilia; flagella; setae
  for defense 212–13
  headlice 306–07
  leaf 208–09
  mammals 202–03
  plant roots 22–23
  in sensory system 124–25
halophiles 278, 324
halteres 166–67
Hamamelis virginiana (witch-hazel) 208
hammered shield lichen 316–17
haptophytes 333
hat-thrower fungi 160, 161
haustoria 35, 317
hawksbeard aphids 243
headlice 306–07
heads (prosomas) 56, 125, 169
hearing 106–7, 108–9
heart 92–93
heat energy 73
Helicobacter pylori 29, 32–33
Helix pomatia (Roman snail) 54–55
Helophilus sp. (hoverfly) 48
hemocoel 157

hemoglobin 90–91
herbivores 54, 68, 292, 297, 314
Hermetia illucens (black soldier fly) 112
heterotrophs 32, 291
hexapods 372–81, 383–84
Hibiscus sp. 208
Hierodula doverii (Asian praying mantis) 47
hindwings 168, 169
Hirundo rustica (swallow) 170–71
hitchhiking 57
hollyhock 237
holoplankton 286
holotrichous isorhizas 42
honeybees 49, 172, 238–39
honeydew 52
horned treehoppers 162–63
hornets 168
horns 162, 184
horseshoe crabs 365
house cricket 106–07
houseflies 48
hoverflies 48
human body 15
  blood 311
  brain 122–23
  guts 308–09
  hair 202–03, 306–07
  heart 75, 92–93
  intestines 60–61
  liver 74–75
  lungs 88
hummingbird hawkmoths 141
hummingbirds 96–97, 116
hyaline cells 204
Hydra sp. 42–43, 226
hydrogen 75
hydroids 42, 138
hydrostatic skeletons 138, 157
hydrozoans 350
hylum tissue 264
Hymanopus coronatus (flower mantis) 194
Hymenoptera 210, 239
hyperthermophiles 278, 325
hyphae
  absorbing food 34–35, 36
  habitats 314, 316, 317
  parasitic 312
  reproduction 230–31
hypogeal germination 265
hypopharynxes 47
hypothecas 178, 184, 185

I

ichneumon wasps 102
ileum 60
immune system 216–17, 276, 308
indumentum 208–09
indusium 260

infections 217, 221
  see also viruses
  malaria 302, 310–11
inflorescence 240, 263
inner ear 108–09
insects
  antennae 102
  aquatic 86–87, 152–53
  classifications of 372–85
  eggs 258–59
  legs 158–59
  sensory systems 104–05, 106–07
  stingers 210–11
  tracheae 84–85
  wings 164–65, 168–69
intestines 30, 60–61, 93, 308
iridescence 116–17, 137
  see also bioluminescence
iridophores 116
iron 76–77, 91
isopods 369
isthmuses 226

J

jaws 141, 201
jejunum 60
jellyfish 42, 351
Joblot, Louis 130
Job's tears 262
jointed legs 158–59
jumping 111, 162–63
jumping spiders 110–11

K

katydids 375
keratin 170, 202
kinetic energy 73
kinorhynchae 299, 363
knee joints 159
knotty shining-claws 56–57

L

labrum 47, 51, 49, 124
lacewings 373
lacteal 60
Lactobacillus 308
lampreys 392
Lamprochernes nodosus (knotty shining-claw) 56–57
Lamproderma arcyrioides 12–13
lancelets 390
lancets 210
larvae
  aquatic 268, 284–85
  blood sucking 51
  caterpillars 64, 212–13, 214–15

diving beetles 140–41
  and galls 300–01
  in metamorphosis 271
  mosquitos 302
  phantom midges 146–47
  plankton 286, 287
  wasps 248
leaf insects 195, 376
leaf-litter see detritivores
leather bugs 48
leaves 94–95, 208–09
  see also detritivores; leaf insects
  carnivorous plants 25
  duckweeds 20–21
  simple and complex 266
Lecanora (lichen) 317
leghemoglobin 26–27
legs 158–59, 293
  arachnids 56, 59, 154–55, 190, 296, 304
  insects 106–07, 140, 150, 270, 273, 307
  for jumping 162, 163
  springtails 189
  velvet worms 156–57
legumes 26
Leishmania 311
lemon balm 208
lens 110, 112, 113
Lepanthes (Colombian orchid) 15
Lepidozia reptans (creeping fingerworts) 266–67
Leptastacus macronyx (copepods) 299
Leptocentrus taurus (horned treehopper) 162–63
Leptospira interrogans 28
lice 306–07, 378
lichens 282–83, 316–17, 347
Licmophora flabellata 290
lift 168
light
  detection 100–01, 110–11, 112
    photoreceptors 100–01, 111
  and production of color 114–15, 137
    bioluminescence 72–73, 118–19
light micrographs (LM) 11
lignin 68, 205
Liliaceae (lily) 236
lily of the valley 94–95
Limacina helicana (sea butterfly) 286
limbs 145
  see also legs
limestone 19
Linnaeus, Carl 322
Liponeura cinerascens (net-winged midge) 284–85
liverworts 266–67
lobe-finned fishes 394
lobsters 370
Locusta migratoria (migratory locust) 48
locusts 47, 48
longhorn beetles 102–03

Loxa viridis (shield bug) 48
luciferin 119
Luidia ciliaris (seven-armed starfish) 148
lungs 82, 88–89, 217
lymphocytes 217
lysosomes 38, 217

M

Macroglossum stellatarum (hummingbird hawkmoth) 141
macroscopic filamentous fungi 346
macronuclei 44
macrophages 133, 217
madreporite 148
maggots 87
malaria 302, 310–11
male reproductive parts (plants) 235
mammals 399
  see also human body
  ears 108
  embryos 257
  hair 202–03
  lungs 88–89
mandibles 46–47, 48, 51, 68, 146
mango seed weevils 192–93
mantises 194, 377
Marchantia sp. (common liverwort) 266
marine plankton 286–87
marjoram 208
mastax 134
maxillae 46–47, 51, 58
mayflies 272–73, 373
meadow grasshoppers 163
Mediterranean dealfish 286
megascleres 186
Megatibicen resh (western dusk singing cicada) 48
meiofauna 298–99
meiosis 233
Melissa officinalis (lemon balm) 208
menaquinone 30
meroplankton 286
merozoites 310
mesophyll 14, 52, 94
metamorphosis 269, 271, 272
  see also larvae; nymphs
methane gas 68
methanogens 324
Micrasterias sp. 290
  M. thomasiana 226–27
  M. truncata 291
microaerophiles 80
Microchrysa sp. (gem soldier fly) 112–13
Micrococcus (bacteria) 277
Microlophium carnosum (common nettle aphids) 52
micronuclei 130
microorganisms
  see also algae; bacteria; fungi

archaeans 12, 13, 29, 68, 80
  in extreme environments 278–79
  lichens 282–83, 316–17
  protozoans 12, 68, 151, 180, 311
    Didinium nastum 40–41
    Plasmodium 244, 245, 302, 310–11
  viruses 11, 220–21, 223
  slime molds 12–13, 244–45
microscleres 186–87
microscopy 11
Microthamnion (branched green alga) 291
microtriches 63
microtrichia 165
microtubules 176, 177
microvilli 45, 60, 177
midges 284–85, 300
migratory locusts 48
millepedes 102, 367
mimicry 194–95
Mimulopsis sp. (acanthus) 236
minnows 142
Misumena vatia (crab spider) 194
mites 11
  defenses 190–91
  feeding 58–59
  habitats 293, 296–97, 304–05
  as parasites 172–73
mitochondria 13, 20, 75, 78, 80, 229
mitosis 254, 255
mole crickets 158–59
mollusks 357–60
monaxon spicules 186
monocot seeds 262–63, 264, 265
Moraxella (bacteria) 276
Morpho sp. (morpho butterfly) 115
mosquito grasses 236
mosquitos 302–03
  flight 164–65, 166
  larvae 151
  and malaria 310, 311
mosses 204–05, 232–33, 266
moths 84, 104, 141, 194, 383
  caterpillars 64, 212–13, 214–15
molds 36, 217, 231, 346
  slime molds 12–13, 244–45
molting 193, 270–71
mouthparts 48–49
  ants 124
  aphids 52
  arachnids 56, 58, 59, 293, 296
  beetles 46–47
  ctenophores 137
  flies 247
  lice 307
  protozoans 40, 41
  rotifers 134
  and sensory systems 104
  springtails 189
  suckers 50, 51, 284
mucilage 18, 24, 33, 61
  mucilaginous leaves 208

multicellular filamentous molds 346
Musca domestica (housefly) 48
muscular systems 138–39, 140–41
  ctenophores 136
  snails 54
  springtails 160, 161
  velvet worms 157
Mustelus antarcticus (gummy shark) 198–99
mycelia 35, 36
mycorrhizae 314
myonemes 138
myosin 141
Myrmarachne formicaria (ant-mimic jumping spider) 154–55
mystacocarids 299
myxamoebas 244, 245
myxobacteria 224–25
Myxococcus xanthus (myxobacteria) 224–25

N

nanostructures 190, 191
natural selection 246
nauplius 145, 288
Navicula sp. (diatom) 178, 286
nectar 49, 51, 97, 104, 239, 302
Neisseria gonorrhoeae 28
nematocysts 42–43
nematodes 217, 294–95, 362
Neottia ovata (orchid) 263
Nephrolepis sp. (sword fern) 260–61
nervous systems 122–23, 125, 138
nettle aphids 52
net-winged midges 284–85
neurons 120, 122
neurotransmitters 122
neuston 151
neutrophils 91, 216, 217
nitrogen 24–25, 26–27, 316
  cycle 26–27, 68
nitrogenase 26–27
nodules 26, 27, 76, 188
nonflowering plants (gymnosperms) 262–63
nonvascular plants 22, 266
nonwoody plants 204–05, 314
Notonecta glauca (common backswimmer) 152–53
nuclei 20, 66, 90, 130, 132, 177
  in cell division 226, 254–55
nursery-web spiders 248–49
nuts 263
nymphs 87, 172, 258
  damselflies 270, 271
  mayflies 272, 273

O

oak apple gall wasps 300, 301
oak trees 300–01
obligate microbes 40, 80
ocean dinoflagellates 118–19
   see also dinoflagellates
ocelli 247
Ocimum basilicum (basil) 208
octopuses 120–21, 360
Octopus vulgaris (common octopus) 120
odontophore 54
offspring 248–49, 272
Old World swallowtail butterflies 214
Olea europaea (olive tree) 208
ommatidia 112, 113, 150
ommochromes 115
onion 262
Onychophora (velvet worm) 157
oocysts 311
oothecae 258
Opercularia 128–29
operculum 306
Ophiocordyceps (fungus) 312–13
opisthosoma 56
oral cilia 45
orange clownfish 256–57
orange potter wasps 248
Orbulina (foram) 180
orb-weaver spiders 64
orchids 15, 262, 263
order (classification) 322
oribatid mites 190–91, 296
Origanum majorana (marjoram) 208
osmeterium 214
osmosis 22, 205
ossicles 108, 196
ostiole 240, 241
ovaries (plants) 235
ovipositors 102, 194, 210, 300
ovum 228–29
oxidation 76, 77
oxygen
   aerobic and anaerobic respiration 80
   air supplies in insects 84–85, 87
   in atmosphere 19
   in circulatory systems 93
   consumption 74–75
   from photosynthesis 94
   in vertebrates 90–91
   from water 86–87

P

painted lady butterflies 49
palisade mesophyll 14
palm trees 262
palps 46–47, 50, 302
   pedipalps 57, 58, 59, 249
Papaver rhoeas (poppy) 262
Papilio machaon (Old World swallowtail butterfly) 214
papillae 156, 188
papulae 196
Paracentrotus lividus (sea urchin) 256
Paracheirodon axelrodi (cardinal tetra) 116–17
Paramacrobiotus kenianus (tardigrade) 281
Paramecium sp. 40–41, 130–31
paramylon 67
parasites
   see also bacteria
   blood suckers 50–51, 302
   copepods 288
   defenses against 216–17
   fungi 35
   of the gut 62–63, 80
   mites 296–97
   nematodes 294
   plankton 290
   Plasmodium 310–11
   of the skin 304–05
parenchyma cells 207
parental care 248–49
Parmelia sulcata (hammered shield lichen) 316–17
parthenogenesis 242, 243
Passiflora caerulea (passion flower) 236
passion flowers 236
passive suspension feeding 44
pathogens 308
paxillae 196
pea plants 26
peanut worms 299
peas 26
pectin 205
pectinelle 41
Pediastrum duplex 290
pedicels 103, 172, 240, 304
Pediculus humanus capitis (human head louse) 306–07
pedipalps 57, 58, 59, 249
   see also palps
pedunculate oaks 236
Pelargonium Crispum (geranium) 208
pellicle 130
pellicule 101
penicillin 36
Penicillium 36–37
pennate diatoms 178–79, 338
Peripatus sp. (velvet worm) 156–57
perithecia 312
peristomes 233
Phacus (algae) 66–67
phage viruses 220–21
phagocytes 91
phagocytosis 38, 217
Phaneus vindex (rainbow scarab beetle) 158
phantom midges 146–47
Phascolion sp. (peanut worm) 299

phialides 230
phloem 52, 206–07
phoresy 172
photoreceptors 100–01, 111
photosynthesis 18–19, 20
   in algae 66, 67, 319
   in lichen 316
   and nitrogen 24–25
   and stomata 94
   sugar production in 314
phyllids 266
Phyllobates terribilis (golden poison frog) 190
phylum (classification) 322
Physarum polycephalum (slime mold) 245
phytoplankton 118–19, 286, 288
pigmentary color 114–15
pili 30, 202, 276
Pilobolus crystallinus (hat-thrower fungus) 160, 161
pincers 56–57, 59
pine nuts 263
pine trees 262
Pinus pinea (pine nut) 263
Pisaura mirabilis (nursery-web spider) 248–49
Pisum sativum (pea) 26
placenta 257
plankton 286–87
   copepods 288
   ctenophores 137
   diatoms 179
   dinoflagellates 184–85
   larvae 268
   phytoplankton 118–19, 286, 288
   pond life 291, 292
   radiolarians 182
   stentors 44–45
Planorbis planorbis (ramshorn snail) 292
plants, classifications of 340–45
plasma 90
Plasmodium 244, 245, 302, 310–11
plasmolysis 205
plates (skeletal) 196
Pleurobrachia pileus (Atlantic ctenophore) 136–37
Poaceae (grass) 236
poisonous animals 190
   see also venom
pollen 235, 239
   grain types 236–37
pollinators 237, 239, 240–41
polychaetes 299
polycystine radiolarians 331
polyextremophiles 278
Polyommatus icarus (common blue butterfly) 10–11
polyps 319
pond life 150–51, 290–91, 292–93
   see also algae
pond skaters 151
poppies 236, 262

pores 236–37
pork tapeworms 62–63
potatoes 15
powdery mildew 34–35
praying mantises 47, 258
pseudoscorpions 56
predatory microbes 40–41
predatory mites 58–59
probiotic bacteria 308
proboscises 49, 51, 52, 104, 302
prokaryotes 13
pronotum 162
prosomas 56
proteins 133, 177
   in eggs 229
   in muscles 141
   in plants 24, 26
   in radiolarians 182
   silk 64–65
proteobacteria 326
prothallus 260
Protoperidinium (dinoflagellate) 184–85
protozoans 12, 68, 151, 180, 311
   Didinium nastum 40–41
   Plasmodium 244, 245, 302, 310–11
Pseudochorthippus parallelus (meadow grasshopper) 163
pseudopods 38, 132–33, 180–81
pseudoscorpions 56–57, 172
Psoroptes sp. (mite) 304–05
psychrophiles 278
ptychocysts 42
pupas 84, 302
Purkinje cells 122–23
Pyrococcus furiosus 278–79
Pyrocystis fusiformis (ocean dinoflagellate) 118–19
Pyrrhosoma nymphula (large red damselfly) 270–71

Q

queens (breeding) 210, 239
Quercus robur (pedunculate oak) 236

R

rabbit ear mites 304–05
rabbit tapeworms 63
rachises 170
radicles 23, 264
radioactivity 77
radiolarians 12, 143, 182–83, 286, 331
radioles 82
radishes 22–23
radulae 54
rainbow scarab beetles 158
rami 143
ramshorn snails 292
Raphanus sativus (radish) 22–23

ray-finned fishes 395
rays 393
red admiral butterflies 114–15
red algae 340
red blood cells 14, 88, 90–91, 93, 310–11
Reddish ground beetle 46–47
red-knee tarantulas 212–13
reindeer lichen 282–83
reindeers 282
reproduction
   see also asexual reproduction; eggs; sexual reproduction; spores
   bacteria 224
   binary fission 130
   budding 78
   fertilization 228–29, 242, 256
   in plants 232, 235
   gametes 232, 235, 263, 310
   pollen 235, 236–37, 239
   of slime molds 244–45
reptiles 397
resilin 107
respiration 84–85, 86–87
   aerobic respiration 75, 78, 80–81
   anaerobic respiration 30, 78, 80–81
   gas exchange 82, 88
retina 110, 111
rheophile 285
Rhizobium bacteria 26
rhizoids 22, 266
Rhyssa persuasoria (ichneumon wasp) 102
ribbon worms 356
RNA (ribonucleic acid) 220, 223
rods (bacteria) 28–29, 76, 277, 308
Roman snails 54–55
roots 206, 264
   fungi 314
   galls on 300
   hairs 22–23
   nitrogen fixing 26
   simple plants 266
rotifers 134–35, 272, 298, 299, 354
roundworms 294

S

Sabellastarte magnifica (feather duster worm) 82–83
Saccharomyces boulardii 308
Saccharomyces cerevisiae (brewer's yeast) 78–79
sago palms 262
Salmonella sp. 32, 308
Salvinia natans (floating fern) 208
sanaidaster spicules 186
sandflies 311
sap 52–53
Sarcoptes sp. (scabies mite) 305

SARS (severe acute respiratory syndrome) 223
saturnid moths 214
scabies mites 305
scabrous leaves 208
scales
   butterflies 114–15
   halteres 166
   leaf 208, 260
   sharks 199
scanning electron microscopy (SEM) 11
scapes 103
Scenedesmus sp. 290
schizonts 310
Schlumbergera sp. (Christmas cactus) 236
scintillons 118, 119
sclerites 50, 193
scolex 62
scolopidium 106
scorpionflies 385
scorpions 366
sea anemones 42, 138, 319, 352
sea butterflies 286
sea cucumbers 148, 196, 387
sea lilies 388
sea potatoes 268–69
sea spiders 365
sea squirts 391
sea stars 388
sea urchins 196, 256, 268–69, 386
seaweeds 339, 340, 341
sedge 262, 263
sediment habitats 298
seedless vascular plants 343
seeds 168, 262–3, 264–65
seed shrimp 292
self-fertilization 235
   see also asexual reproduction
sensilla 104, 164, 165, 166
sensory systems
   algae 100–01
   antennae 102–03
   cnidarians 138
   coordination 124–25
   for flight 165, 166
   hearing 106–07, 108–09
   mouthparts 104, 114–15, 294
   sensilla 104, 164, 165, 166
   tails 272
   vision 110–11
Sepia latimanus (broadclub cuttlefish) 120
Sesamum indicum (sesame) 262
setae
   ants 210
   aquatic animals 142, 144, 145, 146
   arachnids 58, 154, 155, 191
   caterpillars 214
   diatoms 179
   mosquitos 166

sensory 107, 296
   urticating hairs 212–13
   weevils 193
setules 154, 155
sexual attachment (bacteria) 276
sexual dimorphism 247
sexual reproduction 228–29, 243
   see also asexual reproduction
   algae 226
   fungi 36, 230
   plants 234–35
sexual selection 246–47
sharks 198–99, 393
shells
   carapaces 143, 242, 296
   cellulose 184–85
   eggs 258, 259
   forams 180–81
   octopuses 120
shield bugs 48
shoots (seeds) 264
shrews 97
shrimp 144–45, 292, 370
sigma spicules 186
silica 179, 182, 187
silk 56, 64–65, 168, 248
silverfish 372
Sinella curviseta (springtail) 297
skeletal systems
   see also exoskeletons
   echinoderms 196
   muscles 14, 141
   radiolarians 182
   tissue 14
   vertebrates 200–01
skin 14, 97, 176–77
   see also hairs
   color 116–17, 120–21
   parasites 51, 304–05, 307
   sharks 198–99
   springtails 188
slime 25
slime molds 12–13, 244–45, 337
slipper animalcules 130
snails 54–55, 292
Solenopsis diplorotrum (fire ants) 190
Sorex minutus (Eurasian pygmy shrew) 97
sori 260
sound 106–07, 108
species (classification) 322
spermatozoon (sperm) 228–29
sperm nuclei (plants) 235
Sphagnum (bog moss) 204–05
spheres (bacteria) 28–29
spicules 186–87
spiders 366
   see also sea spiders
   aeroplankton 168
   body structures 58, 59, 154–55
   camouflage 194
   sight 110–11

   eggs 248–49
   webs 64–65
spigots 64
spines (defenses)
   caterpillars 214
   echinoderms 149, 196, 268
   parasites 50
   radiolarians 182
   sponges 186
spines (pollen) 237
spines (skeletal) 201
spinnerets 64–65
spiracles 84, 85
spirochete bacteria 28, 32
spirocysts 42
Spirodella sp. (greater duckweed) 20–21
Spirogyra sp. 290
spirrillums bacteria 29, 32
sponges 186–87, 348
spongy bone 14
sporangia 233, 260
   slime molds 244, 245
spores
   dispersal of 160, 161, 224
   ferns 260
   fungi 230, 312, 314
   mosses 232
   slime molds 244
sporophyte 232
sporopollenin 236
sporozoites 310
springtails 160–61, 188–89, 297, 299, 372
squid 360
stalk-eyed flies 246–47
Staphylococcus aureus 29
Staphylococcus epidermidis (bacteria) 276
starfish 148–49, 196, 389
stars, feather and brittle 388
stellate spicules 186
stem cells 15
stems 139
stenoteles 42
stentors 44–45, 67
Sternochetus mangiferae (mango seed weevil) 192–93
stick insects 376
stigma 66, 100, 101, 235
stingers 42, 210–11
stingrays 393
stomach 14, 32–33, 58, 149
stomata 14, 94
stony corals 353
streptobacillus bacteria 28
streptobacillus palisades bacteria 28
Streptococcus 28, 308
   S. pyogenes 28, 221
stromatolites 18–19
strongylaster spicules 186
styles 235
stylets 51, 52
Subbotina (foram) 180

subesophageal ganglion 124
suckers
   larvae 285
   octopuses 121
   parasites 63
sugar 52, 78, 81
   see also glucose
sundews 24–25
sunlight 18, 20, 73
sunstars 196–97
superpositional eyes 112
surface tension 150–51
suspension feeding 44, 45
swallows 170–71
swimming 142, 144–45
   buoyancy 146–47
   drag 199
sword ferns 260–61
syconium wall 240
Symbiodinium 319
symbiotic relationships 67, 300, 314–15, 319
symmetry 178, 226, 227
symphylans 297
synapse 122
Synedra sp. (diatom) 178

T

Taenia pisiformis (rabbit tapeworm) 63
Taenia solium (pork tapeworm) 62–63
tails 96, 147, 170
tapeworms 62–63
tarantulas 212–13
tardigrades 272, 280–81, 298, 364
tarsi 59, 154, 158, 159
   see also feet
taste 47, 104, 294
teeth 54–55, 201
temperature regulation 202
tenaculum 160
tentacles
   anemones 319
   carnivorous plants 24–25
   cnidarians 42, 138
   ctenophores 136, 137
termites 68–69
tetraxon spicules 186
Thalassiosira sp. (diatom) 178
thale cress 209
thallus 266, 267
theca 184
thecogen cells 104
thermophiles 278
thermoregulation 202
thigmotaxic leaves 25
Thiocystis (bacteria) 100
thistle gall flies 300
thoracopods 144, 145
thoraxes
   ants 125
   muscles 140
   pseudoscorpions 56, 59
   thrips 169
threadworms 294
thrips 168–69, 382
thrust 168
Thysia wallichii (flat-faced longhorn beetle) 102–03
tibias 159, 162, 163
Tillandsia sp. (air plant) 208
tissue 14–15
toadstools 35, 230
touch screens 277
toxicysts 40
tracheae 82, 84–85, 106
   aquatic insects 86–87, 146–47
tracheoles 85, 87
Trachipterus trachypterus (Mediterranean dealfish) 286
transmission electron microscopy (TEM) 11
transpiration 94
trap-jawed ant 210
Trebouxia (algae) 316
treehoppers 162–63
trees 205, 262–63
triaxon spicules 186
Triceratium sp. (diatom) 178, 179
Trichia decipiens (slime mold) 244–45
trichobothria 155
Trichodina pediculus 290
trichogens 193, 214
Trichonympha 68
trichromes 208–09
Trifolium sp. (clover) 26–27
Triticum sp. (wheat) 264–65
trophozoites 310
true bugs 48, 258, 379
true flies 384
tube feet 148–49
tube-snout fish 200–01
tubules 176, 177
tubulin 177
tuns 281
turgor pressure 205
turtle mites 172–73
tympanal organs 106–07

U

ultra-small archaeans 325
ultraviolet (UV) light 111
uranium 77
urchins 196, 256, 268–69, 386
urease 32–33
Uroleucon sp. (hawksbeard aphid) 243
Urophora cardui (thistle gall fly) 300
Uropoda sp. (turtle mite) 172–73
urticating hairs 212–13

V

vacuoles 35, 38, 130, 182, 205
valves 92, 178–79
Van der Waals forces 154–55
Vanessa atalanta (red admiral butterfly) 114–15
Vanessa cardui (painted lady butterfly) 49
Varroa destructor (mite) 172
vascular bundles 26, 206, 207
   see also phloem; xylem
vascular plants 14, 22, 266, 343
veins 92, 93
   wings 164, 165
velvet shank mushrooms 230
velvet worms 156–57, 364
venom
   ants 210
   arachnids 56, 58, 59, 111
   backswimmers 152
   caterpillars 214
   cnidarians 42
venule 60, 92
Venus's girdle 136
vertebrates
   see also human body
   circulatory systems 93
   embryos 256–57
   nervous systems 125
   skeletal systems 200–01
vesicles 35, 118, 119, 166, 228
Vespula vulgaris (common wasp) 48
vestibula 130
Vibrio cholerae 28
Viburnum sp. 234–35
villi 60–61
viruses 11, 220–21
   coronavirus 223
vision 110–11
   see also eyes
vitamin K2 30
Volvox sp. (algae) 252–53

W

wasps 380
   body structure 48, 102, 210
   parasitic wasps 240–41, 300–01
   parental care 248
   as prey 111
water absorption (osmosis) 22, 205
water bears (tardigrades) 272, 280–81, 298
water boatman 152
water fleas 142–43, 243, 368
watermeal duckweed 20
water molds 337
water repellency 151, 188, 190, 191
water surface 150–51
water-vascular systems 148
wavelengths
   for color production 115, 116, 137
   in microscopy 11
webs 64–65
weevils 192–93, 312, 322
western dusk singing cicadas 48
wheat 264–65
whirligig beetles 150–51
whiskers 202
white blood cells 90, 91, 133, 217
wings 164–65
   crickets 106
   damselflies 271
   feathers 170
   halteres 166–67
   hummingbirds 96
   mayflies 273
   muscles 141
   thrips 168–69
witch-hazel 208
Wolffia sp. (watermeal duckweed) 20
wood 68
woody plants 205, 206, 315
worms
   annelid 361
   aquatic 82–83, 336, 362, 389
   nematode 294–95, 217, 299
   parasitic 294, 334
   velvet worms 156–57, 364
Wuchereria bancrofti (nematode worm) 217

X

xenosomes 39
xylem 14, 206–07

Y

yeasts 30, 78–79, 245, 308, 347
yolk sacs 256, 257, 258

Z

zona pellucida 228, 229
zooplankton 286
zooxanthellae 318–19
zygotes 310

acknowledgments

DK would like to thank the following people at Smithsonian Enterprises:

Product Development Manager
Kealy Gordon
Director, Licensed Publishing
Jill Corcoran
Vice President, Business Development and Licensing
Brigid Ferraro
President
Carol LeBlanc

DK would also like to extend a special mention to: Martin Oeggerli at *Micronaut* (www.micronaut.ch), Oliver Meckes at *Eye of Science* (www.eyeofscience.de), and Charlotte Peterson-Hill at *Science Photo Library* (www.sciencephoto.com) for their help in supplying and interpreting exceptional scanning electron micrographs.

DK would also like to thank:

Senior Editor:
Helen Fewster
Senior Art Editors:
Duncan Turner
Sharon Spencer
Design assistance:
Jessica Tapolcai
Editorial assistance:
Ankita Gupta
Tina Jindal
Jackets assistance:
Priyanka Sharma
Saloni Singh
Image retouching:
Steve Crozier
Additional illustrations:
Mark Clifton
Dan Crisp
Reference section contributor:
Richard Beatty
Indexer:
Elizabeth Wise
Proofreader:
Richard Gilbert

The publisher would like to thank the following for their kind permission to reproduce their photographs:

(Key: a-above; b-below/bottom; c-center; f-far; l-left; r-right; t-top)

1 Science Photo Library: Steve Gschmeissner. **2-3 Science Photo Library:** Steve Gschmeissner. **4-5 Igor Siwanowicz:** (t). **6-7 Igor Siwanowicz. 8-9 David Liittschwager:** Natural History Photography. **10 Science Photo Library:** Eye Of Science. **11 Science Photo Library:** Wim Van Egmond (tc); Steve Gschmeissner (tr, bc/Mite); Power And Syred (ftr); Gerd Guenther (bc); Eye Of Science (br); Claus Lunau (fbr). **12-13 Andre De Kesel. 14 Science Photo Library:** Wim Van Egmond (bc); Steve Gschmeissner (tl, tc, tr, c, cr); Anne Weston, Em Stp, The Francis Crick Institute (cl); Marek Mis (bl, br). **15 Getty Images:** Science Photo Library / SPL / Steve Gschmeissner (tl). **Science Photo Library:** Daniel Schroen, Cell Applications Inc (cl); Marek Mis (bl). **Sebastian Vieira:** (br, cr). **16-17 Getty Images / iStock:** DigitalVision Vectors / Grafissimo. **18-19 Science Photo Library:** Eye Of Science. **18 Alamy Stock Photo:** Mint Images Limited / Mint Images (cl). **20-21 Science Photo Library:** Biophoto Associates. **20 Dreamstime.com:** Lertwit Sasipreyajun (tl). **22-23 Science Photo Library:** Dennis Kunkel Microscopy. **23 Alamy Stock Photo:** Nigel Cattlin (cla). **24-25 Getty Images / iStock:** CathyKeifer. **24 Dorling Kindersley:** Makoto Honda / www.honda-e.com (t). **Damien L'Hours', Neo-Rajah:** (br). **26 naturepl.com:** Visuals Unlimited (c). **26-27 Science Photo Library:** Eye Of Science. **28 Science Photo Library:** (tr); Eye Of Science (tl, c); Dr Gary Gaugler (tc); SCIMAT (cl); Steve Gschmeissner (cr, br); AMI Images (bl); Dennis Kunkel Microscopy (bc). **29 Science Photo Library:** (tl, bl); Dennis Kunkel Microscopy (cl); Eye Of Science (br). **30 Science Photo Library:** (cb). **30-31 M. Oeggerli:** Micronaut 2008, supported by School of Life Sciences, FHNW. **32-33 M. Oeggerli:** Micronaut 2008, kindly supported by FHNW. **32 Science Photo Library:** Eye Of Science (crb). **34-35 Andre De Kesel. 35 Getty Images / iStock:** Matthew J Thomas (ca). **36 Science Photo Library:** Wim Van Egmond (cb). **36-37 Science Photo Library:** Steve Gschmeissner. **38 Science Photo Library:** Wim Van Egmond (tl). **38-39 Science Photo Library:** Frank Fox. **40-41 Science Photo Library:** Eye Of Science. **40 Science Photo Library:** Biophoto Associates (t). **42 Dr. Robert Berdan:** Science & Art Multimedia, www.canadiannaturephotographer.com (tl). **42-43 Waldo Nell. 44-45 Igor Siwanowicz. 45 Waldo Nell:** (clb). **46-47 Arno van Zon. 47 Science Photo Library:** K Jayaram (cr). **48 Alamy Stock Photo:** alimdi.net (bc); Minden Pictures / Ingo Arndt (tc); Nature Picture Library / Alex Hyde (cl); Phil Degginger (c); Razvan Cornel Constantin (bl). **Dreamstime.com:** Razvan Cornel Constantin (tr); Brett Hondow (tl). **Jan Rosenboom:** (cr, br). **49 Dreamstime.com:** Razvan Cornel Constantin (tr). **Getty Images / iStock:** ConstantinCornel (cl). **naturepl.com:** Klein & Hubert (bl). **50 M. Oeggerli:** Micronaut 2011, with kind support of FHNW. **51 naturepl.com:** Kim Taylor (tr). **52 Dreamstime.com:** Edward Phillips (tl). **52-53 Andre De Kesel. 54 Dreamstime.com:** Travellingtobeprecise (c). **54-55 Science Photo Library:** Eye Of Science. **56-57 Science Photo Library:** Power And Syred. **57 Dorling Kindersley:** Von Reumont, B.M.; Campbell, L.I.; Jenner, R.A. Quo Vadis Venomics? A Roadmap to Neglected Venomous Invertebrates. Toxins 2014, 6, 3488-3551. https://doi.org/10.3390/toxins6123488 (bc). **Dreamstime.com:** Mularczyk (tr). **58-59 M. Oeggerli:** Micronaut 2014, supported by School of Life Sciences, FHNW. **59 Dreamstime.com:** Cathy Keifer (cr). **60 Science Photo Library:** Steve Gschmeissner (tr, cra). **61 Science Photo Library:** Steve Gschmeissner. **62 Science Photo Library:** Steve Gschmeissner. **63 Science Photo Library:** Steve Gschmeissner (b). **64 Science Photo Library:** Dennis Kunkel Microscopy (bc). **65 Alamy Stock Photo:** Minden Pictures / NiS / Jogchum Reitsma. **66 Waldo Nell. 67 Waldo Nell:** (t). **Science Photo Library:** Marek Mis (clb). **68 Science Photo Library:** Michael Abbey (tl). **68-69 eye of science. 70-71 Getty Images / iStock:** DigitalVision Vectors / Nastasic. **72-73 Daniel Kordan (c):** (t). **72 Dreamstime.com:** Cathy Keifer (crb). **74-75 Science Photo Library:** Medimage. **75 Science Photo Library:** Thomas Deerinck, Ncmir (tr). **76-77 Science Photo Library:** Eye Of Science. **77 Science Photo Library:** Eye Of Science (r). **78 Science Photo Library:** Steve Gschmeissner (cra). **78-79 Science Photo Library:** Eye Of Science. **80 Science Photo Library:** AMI Images (c). **80-81 Science Photo Library:** Eye Of Science. **82 Dreamstime.com:** John Anderson (c). **82-83 Getty Images:** Universal Images Group / Wild Horizons. **84 Dreamstime.com:** Raulg2 (clb). **84-85 Science Photo Library:** Steve Gschmeissner. **85 Dreamstime.com:** Roblan (tr). **86-87 Science Photo Library:** Marek Mis. **87 Alamy Stock Photo:** Nature Photographers Ltd / Paul R. Sterry (cr). **88 Science Photo Library:** Zephyr (tl). **88-89 Science Photo Library:** Eye Of Science. **90 Science Photo Library:** Steve Gschmeissner (cra). **90-91 Science Photo Library:** Eye Of Science. **92 Science Photo Library:** Susumu Nishinaga. **93 Science Photo Library:** Lennart Nilsson, TT (tr). **94 Dreamstime.com:** Vaclav Volrab (cl). **94-95 Science Photo Library:** Marek Mis. **96 Alamy Stock Photo:** All Canada Photos / Glenn Bartley. **97 Alamy Stock Photo:** All Canada Photos / Glenn Bartley (cl). **Dreamstime.com:** Rudmer Zwerver (cr). **98-99 Getty Images:** Moment / mikroman6. **100 Waldo Nell:** (bl). **Science Photo Library:** Alfred Pasieka (tc). **100-101 Waldo Nell:** (c). **101 Waldo Nell:** (b). **102 Alamy Stock Photo:** imageBROKER / Volker Lautenbach (cr). **Biodiversity Heritage Library:** Smithsonian Libraries (bc). **102-103 Andre De Kesel. 104-105 Science Photo Library:** Steve Gschmeissner. **104 Dreamstime.com:** Alptraum (clb). **106 Alamy Stock Photo:** Flonline digitale Bildagentur GmbH / Matthias Lenke (bl). **naturepl.com:** MYN / Paul van Hoof (crb). **107 Alamy Stock Photo:** blickwinkel / Lenke. **108 Science Photo Library:** Eye Of Science (bc); Dr David Furness, Keele University (cb). **108-109 Science Photo Library:** Eye Of Science. **110-111 Khairul Bustomi. 111 Dreamstime.com:** Dwi Yulianto (tl). **112 Science Photo Library:** Javier Torrent, Vw Pics (tl). **112-113 Getty Images:** 500px / Voicu Iulian. **114-115 Andre De Kesel. 115 Science Photo Library:** Eye Of Science (c). **116 Dreamstime.com:** Cătălin Gagiu (clb). **116-117 Waldo Nell. 118-119 Science Photo Library:** Gerd Guenther. **119 Alamy Stock Photo:** Eric Nathan (tc). **120 Alamy Stock Photo:** Nature Picture Library / Georgette Douwma (tr). **Alexander Semenov:** (b). **121 Alexander Semenov:** (t). **122 Dreamstime.com:** Piyapong Thongdumhyu (cra). **123 Science Photo Library:** Thomas Deerinck, NCMIR. **124-125 Getty Images:** 500Px Plus / Yudy Sauw. **125 Alamy Stock Photo:** Nature Picture Library / MYN / Lily Kumpe (tr). **126-127 Dreamstime.com:** Evgeny Turaev. **128-129 Science Photo Library:** Frank Fox. **130 Alamy Stock Photo:** blickwinkel / Fox (cb). **130-131 Alamy Stock Photo:** Panther Media GmbH / Kreutz. **132 Alamy Stock Photo:** blickwinkel / Guenther (tl); Panther Media GmbH / Kreutz (bl). **Science Photo Library:** Gerd Guenther (c). **132-133 Getty Images / iStock:** micro_photo (tc). **133 Science Photo Library:** David M. Phillips (tr). **134-135 Waldo Nell. 134 Biodiversity Heritage Library:** Smithsonian Libraries (tc). **135 Science Photo Library:** Rogelio Moreno (tr). **136 Alamy Stock Photo:** Biosphoto / Christoph Gerigk (crb). **Science Photo Library:** Wim Van Egmond (c). **137 Science Photo Library:** Wim Van Egmond. **138 Science Photo Library:** Jannicke Wiik-Nielsen (bc). **138-139 Science Photo Library:** Jannicke Wiik-Nielsen. **140-141 Waldo Nell. 141 Dreamstime.com:** Vlasto Opatovsky (cr). **142-143 Science Photo Library:** Rogelio Moreno. **143 Science Photo Library:** Wim Van Egmond (crb). **144-145 Igor Siwanowicz. 145 Science Photo Library:** Frank Fox (cr). **146 Science Photo Library:** Frank Fox. **147 Alamy Stock Photo:** blickwinkel / Hartl (cr). **Science Photo Library:** Frank Fox (tc). **148 Alamy Stock Photo:** agefotostock / Marevision (bl). **148-149 Evan Darling:** Slide preparation and imaging by Evan Darling, whos other image making can be seen on Instagram @lambdascans. **150-151 Alamy Stock Photo:** blickwinkel / Lenke. **151 Getty Images / iStock:** JanMiko (cr). **152 naturepl.com:** Jan Hamrsky (cb). **152-153 naturepl.com:** Jan Hamrsky. **154-155 Science Photo Library:** Eye Of Science. **154 Dorling Kindersley:** Jonas O. Wolff ,Wolfgang Nentwig,Stanislav N. Gorb, doi: https://doi.org/10.1371/journal.pone.0062682.g001 (cla). **Shamsul Hidayat:** (c). **156-157 Science Photo Library:** Dr Morley Read. **157 Science Photo Library:** Alex Hyde (tl). **158 Alamy Stock Photo:** Nature Picture Library (br). **Andre De Kesel:** (bl). **159 Dreamstime.com:** Angel Luis Simon Martin (bl). **Andre De Kesel:** (br). **160-161 Science Photo Library:** Eye Of Science. **160 Dorling Kindersley:** Sakes, Aimee & Wiel, Marleen & Henselmans, Paul & Van Leeuwen, Johan L & Dodou, Dimitra & Breedveld, Paul. (2016). Shooting Mechanisms in Nature: A Systematic Review. PLOS ONE. 11. e0158277. 10.1371 / journal.pone.0158277. (cla). **161 Science Photo Library:** Wim Van Egmond (tr). **162-163 Andre De Kesel**. **163 Alamy Stock Photo:** Minden Pictures / Stephen Dalton (tc). **164-165 Science Photo Library:** Marek Mis. **165 Dreamstime.com:** Romantiche (cr). **166 Science Photo Library:** Steve Gschmeissner (tl). **167 Andre De Kesel**. **168-169 Science Photo Library:** Steve Gschmeissner. **168 Alamy Stock Photo:** Nature Picture Library / Michael Hutchinson (tl). **Science Photo Library:** Claude Nuridsany & Marie Perennou (tc). **170 Paul Hollingworth Photography:** (b). **171 Science Photo Library:** Power And Syred. **172-173 John Hallmen**. **172 Getty Images / iStock:** Bee-individual (tl). **174-175 Library of Congress, Washington, D.C.:** Giltsch, Adolf, 1852-1911, LC-DIG-ds-07542. **176-177 Science Photo Library:** Dr Torsten Wittmann. **177 Science Photo Library:** Don W. Fawcett (cr). **178 Science Photo Library:** Dennis Kunkel Microscopy (bl); Steve Gschmeissner (tl, tc, tr, cl, c, cr, bc, br). **179 Science Photo Library:** Steve Gschmeissner (tl, tr, cl, bl). **180-181 Science Photo Library:** Steve Gschmeissner. **180 Science Photo Library:** Steve Gschmeissner (tl, tc, tr). **182 Science Photo Library:** Wim Van Egmond (tl). **183 Biodiversity Heritage Library:** Harvard University, Museum of Comparative Zoology, Ernst Mayr Library. **184 Science Photo Library:** Steve Gschmeissner (ca). **185 Science Photo Library:** Steve Gschmeissner. **186 Science Photo Library:** Eye Of Science (br); Steve Gschmeissner (tl, tc, cl, c, cr, bl, bc); Ikelos GMBH / Dr. Christopher B. Jackson (tr). **187 Science Photo Library:** Steve Gschmeissner (cl, bl); Natural History Museum, London (r). **188 Science Photo Library:** Eye Of Science (cl). **188-189 Science Photo Library:** Eye Of Science. **190-191 M. Oeggerli:** Micronaut 2013. With kind support by School of Life Sciences, FHNW. **191 Science Photo Library:** Steve Gschmeissner (crb). **192 © State of Western Australia (Department of Primary Industries and Regional Development, WA):** Pia Scanlon. **193 Greg Bartman:** USDA APHIS PPQ, Bugwood.org (tr). **194-195 SuperStock:** Ch'ien Lee. **194 Alamy Stock Photo:** Biosphoto / Husni Che Ngah (cla). **Dreamstime.com:** Cathy Keifer (c); Lidia Rakcheeva (tl); Mohd Zaidi Abdul Razak (tc). **SuperStock:** Minden Pictures (ca); Minden Pictures / Piotr Naskrecki (cl). **196 Alamy Stock Photo:** Islandstock (cl). **196-197 Alexander Semenov**. **198-199 Science Photo Library:** Ted Kinsman. **199 OceanwideImages.com:** Gary Bell (tc). **Science Photo Library:** Ted Kinsman (cra). **200-201 Dr. Adam P. Summers**. **201 Dreamstime.com:** Awcnz62 (c). **202-203 Science Photo Library:** Steve Gschmeissner. **204-205 Waldo Nell**. **204 Science Photo Library:** Marek Mis (br). **206-207 Science Photo Library:** Steve Gschmeissner. **206 123RF.com:** Gabriele SiebenhÃ¼hner (crb). **Dreamstime.com:** Svetlana Foote (c). **208 Science Photo Library:** Eye Of Science (tr, cl, cr, bl, bc, br); Martin Oeggerli (tl); Steve Gschmeissner (tc). **SuperStock:** Minden Pictures / Albert Lleal (c). **209 Science Photo Library:** Stefan Diller (tl); Visuals Unlimited, Inc. / Dr. Stanley Flegler (cl); Eye Of Science (bl); Science Source / Ted Kinsman (r). **210 Alamy Stock Photo:** Minden Pictures / Mark Moffett (tl). **210-211 M. Oeggerli:** Micronaut 2018, supported by Pathology, University Hospital Basel, and School of Life Sciences, FHNW. **212 Shutterstock.com:** pets in frames (cl). **212-213 Science Photo Library:** Power And Syred (c). **213 Science Photo Library:** Power And Syred. **214 Dreamstime.com:** Juan Francisco Moreno GÃmez (cl). **214-215 Igor Siwanowicz**. **216-217 Science Photo Library**. **217 Science Photo Library:** Eye Of Science (tr). **218-219 Alamy Stock Photo:** Quagga Media. **220-221 Science Photo Library**. **222-223 Science Photo Library:** NIAID / National Institutes Of Health. **223 Science Photo Library:** KTSDESIGN (cb). **224 Dreamstime.com:** Iuliia Morozova (cl). **224-225 Science Photo Library:** Eye Of Science. **226-227 Science Photo Library:** Wim Van Egmond (t). **226 Avalon:** M I Walker (crb). **228 Science Photo Library:** Lennart Nilsson, TT (tr). **229 Science Photo Library:** Eye Of Science. **230 Science Photo Library:** Nature Picture Library / Alex Hyde (cl). **230-231 Science Photo Library:** Eye Of Science. **232 Getty Images / iStock:** Anest (c). **Science Photo Library:** Eye Of Science (br). **233 Science Photo Library:** Eye Of Science. **234-235 Martin Oeggerli:** Micronaut. **235 Dreamstime.com:** Alfio Scisetti (c). **236 Science Photo Library:** Eye Of Science (cl, br); Steve Gschmeissner (tl, tr, c, cr, bc); Power And Syred (tc); Ikelos Gmbh / Dr. Christopher B. Jackson (bl). **237 Dorling Kindersley:** Sue Barnes / EMU Unit of the Natural History Museum, London (bl). **Science Photo Library:** Eye Of Science (cl); Steve Gschmeissner (tr); Linear Imaging / Linnea Rundgren (tr). **238-239 Getty Images:** John Kimbler. **239 Science Photo Library:** Susumu Nishinaga (cb). **240 Science Photo Library:** Steve Lowry. **241 Alamy Stock Photo:** Minden Pictures / Mark Moffett (tr). **242-243 Science Photo Library:** Marek Mis. **243 naturepl.com:** Doug Wechsler (c). **244-245 Science Photo Library:** Eye Of Science. **245 Alamy Stock Photo:** Nature Picture Library (tr). **246-247 M. Oeggerli:** Micronaut 2013, supported by School of Life Sciences, FHNW. **247 Biodiversity Heritage Library:** Smithsonian Libraries (br). **naturepl.com:** Mark Moffett (bc). **Martin Oeggerli:** Micronaut 2013, kindly supported by School of Life Sciences, FHNW (cr); Micronaut 2013, supported by School of Life Sciences, FHNW (crb). **248 Dreamstime.com:** Kengriffiths6 (tl, tc, tr). **248-249 Andre De Kesel**. **250-251 Alamy Stock Photo:** Patrick Guenette. **252-253 Science Photo Library:** Frank Fox. **252 Science Photo Library:** AMI Images (tc). **254 Science Photo Library:** Biophoto Associates (tr). **254-255 Science Photo Library:** Steve Gschmeissner (b). **256 Science Photo Library:** Gerard Peaucellier, ISM (tr). **256-257 Daniel Knop:** (b). **258 Alamy Stock Photo:** bilwissedition Ltd. & Co. KG (cl); Nik Bruining (bl). **naturepl.com:** Hans Christoph Kappel (br). **Science Photo Library:** Nicolas Reusens (bc). **258-259 Martin Oeggerli:** Micronaut. **260 Alamy Stock Photo:** Chris Lloyd (c). **260-261 Science Photo Library:** Rogelio Moreno. **262 Alamy Stock Photo:** Hans Stuessi (cl). **Dreamstime.com:** Cl2004lhy (tr); Antonio Ribeiro (tl); Tatsuya Otsuka (tc). **Science Photo Library:** Steve Gschmeissner (bl); Gerd Guenther (c); Petr Jan Juracka (cr); Power And Syred (bc, br). **263 Science Photo Library:** Eye Of Science (cl); Th Foto-Werbung (tl); SCIMAT (bl); US Geological Survey (br). **264-265 Science Photo Library:** Eye Of Science. **266 Dreamstime.com:** Callistemon3 (c). **266-267 Science Photo Library:** Magda Turzanska. **268 Biodiversity Heritage Library:** Smithsonian Libraries (bl). **268-269 Science Photo Library:** Wim Van Egmond. **269 Alamy Stock Photo:** Steve. Trewhella (tr). **270 Science Photo Library:** Nature Picture Library / / 2020vision / Ross Hoddinott (bl, br). **271 Science Photo Library:** Nature Picture Library / / 2020vision / Ross Hoddinott (bl, br). **272-273 Andre De Kesel**. **273 Science Photo Library:** Nature Picture Library / Jan Hamrsky (tr). **274-275 Getty Images / iStock:** DigitalVision Vectors / ilbusca. **276-277 Science Photo Library:** Steve Gschmeissner. **276 Science Photo Library:** Steve Gschmeissner (tl). **278-279 Science Photo Library:** Eye Of Science. **280-281 eye of science**. **281 Science Photo Library:** Eye Of Science (cb). **282 Dreamstime.com:** Anna Krivitskaia (c). **282-283 Science Photo Library:** Edward Kinsman. **284-285 eye of science**. **285 Dreamstime.com:** Wirestock (br). **286 Science Photo Library:** John Burbidge (bc); Alexander Semenov (br, fbr). **287 Getty Images / iStock:** ZU_09. **288 Waldo Nell:** (r). **Science Photo Library:** Teresa Zgoda (cb). **289 Waldo Nell:** (b, tr). **290 Science Photo Library:** Wim Van Egmond (tl, tc, cl, c, cr, bc); Rogelio Moreno (tr); Gerd Guenther (bl); Frank Fox (br). **291 Science Photo Library:** Wim Van Egmond (tl, cl); Rogelio Moreno (bl); Gerd Guenther (tr). **292 Alamy Stock Photo:** RGB Ventures / Charles Krebs (tl). **naturepl.com:** Jan Hamrsky (tc, tc/Diving Beetle). **293 M. Oeggerli:** Micronaut 2010, kindly supported by FHNW, Muttenz. **294-295 Science Photo Library:** Steve Gschmeissner. **294 Science Photo Library:** Dennis Kunkel Microscopy (cb). **296-297 M. Oeggerli:** Micronaut 2014, supported by School of Life Sciences, FHNW. **297 naturepl.com:** Doug Wechsler (tc). **Science Photo Library:** Wim Van Egmond (tr). **298 Science Photo Library:** Eye Of Science. **299 Science Photo Library:** Dennis Kunkel Microscopy (cl, bc); Eye Of Science (tl); Steve Gschmeissner (tc, cr); David Scharf (tr, c, bl, br). **300 Dreamstime.com:** Henrikhl (tr); Paul Reeves (tc). **Science Photo Library:** Nature Picture Library / Solvin Zankl (tl). **301 Alamy Stock Photo:** Science History Images. **302-303 Science Photo Library:** Eye Of Science. **302 Science Photo Library:** Claude Nuridsany & Marie Perennou (cb). **304-305 M. Oeggerli:** Micronaut 2014, supported by School of Life Sciences, FHNW. **305 Science Photo Library:** Steve Gschmeissner (tl). **306 Alamy Stock Photo:** blickwinkel / H. Bellmann / F. Hecker (clb). **306-307 M. Oeggerli:** Micronaut 2007. **308 Science Photo Library:** Steve Gschmeissner (c). **308-309 M. Oeggerli:** Micronaut, 2012, kindly supported by School of Life Sciences, FHNW. **310 Science Photo Library:** Eye Of Science (tl, c). **311 Science Photo Library:** Eye Of Science (cl, clb, bc, br, cr). **312 Alamy Stock Photo:** Biosphoto / Frank Deschandol & Philippe Sabine (b). **313 John Hallmen**. **314 Dreamstime.com:** Heinz Peter Schwerin (cl). **314-315 Science Photo Library:** Eye Of Science. **316-317 Science Photo Library:** Eye Of Science. **318-319 Waldo Nell**. **319 Dreamstime.com:** Andy Nowack (cr). **320-321 Dreamstime.com:** Patrick Guenette. **322 Andre De Kesel**. **323 Science Photo Library**. **324 Science Photo Library:** Eye Of Science (crb); Power And Syred (cra). **325 Berkeley Lab:** (crb). **Science Photo Library:** Eye Of Science (cra). **326 Getty Images / iStock:** powerofforever (cra). **Wellcome Collection:** Tuberculosis and anthrax bacilli, actinomyces and micrococci, seen through a microscope. Five photographs. Wellcome Collection. Public Domain Mark (crb). **327 Science Photo Library:** Dennis Kunkel Microscopy (cra); Claire Ting (crb). **328 Alamy Stock Photo:** BSIP SA / BOSCARIOL (cra). **Pacific Northwest National Laboratory:** TEM micrograph of Deinococcus radiodurans. Image courtesy Alice Dohnalkova, Pacific Northwest National Laboratory (bc). **329 Science Photo Library:** CNRI (cra, crb). **330 Alamy Stock Photo:** Alpha Stock (crb). **Wellcome Collection:** Trypanosomes at 1000X magnification. Coloured process print after R. D. Muir. Wellcome Collection. Public Domain Mark (cra). **331 Biodiversity Heritage Library:** Harvard University, Museum of Comparative Zoology, Ernst Mayr Library (cra, crb). **332 Biodiversity Heritage Library:** MBLWHOI Library (cr). **NOAA:** Image courtesy of IFE, URI-IAO, Lost City science party, and NOAA. (bc/ Xenophyophorea). **Science Photo Library:** Wim Van Egmond (bl, br); Steve Gschmeissner (bc). **333 Getty Images:** De Agostini Editorial / De Agostini Picture Library (bl). **Science Photo Library:** Steve Gschmeissner (cr, bc, br). **334 Dorling Kindersley:** David Patterson / Bob Andersen (br). **Getty Images / iStock:** itsme23 (cr); Sinhyu (bc); Oxford Scientific (bc/ Pyrocystis). **SuperStock:** Minden Pictures /

Albert Lleal (bl). **335 artscult.com:** (cra). **Getty Images / iStock:** powerofforever (crb). **336 Biodiversity Heritage Library:** Cornell University Library (cr). **Getty Images / iStock:** micro_photo (bc). **Science Photo Library:** Gerd Guenther (bc/Lesquereusia, br); Marek Mis (bl). **337 Alamy Stock Photo:** Glasshouse Images / JT Vintage (bc). **Biodiversity Heritage Library:** University of Toronto - Gerstein Science Information Centre (cra). **338 Biodiversity Heritage Library:** New York Botanical Garden, LuEsther T. Mertz Library (cra); Smithsonian Libraries (br). **339 Alamy Stock Photo:** AF Fotografie (cra). **artscult.com:** (br). **340 Alamy Stock Photo:** Nature Picture Library / Sue Daly (bc/Dulse); Nick Upton (bl); Roberto Nistri (bc). **Biodiversity Heritage Library:** Smithsonian Libraries (cr). **Dreamstime.com:** Sandyloxton (br). **341 Biodiversity Heritage Library:** MBLWHOI Library (cr). **Dreamstime.com:** Ashish Kumar (bc); Wrangel (bc/Sea lettuce); Mps197 (br). **Shutterstock.com:** Lebendkulturen.de (bl). **342 Alamy Stock Photo:** Chris Mattison (bc/Red Bog Moss). **Biodiversity Heritage Library:** University Library, University of Illinois Urbana Champaign (cr). **Dreamstime.com:** Dmitry Maslov (bc); Ian Redding (br). **Shutterstock.com:** Gondronx Studio (bl). **343 Biodiversity Heritage Library:** Cornell University Library (cr). **Dreamstime.com:** Steve Callahan (bc); Mkojot (bc/Delta Maidenhair); Orest Lyzhechka (bl); Karin De Mamiel (br). **344 Biodiversity Heritage Library:** New York Botanical Garden, LuEsther T. Mertz Library (cr). **Dreamstime.com:** Bennnn (bl); Adam Radosavljevic (br); Simona Pavan (bc/Ginkgo); Puripat Penpun (bc). **345 Biodiversity Heritage Library:** New York Botanical Garden, LuEsther T. Mertz Library (cr). **Dreamstime.com:** Cindy Daly (br); Lukich (bc); Chris Moncrieff (bc/Southern Marsh Orchid); Twilightartpictures (bl). **346 Biodiversity Heritage Library:** MBLWHOI Library (crb). **Dreamstime.com:** Piyapong Thongdumhyu (cra). **347 Biodiversity Heritage Library:** New York Botanical Garden, LuEsther T. Mertz Library (crb). **Wellcome Collection:** Candida albicans. Wellcome Collection. CC0 1.0 Universal (cra). **348 Biodiversity Heritage Library:** American Museum of Natural History Library (cr). **Dreamstime.com:** John Anderson (bl); Christian Weiß (bl); Kevin Panizza / Kpanizza (bc/Yellow Tube Sponge); Seadam (br). **349 Alamy Stock Photo:** WaterFrame / WaterFrame_fba (bl). **Biodiversity Heritage Library:** Smithsonian Libraries (cr). **Dreamstime.com:** Tignogartnahc (bc); Vojcekolevski (br). **350 Biodiversity Heritage Library:** Smithsonian Libraries (cr). **Dreamstime.com:** Aldorado10 (bl); Daniel Poloha (bc/Hydroid Fan); Jeremy Brown (br). **Getty Images / iStock:** micro_photo (bc). **351 Alamy Stock Photo:** Andrey Nekrasov (br). **Biodiversity Heritage Library:** University Library, University of Illinois Urbana Champaign (cr). **Dreamstime.com:** Pniesen (bc/Cyanea capillata); Koval Viktoria (bl); Lynda Dobbin Turner (bc). **352 Biodiversity Heritage Library:** Harvard University, Museum of Comparative Zoology, Ernst Mayr Library (cr). **Dreamstime.com:** Deeann Cranston (br); Vojcekolevski (bl); Steve Estvanik (bc/Plumose Anemone). **naturepl.com:** Chris Newbert (bc). **353 Alamy Stock Photo:** Florilegius (br). **Dreamstime.com:** John Anderson (bl, br); Cigdem Sean Cooper (bc); Vojcekolevski (bc/Torch Coral). **354 Biodiversity Heritage Library:** Smithsonian Libraries (cra, crb). **355 Alamy Stock Photo:** agefotostock / José Antonio Hernaiz (br). **Biodiversity Heritage Library:** Smithsonian Libraries (cr). **Science Photo Library:** Wim Van Egmond (bl). **356 Biodiversity Heritage Library:** Smithsonian Libraries (cra, crb). **357 Alamy Stock Photo:** The Book Worm (crb). **Biodiversity Heritage Library:** Smithsonian Libraries (cra). **358 Biodiversity Heritage Library:** Smithsonian Libraries (cr). **Dreamstime.com:** Mariuszks (bc/Blue Mussels); Seadam (bl, bc); Slowmotiongli (br). **359 Biodiversity Heritage Library:** Harvard University, Museum of Comparative Zoology, Ernst Mayr Library (cr). **Dreamstime.com:** Martin Pelanek (bl); Suwat Sirivutcharungchit (bc); Seadam (br). **360 Biodiversity Heritage Library:** Smithsonian Libraries (cr). **Dreamstime.com:** Izanbar (bc); Peterclark1985 (bl); Rodho (bc/Nautilus); Tignogartnahc (br). **361 Alamy Stock Photo:** blickwinkel / Hecker (bc); Nature Photographers Ltd / Paul R. Sterry (bc/Peacock Worm). **Biodiversity Heritage Library:** Smithsonian Libraries (cr). **Dreamstime.com:** Koolsabuy (br). **Getty Images / iStock:** ePhotocorp (bl). **362 Alamy Stock Photo:** Archive PL (crb). **Dreamstime.com:** Olga Rudneva (cr). **363 Science Photo Library:** Claude Nuridsany & Marie Perennou (cra); David Scharf (crb). **364 Alamy Stock Photo:** Darling Archive (cra). **Dreamstime.com:** Patrick Guenette (crb). **365 Alamy Stock Photo:** Paul Fearn (cra). **Biodiversity Heritage Library:** Smithsonian Libraries (crb). **366 Biodiversity Heritage Library:** Museums Victoria (cr). **Dorling Kindersley:** Paolo Mazzei (br). **Dreamstime.com:** Geza Farkas (bc); Volodymyr Melnyk (bl); Conny Skogberg (bc/House Pseudoscorpion). **367 Alamy Stock Photo:** Oliver Thompson-Holmes (bc); mauritius images GmbH / Hana und Vladimir Motycka (br). **Biodiversity Heritage Library:** Smithsonian Libraries (cr). **Dreamstime.com:** James Jacob (bl); Mikelane45 (bc/Scolopendra Cingulata). **368 Biodiversity Heritage Library:** Harvard University, Museum of Comparative Zoology, Ernst Mayr Library (cra); Smithsonian Libraries (crb). **369 Biodiversity Heritage Library:** MBLWHOI Library (cra); Smithsonian Libraries (crb). **370 Biodiversity Heritage Library:** Smithsonian Libraries (cr). **Dreamstime.com:** Nicefishes (bc/Blue Crayfish); Piboon Srimak (bl); Vojcekolevski (bc). **Getty Images:** The Image Bank / Stephen Frink (br). **371 Alamy Stock Photo:** Michael Patrick O'Neill (bl). **Biodiversity Heritage Library:** Smithsonian Libraries (cr). **Dreamstime.com:** Tristan Barrington (bc/Sally Lightfoot Crab); Edwin Butter (bc); Ralf Lehmann (br). **372 artscult.com:** (cra). **Getty Images:** De Agostini Editorial / De Agostini Picture Library (crb). **373 Biodiversity Heritage Library:** Cornell University Library (crb). **Getty Images:** De Agostini Editorial / De Agostini Picture Library (cra). **374 Biodiversity Heritage Library:** Smithsonian Libraries (cr). **Dreamstime.com:** Alslutsky (bc); Imphilip (bl); Stuartan (bc/Banded Demoiselle); Paul Sparks (br). **375 Biodiversity Heritage Library:** Smithsonian Libraries (cr). **Dreamstime.com:** Siedykholena (br); Rudmer Zwerver (bl). **naturepl.com:** Ingrid Visser (bc). **376 Biodiversity Heritage Library:** NCSU Libraries (archive.org) (cr). **Dreamstime.com:** Lukas Blazek (bl); Klomsky (bc/Peruphasma Schultei); Simon Shim (br). **naturepl.com:** Emanuele Biggi (bc). **377 Alamy Stock Photo:** Florilegius (bc). **Wellcome Collection:** Ten insects, including the earwig, cockroach, grasshopper, praying mantis, cricket and mole cricket. Coloured engraving by J. W. Lowry after C. Bone (cra). **378 Biodiversity Heritage Library:** Cornell University Library (crb); Smithsonian Libraries (cra). **379 123RF.com:** Zhang YuanGeng (bc). **Biodiversity Heritage Library:** Smithsonian Libraries (cr). **Dreamstime.com:** Darius Baužys (br); Paul Lemke (bl); Julie Feinstein (bc/Periodical Cicada). **380 Alamy Stock Photo:** Zoonar GmbH / Alfred Schauhuber (bl). **artscult.com:** (br). **Dreamstime.com:** Cbechinie (bc/Weaver Ant); Tomas1111 (br). **381 Alamy Stock Photo:** Hakan Soderholm (bc). **Biodiversity Heritage Library:** Smithsonian Libraries (cr). **Dreamstime.com:** Oleksii Kriachko (br); Oskanov (bl); Sovpag (bc/Ladybird). **382 Alamy Stock Photo:** Florilegius (crb). **Biodiversity Heritage Library:** NCSU Libraries (archive.org) (cra). **383 Dreamstime.com:** Samuel Areny (bc/Common Brimstone); Ian Redding (bl, br); Glomnakub (bc). **Rijksmuseum, Amsterdam:** C. Mensing, 2020, 'workshop of Maria Sibylla Merian, Metamorphosis of a Small Emperor Moth, after 1679', in J. Turner (ed.), Dutch Drawings of the Seventeenth Century in the Rijksmuseum, online coll. cat. Amsterdam: hdl.handle.net / 10934 / RM0001.COLLECT.55730 (accessed 27 May 2021). (cr). **384 Biodiversity Heritage Library:** Smithsonian Libraries (cr). **Dreamstime.com:** FlorianAndronache (bc); Marcouliana (bl); Ava Peattie (bc/Beefly); Dennis Jacobsen (br). **385 Biodiversity Heritage Library:** Smithsonian Libraries (cr). **Dreamstime.com:** Pavel Kusmartsev (crb). **386 Alamy Stock Photo:** Keith Corrigan (cr). **Dreamstime.com:** Marie Debs (bc); Mollynz (bl); Ashish Kumar (bc/Diadema); Wrangel (br). **387 Biodiversity Heritage Library:** Harvard University, Museum of Comparative Zoology, Ernst Mayr Library (cr). **Dreamstime.com:** Ethan Daniels (br); Asther Lau Choon Siew (bl); Vojcekolevski (bc/Yellow Sea Cucumber). **388 Biodiversity Heritage Library:** Smithsonian Libraries (cra, crb). **389 Alamy Stock Photo:** Alpha Stock (crb). **Biodiversity Heritage Library:** University Library, University of Illinois Urbana Champaign (cra). **390 Alamy Stock Photo:** Archive PL (crb). **Biodiversity Heritage Library:** Smithsonian Libraries (cra). **391 Biodiversity Heritage Library:** MBLWHOI Library (cr). **Dreamstime.com:** James Dvorak (br); Natalie11345 (bl); Francofirpo (bc); Seadam (bc/Clavelina). **392 Biodiversity Heritage Library:** Naturalis Biodiversity Center (crb). **The New York Public Library:** Rare Book Division, The New York Public Library. "Great, or Sea Lamprey, Petromyzon marinus." The New York Public Library Digital Collections. 1806. https://digitalcollections.nypl.org/items/510d47da-6343-a3d9-e040-e00a18064a99 (cra). **393 Biodiversity Heritage Library:** Naturalis Biodiversity Center (cra). **Dreamstime.com:** Richard Carey (br); Jill Lang (bc). **Getty Images / iStock:** E+ / RainervonBrandis (bl). **394 Alamy Stock Photo:** Pally (bc); Pally (br). **Dreamstime.com:** Slowmotiongli (bl). **Getty Images / iStock:** DigitalVision Vectors / ilbusca (cr). **395 Biodiversity Heritage Library:** Harvard University, Museum of Comparative Zoology, Ernst Mayr Library (cr). **Dreamstime.com:** Lukas Blazek (br); Michael Siluk (bl); Slowmotiongli (bc); Bill Kennedy (bc/Orange Seahorse). **396 Biodiversity Heritage Library:** Smithsonian Libraries (cr). **Dreamstime.com:** Dirk Ercken (bl, bc); Ondřej Prosický (bc/Fire Salamander). **Getty Images:** Photodisc / R. Andrew Odum (br).

397 Biodiversity Heritage Library: Harvard University, Museum of Comparative Zoology, Ernst Mayr Library (cr). **Dreamstime.com:** Cleanylee (bc/Rat Snake); Johannes Gerhardus Swanepoel (bl); Slowmotiongli (bc); Isabellebonaire (br). **398 Biodiversity Heritage Library:** Smithsonian Libraries (cr). **Dreamstime.com:** K Quinn Ferris (bc); Petar Kremenarov (bl); Vladimir Seliverstov (bc/Emperor Penguin); Svetlana Foote (br). **399 Biodiversity Heritage Library:** Smithsonian Libraries (cr). **Dreamstime.com:** Paul Banton (bc/Giraffe); Beehler (bl); Vishwa Kiran (br); Photowitch (bc)

Endpaper images: *Front and Back:* **Science Photo Library:** Dennis Kunkel Microscopy

All other images © Dorling Kindersley
For further information see:
www.dkimages.com